Studies in Systems, Decision and Control

Volume 570

AF816326

The series "Studies in Systems, Decision and Control" (SSDC) covers both new developments and advances, as well as the state of the art, in the various areas of broadly perceived systems, decision making, and control–quickly, up to date and with a high quality. The intent is to cover the theory, applications, and perspectives on the state of the art and future developments relevant to systems, decision making, control, complex processes, and related areas, as embedded in the fields of engineering, computer science, physics, economics, social and life sciences, as well as the paradigms and methodologies behind them. The series contains monographs, textbooks, lecture notes, and edited volumes in systems, decision making, and control spanning the areas of Cyber-Physical Systems, Autonomous Systems, Sensor Networks, Control Systems, Energy Systems, Automotive Systems, Biological Systems, Vehicular Networking and Connected Vehicles, Aerospace Systems, Automation, Manufacturing, Smart Grids, Nonlinear Systems, Power Systems, Robotics, Social Systems, Economic Systems and other. Of particular value to both the contributors and the readership are the short publication timeframe and the worldwide distribution and exposure which enable both a wide and rapid dissemination of research output.

Indexed by SCOPUS, DBLP, WTI Frankfurt eG, zbMATH, SCImago.

All books published in the series are submitted for consideration in Web of Science.

Peter H. M. P. Roelofsma · Fakhra Jabeen ·
H. Rob Taal · Jan Treur
Editors

Using Shared Mental Models and Organisational Learning to Support Safety and Security Through Cyberspace: A Computational Analysis Approach

Springer

Editors
Peter H. M. P. Roelofsma
Risk Management and Cybersecurity Group
Centre of Expertise Cybersecurity
The Hague University of Applied Sciences
The Hague, The Netherlands

Division of Neonatology
Department of Neonatal and Paediatric
Intensive Care
Erasmus University Medical Center
Rotterdam, The Netherlands

H. Rob Taal
Division of Neonatology
Department of Neonatal and Paediatric
Intensive Care
Erasmus University Medical Center
Rotterdam, The Netherlands

Fakhra Jabeen
Faculty of Social Sciences
Communication and Media
Radboud University
Nijmegen, The Netherlands

Jan Treur
Department of Computer Science
Social AI Group
Vrije Universiteit Amsterdam
Amsterdam, The Netherlands

Risk Management and Cybersecurity Group
Centre of Expertise Cybersecurity
The Hague University of Applied Sciences
The Hague, The Netherlands

ISSN 2198-4182 ISSN 2198-4190 (electronic)
Studies in Systems, Decision and Control
ISBN 978-3-031-72077-2 ISBN 978-3-031-72075-8 (eBook)
https://doi.org/10.1007/978-3-031-72075-8

This Springer imprint is published by the registered company Springer Nature Switzerland AG
The registered company address is: Gewerbestrasse 11, 6330 Cham, Switzerland

To Christa Eleonora, in memory of all our times, and still missing you at every single moment.

Preface

In an era defined by rapid technological advancements and an ever-evolving digital landscape, the importance of safety and security in cyberspace cannot be overstated. As organizations increasingly rely on digital infrastructures, the need for robust strategies to safeguard these systems becomes paramount. This book, *Using Shared Mental Models and Organisational Learning to Support Safety and Security Through Cyberspace: A Computational Analysis Approach*, addresses this critical need by exploring the integration of shared mental models and organizational learning to enhance safety and security measures.

The concept of shared mental models refers to the common understanding and shared knowledge among team members, which is crucial for coordinated action and decision-making. When applied to safety and security, shared mental models enable teams to anticipate potential threats, respond swiftly to incidents, and implement preventive measures effectively. This book delves into the mechanisms through which shared mental models can be cultivated within organizations, fostering a culture of vigilance and proactive defense.

Organizational learning, on the other hand, involves the continuous process of acquiring, sharing, and utilizing knowledge to adapt and improve. In the context of safety and security, through cyberspace this means learning from past incidents and mistakes, staying abreast of emerging threats, and continuously updating security protocols developing shared mental models and make individuals learn from them. By examining case studies and simulations of real-world scenarios, this book illustrates how organizations can leverage organizational learning to stay ahead of cyber adversaries and build resilient digital infrastructures.

Many examples of the book are from the medical domains of anesthesiology and neonatology. The reason for this is twofold. First, in these fields teamwork and the maintaining of a shared mental model form a crucial part in successful medical performance. Second, the protocols of anesthesiologists and neonatologists appeared in general relatively more precise than others so as to form a secure basis from which we could start our enterprise. Our work, however, is generalizable to any other domain of team and organizational performance.

Ensuring patient safety and security through cyberspace requires that all care professionals operate as a team and community. In order to be successful, it is of paramount importance that all members of the team have a shared understanding of the diagnosis, the condition of the patient, the secure use of medical devices, and the plan of action.

At present, to ensure that all members have this 'shared mental model', members communicate and observe each other's actions. Based upon this information, members should be able to confirm if indeed all have a shared mental model and speak up when deviations in one or more members and/or processes are suspected. From a group dynamical and information processing perspective, this verification process is known to be very vulnerable: how can red flags be detected in complicated surgery settings and do members feel psychologically safe enough to speak up when they have concerns about being on the same page as the rest of their team?

This book presents a new approach for safety and security through cyberspace through introducing a concept of co-designed clinical pathways supported by the AI coach. The AI coach will be an intervention for both improving hospital-wide safety and security through cyberspace by exemplary clinical pathways in exceptionally vulnerable patients. It will empower users by supporting and facilitating the development of a shared mental model for team and organisational learning. The AI coach will function as an information, communication, cooperation and decision support system.

A unique feature of this book is its emphasis on computational analysis. Computational tools and techniques offer powerful means to identify patterns and predict potential vulnerabilities. This book presents a comprehensive overview of the applicability and use of network-oriented modeling and analysis. By integrating these methods with shared mental models and organizational learning, this book proposes a holistic approach to safety and security in cyberspace that is both innovative and practical.

The journey through the pages of this book will equip readers with the theoretical insights and practical tools needed to enhance safety and security in their organizations.

We would like to express our gratitude to the many experts and practitioners who have contributed their insights and experiences to this work. Their collective wisdom has enriched the content and ensured its relevance to the current cybersecurity landscape. We are also indebted to our colleagues and authors of chapters whose guidance, support, and work have been invaluable in bringing this book to fruition.

The material at the basis of this book has been presented in recent years as original work with publications by Springer Nature in conferences such as ICCCI'21, ICICT'22, SocialCom'22, BICA*AI'22, BICA*AI'23, and ComplexNetworks'23, and in journals such as Cognitive Systems Research from Elsevier and Journal of Information and Telecommunication from Taylor and Francis.

As you embark on this exploration of shared mental models, organizational learning, and computational analysis, we hope you find the inspiration and knowledge to make meaningful strides. Together, we can build a safer and more secure cyberspace for all.

The Hague, The Netherlands
June 2024

Peter H. M. P. Roelofsma

Contents

Introduction

Ensuring patient safety and security in healthcare settings is increasingly complex, especially with the growing reliance on digital technologies. The introduction chapter of "Computational Analysis to Support Safety and Security Through Cyberspace Using Shared Mental Models and Organisational Learning" outlines the critical need for effective communication and shared understanding among healthcare professionals. The book posits that achieving a shared mental model, where all team members have a consistent understanding of patient diagnosis, condition, and treatment plans, is fundamental to ensuring patient safety and the secure use of medical devices.

The traditional method for developing a shared mental model involves team members communicating and observing each other's actions. However, this verification process is fraught with vulnerabilities, particularly in high-stress, complicated surgical environments where red flags can be easily overlooked, and psychological safety to speak up is not always present. To address these challenges, the book introduces an innovative approach involving the co-design of clinical pathways supported by an AI coach. This AI coach is envisioned as an intervention to enhance hospital-wide safety and security, particularly in scenarios involving exceptionally vulnerable patients.

The AI coach aims to empower healthcare professionals by facilitating the development of shared mental models and fostering team and organizational learning. It functions as an integrated system providing information, communication, cooperation, and decision support, thereby addressing the identified vulnerabilities in the traditional verification processes. This introduction sets the stage for a comprehensive exploration of how AI and network-oriented modeling can transform safety and security in healthcare.

What Will be Addressed in This Part

The introduction chapter by Roelofsma and Treur delves into the core contributions and implications of integrating an AI coach into healthcare settings. The AI coach's ability to support the development of shared mental models among healthcare teams addresses a significant gap in current practices. By ensuring that all team members have a consistent understanding of patient care processes, the AI coach reduces the risk of miscommunication and enhances overall patient safety.

One of the primary advantages of the AI coach is its ability to provide real-time information and decision support. This feature is particularly crucial in high-stakes environments like surgery, where the timely identification and response to red flags can mean the difference between a successful outcome and a critical failure. The AI coach's continuous monitoring and analysis capabilities help in early detection of potential issues, allowing for prompt intervention.

Moreover, the AI coach promotes psychological safety among team members by creating an environment where speaking up about concerns is encouraged and facilitated. This aspect is vital for effective teamwork, as it ensures that all voices are heard and considered, thereby improving the decision-making and learning processes. The AI coach's role in fostering a culture of open communication and mutual support is a significant step forward in addressing the challenges identified in the traditional verification process.

The integration of AI in healthcare also raises important considerations regarding implementation and adaptation. The book discusses the need for a co-designed approach, where clinical pathways are developed collaboratively with input from various stakeholders. This collaborative approach ensures that the AI coach is tailored to the specific needs and contexts of different healthcare settings, enhancing its effectiveness and acceptance among users.

Additionally, the AI coach's impact on organizational learning cannot be overstated. By systematically collecting and analyzing data on team interactions and patient outcomes, the AI coach provides valuable insights that can inform continuous improvement efforts. This data-driven approach to learning helps healthcare organizations to identify best practices, address recurring issues, and adapt to evolving challenges.

Conclusion

In conclusion, the introduction of an AI coach for enhancing safety and security through cyberspace represents a significant advancement in healthcare. The AI coach addresses critical vulnerabilities in the traditional process of developing shared mental models among healthcare teams, offering a robust solution that integrates information, communication, cooperation, and decision support. By facilitating the development of shared mental models, promoting psychological safety, and supporting organizational learning, the AI coach has the potential to transform safety and security practices in healthcare settings.

This volume *Using Shared Mental Models and Organisational Learning to Support Safety and Security Through Cyberspace: A Computational Analysis Approach* provides a comprehensive exploration of this innovative approach, highlighting its theoretical foundations, practical applications, and potential impact. The proposed AI coach not only enhances current safety and security measures but also paves the way for future innovations in healthcare risk management.

As healthcare continues to evolve in the digital age, the integration of AI technologies like the AI coach will be crucial in ensuring that patient safety and security remain paramount. The collaborative, adaptive, and user-centered nature of the AI coach ensures that it can effectively meet the diverse needs of healthcare organizations, making it a versatile tool for global implementation. The next steps in research and development will be critical in realizing the full potential of the AI coach, and this book lays the groundwork for these future endeavors.

Computational Analysis to Support Safety and Security Through Cyberspace Using Shared Mental Models and Organisational Learning

Peter H. M. P. Roelofsma and Jan Treur

Abstract This chapter presents an introduction to this book on a computational analysis approach for safety and security through cyberspace. It briefly introduces main concepts such as mental models, shared mental models, organizational learning, and how dynamics and adaptivity can be modeled by adaptive networks. Finally, it will provide an overview of the parts and chapters of the book.

1 Introduction

Ensuring patient safety and security through cyberspace requires that all care professionals operate as a team and community. In order to be successful, it is of paramount importance that all members of the team have a shared understanding of the diagnosis, the condition of the patient, the secure use of medical devices and the plan of action. At present, to ensure that all members have this 'shared mental model', members communicate and observe each other's actions. Based upon this information, members should be able to confirm if indeed all have a shared mental model and speak up when deviations in one or more members and/or processes are suspected. From a group dynamical and information processing perspective, this verification

P. H. M. P. Roelofsma · J. Treur (✉)
Risk Management and Cybersecurity Group, Centre of Expertise Cybersecurity, The Hague University of Applied Sciences, The Hague, The Netherlands
e-mail: j.treur@vu.nl; j.treur@hhs.nl

P. H. M. P. Roelofsma
e-mail: p.h.m.p.roelofsma@hhs.nl

P. H. M. P. Roelofsma
Department of Pediatrics, Division Neonatology, Erasmus University Medical Centre, Erasmus MC, University Medical Center-Sophia Children's Hospital, Rotterdam, The Netherlands

J. Treur
Social AI Group, Department of Computer Science, Vrije Universiteit Amsterdam, Amsterdam, The Netherlands

P. H. M. P. Roelofsma et al. (eds.), *Using Shared Mental Models and Organisational Learning to Support Safety and Security Through Cyberspace: A Computational Analysis Approach*, Studies in Systems, Decision and Control 570,
https://doi.org/10.1007/978-3-031-72075-8_1

process is known to be very vulnerable: how can red flags be detected in complicated surgery settings and do members feel psychologically safe enough to speak up when they have concerns about being on the same page as the rest of their team? This book presents a new approach for safety and security through cyberspace through introducing a concept of co-designed clinical pathways supported by the AI coach. The AI coach will be an intervention for both improving hospital-wide safety and security through cyberspace by exemplary clinical pathways in exceptionally vulnerable patients. It will empower users by supporting and facilitating the development of a shared mental model for team and organisational learning. The AI coach will function as an information, communication, cooperation and decision support system.

2 Shared Mental Models

Historically, Kenneth Craik is considered a central person who introduced the concept of mental model. In Craik (1943) he considers a mental model as a *small-scale model* that is carried by an organism within its head as follows:

> If the organism carries a "small-scale model" of external reality and of its own possible actions within its head, it is able to try out various alternatives, conclude which is the best of them, react to future situations before they arise, utilise the knowledge of past events in dealing with the present and future, and in every way to react in a much fuller, safer, and more competent manner to the emergencies which face it. (Craik 1943, p. 61)

Note how he relates mental models to safety in how we respond to what happens in the world. Such internal models use certain types of *relation-structure* making a mental model work similarly to how the real world works:

> By "relation-structure" I do not mean some obscure non-physical entity which attends the model, but the fact that it is a physical working model which works in the same way as the process it parallels … (Craik 1943, p. 51).

The similarity he mentions is depicted in Fig. 1, where the relation $M \rightarrow M'$ within the mental model corresponds to a similar relation $W \rightarrow W'$ in the world, see also (Williams 2018a, b).

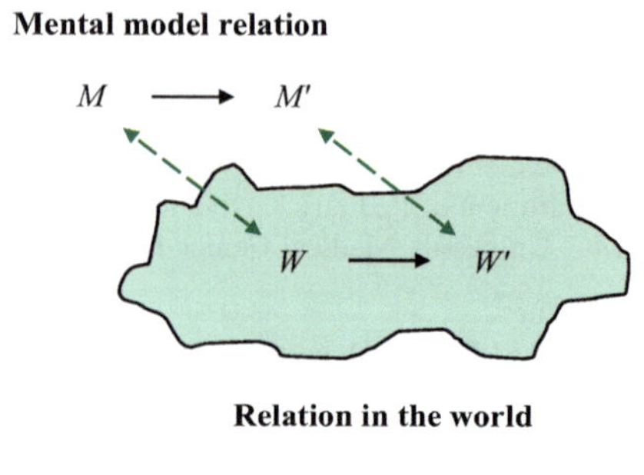

Fig. 1 A relation $W \rightarrow W'$ in the world has a corresponding relation $M \rightarrow M'$ in a mental model

This relational view in principle provides possibilities for network-oriented modeling of mental models where relations can be described by connections. However, as mental models are usually not static structures but are used for dynamics and adaptivity, a deeper analysis is required: mental models can play a role in different types of processes. Based on an analysis of several types of case studies, the following overview of these processes has been made and used as a basis for cognitive computational modeling in Van Ments and Treur (2021), Treur and Van Ments (2022):

Dynamics: internal simulation based on mental models

Mental models are often used for internal or mental simulation to relate what is known to what is yet unknown about world or mental states or processes. For example, in planning, visualisation in sport, flashback movies in PTSD, dreaming, and reasoning.

Adaptation: mental models can be adapted by learning or forgetting

Mental models can be formed or learned and they can be revised or forgotten, e.g., Piaget (1954).

Control: mental models can be controlled

Using mental models and adapting them is in principle not random but applies some form of control.

Based on this analysis, in Van Ments and Treur (2021) the cognitive architecture shown in Fig. 2 has been proposed and used as a basis for computational modeling of various case studies in Treur and Van Ments (2022). Here at the lowest level (pink plane) the structure of a mental model by its relations is used for internal simulations, at the middle level (blue plane) learning and forgetting takes place, and at the highest level (purple plane) these processes are controlled.

A shared team mental model is considered very important for high performing teams in order to perform complex tasks well, e.g., Burtscher and Manser (2012), Langan-Fox et al. (2000), Mohammed et al. (2010). This provides team members with a shared understanding of the relevant elements to perform a specific task. A team mental model originates in attuning the individual team member mental models

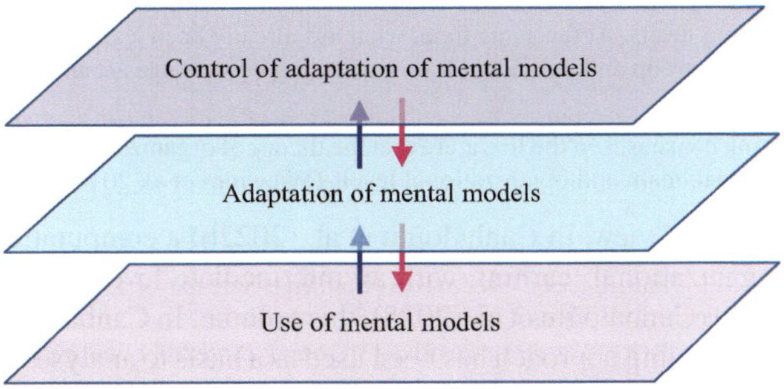

Fig. 2 Cognitive architecture for mental models based on three levels of mental processing

(DeChurch and Mesmer-Magnus 2010). The main functions of team mental models are improved planning, coordination and alignment (Nini 2019). Two different types of team mental models mentioned in Mohammed et al. (2010) are task-related team mental models and team-related team mental models. In Van Ments et al. (2021) a network modeling approach for handling a team mental model in a medical context has been introduced; see also Treur and Van Ments (2022), Chap. 14.

3 Organisational Learning

From a socio-cognitive perspective on learning, Kim (1993) claims that learning initiates within individuals who adjust their mental models when they are exposed to new experiences or process new information. By sharing their mental models with others via discussions or joint problem solving, they transfer them to other individuals and groups. After that learning is institutionalized it becomes embedded in organizational routines.

> Organizational learning is dependent on individuals improving their mental models; making those mental models explicit is crucial to developing new shared mental models. This process allows organizational learning to be independent of any specific individual. Why put so much emphasis on mental models? Because the mental models in individuals' heads are where a vast majority of an organization's knowledge (both know-how and know-why) lies. (Kim 1993), p. 44

Based on this view, in Canbaloğlu et al. (2022a) a computational modeling approach to organizational learning has been introduced.

Organizational learning is not just a collection of individual learnings, but can remain within the organization, stored in systems, manuals, and policies, also after individuals leave. Organizational learning is dynamic because it involves individuals and their learning, it is multilevel because it also extends to a team or organizational level, and it is non-linear because it has multiple feedback mechanisms which, for example, allows organization to learn from individuals (feed forward learning) and the individuals to learn from the organization (feedback learning); see also Fig. 3 and Crossan et al. (1999), Wiewiora et al. (2019):

> Through feed-forward processes, new ideas and actions flow from the individual to the group to the organization levels. At the same time, what has already been learned feeds back from the organization to group and individual levels, affecting how people act and think. (Crossan et al. 1999), p. 532.

> There is growing consensus in the literature that the theory of organizational learning should consider individual, team and organizational levels (Wiewiora et al. 2019), p. 94

Based on the latter view, in Canbaloğlu et al. (2022b) a computational modeling approach to organizational learning with an intermediate level for teams has been introduced see also Canbaloğlu et al. (2025), this volume. In Canbaloğlu et al. (2023) this introduced modeling approach has been used as a basis to analyse various organizational learning processes and various contextual aspects influencing that learning, such as organizational culture, leadership, and other contextual elements.

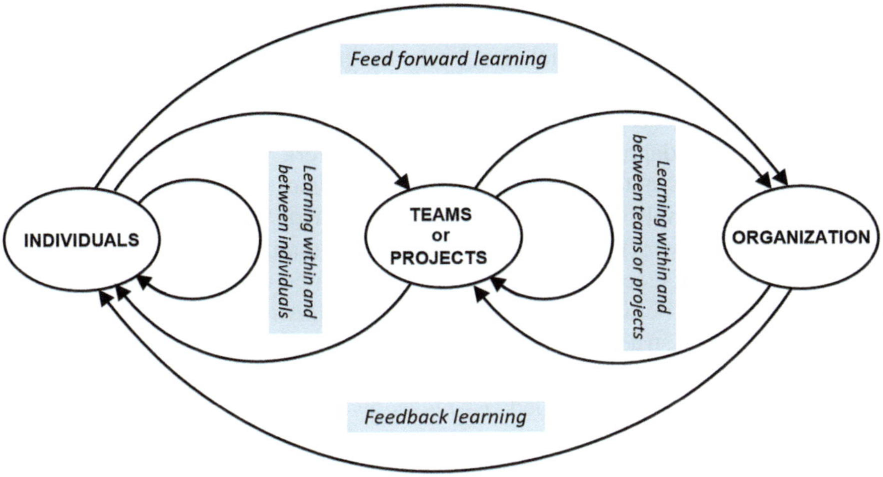

Fig. 3 Multilevel organizational learning: multiple levels and nested cycles

4 Network Modeling of Complex Adaptive Dynamical Systems

In the sections above it has been discussed how from the perspective of cognitive science and management science the use of mental models, sharing them, and organisational learning based on them, involve complex nonlinear dynamical and adaptive processes. This presents us with serious challenges when computational modelling is aimed for. However, in the current section it is discussed how higher-order adaptive dynamical systems can successfully be used to obtain computational models for these challenging processes. Moreover, it will be discussed how following Treur (2020a, b) such higher-order adaptive dynamical systems can be designed as network models. Here, for socalled temporal-causal network models their structure can be described by the following network characteristics:

- *Connectivity characteristics*

  Connections $X \rightarrow Y$ from a state X to a state Y and their weights $\boldsymbol{\omega}_{X,Y}$
- *Aggregation characteristics*

  For any state Y, a combination function $\mathbf{c}_Y$ defines the aggregation that is applied to the impacts on Y from its incoming connections from states X
- *Timing characteristics*

  Each state Y has a speed factor $\boldsymbol{\eta}_Y$ defining how fast it changes for given aggregated causal impact

Based on this network structure, by the software environment described in Treur (2020b, Chap. 9) in a standard manner difference equations can be obtained defining the network states' dynamics. This can be used among others to model internal simulation based on given mental models: the lower (pink) plane in Fig. 2. But when the network connections are static, this does not allow to make the mental models adaptive, what is needed to learn (or forget) them.

To also model adaptive processes, adaptive network models can be used: then not only a network's states but also (some of) its network characteristics can change over time, for example the connections. By using a *self-modelling network* (also called a *reified* network), a network-oriented conceptualization can also be applied to adaptive networks; see Treur (2020a, b). This can be done through the addition of new states to the network (called *self-model* or *reification* states) which describe part of the network structure by representing (adaptive) network characteristics. In the graphical 3D-format as shown, for example, in Fig. 2, such additional states are depicted at the middle level (called *self-model* level), where the original network is at the *base level.* As an example, the weight $\boldsymbol{\omega}_{X,Y}$ of a connection $X \rightarrow Y$ within a mental model at the base level (the lower, pink plane in Fig. 2) can be represented (at a next self-model level: the middle, blue plane in Fig. 2) by a self-model state named $\mathbf{W}_{X,Y}$. As these self-model states $\mathbf{W}_{X,Y}$ can have their own dynamics (like any state has, as pointed out above), in this way the relations defining a mental model do not need to be static anymore but can become adaptive: they can change over time. This enables computation modelling of learning and forgetting of mental models as well.

As the self-modeling principle can be applied in an iterative manner as has been shown in detail in Treur (2020b, Chap. 10), also the control over the learning or forgetting can be modeled in a similar way, this time by second-order self-model states (in the upper, purple plane in Fig. 2). The obtained network-oriented modeling approach to mental models has been introduced in detail and applied to many case studies in the work underlying (Treur and Van Ments 2022) and for shared mental models in particular also to a case study for teamwork in a hospital in Van Ments et al. (2021). Moreover, this approach has also been used in Canbaloğlu et al. (2022a, b) to introduce a computational modeling approach to organizational learning and is also the basis for the work in Canbaloğlu et al. (2023).

In line with (Ashby 1960), in Treur (2021), Hendrikse et al. (2023), it is shown that any smooth dynamical system has a canonical representation as a temporal-causal network and any smooth (multi-order) adaptive dynamical system has a canonical representation as a (multi-order) self-modeling temporal-causal network. Therefore, compared to adaptive dynamical systems in general, the network-oriented modeling approach used here does not introduce any fundamental limitations concerning what it can model. This has also been confirmed by applications to case studies in practice, e.g., Treur and Van Ments (2022), Canbaloğlu et al. (2023), Hosseini et al. (2025).

5 Overview of This Book

The book is composed of six parts that each will be discussed in this section:

I. Introduction
II. Basics for the Area: Shared Mental Models and Organisational Learning
III. Supporting Safety and Security in Medical Teamwork in the Neonatal Domain through Cyberspace
IV. Supporting Safety and Security for Communication with Patients and Colleagues through Cyberspace
V. Organisational Learning for Safety and Security through Cyberspace
VI. Results, Further Perspectives, and Discussion.

5.1 Part I. Introduction

In this chapter we present the overview of the research program and the book (Roelofsma and Treur 2025a). The research program addresses a computational analysis leading to a proposed Artificially Intelligent (AI) coach who collects and combines situational, behavioural and verbal information through sensors and microphones and an interactive dashboard, and who processes this using human-like mental processes and learning algorithms. The AI coach can ask relevant questions to verify that all members have a shared mental model and intervene if necessary, using mediated communication through the interactive dashboard and artificial speech. Such a system would have superior diagnostic capabilities and no concerns about speaking up.

The overarching aim of this project is to explore: How can this AI coach contribute in the hospital to improving patient safety and security through cyberspace by creating an adaptive shared mental model for team and supporting organisational learning? Our research program will develop the underlying models and algorithms of an AI-coach to enable safety and security through cyberspace.

The vision is that the AI Coach will interact with users (patients and clinical and non-clinical hospital staff, the organisation) in a personalised and motivating way and guide the organisation towards improvement of patient safety and security through cyberspace. The project aims to optimise the utilisation of all available knowledge and the professional, personal and social skills of all team members facilitated by the AI-coach.

The AI coach will stimulate hospital staff by giving suggestions for initiatives and opportunities to improve patient safety and security through cyberspace. This will be done by stimulating and implementing, e.g., behavioural change within the organisation, elements of safety and shared mental models. The research program is urgent since the available knowledge and current recommendations exceed the cognitive bandwidth of the average healthcare provider and system user, especially in pathways where multidisciplinary care is provided. AI provides important solutions for challenges such as contradictory recommendations and innovations in this area.

5.2 Part II Basics for the Area: Shared Mental Models, Organisational Learning, and Adaptive Network Modeling

Part II of the book consists of Chaps. 2 and 3 which describe the basics for the area.

In Chap. 2 a second-order adaptive network model (designed according to the architecture depicted in Fig. 2) is described for mental processes making use of shared mental models for team performance (Van Ments et al. 2025). The chapter illustrates the value of adequate shared mental models for safe and efficient team performance and in cases of imperfections of such shared team models how this complicates team performance. It is illustrated for a context of a medical team performing a tracheal intubation. Simulations illustrate how the adaptive network model is able to address the type of complications that can occur in realistic scenarios.

In Chap. 3 multilevel organizational learning is described that concerns an interplay of different types of learning at individual, team, and organizational levels (Canbaloğlu et al. 2025). These processes use complex dynamic and adaptive mechanisms (addressed like depicted in Fig. 3). A second-order adaptive network model for this is introduced (again designed according to the architecture described in Fig. 2) in this chapter and illustrated.

5.3 Part III Supporting Safety and Security in Medical Teamwork in the Neonatal Domain Through Cyberspace

Chapters 4–7 are Part III of the book. These chapters focus in particular on the use of shared mental models and organisational learning for medical teamwork in the neonatal domain and an AI Coach in cyberspace to support safety and security.

Chapter 4 shows how second-order adaptive agent-based network models can be used to model a medical team supported by a virtual AI Coach (Xu et al. 2025). It is illustrated for the case of a newborn baby in danger. The design of these computational agent models is based on an adaptive self-modeling network modeling approach. This also addresses how the AI Coach can play a central role in organizational learning, following Canbaloğlu et al. (2025).

Chapter 5 presents the use of second-order adaptive network models of hospital teams consisting of doctors and nurses, interacting together (Mokadem et al. 2025). A variety of scenarios are modelled and simulated, in relation with respiratory distress of a neonate, along with the integration of an AI-Coach for monitoring and support of such teams and of organizational learning. The research highlights the benefits of introducing a virtual AI-Coach in a hospital setting. The practical application setting revolves around a medical team responsible for managing neonates with respiratory distress. In this setting an AI-Coach acts as an additional team member, to ensure correct execution of medical procedure.

Chapter 6 shows how second-order adaptive agent-based network models can be used to support a medical team in healthcare institutions to adhere to specific neonatal hypoglycemia and neonatal hyperbilirubinemia treatment guidelines through the integration of an Artificial Intelligence (AI) Virtual Coach (Chawla et al. 2025). The proposed AI Coach is designed to provide timely interventions and correct deviations when lapses in the health care practitioner's internal mental states occur.

Chapter 7 presents an approach to enhancing neonatal care through the application of Artificial Intelligence (AI) (Sebahi et al. 2025). Utilizing network-oriented modeling methodologies, the study aims to develop a network model in particular to improve outcomes in neonatal respiratory support. A significant contribution is the introduction of an AI coach, which serves as a real-time intervention mechanism to fill in the doctor's knowledge gaps. The research serves as a seminal work in the intersection of artificial intelligence and healthcare, demonstrating the potential of AI-driven network modeling in improving patient outcomes and streamlining healthcare protocols.

5.4 Part IV Supporting Safety and Security for Communication with Patients and Colleagues Through Cyberspace

The focus of part IV is in particular on the use of an AI Coach in cyberspace to support safety and security in the neotal domain concerning communication between colleagues and with patients.

Chapter 8 describes an extension of a safety culture within hospital organizations providing more transparency and acknowledgement of all actors, and in particular the parents (Weigl et al. 2025). It contributes a model architecture to support a hospital to develop such an extended safety culture. It is illustrated for prevention of postpartum depression. The chapter illustrates the benefits of simple support in terms of communication during childbirth, which has lasting effects, even outside the hospital. For the impact of additional communication, a Virtual Safety Coach is designed that intervenes when necessary to provide support, i.e., when a health care practitioner doesn't. Moreover, organizational learning is also modelled to improve the mental models of both the Safety Coach and the Health Care Practitioner.

Chapter 9 This chapter explores a computational model concerning the mechanisms known from psychological, observational, and medical literature which underlie the speaking up behaviour of a health care professional (Doornkamp et al. 2025). It also addresses how a doctor may respond to a communicated message about a medical error made. Through several scenarios we illustrate what pattern of factors causes a healthcare professional to speak up when witnessing a (near) medical error. We moreover demonstrate how introducing an observant agent can facilitate effective communication and help to ensure patient safety through speaking up when a nurse cannot.

5.5 Part V Organisational Learning for Safety and Security Through Cyberspace

Part V of the book addresses in particular the role of mistakes and learning from them and how an organisation-wide strategy can be followed to support safety and security.

Chapter 10 exposes the consequences of types of attitudes of leaders in organizations (Hosseini et al. 2025). One is the mindset of a boss who dominantly believes that mistakes usually have negative consequences and therefore avoids them by only accepting simple tasks. A second attitude concerns bosses who select simple tasks to be always seen as perfect. This chapter exposes the consequences of these attitudes for the organizations.

Chapter 11 focusses on using a self-modelling network modeling approach to analyse and predict the behavior of people within organizations, specifically on the implementation of EHS standards (Van den Hurk et al. 2024). The study focusses on a real-world scenario in which EHS standards are implemented in a multi-level organizational learning scenario. The study shows the possibility of using a self-modelling social network in the execution of EHS activities. It demonstrates the importance of cooperation and open communication between the system owner, EHS expert and department. This is needed to be able to learn about processes and efficiently safeguard the organization and its members.

5.6 Part VI Results, Further Perspectives, and Discussion

Part VI summarizes the obtained results and points out what can be put on a future research agenda.

Chapter 12 describes the specific effects of AI for cybersecurity. It uses a cyber risk management model for anticipation, monitoring, reacting and learning in the context of phishing emails (Van den Hurk et al. 2024). It particular, the chapter discusses the importance of AI for cyber risk management and provides relevant future research questions for this field.

Finally, Chap. 13 summarizes the integrated results of the book and delineates what the next steps for the research program should be (Roelofsma and Treur 2025b).

6 Discussion

This book highlights the critical role of shared mental models in ensuring effective safe and secure teamwork and communication within high-stakes environments, such as healthcare. The concept, rooted in the early work of Kenneth Craik, emphasizes the importance of internal simulations, adaptation, and control in managing complex tasks and scenarios. The application of these principles to computational modeling allows for the creation of adaptive systems capable of learning and evolving over time.

One of the key contributions of this work is the introduction of the AI coach, designed to facilitate the development of shared mental models within medical teams. This tool not only supports the secure use of medical devices but also enhances team coordination and communication, ultimately leading to improved patient outcomes. The AI coach serves as a decision support system, offering real-time insights and fostering a psychologically safe and secure environment where team members can voice concerns and ensure alignment in their understanding and actions.

The inclusion of organizational learning further enriches the framework by integrating individual and collective learning processes. The multilevel and nonlinear nature of organizational learning underscores the dynamic interplay between individuals, teams, and the broader organization. The computational modeling approaches discussed provide a structured way to analyze and enhance these learning processes, thereby taking into account various contextual factors such as organizational culture and leadership.

The network-oriented modeling of complex adaptive dynamical systems, as presented in this book, addresses the challenges inherent in modeling such intricate systems. By employing higher-order adaptive systems represented as adaptive temporal-causal networks, the authors offer a comprehensive methodology for capturing the dynamics of mental models and organizational learning. This approach not only supports the modeling of static relationships but also accommodates the adaptivity required for learning and forgetting, thus providing a more realistic representation of cognitive and organizational processes.

The text provides in several steps a comprehensive overview of a multi-faceted approach to enhancing safety and security in healthcare and organizational settings through the use of adaptive network models, shared mental models, and AI-assisted learning. The integration of these models and technologies addresses various dimensions of team performance, organizational learning, and risk management.

First, we lay the foundational concepts by introducing adaptive network models for mental processes and organizational learning. The use of shared mental models is crucial for team performance, as demonstrated through the case study of a medical team performing a tracheal intubation. The simulations illustrate the potential of these models to mitigate complications arising from imperfect shared mental models. The book expands on organizational learning by detailing a multi-level adaptive network model, highlighting the complexity and dynamism of learning processes at individual, team, and organizational levels.

Second, the focus shifts to practical applications in the neonatal domain. The book emphasizes the role of AI coaches in supporting medical teams. In Chaps. 4–7 we demonstrate how second-order adaptive agent-based network models can enhance team performance and adherence to protocols through real-time interventions and support. The use of AI coaches is shown to be effective in scenarios involving neonatal respiratory distress and adherence to treatment guidelines for conditions like neonatal hypoglycemia and hyperbilirubinemia. The AI coach's ability to fill knowledge gaps and ensure protocol compliance is a significant advancement in improving patient safety outcomes.

Third, we extend the application of AI Coaches to enhance communication within hospital organizations. We highlight the development of a safety culture that includes transparent communication with parents, aiming to prevent postpartum depression. The inclusion of a virtual AI coach to support healthcare practitioners in communication scenarios is a novel approach to improving long-term outcomes. We have delved into the psychological and observational mechanisms underlying healthcare professionals' speaking-up behavior, showcasing the role of an observant virtual agent in facilitating effective communication and ensuring patient safety.

Next, we address the role of organizational learning in safety and security, with a particular focus on learning from mistakes. We discussed the impact of leadership attitudes towards mistakes on organizational performance. It emphasizes the negative consequences of a risk-averse mindset and the importance of embracing mistakes as learning opportunities. We also illustrate the implementation of Environment, Health, and Safety (EHS) standards through adaptive network models, highlighting the importance of cooperation and open communication for effective organizational learning.

Finally, in the last step we summarize the results and outline future research directions. We emphasize the significance of AI in cybersecurity, particularly in managing cyber risks such as phishing emails. It calls for further research into AI-driven cyber risk management models. We then consolidate our findings and propose next steps for advancing the research program in delineating the Cybersecurity Living Lab concept.

In conclusion, this book offers a thorough exploration of the theoretical foundations and practical applications of computational analysis in enhancing safety and security through cyberspace. The integration of shared mental models, organizational learning, and adaptive network modeling presents a cohesive framework that holds significant promise for future research and development in the field. By bridging cognitive science, management science, and computational modeling, this work paves the way for innovative solutions to complex challenges in high-stakes environments. The multi-level approach, encompassing individual, team, and organizational learning, demonstrates the potential of these models to improve performance, adherence to protocols, and communication within teams. The practical applications in the neonatal domain and the use of AI coaches highlight significant advancements in patient care and organizational safety culture.

Future research should focus on further refining these models and exploring their applications in other domains. The continuous development of AI-driven interventions and the integration of adaptive learning mechanisms will be crucial in addressing emerging challenges in safety and security. This comprehensive approach, combining theoretical foundations with practical applications, sets a strong precedent for the ongoing evolution of safety and security management in complex systems.

References

Abraham, W.C., Bear, M.F.: Metaplasticity: the plasticity of synaptic plasticity. Trends Neurosci. **19**(4), 126–130 (1996). https://doi.org/10.1016/s0166-2236(96)80018-x. PMID: 8658594

Ashby, W.R.: Design for a Brain. Chapman and Hall, London (second extended edition) (1960)

Burtscher, M.J., Manser, T.: Team mental models and their potential to improve teamwork and safety. A review and implications for future research in healthcare. Saf. Sci. **50**(5), 1344–1354 (2012). https://doi.org/10.1016/j.ssci.2011.12.033

Canbaloğlu, G., Treur, J., Roelofsma, P.H.M.P.: Computational modeling of organisational learning by self-modeling networks. Cogn. Syst. Res. **73**, 51–64 (2022a)

Canbaloğlu, C., Treur, J., Roelofsma, P.H.M.P.: An adaptive self-modeling network model for multilevel organizational learning. In: Proceedings of the Seventh International Congress on Information and Communication Technology, ICICT 2022, Volume 2. Lecture Notes in Networks and Systems, vol. 448, pp. 179–191. Springer Nature (2022b)

Canbaloğlu, G., Treur, J., Wiewiora, A. (eds.): Computational Modeling of Multilevel Organisational Learning and Its Control Using Self-modeling Network Models. Springer Nature (2023)

Canbaloğlu, G., Treur, J., Roelofsma, P.H.M.P.: Learning of safety and security management through cyberspace: an adaptive self-modeling network model for multilevel organizational learning. In: Roelofsma, P.H.M.P., Jabeen, F., Taal, H.R., Treur, J. (eds.) Using Shared Mental Models and Organisational Learning to Support Safety and Security through Cyberspace: A Computational Analysis Approach. Springer Nature, Cham, Switzerland (this volume) (2025)

Chawla, R., Jabeen, F., Treur, J., Taal, H.R., Roelofsma, P.H.M.P.: Supporting risk management through cyberspace: an adaptive network model simulating AI coach effects by inducing adherence to guidelines in neonatal medical protocols. In: Roelofsma, P.H.M.P., Jabeen, F., Taal, H.R., Treur, J. (eds.) Using Shared Mental Models and Organisational Learning to Support Safety and Security through Cyberspace: A Computational Analysis Approach. Springer Nature, Cham, Switzerland (this volume) (2025)

Craik, K.J.W.: The Nature of Explanation. University Press, Cambridge, MA (1943)

Crossan, M.M., Lane, H.W., White, R.E.: An organizational learning framework: from intuition to institution. Acad. Manag. Rev. **24**, 522–537 (1999)

DeChurch, L.A., Mesmer-Magnus, J.R.: Measuring shared team mental models: a meta-analysis. Group Dyn. Theory Res. Practice **14**(1), pp. 1–14 (2010). https://doi.org/10.1037/a0017455

Doornkamp, S., Jabeen, F., Treur, J., Taal, H.R., Roelofsma, P.H.M.P.: Optimizing speaking up behavior for safety and security through cyberspace: a controlled adaptive computational network model of a virtual coach. In: Roelofsma, P.H.M.P., Jabeen, F., Taal, H.R., Treur, J. (eds.) Using Shared Mental Models and Organisational Learning to Support Safety and Security through Cyberspace: A Computational Analysis Approach. Springer Nature, Cham, Switzerland (this volume) (2025)

Gentner, D., Stevens, A.L. (eds.): Mental Models. Lawrence Erlbaum Associates, Hillsdale NJ (1983)

Hendrikse, S.C.F., Treur. J., Koole. S.L.: Modeling emerging interpersonal synchrony and its related adaptive short-term affiliation and long-term bonding: a second-order multi-adaptive neural agent model. Int. J. Neural Syst. **33**(7), 2350038, pp. 1–41 (2023). https://doi.org/10.1142/S0129065723500387

Hendrikse, S.C.F., Treur, J., Koole, S.L. (eds.): New Analysis and Modeling Directions in Social Interaction Science: Emergent Multimodal Interpersonal Synchrony, Affiliation, and Bonding. Springer Nature, to appear (2025)

Hosseini, M., Treur, J., Kucharska, W.: Learning from mistakes within organizations: An adaptive network-oriented model for a double bias perspective for safety and security through cyberspace. In: Roelofsma, P.H.M.P., Jabeen, F., Taal, H.R., Treur, J. (eds.) Using Shared Mental Models and Organisational Learning to Support Safety and Security through Cyberspace: A Computational Analysis Approach. Springer Nature, Cham, Switzerland (this volume) (2025)

Kim, D.H.: The link between individual and organizational learning. In: Sloan Management Review, Fall, pp. 37–50 (1993). Also in: Klein, D.A. (ed.) The Strategic Management of Intellectual Capital. Routledge-Butterworth-Heinemann, Oxford (1993)

Langan-Fox, J., Code, S., Langfield-Smith, K.: Team mental models: techniques, methods, and analytic approaches. Hum. Factors **42**(2), 242–271 (2000). https://doi.org/10.1518/001872000779656534

Mohammed, S., Ferzandi, L., Hamilton, K.: Metaphor no more: A 15-year review of the team mental model construct. J. Manag. **36**(4), 876–910 (2010)

Mokadem, N., Jabeen, F., Treur, J., Taal, H.R., Roelofsma, P.H.M.P.: Increasing safety and security through cyberspace by an adaptive network model for AI-assisted risk management of neonatal respiratory distress. In: Roelofsma, P.H.M.P., Jabeen, F., Taal, H.R., Treur, J. (eds.) Using Shared Mental Models and Organisational Learning to Support Safety and Security through Cyberspace: A Computational Analysis Approach. Springer Nature, Cham, Switzerland (this volume) (2025)

Nini, M.: All on the same page: How Team Mental Models (TMM) increase team performance. CQ Net (2019). https://www.ckju.net/en/dossier/team-mental-models-increase-team-performance

Piaget, J.: The Construction of Reality in the Child. Basic Books Inc., New York (1954)

Port, R.F., van Gelder, T. (eds.): Mind as Motion: Explorations in the Dynamics of Cognition. MIT Press, Cambridge, MA (1995)

Roelofsma, P.H.M.P., Treur, J.: Shared mental models and organisational learning for safety and security through cyberspace: a computational analysis approach. In: Roelofsma, P.H.M.P., Jabeen, F., Taal, H.R., Treur, J. (eds.) Using Shared Mental Models and Organisational Learning to Support Safety and Security through Cyberspace: A Computational Analysis Approach. Springer Nature, Cham, Switzerland (this volume) (2025a)

Roelofsma, P.H.M.P., Treur, J.: Towards an AI coach for risk management and cybersecurity in health systems. In: Roelofsma, P.H.M.P., Jabeen, F., Taal, H.R., Treur, J. (eds.) Using Shared Mental Models and Organisational Learning to Support Safety and Security through Cyberspace: A Computational Analysis Approach. Springer Nature, Cham, Switzerland (this volume) (2025b)

Sebahi, Y., Jabeen, F., Treur, J., Taal, H.R., Roelofsma, P.H.M.P.: Improving risk management through cyberspace: Optimizing neonatal respiratory support through network oriented modeling. In: Roelofsma, P.H.M.P., Jabeen, F., Taal, H.R., Treur, J. (eds.) Using Shared Mental Models and Organisational Learning to Support Safety and Security through Cyberspace: A Computational Analysis Approach. Springer Nature, Cham, Switzerland (this volume) (2025)

Seel, N.M.: Mental models in learning situations. In: Advances in Psychology, vol. 138, pp. 85–107. North-Holland, Amsterdam (2006)

Treur, J.: Network reification as a unified approach to represent network adaptation principles within a network. In: Proceedings of the 7th International Conference on Theory and Practice of Natural Computing, TPNC'18. Lecture Notes in Computer Science, vol. 11324, pp. 344–358. Springer Publishers (2018)

Treur, J.: Modeling higher-order adaptivity of a network by multilevel network reification. Netw. Sci. **8**, 110–144 (2020a)

Treur, J.: On the dynamics and adaptivity of mental processes: relating adaptive dynamical systems and self-modeling network models by mathematical analysis. Cogn. Syst. Res. **70**, 93–100 (2021)

Treur, J.: Network-Oriented Modeling for Adaptive Networks: Designing Higher-Order Adaptive Biological, Mental and Social Network Models. Springer Nature (2020b)

Treur, J., Van Ments, L. (eds.): Mental Models and Their Dynamics, Adaptation, and Control: A Self-Modeling Network Modeling Approach. Springer Nature (2022)

Van den Hout, N.J., Bouma, D., Hoffmans, C., Zwarts, N., Treur, J., Roelofsma, P.H.M.P.: Addressing organizational cyber security challenges in healthcare using adaptive network-oriented modeling. In: Roelofsma, P.H.M.P., Jabeen, F., Taal, H.R., Treur, J. (eds.): Using Shared Mental Models and Organisational Learning to Support Safety and Security through Cyberspace: A Computational Analysis Approach. Springer Nature, Cham, Switzerland (this volume) (2025)

Van den Hurk, N., Treur, J., Roelofsma, P.H.M.P.: Organisational learning for safety and security through cyberspace: adaptive modelling of the implementation of environment health and safety standards. In: Roelofsma, P.H.M.P., Jabeen, F., Taal, H.R., Treur, J. (eds.) Using Shared Mental Models and Organisational Learning to Support Safety and Security through Cyberspace: A Computational Analysis Approach. Springer Nature, Cham, Switzerland (this volume) (2024)
Van Ments, L., Treur, J., Klein, J., Roelofsma, P.H.M.P.: A second-order adaptive network model for shared mental models in hospital teamwork. In: Nguyen, N.T., et al. (eds.) Proceedings of the 13th International Conference on Computational Collective Intelligence, ICCCI'21. Lecture Notes in AI, vol. 12876, pp. 126–140. Springer Nature (2021)
Van Ments, L., Treur, J., Klein, J. Roelofsma, P.H.M.P.: Developing a safety and security AI coach: a second-order adaptive network model of shared mental models in hospital teamwork. In: Roelofsma, P.H.M.P., Jabeen, F., Taal, H.R., Treur, J. (eds.) Using Shared Mental Models and Organisational Learning to Support Safety and Security through Cyberspace: A Computational Analysis Approach. Springer Nature, Cham, Switzerland (this volume) (2025)
Van Ments, L., Treur, J.: Reflections on dynamics, adaptation and control: a cognitive architecture for mental models. Cogn. Syst. Res. **70**, 1–9 (2021)
Weigl, L., Jabeen, F., Treur, J., Taal, H.R., Roelofsma, P.H.M.P.: Learning for a better safety and security culture within an organization: Reducing the risk in communication with AI Coaching for Security Communication through Cyberspace. In: Roelofsma, P.H.M.P., Jabeen, F., Taal, H.R., Treur, J. (eds.), Using Shared Mental Models and Organisational Learning to Support Safety and Security through Cyberspace: A Computational Analysis Approach. Springer Nature, Cham, Switzerland (this volume) (2025)
Wiewiora, A., Smidt, M., Chang, A.: The 'How' of multilevel learning dynamics: a systematic literature review exploring how mechanisms bridge learning between individuals, teams/projects and the organization. Eur. Manag. Rev. **16**, 93–115 (2019)
Williams, D.: Predictive minds and small-scale models: Kenneth Craik's contribution to cognitive science. Philos. Explor. **21**(2), 245–263 (2018a)
Williams, D.: The Mind as a Predictive Modelling Engine: Generative Models, Structural Similarity, and Mental Representation. Ph.D. Thesis, University of Cambridge, UK (2018b)
Xu, Y., Jabeen, F., Treur, J., Taal, H.R., Roelofsma, P.H.M.P.: Supporting medical team decision in the neonatal domain: how safety and security risk management through cyberspace can be supported by adaptive agent network models with internal mental models. In: Roelofsma, P.H.M.P., Jabeen, F., Taal, H.R., Treur, J. (eds.) Using Shared Mental Models and Organisational Learning to Support Safety and Security through Cyberspace: A Computational Analysis Approach. Springer Nature, Cham, Switzerland (this volume) (2025)

Basics for the Area: Shared Mental Models, Organisational Learning, and Adaptive Network Modeling

In the intricate and high-stakes environment of healthcare, teamwork and communication are pivotal for ensuring safety and efficiency. This second part of this book delves into foundational concepts that underpin these critical aspects, focusing on shared mental models, organizational learning, and adaptive network modeling. These concepts are essential for understanding how healthcare teams can function optimally, adapt to new challenges, and continually improve their practices. This part introduces key methodologies for modeling and analyzing these processes, providing a robust framework for enhancing team performance and organizational learning in healthcare settings.

What Will be Addressed in This Part

Shared Mental Models and Adaptive Network Modeling
The concept of shared mental models is integral to effective teamwork, particularly in high-pressure environments such as hospitals. Chapter "Developing a Safety and Security AI Coach: A Second-Order Adaptive Network Model of Shared Mental Models in Hospital Teamwork" by Van Ments et al. presents a second-order adaptive network model that captures the dynamics of mental processes within hospital teams. By focusing on shared mental models, this research highlights how a well-aligned understanding among team members can lead to safer and more efficient performance. The study uses the example of a medical team performing a tracheal intubation to demonstrate how shared mental models facilitate coordination and reduce errors. Through simulations, the chapter illustrates how the adaptive network model can address complications arising from imperfect mental models, offering insights into improving team performance under various scenarios.

Organisational Learning and Adaptive Network Modeling
Building on the theme of organizational learning, Canbaloğlu et al. explore in Chapter "Learning of Safety and Security Management Through Cyberspace: An Adaptive Self-Modeling Network Model for Multilevel Organizational Learning" the multilevel nature of learning within healthcare organizations. Their work introduces an

adaptive self-modeling network model that encompasses individual, team, and organizational learning processes. This multilevel approach acknowledges the complexity of learning mechanisms and their interdependencies. By modeling these processes computationally, the chapter provides a detailed analysis of how different types of learning interact and support each other. The adaptive network model is designed to facilitate continuous improvement and adaptation, enabling healthcare organizations to evolve and respond to new challenges effectively.

Shared Mental Models and Organisational Learning in Health Care
Both chapters underscore the importance of adaptive network modeling in understanding and enhancing the cognitive and learning processes within healthcare teams and organizations. These models provide a computational framework for analyzing how shared mental models and organizational learning contribute to safety and performance. By simulating realistic scenarios, the research demonstrates the practical applications of these models in addressing common challenges in healthcare settings. The insights gained from these studies are crucial for developing strategies to improve teamwork, communication, and organizational learning, ultimately leading to better patient outcomes and a safer healthcare environment.

Conclusion
The second part of this book lays the groundwork for understanding the complex dynamics of teamwork and organizational learning in healthcare. Through the exploration of shared mental models and adaptive network modeling, this part provides essential insights into how healthcare teams can function more effectively and safely. The two chapters by Van Ments et al. and Canbaloğlu et al. offer valuable contributions to this field, demonstrating the practical applications of these concepts in enhancing team performance and facilitating organizational learning. As healthcare continues to face new challenges, the methodologies and frameworks presented in this part will be instrumental in fostering a culture of continuous improvement and adaptation, ensuring that healthcare teams can deliver the highest quality of care to their patients.

Developing a Safety and Security AI Coach: A Second-Order Adaptive Network Model of Shared Mental Models in Hospital Teamwork

Laila Van Ments, Jan Treur, Jan Klein, and Peter H. M. P. Roelofsma

Abstract This chapter describes a second-order adaptive network model for mental processes making use of shared mental models (SMM) for team performance. The chapter illustrates on the one hand the value of adequate SMM's for safe and efficient team performance and on the other hand in cases of imperfections of such shared team models how this complicates the team performance. To this end, the adaptive network model covers use, adaptation and control of the shared mental model. It is illustrated for an application context of a medical team performing a tracheal intubation, executed by a nurse and a medical specialist. Simulations illustrate how the adaptive network model is able to address the type of complications that can occur in realistic scenarios.

Keywords Shared mental model · Second-order network model · Hospital · Team performance · Healthcare safety

---

L. Van Ments
Fullpath, Jerusalem, Israel
e-mail: lailavanments@hotmail.com; laila@fullpath.com

J. Treur (✉)
Social AI Group, Vrije Universiteit Amsterdam, Amsterdam, The Netherlands
e-mail: j.treur@vu.nl; j.treur@hhs.nl

J. Treur · P. H. M. P. Roelofsma
Risk Management and Cybersecurity Group, Centre of Expertise Cybersecurity, The Hague University of Applied Sciences, The Hague, The Netherlands
e-mail: p.h.m.p.roelofsma@hhs.nl

J. Klein
Delft University of Technology, Delft, The Netherlands
e-mail: j.klein@tudelft.nl

P. H. M. P. Roelofsma
Department of Pediatrics, Division Neonatology, Erasmus University Medical Centre, Erasmus MC, University Medical Center-Sophia Children's Hospital, Rotterdam, The Netherlands

P. H. M. P. Roelofsma et al. (eds.), *Using Shared Mental Models and Organisational Learning to Support Safety and Security Through Cyberspace: A Computational Analysis Approach*, Studies in Systems, Decision and Control 570,
https://doi.org/10.1007/978-3-031-72075-8_2

1 Introduction

The concept of a shared mental model (SMM) has recently received increased attention in medical team performance literature as well as in other domains. SMM's are often brought in relation to the quality of team performance and safety (Burthscher et al. 2011; Wilson 2019; Todd 2018; Higgs et al. 2018; Seo et al. 2021). A team has a shared mental model when relevant knowledge structures concerning how reality works or should work are held by all team members and when there is sufficient alignment in the internal representations of these knowledge structures (Fischhof and Johnson 1997; Jones and Roelofsma 2000; Mathieu et al. 2000). Like mental models in general, shared mental models are used in mental processes for internal mental simulation and decision making based on their outcomes; e.g., Craik (1943). Moreover, they often are adaptive in the sense that they can be learnt or forgotten, and for such adaptation usually a form of control is applied. These aspects of shared mental models are all addressed in the current chapter.

In this chapter, it is shown how the adaptive network-oriented modeling approach from Treur (2020) and the three-level cognitive architecture for mental models described in Van Ments and Treur (2021c) can be used to model second-order adaptive mental processes involving shared mental models. It is illustrated in particular for the mental processes of members of a medical team. The real-world challenge addressed here is to cover (1) the errors and other imperfections that are daily practice in such teams and (2) the way in which such teams handle them.

In Sect. 2, a general introduction and background is described. This section also describes the domain specifics of the example scenario: the use case. This use case is a tracheal intubation performed by a medical team consisting of a specialist and a nurse. In Sect. 3 the design of the adaptive network model for this type of shared mental model is presented. Section 4, then, describes the illustrative simulation examples. Section 5 gives the chapter's main conclusions and provides a discussion for further extensions of the adaptive network model using SMM to support team performance and healthcare safety.

2 Background

The second-order adaptive network model introduced here integrates knowledge of mental models from psychology, team mental models from social sciences, hospital protocols from medical- and safety sciences, and the AI-domain of network-oriented modeling.

2.1 Mental Models

For the history of the mental model area, often Kenneth Craik is mentioned as a central person. In his book (Craik 1943), he describes a mental model as a *small-scale model* that is carried by an organism within its head as follows; see also Williams (2018):

> If the organism carries a "small-scale model" of external reality and of its own possible actions within its head, it is able to try out various alternatives, conclude which is the best of them, react to future situations before they arise, utilize the knowledge of past events in dealing with the present and future, and in every way to react in a much fuller, safer, and more competent manner to the emergencies which face it. (Craik 1943, p. 61)

Note that this quote covers both the usage of a mental model based on so-called internal mental simulation ('try out various alternatives') and the learning of it ('utilize the knowledge of past events'). Other authors also have formulated what mental models are. For example, with an emphasis on causal relations, Shih and Alessi (1993, p. 157) explain that

> By a mental model we mean a person's understanding of the environment. It can represent different states of the problem and the causal relationships among states.

De Kleer and Brown (1983) describe a mental model as the envisioning of a system, including a topological representation of the system components, the possible states of each of the components, and the structural relations between these components, the running or execution of the causal model based on basic operational rules and on general scientific principles.

2.2 Shared Mental Models

Team errors have often been linked to inadequacies of the shared mental model and the lack of adaptivity of it Fisschoff and Johnson (1997), Jones and Roelofsma (2000), Mathieu et al. (2000), Burthscher et al. (2011), Wilson (2019), Todd (2018). This has major implications for health care and patent safety in the operation room, e.g., concerning open heart operation and tracheal intubation (Higgs et al. 2018; Seo et al. 2021). Jones and Roelofsma (2000) discuss four types of team errors resulting from inadequate shared mental models.

The first is called 'false consensus'. The false consensus effect (Ross et al. 1977; Krueger 1998) refers to the tendency to overestimate the degree of similarity between self and others team members and this may result in biased judgements or team decisions. It is often described as people's tendency to 'see their own behavioural choices and judgements as relatively common and appropriate to existing circumstances while viewing alternative responses as uncommon, deviant, or inappropriate'.

A second type of team error and perhaps the most well-known is 'groupthink'; e.g., Janis (1972), Kleindorfer et al. (1993). It is often described as a mode of thinking that people engage in when they are deeply involved in a cohesive in-group, when the members' striving for unanimity overrides their motivation to realistically appraise alternative courses of action. Groupthink refers to a deterioration of mental efficiency and reality testing that results from in-group pressures.

A third type of team error resulting from inadequate shared mental model is group polarization; e.g., Isenberg (1986). This refers to the phenomenon that occurs when the position that is held on an issue by the majority of the group members is intensified as a result of discussion. For example, if group members are initially generally in favour of a particular preference of action, then group discussion will further enhance the favorability of this preference at an individual level. There are two special cases of group polarization. One is termed risky shift and occurs when a group, overall, becomes more risk seeking than the initial average risk seeking tendencies of the individual members. The other is termed cautious shift and occurs when groups become more risk averse than the initial average risk averse tendencies of the individual members. In both cases the average response of the individual group members is more extreme after discussion. Such shifts in preference have been demonstrated by an overwhelming number of studies.

A fourth team error is labelled escalation of commitment; e.g., Bazerman et al. (1984), Staw (1976). This refers to the tendency for individuals or groups to continue to support a course of action despite evidence that it is failing. In other words, it is the tendency for a decision to support a previous decision for which there was a negative outcome. The specific concern is with non-rational escalation of commitment with a degree to which an individual escalates commitment to a previously selected course of action beyond that which a rational.

2.3 Case Description

The general setting of the addressed case is an emergency department where an emergency team is coming together for preparing to intubate a critically ill patient with deteriorating conscious state. The airway has been assessed as being normal and there is no expectation that there are going to be any difficulties with intubation. A doctor (D) is called in to perform a tracheal intubation in collaboration with a nurse (N). In general, a tracheal intubation induces stress for D and A. The call of the doctor triggers the activation of the initial state of a shared mental model with separate roles and activities for the tracheal intubation for the D and N. The roles and activities are unique for D and N. The roles for the doctor are:

- team leader
- prepare team
- prepare for difficulties
- intubator.

The roles for the nurse are:

- intubator's assistant
- prepare patient
- prepare equipment
- prepare drugs
- give drugs
- monitoring the patient
- cricoid force
- runner for help and/or additional equipment.

In addition to the allocation of roles, the shared mental model contains the corresponding (temporal) sequence of activities for D and N. For the chosen example scenario based on an imperfect shared mental model considered here, this consists of the following sequence.

- The nurse prepares the patient.
- According to the protocol she should then have performed the preparation of the equipment; but she forgets this and goes on to perform the preparation of the drugs.
- The doctor executes pre oxygenation and
- starts with the preparation of the team and
- the preparation for difficulties.
- The nurse listens to and observes the doctor's team preparation.
- The nurse gives drugs to the patient and
- applies cricoid to the patient.
- Then the doctor initiates the executing of plan A Larynscopy and
- starts the first intubation attempt.
- The nurse assists the doctor in the intubation attempt.
- The nurse monitors the patient.
- When the first attempt is finished, the nurse seeks confirmation of its success by monitoring the capnograph. Then N realizes the earlier omission and sees that the capnograph is not active.

The intubation attempt is repeated with the exclusion of the preparation and the giving of the drugs to the patient by N. Also the preparation of the team for the intubation and for difficulties are not performed anymore by D. All other tasks are repeated in a second round and when this is not successful also in a third attempt. According to the protocol the doctor should have asked for help when the third attempt is not successful. But she does not do this.

2.4 Network-Oriented Modeling

The Network-Oriented Modelling approach based on temporal-causal networks from Treur (2016, 2020) is a suitable modeling approach to represent causal relations and the way they can be processed to generate mental processes, as needed for the use of shared mental models as described above. In particular, in Treur (2020) it is described how adaptive networks of different orders can be modelled relatively easily. Therefore, this approach was used to design a second-order adaptive network model for using shared mental models in a team member's mental processing and acting.

Network nodes X have state values indicated by real numbers $X(t)$ that vary over time t; nodes are also called states. The characteristics defining a network model are:

- **Connectivity characteristics**:
  Connections from states X to Y, having *connection weights* $\boldsymbol{\omega}_{X,Y}$ specifying their strengths
- **Aggregation characteristics**:
  Each state Y has a *combination function* $\mathbf{c}_Y$ that specifies how impact from all incoming connections on Y is aggregated. Based on a list of basic combination functions $\mathbf{bcf}_i$ (each with some parameters) provided by an available library, such a combination function can be specified by weights $\boldsymbol{\gamma}_i$ and parameters $\boldsymbol{\pi}_{i,j}$ for these basic combination functions $\mathbf{bcf}_i$
- **Timing characteristics**:
  Each state Y has a *speed factor* $\boldsymbol{\eta}_Y$ specifying how fast Y changes.

The numerical representation created by the available dedicated software environment is based on the following equations based on the above network characteristics (where $X_1, \ldots, X_k$ are the states from which state Y gets incoming connections):

$$Y(t + \Delta t) = Y(t) + \boldsymbol{\eta}_Y\left[\mathbf{c}_Y\left(\boldsymbol{\omega}_{X_1,Y}X_1(t), \ldots, \boldsymbol{\omega}_{X_k,Y}X_k(t)\right) - Y(t)\right]\Delta t \qquad (1)$$

A computational network engine developed within this software environment based on the generic Eq. (1) takes care for the processing of all network states thereby using their connections and other network characteristics.

2.5 *Self-modeling Networks to Model Adaptivity and Control*

First-order adaptation (also called plasticity) will be applied to the strength of the relations (connections) within a mental model: by Hebbian learning (Hebb 1949) they can become stronger and by extinction (forgetting) they can become weaker. In addition, second-order adaptation will be applied to model as a form of meta-plasticity (Abraham and Bear 1996; Garcia 2002), for the control effect of the contextual stress on the first-order adaptation process of learning and forgetting. To design network models that are adaptive, the concept *self-modeling network* (also called reified network) introduced in Treur (2020) has turned out very useful. A self-modeling network is obtained if for some of the network characteristics $\boldsymbol{\omega}$, $\mathbf{c}$, $\boldsymbol{\eta}$ as introduced above, network states are added to the network that represent their value. For example, for a connection weight $\boldsymbol{\omega}_{X,Y}$ an additional state $\mathbf{W}_{X,Y}$ (called *self-model state*) is added to the network that represents this weight and is used for that weight in the processing. Also, notations such as **IW**, **LW** and **RW** can be used instead of **W**, to distinguish different types of connection weights (see Sect. 3 for examples). For such an additional network **W**-state, also additional network characteristics are added to get an adequate embedding in the obtained self-modeling network. Next, as another example, for the combination function of a self-model state $\mathbf{W}_{X,Y}$, a persistence parameter $\boldsymbol{\mu}_{\mathbf{W}_{X,Y}}$ can be used that is represented by another self-model state $\mathbf{M}_{\mathbf{W}_{X,Y}}$. The latter network state is a *second-order self-model state* as it represents a network characteristic related to (first-order) self-model state $\mathbf{W}_{X,Y}$. These two types of self-model states will be used in the adaptive network model introduced in Sect. 3 to model learning of (shared) mental models and forgetting them under stressful circumstances.

3 The Adaptive Network Model Using a Shared Mental Model

As also indicated in the quote from Craik (1943) in Sect. 2, mental models are not only used by a form of internal mental simulation, but are also adaptive in that over time they are learned and can be forgotten. Moreover, cognitive control is exerted over these processes. To cover these three types of mental processes, the second-order adaptive network model introduced here follows the generic three-level cognitive architecture depicted in Fig. 1, as described in Van Ments and Treur (2021c). For more details and several case studies, see (Treur and Van Ments, 2022) and for applications to multilevel organisational learning, see (Canbaloğlu et al. 2023). Here the base level (lower, pink plane) models the use of a mental model by internal simulation, the adaptation level (middle, blue plane) the learning (and forgetting) of a mental model, and the control level (upper, purple plane) models the control of these processes.

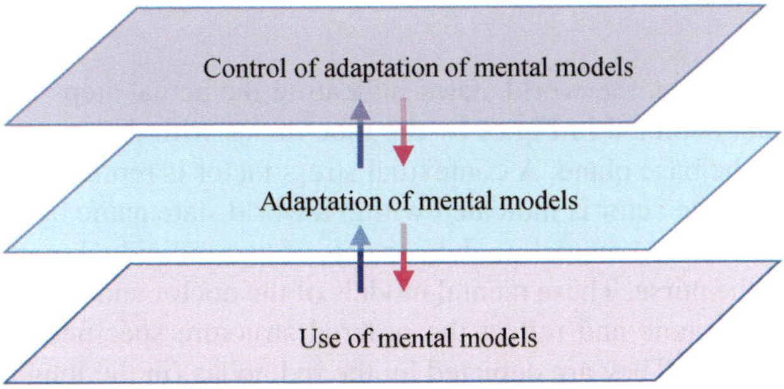

Fig. 1 Cognitive architecture with three levels of mental processing for usage (by internal mental simulation), adaptation (by learning and forgetting), and (cognitive) control of mental model adaptation

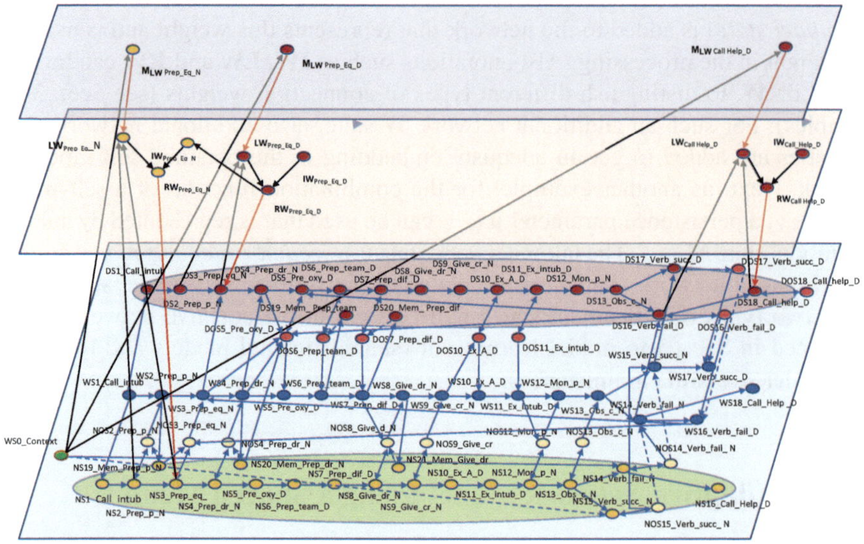

Fig. 2 Connectivity of the designed adaptive network model for the shared mental model. It includes the two mental models of the nurse (long yellow oval) and of the doctor (long red oval) and the self-models for the first-order (the pink plane) and second-order (the purple plane) adaptation. Dashed connections indicate connections with negative weights

The adaptive network model introduced here is a refinement and instantiation of this generic architecture for the scenario described in Sect. 2. It has connectivity as depicted in Fig. 2; for an explanation of the most important states, see Tables 1, 2 and 3. The scenario concerns a sequence of actions with actors assigned performing them and their temporal order, according to the realistic example scenario as described in Sect. 2.

3.1 Base Level: Overview

Within the base plane, the world states indicating the actual steps in the world for this scenario are depicted in Fig. 2 by the blue nodes with their connections in the middle area of the base plane. A contextual stress factor is represented by the green node on the left. The actor is indicated within a world state name by D for doctor or N for nurse. The shared mental model consists of two individual mental models for the doctor and the nurse. These mental models of the doctor and the nurse are shown in the base level plane and reflect the ordered structure specified in the use case discussed in Sect. 2. They are depicted by the red nodes (in the long light-red oval) and yellow nodes (in the yellow-green oval) in the base plane and their connections, respectively.

Table 1 Overview of the world states (WS) and the mental model states for the doctor (DS) and nurse (NS) reflecting these world states

World, Doctor and Nurse				Explanation
WS0			Context	Contextual stress factor
WS1	DS1	NS1	Call_intub	External call for intubation
WS2	DS2	NS2	Prep_p_N	Preparation of the patient by the nurse
WS3	DS3	NS3	Prep_eq_N	Preparation of the intubation equipment by the nurse
WS4	DS4	NS4	Prep_dr_N	Nurse prepares drugs for the patient
WS5	DS5	NS5	Pre_oxy_D	Doctor executes pre oxygenation
WS6	DS6	NS6	Prep_team_D	Doctor prepares the team for intubation
WS7	DS7	NS7	Prep_dif_D	Doctor prepares the team for difficulties
WS8	DS8	NS8	Give_dr_N	Nurse gives the patient drugs
WS9	DS9	NS9	Give_cr_N	Nurse applies cricoid to the patient
WS10	DS10	NS10	E_A_D	Doctor executes plan A Laryngoscopy
WS11	DS11	NS11	E_intub_D	Doctor intubates the patient
WS12	DS12	NS12	Mon_p_N	Nurse monitors patient
WS13	DS13	NS13	Obs_c_N	Nurse observes capnograph
WS14		NS14	Verb_fail_N	Nurse verbalizes failure of intubation
WS15		NS15	Verb_succ_N	Nurse verbalizes success of intubation
WS16	DS16		Verb_fail_D	Doctor verbalizes failure of intubation
WS17	DS17		Verb_succ_D	Doctor verbalizes success of intubation
WS18	DS18	NS18	Call_help_D	Doctor calls for help

Table 2 Overview of the memory states and ownership states for the doctor and nurse

Name		Explanation
DS19	Mem for Prep team D	Memory state of Doctor for the action of preparing the team
DS20	Mem for Prep dif D	Memory state of Doctor for the action of preparing the team for difficulties
DOS5	DOS for Pre_oxy_D	Ownership state for the action of preoxygenation
DOS6	DOS for Prep_team_D	Ownership state for the action of preparing the team
DOS7	DOS for Prep_dif_D	Ownership state for the action of preparing the team for difficulties
DOS10	DOS for E_A_D	Ownership state for the action of plan A Laryngoscopy by doctor
DOS11	DOS for E_intub_D	Ownership state for the action of intubating first attempt by doctor
DOS16	DOS for Verb_fail_D	Ownership state for the action of verbalizing that attempt has failed by doctor
DOS17	DOS for Verb_succ_D	Ownership state for the action of verbalizing that attempt has succeeded by doctor
DOS18	DOS for Call_help_D	Ownership state for the action of call for help, by doctor
NS19	Mem for Prep_p N	Memory state of Nurse for the action of preparing the patient
NS20	Mem for Prep_dr N	Memory state of Nurse for the action of preparing the drugs
NS21	Mem for Give_dr N	Memory state of Nurse for the action of giving the drugs
NOS2	NOS for Prep_N	Nurse Ownership State for Preparation patient
NOS3	NOS for Prep_eq_N	Nurse Ownership State for Preparation equipment
NOS4	NOS for Prep_dr_N	Nurse Ownership State for preparing drugs
NOS8	NOS for Give_d_N	Nurse Ownership State for Nurse gives drugs
NOS9	NOS for Give_cr_N	Nurse Ownership State for Nurse gives cricoid
NOS12	NOS for Mon_p_N	Nurse Ownership State for Nurse monitors patient
NOS13	NOS for Obs_c_N	Nurse Ownership State for observing capnograph
NOS14	NOS for Verb_fail_N	Nurse Ownership State for verbalizing that attempt has failed
NOS15	NOS for Verb_succ_N	Nurse Ownership State for verbalizing that attempt has succeeded

Table 3 Overview of the first-and second-order self-model states

Name		Explanation
W1	$\mathbf{LW}_{\text{Prep p Nurse} \rightarrow \text{Prep eq Nurse}}$	First-order self-model state for the Nurse's weight of the connection from the preparing the patient mental model state to the preparing the equipment mental model state as learnt by Hebbian learning
W2	$\mathbf{IW}_{\text{Prep p Nurse} \rightarrow \text{Prep eq Nurse}}$	First-order self-model state for the Nurse's weight of the connection from preparing the patient mental model state to preparing the equipment mental model state as learnt from instruction by the doctor
W3	$\mathbf{RW}_{\text{Prep p Nurse} \rightarrow \text{Prep eq Nurse}}$	First-order self-model state for the Nurse's overall weight of the connection from preparing the patient mental model state to preparing the equipment mental model state
W4	$\mathbf{LW}_{\text{Prep p N D} \rightarrow \text{Prep eq N D}}$	First-order self-model state for the Doctor's weight of the connection from preparing the patient by the Nurse mental model state to preparing the equipment by the Nurse mental model state as learnt by Hebbian learning
W5	$\mathbf{IW}_{\text{Prep p N D} \rightarrow \text{Prep eq N D}}$	First-order self-model state for the Doctor's weight of the connection from preparing the patient by the Nurse mental model state to preparing the equipment by the Nurse mental model state as known to the Doctor
W6	$\mathbf{RW}_{\text{Prep p N D} \rightarrow \text{Prep eq N D}}$	First-order self-model state for the Doctor's overall weight of the connection from preparing the patient by the Nurse mental model state to preparing the equipment by the Nurse mental model state
W7	$\mathbf{LW}_{\text{Verb fail D} \rightarrow \text{Call help D}}$	First-order self-model state for the Doctor's weight of the connection from verbalisation of failure mental model state to call for help mental model state as learnt by Hebbian learning
W8	$\mathbf{IW}_{\text{Verb fail D} \rightarrow \text{Call help D}}$	First-order self-model state for the Doctor's weight of the connection from verbalisation of failure mental model state to call for help mental model state as known to the Doctor
W9	$\mathbf{RW}_{\text{Verb fail D} \rightarrow \text{Call help D}}$	First-order self-model state for the Doctor's overall weight of the connection from verbalisation of failure mental model state to call for help mental model state
M1	$\mathbf{M}_{\mathbf{LW}_{\text{Prep p Nurse} \rightarrow \text{Prep eq Nurse}}}$	Second-order self-model state for the persistence factor of the Nurse's weight of the connection from preparing the patient mental model state to preparing the equipment mental model state as learnt by Hebbian learning
M2	$\mathbf{M}_{\mathbf{LW}_{\text{Prep p N D} \rightarrow \text{Prep eq N D}}}$	Second-order self-model state for the persistence factor of the Doctor's weight of the connection from preparing the patient by the Nurse mental model state to preparing the equipment by the Nurse mental model state as learnt by Hebbian learning
M3	$\mathbf{M}_{\mathbf{LW}_{\text{Verb fail D} \rightarrow \text{Call help D}}}$	Second-order self-model state for the persistence factor of the Doctor's weight of the connection from preparing the patient by the Nurse mental model state to preparing the equipment by the Nurse mental model state as learnt by Hebbian learning

The states within the mental models refer to the world states they model and like these world states they also specify an actor, indicated by D for doctor or N for nurse. The two individual mental models are two instances of an overall team mental model incorporating both the course of actions and the roles of the different team members for these actions. These individual instances of the team mental model can have differences, as in general not all team members will possess one and the same perfect team mental model.

3.2 *Base Level: Memory States in the Mental Models*

Within the mental models some specific states enable to take into account what has occurred in the past; these mental model states are called *memory states*. These are particularly useful if parts of the processes have to be repeated because of failures. Usually then only some of the actions have to be redone, while other actions can be skipped, as is illustrated in the addressed scenario. For example, preparation of the patient does not need to be redone, but preparation of the equipment has to be redone when the process has to be repeated. The memory states within the mental models are a crucial element to obtain this form of flexibility as they enable to model such issues in a context-sensitive manner taking into account the history of the process.

3.3 *Base Level: Action Ownership States*

By each of the two team members, their own mental model is used to determine their actions in the world. This goes through the member's action ownership states (indicated in light red for the doctor and in light yellow for the nurse). These ownership states are mental states but are not considered to be part of the mental models. Instead, they use input from the mental models and realise a form mediation from mental model to the real world by initiating the execution of the indicated actions, which leads to affecting the related world states. In this way, the mental models affect the decisions for actions activating the world states. Conversely, connections from world states to corresponding mental model states are (at some points) used to feed information about the world into the mental models.

3.4 *Middle Level: Adaptation of the Mental Models (Plasticity)*

The middle (blue) plane addresses the mental processes for learning and forgetting of the mental models. In particular, this addresses the connection within the nurse's mental model from the mental model state for preparation of the patient to the mental model state for preparation of the equipment. Inspired by Bhalwankar and Treur (2021a, b), where it was shown how instructional learning and observational learning of mental models can be integrated, in a similar manner two types of learning are covered here:

- Learning by instruction from the Doctor (modelled by the Nurse's **IW**-state and its incoming connection from the Doctor's **RW**-state)
- Hebbian learning (Hebb 1949) based on running the own internal simulation, among others triggered by observation (modelled by the Nurse's **LW**-state with its incoming connections from the two relevant Nurse's mental model states)

Table 4 Combination functions from the library used in the introduced network model

	Notation	Formula	Parameters
Steponce	**steponce**(V)	1 if $\alpha \le t \le \beta$, else 0	α start, β end time
Scalemap	**scalemap**$_{\lambda,\upsilon}$(V)	$\lambda + (\upsilon - \lambda)\, V$	Lower bound λ; Upper bound υ
Advanced logistic sum	**alogistic**$_{\sigma,\tau}$($V_1, \ldots, V_k$)	$[\frac{1}{1+e^{-\sigma(V_1+\cdots+V_k-\tau)}} - \frac{1}{1+e^{\sigma\tau})}]\,(1 + e^{-\sigma\tau})$	Steepness $\sigma > 0$ Excitability threshold τ
Hebbian learning	**hebb**$_{\mu}$(V_1, V_1, W)	$V_1V_2\,(1-W) + \mu\, W$	V_1,V_2 activation levels of the connected states; W activation level of the self-model state for the connection weight μ persistence factor

The values of these two states are integrated in the **RW**-state, which represents the overall value that is actually used as connection weight in the internal simulation at the base level. The Hebbian learning applied for the **LW**-state is modeled by applying combination function **hebb**$_{\mu}$ (shown in Table 4) for aggregation for the **LW**-states; this learning is not fully persistent as this combination function **hebb**$_{\mu}$ used includes a persistence factor μ that represents the fraction (of the learnt value) that persists per time unit. For example, if μ is 0.9, then every time unit 10% of the learnt value is lost (forgotten). In this way, the Hebbian learning also covers extinction or forgetting. This combination function parameter μ is one of the aggregation network characteristics of the (first-order) self-model represented in the middle plane.

3.5 *Upper Level: Control of the Adaptation of Mental Models (Metaplasticity)*

Within the adaptive network model introduced here, the extent μ of persistence is not assumed constant but depends on circumstances. This means that this network characteristic of the first-order self-model is adaptive, which is modelled by including a second-order self-model **M**-state within the upper-level plane that represents the adaptive value of μ. This enables modelling metaplasticity (Abraham and Bear 1996; Garcia 2002) of mental processes, which is plasticity of the plasticity. For the considered scenario, it is assumed that in particular a high stress level leads to a decreased value of the **M**-state; in this way forgetting due to stressful circumstances is modelled, in line with (Garcia 2002). This is specified by the (suppressing) upward connections from the stressful context state in the base level to the **M**-states.

The combination functions from the combination function library available within the software environment used here are shown in Table 4.

4 Simulation for the Example Scenario

Recall from the introduction that the main real-world challenge addressed for the designed adaptive network model is that it is able to cover.

(1) the errors and other imperfections that are daily practice in medical teams and
(2) the way in which such teams handle them.

This can be considered a performance indicator against which the model can be validated. In this section, it will be shown by the realistic example simulation scenario from Sect. 2 how the model indeed satisfies this performance indicator. In this simulation, a repeatedly unsuccessful intubation process is shown.

The network characteristics defining the network model introduced above have been specified in a standard table format (called role matrices) that can be used as input for the available dedicated software environment; see also the Appendix Sect. 6. When transferred to this software environment, these tables with network characteristics are automatically used by the incorporated generic differential Eq. (1) when running simulations. The example simulation discussed here was run over a time interval of 0 to 180 with step size $\Delta t = 0.5$. This provides us with graphs of simulations based on the values chosen for the network characteristics. In this simulation, a repeatedly unsuccessful intubation process is shown. The contextual stress level has been set relatively high (0.5). For reasons of clarity, the figures have split the world states (Fig. 3), the nurse's states (Fig. 5), the doctor's states (Fig. 4), and the adaptivity states (Fig. 6), but they all happen in the same simulation at the indicated time points.

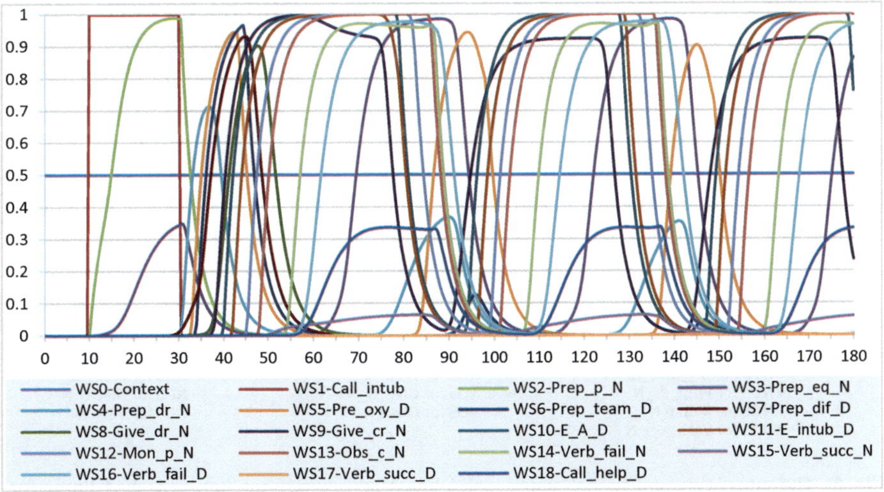

Fig. 3 World states of a repeated failing intubation process

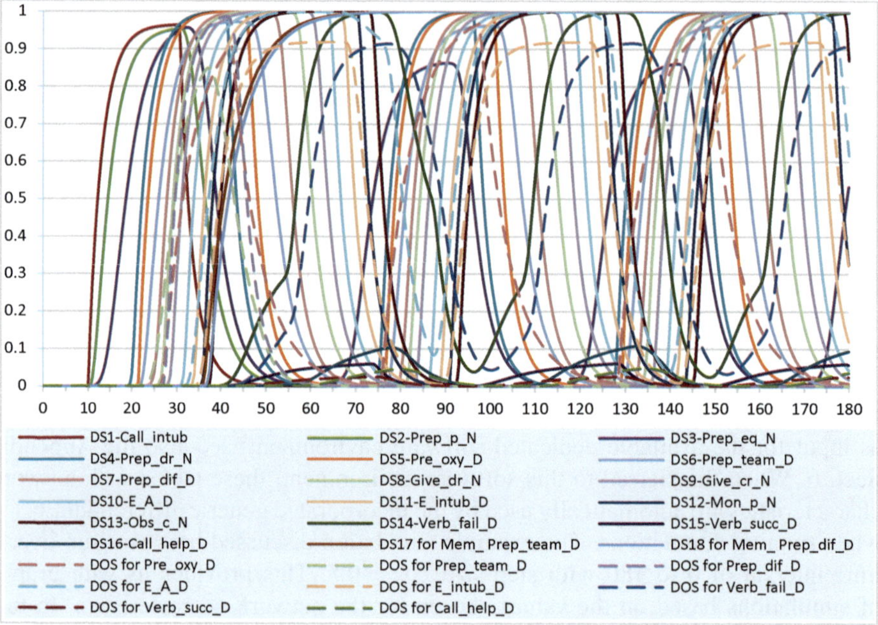

Fig. 4 The doctor's mental model states (solid lines) and ownership states (dashed lines) for a repeated failing process

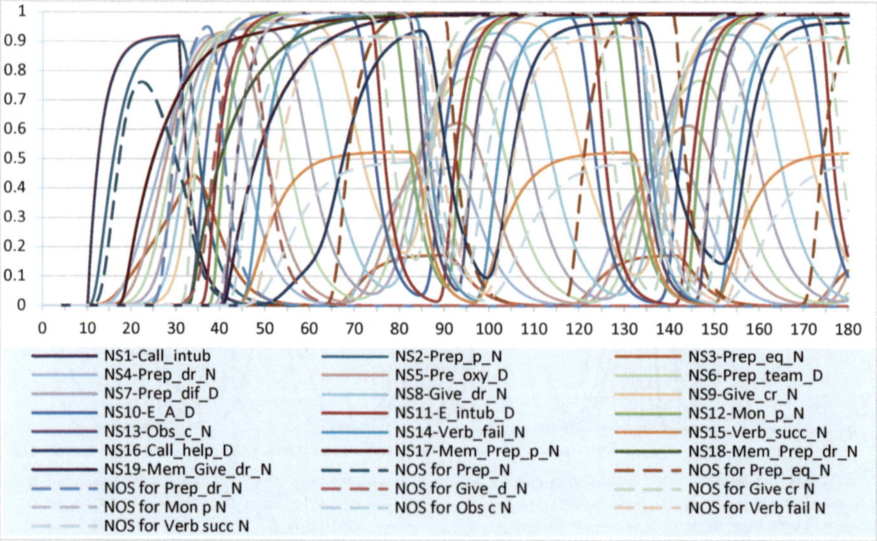

Fig. 5 The nurse's mental model states (solid lines) and ownership states (dashed lines) for a repeated failing intubation process

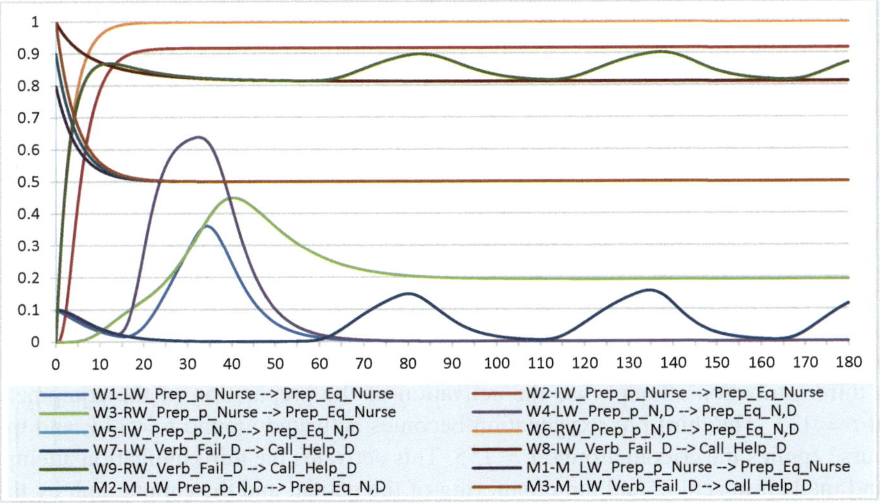

Fig. 6 The first- and second-order self-model states for adaptation (learning and forgetting) and control of it (stress leading to forgetting)

4.1 The World States

Figure 3 shows the simulation output for the world states. This shows how the actual process in the world proceeds. In time period $t = 10$–30 a call for intubation takes place, which sets in motion the intubation sequence for the scenario indicated by the use case described in Sect. 2. A bit after the call for intubation, the Nurse starts preparing the patient (the light green line). In this scenario, the purple line indicating the preparation of the equipment starting at time $t = 15$ does not reach an adequate level of activation, only around 0.375, meaning that this preparation of equipment is not (sufficiently) executed in the world. However, the next step in the scenario of preparing the drugs becomes active around $t = 33$ and does get activated enough. Subsequently, also the rest of the steps in the scenario continue as prescribed by the shared mental model. Between $t = 40$ and $t = 45$, the doctor's first actions all become activated: the pre-oxygenation of the patient, the preparation of the team and the preparation for difficulties. After this, the nurse continues with giving the drugs to the patient (dark green around $t = 47$), and applies the cricoid force right after. Now, the execution of the attempt laryngoscopy A, and the intubation action itself both become activated between $t = 47$ and $t = 50$. This also activates the nurse's actions to monitor the patient, around $t = 55$, and to observe the capnograph equipment around $t = 60$. Around this time point, the nurse will realize she did not prepare the equipment (remember the non-activated prepare equipment state), and verbalize the failed intubation attempt as a result, around $t = 67$. Soon after, around $t = 69$, the doctor confirms this by also verbalizing the failed intubation attempt.

After this verbalization of failure, the doctor and nurse will start their second attempt. This time, this starts with the nurse preparing the equipment; see how the light purple prepare equipment line now does reach activation around $t = 77$. Now, the preparation of the patient and drugs are skipped by the nurse because they do not need to happen more than once. These states do not get activated in the simulation and there is a slight gap, until the pre-oxygenation gets activated around $t = 93$: the orange line. After this, the doctor skips the preparation of the team and for difficulties, because these steps already happened and do not need to be repeated. Around $t = 100$, the nurse gives cricoid force, and the doctor starts the second intubation attempt. Again, the nurse monitors the patient, and the capnograph, but unfortunately also this intubation attempt fails. The nurse verbalizes this failed attempt around $t = 120$, and the doctor verbalizes the failure around $t = 125$. The team now continues with a third intubation attempt, see the activation of the preparation of the equipment at $t = 133$. The third pre-oxygenation becomes activated around $t = 145$, and the nurse applies the cricoid around $t = 155$. This activates the third intubation attempt and intubation action, and the monitoring of the patient and the capnograph by the nurse. Note that also in this attempt the same actions are skipped as in the second attempt, because they do not need to happen again. This attempt fails too and this is verbalized by the nurse at $t = 170$, and right after by the doctor as well. In the simulation the same pattern keeps on repeating after this time point. Note that after the failure verbalisations the 'call for help' state (the dark blue line) gets a small level of activation, up to around 0.35, but this is not enough to actually happen, so no help is called in this simulation scenario.

Figures 4 and 5 show for the addressed intubation scenario, what precedes the world state activations described above: respectively the internal simulations by the doctor and nurse of their own mental model and activating accordingly their ownership states for the actions they execute in the world.

4.2 The Doctor's Mental Processes Based on Her Mental Model

Figure 4 shows the doctor's mental model states and the doctor's ownership states simulated over time. At $t = 10$, the world state for the call for intubation becomes activated, hereby activating the sequence of events in the mental model of intubation of the doctor. Right after the call for intubation, the doctor's mental model for the nurse preparing the patient, equipment and drugs get activated at $t = 15$ (note that at this point this action only happens in the doctor's mental model, but not in the real world). Then the mental model states for the doctors first 'own' actions get activated: to pre oxygenate the patient, prepare the team and for difficulties, around $t = 20$. Now, the doctor's ownership states for the doctors' actions (pre oxygenate, preparing the team and for difficulties) get activated at $t = 30$, which will ultimately lead to the corresponding real-world actions. Around $t = 32$ some mental model states of actions the nurse has to do become activated, namely, to give the patient drugs and to apply cricoid force. This triggers the doctor's mental model state of starting plan A of intubation, and the actual intubation, also around $t = 32$, and slightly after that around $t = 35$ the ownership states for these actions. At around $t = 37$, the nurse's actions activate in the doctor's mental model: to monitor the patient and to observe the capnograph. This leads to the doctor to verbalize failure in her mental model around $t = 50$, and to develop activation of ownership of this verbalisation action a bit after that. The call for help does not get proper activation. This round ends around $t = 75$, after which the next round starts immediately (as an emergent process; there is no round information or control of rounds in the model).

4.3 The Nurse's Mental Processes Based on Her Mental Model

Figure 5 shows the nurse's mental model states and the nurse's ownership states simulated over time. At $t = 10$, the world state for the call for intubation becomes activated, hereby activating the sequence of events in the mental model of intubation of the nurse. Right after the call for intubation, the nurse's mental model for herself preparing the patient gets activated at $t = 15$, and right after also the ownership state for the first action gets activated, meaning the nurse executes the preparation of the patient. At around $t = 25$, the memory state of the nurse for preparing the patient reaches activation, meaning that the nurse can remember that she did this and does not have to repeat this action. Around this time, the prepare equipment mental model state reaches partial activation, but not enough to activate the ownership state for this action. From $t = 24$ until around $t = 37$, the mental model states for the preparation and execution actions of the intubation get activated, with the mental model state for the intubation reaching activation around $t = 39$. Also, the ownership states for most actions become activated in this time period, although the ownership state for 'prepare equipment' does not become activated, indicating that the nurse does not execute this action. At $t = 40$, the mental model state for monitoring the patient, and around $t = 44$ the mental model state for observing the capnograph become activated. At $t = 49$ the nurse's mental model state for verbalizing failure gets activation, and interestingly at $t = 69$, the nurse's mental model state for calling for help gets activated, even though this does not get executed by the doctor in the real world.

Note how also the memory states for preparing the patient, preparing the drugs and giving the drugs become activated at respectively $t = 25$, $t = 45$ and $t = 54$, after the ownership states for the same actions, causing the nurse to remember and not execute these actions in following rounds. At $t = 70$, the ownership state for preparing the equipment becomes activated, indicating the start of the second attempt at intubation. Note that the mental model state for the equipment preparation does not reach activation, showing that the nurse gets this input from an external source (the verbalization of failure of the intubation, by the doctor).

4.4 The Learning and Forgetting States

Figure 6 shows the activation levels of the states involved in adaptation (learning and forgetting) of the mental models, as shown in the first- and second-order self-model levels in Fig. 2. In the current model, for the sake of simplicity there are only three places in the model where learning and forgetting have been incorporated: in the nurse's mental model between the 'prepare equipment' state and the 'prepare drugs' state, in the doctor's mental model between the 'prepare equipment' state and the 'prepare drugs' state and finally in the doctor's mental model between the 'verbalization of failure' state and the 'call for help' state. As also discussed in Sect. 3, in each of these cases, the applied adaptation mechanism was built upon three sub-mechanisms, modelled in a network-oriented manner by additional states and connections:

- **LW**-states, representing the Hebbian learning. This means that the person learns by activation of connected states of the mental model, for example by using the mental model for internal simulation or triggered by observing the corresponding states in the real world. In the model, an **LW**-state is activated from the source and destination mental model states of the learnt connection. The persistence involved in the adaptation represented by an **LW**-state is controlled by an **M**-state in the second-order adaptation level (which represents the persistence factor μ).
- **IW**-states, representing instructional learning. This means that the agent learns by getting some sort of instruction, either during the process or before. In the model, the nurse learns from the doctor's instructions. The doctor applies previously acquired knowledge for this.
- **RW**-states, which models just a combination of the above two states.

In Fig. 6, the above three mechanisms are shown in a simulation graph. All three **M**-states start at a high level (0.8, 0.9 or 1) and due to the high stress level drop to 0.5 around $t = 10$. There are a few states that get activated around $t = 15$: **RW** for preparing the equipment in the doctor's mental model (W6), **RW** for the verbalization of failure by the doctor (W9) and **IW** for calling for help by the doctor (W8).

Besides that, none of the learning states really reach proper activation. Therefore, while the learning mechanisms in principle are working as can be seen from the changing activation levels, they do not have a strong enough positive learning impact but instead an overall negative impact, due to the forgetting that is induced by the high stress level (Garcia 2002), making the persistence factor representations **M** low. This negative effect contributes to the omission of the preparation of the equipment by the nurse in the first round and also to not calling for help after each failed round.

5 Discussion

In this chapter, a quite flexible and second-order adaptive computational network model was introduced enabling simulation of mental processes involving a shared mental model for teams, illustrated for a doctor and a nurse performing tracheal intubation of a patient. It is mainly based on material from Van Ments et al. (2021b). The model allows for the representation and processing of the actions in the world, the preceding internal simulation of the two mental models of the nurse and doctor and the dynamics of the interactions between them via the ownership states that represent how the actors actually decide based on the internal simulation and perform the actions. A contextual stress factor is included that determines the effects of stress on these mental processes, in particular forgetting parts of a mental model as a negative effect of metaplasticity (Abraham and Bear 1996) as described in more detail in Garcia (2002). Accordingly, in simulation experiments it was shown how learning and forgetting of shared mental models can happen and how failing team processes and redoing them can be modelled in a context-sensitive and flexible manner.

The computational model was developed based on the network-oriented modeling approach described in Treur (2016, 2020) and its dedicated software environment described in Treur (2020, Chap. 9). In earlier work it has been shown how this modeling approach enables modeling of different types of mental models, for example, for mental models representing flashback experiences in PTSD (van Ments and Treur 2021a), for joint decision making based on certain metaphors (Van Ments and Treur 2021b), and for how a mental God-model can affect empathic and disempathic human behaviour (Van Ments et al. 2016, 2018, 2022). Other computational approaches such as described in Dionne et al. (2010), Outland (2019), Scheutz (2013), use agent-based models (which usually brings more added complexity), dynamical system models or program code. This lacks a well-defined description at a modelling level and makes it hard if not impossible to incorporate second-order adaptation in a transparent manner in the model, as needed here. Otherwise it is hard to cover the positive and negative effects of metaplasticity as described in Garcia (2002). In contrast, the current chapter describes at a modelling level a very flexible second-order adaptive network model. It addressing shared mental models for teamwork and illustrates this by a hospital teamwork scenario.

A less flexible precursor of the second-order adaptive network model introduced in the current chapter was described in Van Ments et al. (2021a, b). The latter network model only addressed parts (not including memory states) of the base level. Therefore, it was nonadaptive and also did not cover errors and other imperfections of the team members occurring in their daily practice. For example, it was shown that the adaptive network model introduced here is able to model forgetting part of a shared mental model as illustrated in the simulated scenario, failure of the action and redoing the process in a context-sensitive manner after it has turned out to fail. In this way as a form of validation, the current model has be shown to be much closer to real-world team processes.

To achieve this, the two levels of self-models for first-order and second-order adaptation are new in the current model; this enabled modelling the positive and negative effects of metaplasticity as described in Garcia (2002) in the form of learning and forgetting parts of a mental model. In addition, also the use of memory states to be able to redo a failed attempt in a history-context-sensitive manner is new, providing a mechanism for only redoing the actions that are needed and skipping the ones that are not needed again as can be observed naturally in practice; the precursor model from Van Ments et al. (2021a, b) is much more rigid and lacks also this type of flexibility.

A next step will be to model the occurrence of a wider variety of errors and incidents—and their solutions—that are specific for team and group performance. Examples are: false consensus, group think, escalation of commitment and group polarization (Jones and Roelofsma 2000). Another relevant issue would be to examine the effect of group dynamics as a function of the team size. Often it is suggested that increasing the team, would lead to more safety and efficiency (Higgs et al. 2018) but increasing group size also leads to new group dynamics with corresponding potential problems. As mentioned, shared mental models are used in a variety of safety-related situations such as aviation, firefighting teams, dealing rooms, shipping control, etc. An important line for future research is to examine the descriptive validity of our model and further extensions of it for such domains.

Appendix: Specification of the Network Model

In this section the full specification of the adaptive network model is given by role matrices as explained in Treur (2020, Chap. 9) and the initial values used are listed. The role matrices provide a good basis for communication between modellers, and they can be used as input for the software environment to run simulations based on them.

Role matrices provide a compact standardised and structured table format that can be used to specify the network charactistics $\omega_{X,Y}$, $\gamma_{j,Y}$, $\pi_{i,j,Y}$, η_Y that define a design of a (self-modeling) network model. As discussed in Sect. 4, the three types of characteristics are:

Connectivity specified in role matrices **mb** (for base connections $X \rightarrow Y$) and **mcw** (for connection weights $\omega_{X,Y}$).

Aggregation specified by role matrices **mcfw** (for combination function weights $\gamma_{j,Y}$) and **mcfp** (for combination function parameters $\pi_{i,j,Y}$).

Timing specified by role matrix **ms** (for speed factors η_Y).

Role matrices have rows for all of the states in the network model, indicated by the state names X_i on the left side (which is also followed by a more informative name).

For *connectivity* characteristics, in role matrix **mb**, in each row it is listed which are the states X_j from which X_i has incoming connections. For example in the row for X_8 it is indicated that state X_8 (which is also named WS3-Prep_eq_N) has incoming connections from states X_7, X_{70} and X_{21}. In role matrix **mcw**, for each of these connections weights are specified in the corresponding cell. For example, in the row for X_8, in the first column it is indicated that the the weights from all connections to X_8 are 1. In this way, a compact overview is obtained for all connection weights $\omega_{X,Y}$ of the network model. Note that in some of the (peach-red shaded) cells of **mcw** no numbers are specified but state names X_k. This is the way in which it is indicated that that X_k is a self-model state which plays the role of the connection weight for the cell in which it is specified. In a computational sense, this means that at any time in computations the value of that state is used for the concerning connection weight. This makes the adaptation of the connection weights happen. Similarly, for *timing* characteristics, in role matrix **ms** a list of speed factors for all states is given.

For *aggregation* characteristics, in role matrix **mcfw** the combination function weights are specified. In addition, in role matrix **mcfp** the parameters for the combination functions are specified. For example, in **mcfw** it is indicated that state X_8 uses the advanced logistic sum function **alogistic** (see Table 4) with steepness parameter $\sigma = 5$ and threshold parameter $\tau = 1.1$ indicated in **mcfp**. Also in some of the (peach-red shaded) cells of **mcfp**, no numbers are specified but state names X_k. In particular, this holds for the first-order self-model **LW**-states X_{78}, X_{81}, and X_{84} which model Hebbian learning; it is indicated in **mcfp** that their persistence factors are adaptive and represented by second-order self-model $\mathbf{M_{LW}}$-states X_{87}, X_{88}, and X_{89}, respectively.

Finally, also a list of initial values is given (Figs. 7, 8, 9, 10, 11 and 12).

mb	base-connectivity	1	2	3	4	5
X_1						
X_2						
X_3						
X_4						
X_5	WS0-Context	X_5				
X_6	WS1-Call_intub	X_6				
X_7	WS2-Prep_p_N	X_6	X_{69}			
X_8	WS3-Prep_eq_N	X_7	X_{70}	X_{21}		
X_9	WS4-Prep_dr_N	X_8	X_{71}			
X_{10}	WS5-Pre_oxy_D	X_9	X_{42}			
X_{11}	WS6-Prep_team_D	X_{10}	X_{43}			
X_{12}	WS7-Prep_dif_D	X_{11}	X_{44}			
X_{13}	WS8-Give_dr_N	X_{12}	X_{72}			
X_{14}	WS9-Give_cr_N	X_{13}	X_{73}			
X_{15}	WS10-E_A_D	X_{14}	X_{45}			
X_{16}	WS11-E_intub_D	X_{15}	X_{46}			
X_{17}	WS12-Mon_p_N	X_{16}	X_{74}			
X_{18}	WS13-Obs_c_N	X_{17}	X_{75}			
X_{19}	WS14-Verb_fail_N	X_{18}	X_{76}	X_{77}		
X_{20}	WS15-Verb_succ_N	X_{18}	X_{77}	X_{76}		
X_{21}	WS16-Verb_fail_D	X_{19}	X_{47}	X_{48}		
X_{22}	WS17-Verb_succ_D	X_{20}	X_{48}	X_{47}		
X_{23}	WS18-Call_help_D	X_{19}				
X_{24}	DS1-Call_intub	X_6				
X_{25}	DS2-Prep_p_N	X_{24}				
X_{26}	DS3-Prep_eq_N	X_{25}	X_8			
X_{27}	DS4-Prep_dr_N	X_{26}	X_9			
X_{28}	DS5-Pre_oxy_D	X_{27}	X_{42}			
X_{29}	DS6-Prep_team_D	X_{28}	X_{43}			
X_{30}	DS7-Prep_dif_D	X_{29}	X_{44}			
X_{31}	DS8-Give_dr_N	X_{30}				
X_{32}	DS9-Give_cr_N	X_{31}				
X_{33}	DS10-E_A_D	X_{32}	X_{45}			
X_{34}	DS11-E_intub_D	X_{33}	X_{46}			
X_{35}	DS12-Mon_p_N	X_{34}				
X_{36}	DS13-Obs_c_N	X_{35}				

Fig. 7 Connectivity characteristics: role matrix **mb**

X_{37}	DS16-Verb_fail_D	X_{36}	X_{47}	X_{19}		
X_{38}	DS17-Verb_succ_D	X_{36}	X_{48}	X_{20}		
X_{39}	DS18-Call_help_D	X_{37}	X_{49}			
X_{40}	DS19-Mem-Prep_team_D	X_{11}	X_{40}			
X_{41}	DS20-Mem-Prep_dif_D	X_{12}	X_{41}			
X_{42}	DOS5-Pre_oxy_D	X_{9}	X_{28}	X_{37}	X_{27}	X_{8}
X_{43}	DOS6-Prep_team_D	X_{29}	X_{40}			
X_{44}	DOS7-Prep_dif_D	X_{30}	X_{41}			
X_{45}	DOS10-E_A_D	X_{14}	X_{33}			
X_{46}	DOS11-E_intub_D	X_{34}				
X_{47}	DOS16-Verb_fail_D	X_{37}				
X_{48}	DOS17-Verb_succ_D	X_{38}				
X_{49}	DOS18-Call_help_D	X_{39}				
X_{50}	NS1-Call_intub	X_{6}				
X_{51}	NS2-Prep_p_N	X_{50}	X_{69}			
X_{52}	NS3-Prep_eq_N	X_{51}	X_{70}			
X_{53}	NS4-Prep_dr_N	X_{52}	X_{71}			
X_{54}	NS5-Pre_oxy_D	X_{53}				
X_{55}	NS6-Prep_team_D	X_{54}				
X_{56}	NS7-Prep_dif_D	X_{55}				
X_{57}	NS8-Give_dr_N	X_{56}	X_{72}			
X_{58}	NS9-Give_cr_N	X_{57}	X_{73}			
X_{59}	NS10-E_A_D	X_{58}				
X_{60}	NS11-E_intub_D	X_{59}				
X_{61}	NS12-Mon_p_N	X_{60}	X_{74}			
X_{62}	NS13-Obs_c_N	X_{61}	X_{75}			
X_{63}	NS14-Verb_fail_N	X_{62}	X_{5}	X_{76}		
X_{64}	NS15-Verb_succ_N	X_{62}	X_{5}	X_{77}		
X_{65}	NS18-Call_help_D	X_{63}	X_{3}			
X_{66}	NS19-Mem_Prep_p_N	X_{7}	X_{66}			
X_{67}	NS20-Mem_Prep_dr_N	X_{9}	X_{67}			
X_{68}	NS21-Mem_Give_dr_N	X_{13}	X_{68}			
X_{69}	NOS2-Prep_p_N	X_{51}	X_{66}			
X_{70}	NOS3-Prep_eq_N	X_{52}	X_{21}	X_{66}		
X_{71}	NOS4-Prep_dr_N	X_{53}	X_{67}			
X_{72}	NOS8-Give_dr_N	X_{12}	X_{57}	X_{68}		
X_{73}	NOS9-Give_cr_N	X_{58}				
X_{74}	NOS12-Mon_p_N	X_{16}	X_{61}			
X_{75}	NOS13-Obs_c_N	X_{62}				
X_{76}	NOS14-Verb_fail_N	X_{63}				
X_{77}	NOS15-Verb_succ_N	X_{64}				
X_{78}	W1-**LW**_Prep_p_Nurse→Prep_eq_Nurse	X_{51}	X_{52}	X_{78}		
X_{79}	W2-**IW**_Prep_p_Nurse→Prep_eq_Nurse	X_{83}				
X_{80}	W3-**RW**_Prep_p_Nurse→Prep_eq_Nurse	X_{78}	X_{79}			
X_{81}	W4-**LW**_Prep_p_N,D→Prep_eq_N,D	X_{25}	X_{26}	X_{81}		
X_{82}	W5-**IW**_Prep_p_N,D→Prep_eq_N,D	X_{82}				
X_{83}	W6-**RW**_Prep_p_N,D→Prep_eq_N,D	X_{81}	X_{82}			
X_{84}	W7-**LW**_Verb_fail_D→Call_help_D	X_{37}	X_{39}	X_{84}		
X_{85}	W8-**IW**_Verb_fail_D→Call_help_D	X_{85}				
X_{86}	W9-**RW**_Verb_fail_D→Call_help_D	X_{84}	X_{85}			
X_{87}	M1-**M_LW**_Prep_p_Nurse→Prep_eq_Nurse	X_{5}				
X_{88}	M2-**M_LW**_Prep_p_N,D→Prep_eq_N,D	X_{5}				
X_{89}	M3-**M_LW**_Verb_fail_D→Call_help_D	X_{5}				

Fig. 7 (continued)

mcw	connection weights	1	2	3	4	5
X_1						
X_2						
X_3						
X_4						
X_5	WS0-Context	1				
X_6	WS1-Call_intub	1				
X_7	WS2-Prep_p_N	1	1			
X_8	WS3-Prep_eq_N	1	1	1		
X_9	WS4-Prep_d_N	1	1			
X_{10}	WS5-Pre_oxy_D	1	1			
X_{11}	WS6-Prep_team_D	1	1			
X_{12}	WS7-Prep_dif_D	1	1			
X_{13}	WS8-Give_d_N	1	1			
X_{14}	WS9-Give_cr_N	1	1			
X_{15}	WS10-E_A_D	1	1			
X_{16}	WS11-E_intub_D	1	1			
X_{17}	WS12-Mon_p_N	1	1			
X_{18}	WS13-Obs_c_N	1	1			
X_{19}	WS14-Verb_fail_N	1	1	-1		
X_{20}	WS15-Verb_succ_N	1	1	-1		
X_{21}	WS16-Verb_fail_D	1	1	-1		
X_{22}	WS17-Verb_succ_D	1	1	-1		
X_{23}	WS18-Call_help_D	1				
X_{24}	DS1-Call_intub	1				
X_{25}	DS2-Prep_p_N	1				
X_{26}	DS3-Prep_eq_N	X_{83}	1			
X_{27}	DS4-Prep_d_N	1	1			
X_{28}	DS5-Pre_oxy_D	1	0.1			
X_{29}	DS6-Prep_team_D	1	0.1			
X_{30}	DS7-Prep_dif_D	1	0.1			
X_{31}	DS8-Give_d_N	1				
X_{32}	DS9-Give_cr_N	1				
X_{33}	DS10-E_A_D	1	0.1			
X_{34}	DS11-E_intub_D	1	0.1			
X_{35}	DS12-Mon_p_N	1				
X_{36}	DS13-Obs_c_N	1				
X_{37}	DS16-Verb_fail_D	1	0.1	1		
X_{38}	DS17-Verb_succ_D	1	0.1	1		
X_{39}	DS18-Call_help_D	X_{86}	0.1			
X_{40}	DS19-Mem-Prep_team_D	1	1			
X_{41}	DS20-Mem-Prep_dif_D	1	1			
X_{42}	DOS5-Pre_oxy_D	1	1	1	1	1

Fig. 8 Connectivity characteristics: role matrix **mcw** for connection weights

X_{43}	DOS6-Prep_team_D	1	-1	
X_{44}	DOS7-Prep_dif_D	1	-1	
X_{45}	DOS10-E_A_D	1	1	
X_{46}	DOS11-E_intub_D	1		
X_{47}	DOS16-Verb_fail_D	1		
X_{48}	DOS17-Verb_succ_D	1		
X_{49}	DOS18-Call_help_D	1		
X_{50}	NS1-Call_intub	1		
X_{51}	NS2-Prep_p_N	1	0.1	
X_{52}	NS3-Prep_eq_N	X_{80}	0.1	
X_{53}	NS4-Prep_d_N	1	0.1	
X_{54}	NS5-Pre_oxy_D	1		
X_{55}	NS6-Prep_team_D	1		
X_{56}	NS7-Prep_dif_D	1		
X_{57}	NS8-Give_d_N	1	0.1	
X_{58}	NS9-Give_cr_N	1	0.1	
X_{59}	NS10-E_A_D	1		
X_{60}	NS11-E_intub_D	1		
X_{61}	NS12-Mon_p_N	1	0.1	
X_{62}	NS13-Obs_c_N	1	0.1	
X_{63}	NS14-Verb_fail_N	1	1	0.1
X_{64}	NS15-Verb_succ_N	1	-1	0.1
X_{65}	NS18-Call_help_D	1	1	
X_{66}	NS19-Mem_Prep_p_N	1	1	
X_{67}	NS20-Mem_Prep_dr_N	1	1	
X_{68}	NS21-Mem_Give_dr_N	1	1	
X_{69}	NOS2-Prep_p_N	1	-1	
X_{70}	NOS3-Prep_eq_N	1	1	1
X_{71}	NOS4-Prep_dr_N	1	-1	
X_{72}	NOS8-Give_d_N	1	1	-1
X_{73}	NOS9-Give_cr_N	1		
X_{74}	NOS12-Mon_p_N	1	1	
X_{75}	NOS13-Obs_c_N	1		
X_{76}	NOS14-Verb_fail_N	1		
X_{77}	NOS15-Verb_succ_N	1		
X_{78}	W1-**LW**_Prep_p_Nurse→Prep_eq_Nurse	1	1	1
X_{79}	W2-**IW**_Prep_p_Nurse→Prep_eq_Nurse	1		
X_{80}	W3-**RW**_Prep_p_Nurse→Prep_eq_Nurse	1	1	
X_{81}	W4-**LW**_Prep_p_N,D→Prep_eq_N,D	1	1	1
X_{82}	W5-**IW**_Prep_p_N,D→Prep_eq_N,D	1		
X_{83}	W6-**RW**_Prep_p_N,D→Prep_eq_N,D	1	1	
X_{84}	W7-**LW**_Verb_fail_D→Call_help_D	1	1	1
X_{85}	W8-**IW**_Verb_fail_D→Call_help_D	1		
X_{86}	W9-**RW**_Verb_fail_D→Call_help_D	1	1	
X_{87}	M1-**M_LW**_Prep_p_Nurse→Prep_eq_Nurse	1		
X_{88}	M2-**M_LW**_Prep_p_N,D→Prep_eq_N,D	1		
X_{89}	M3-**M_LW**_Verb_fail_D→Call_help_D	1		

Fig. 8 (continued)

Fig. 9 Timing characteristics: role matrix **ms** for speed factors

ms	speed factors	1
X_1		
X_2		
X_3		
X_4		
X_5	WS0-Context	0
X_6	WS1-Call_intub	2
X_7	WS2-Prep_p_N	0.3
X_8	WS3-Prep_eq_N	0.3
X_9	WS4-Prep_d_N	0.3
X_{10}	WS5-Pre_oxy_D	0.3
X_{11}	WS6-Prep_team_D	0.3
X_{12}	WS7-Prep_dif_D	0.3
X_{13}	WS8-Give_d_N	0.3
X_{14}	WS9-Give_cr_N	0.3
X_{15}	WS10-E_A_D	0.3
X_{16}	WS11-E_intub_D	0.3
X_{17}	WS12-Mon_p_N	0.3
X_{18}	WS13-Obs_c_N	0.3
X_{19}	WS14-Verb_fail_N	0.3
X_{20}	WS15-Verb_succ_N	0.3
X_{21}	WS16-Verb_fail_D	0.3
X_{22}	WS17-Verb_succ_D	0.3
X_{23}	WS18-Call_help_D	0.3
X_{24}	DS1-Call_intub	0.3
X_{25}	DS2-Prep_p_N	0.3
X_{26}	DS3-Prep_eq_N	0.3
X_{27}	DS4-Prep_d_N	0.3
X_{28}	DS5-Pre_oxy_D	0.3
X_{29}	DS6-Prep_team_D	0.3
X_{30}	DS7-Prep_dif_D	0.3
X_{31}	DS8-Give_d_N	0.3
X_{32}	DS9-Give_cr_N	0.3
X_{33}	DS10-E_A_D	0.3
X_{34}	DS11-E_intub_D	0.3
X_{35}	DS12-Mon_p_N	0.3
X_{36}	DS13-Obs_c_N	0.3
X_{37}	DS16-Verb_fail_D	0.3
X_{38}	DS17-Verb_succ_D	0.3
X_{39}	DS18-Call_help_D	0.3
X_{40}	DS19-Mem-Prep_team_D	0.15
X_{41}	DS20-Mem-Prep_dif_D	0.15
X_{42}	DOS5-Pre_oxy_D	0.15
X_{43}	DOS6-Prep_team_D	0.3
X_{44}	DOS7-Prep_dif_D	0.3
X_{45}	DOS10-E_A_D	0.3
X_{46}	DOS11-E_intub_D	0.3
X_{47}	DOS16-Verb_fail_D	0.3

Fig. 9 (continued)

X_{48}	DOS17-Verb_succ_D	0.3
X_{49}	DOS18-Call_help_D	0.3
X_{50}	NS1-Call_intub	0.3
X_{51}	NS2-Prep_p_N	0.3
X_{52}	NS3-Prep_eq_N	0.3
X_{53}	NS4-Prep_d_N	0.3
X_{54}	NS5-Pre_oxy_D	0.3
X_{55}	NS6-Prep_team_D	0.3
X_{56}	NS7-Prep_dif_D	0.3
X_{57}	NS8-Give_d_N	0.3
X_{58}	NS9-Give_cr_N	0.3
X_{59}	NS10-E_A_D	0.3
X_{60}	NS11-E_intub_D	0.3
X_{61}	NS12-Mon_p_N	0.3
X_{62}	NS13-Obs_c_N	0.3
X_{63}	NS14-Verb_fail_N	0.3
X_{64}	NS15-Verb_succ_N	0.3
X_{65}	NS18-Call_help_D	0.3
X_{66}	NS19-Mem_Prep_p_N	0.1
X_{67}	NS20-Mem_Prep_dr_N	0.1
X_{68}	NS21-Mem_Give_dr_N	0.1
X_{69}	NOS2-Prep_p_N	0.3
X_{70}	NOS3-Prep_eq_N	0.3
X_{71}	NOS4-Prep_dr_N	0.3
X_{72}	NOS8-Give_d_N	0.3
X_{73}	NOS9-Give_cr_N	0.3
X_{74}	NOS12-Mon_p_N	0.3
X_{75}	NOS13-Obs_c_N	0.3
X_{76}	NOS14-Verb_fail_N	0.3
X_{77}	NOS15-Verb_succ_N	0.3
X_{78}	W1-**LW**_Prep_p_Nurse→Prep_eq_Nurse	0.3
X_{79}	W2-**IW**_Prep_p_Nurse→Prep_eq_Nurse	0.3
X_{80}	W3-**RW**_Prep_p_Nurse→Prep_eq_Nurse	0.1
X_{81}	W4-**LW**_Prep_p_N,D→Prep_eq_N,D	0.3
X_{82}	W5-**IW**_Prep_p_N,D→Prep_eq_N,D	0.3
X_{83}	W6-**RW**_Prep_p_N,D→Prep_eq_N,D	0.3
X_{84}	W7-**LW**_Verb_fail_D→Call_help_D	0.3
X_{85}	W8-**IW**_Verb_fail_D→Call_help_D	0.3
X_{86}	W9-**RW**_Verb_fail_D→Call_help_D	0.3
X_{87}	M1-**M_LW**_Prep_p_Nurse→Prep_eq_Nurse	0.7
X_{88}	M2-**M_LW**_Prep_p_N,D→Prep_eq_N,D	0.3
X_{89}	M3-**M_LW**_Verb_fail_D→Call_help_D	0.3

mcfw	combination function weights	1 alogistic	2 hebb	3 scalemap	4 steponce
X_1					
X_2					
X_3					
X_4					
X_5	WS0-Context	1			
X_6	WS1-Call_intub				1
X_7	WS2-Prep_p_N	1			
X_8	WS3-Prep_eq_N	1			
X_9	WS4-Prep_d_N	1			
X_{10}	WS5-Pre_oxy_D	1			
X_{11}	WS6-Prep_team_D	1			
X_{12}	WS7-Prep_dif_D	1			
X_{13}	WS8-Give_d_N	1			
X_{14}	WS9-Give_cr_N	1			
X_{15}	WS10-E_A_D	1			
X_{16}	WS11-E_intub_D	1			
X_{17}	WS12-Mon_p_N	1			
X_{18}	WS13-Obs_c_N	1			
X_{19}	WS14-Verb_fail_N	1			
X_{20}	WS15-Verb_succ_N	1			
X_{21}	WS16-Verb_fail_D	1			
X_{22}	WS17-Verb_succ_D	1			
X_{23}	WS18-Call_help_D	1			
X_{24}	DS1-Call_intub	1			
X_{25}	DS2-Prep_p_N	1			
X_{26}	DS3-Prep_eq_N	1			
X_{27}	DS4-Prep_d_N	1			
X_{28}	DS5-Pre_oxy_D	1			
X_{29}	DS6-Prep_team_D	1			
X_{30}	DS7-Prep_dif_D	1			
X_{31}	DS8-Give_d_N	1			
X_{32}	DS9-Give_cr_N	1			
X_{33}	DS10-E_A_D	1			
X_{34}	DS11-E_intub_D	1			
X_{35}	DS12-Mon_p_N	1			
X_{36}	DS13-Obs_c_N	1			
X_{37}	DS16-Verb_fail_D	1			
X_{38}	DS17-Verb_succ_D	1			
X_{39}	DS18-Call_help_D	1			
X_{40}	DS19-Mem-Prep_team_D	1			
X_{41}	DS20-Mem-Prep_dif_D	1			
X_{42}	DOS5-Pre_oxy_D	1			
X_{43}	DOS6-Prep_team_D	1			
X_{44}	DOS7-Prep_dif_D	1			
X_{45}	DOS10-E_A_D	1			
X_{46}	DOS11-E_intub_D	1			
X_{47}	DOS16-Verb_fail_D	1			

Fig. 10 Aggregation characteristics: role matrix **mcfw** for combination function weights

X_{48}	DOS17-Verb_succ_D	1		
X_{49}	DOS18-Call_help_D	1		
X_{50}	NS1-Call_intub	1		
X_{51}	NS2-Prep_p_N	1		
X_{52}	NS3-Prep_eq_N	1		
X_{53}	NS4-Prep_d_N	1		
X_{54}	NS5-Pre_oxy_D	1		
X_{55}	NS6-Prep_team_D	1		
X_{56}	NS7-Prep_dif_D	1		
X_{57}	NS8-Give_d_N	1		
X_{58}	NS9-Give_cr_N	1		
X_{59}	NS10-E_A_D	1		
X_{60}	NS11-E_intub_D	1		
X_{61}	NS12-Mon_p_N	1		
X_{62}	NS13-Obs_c_N	1		
X_{63}	NS14-Verb_fail_N	1		
X_{64}	NS15-Verb_succ_N	1		
X_{65}	NS18-Call_help_D	1		
X_{66}	NS19-Mem_Prep_p_N	1		
X_{67}	NS20-Mem_Prep_dr_N	1		
X_{68}	NS21-Mem_Give_dr_N	1		
X_{69}	NOS2-Prep_p_N	1		
X_{70}	NOS3-Prep_eq_N	1		
X_{71}	NOS4-Prep_dr_N	1		
X_{72}	NOS8-Give_d_N	1		
X_{73}	NOS9-Give_cr_N	1		
X_{74}	NOS12-Mon_p_N	1		
X_{75}	NOS13-Obs_c_N	1		
X_{76}	NOS14-Verb_fail_N	1		
X_{77}	NOS15-Verb_succ_N	1		
X_{78}	W1-**LW**_Prep_p_Nurse→Prep_eq_Nurse		1	
X_{79}	W2-**IW**_Prep_p_Nurse→Prep_eq_Nurse	1		
X_{80}	W3-**RW**_Prep_p_Nurse→Prep_eq_Nurse	1		
X_{81}	W4-**LW**_Prep_p_N,D→Prep_eq_N,D		1	
X_{82}	W5-**IW**_Prep_p_N,D→Prep_eq_N,D	1		
X_{83}	W6-**RW**_Prep_p_N,D→Prep_eq_N,D	1		
X_{84}	W7-**LW**_Verb_fail_D→Call_help_D		1	
X_{85}	W8-**IW**_Verb_fail_D→Call_help_D	1		
X_{86}	W9-**RW**_Verb_fail_D→Call_help_D	1		
X_{87}	M1-**M_LW**_Prep_p_Nurse→Prep_eq_Nurse			1
X_{88}	M2-**M_LW**_Prep_p_N,D→Prep_eq_N,D			1
X_{89}	M3-**M_LW**_Verb_fail_D→Call_help_D			1

Fig. 10 (continued)

mcfp	combination function parameters	1 alogistic		2 hebb	3 scalemap	4 steponce	
X_1							
X_2							
X_3							
X_4							
X_5	WS0-Context	5	0.5				
X_6	WS1-Call_intub					10	30
X_7	WS2-Prep_p_N	8	1				
X_8	WS3-Prep_eq_N	5	1.1				
X_9	WS4-Prep_d_N	30	1				
X_{10}	WS5-Pre_oxy_D	25	1				
X_{11}	WS6-Prep_team_D	25	1				
X_{12}	WS7-Prep_dif_D	5	1.1				
X_{13}	WS8-Give_d_N	5	1.1				
X_{14}	WS9-Give_cr_N	25	0.9				
X_{15}	WS10-E_A_D	25	1.1				
X_{16}	WS11-E_intub_D	25	1.1				
X_{17}	WS12-Mon_p_N	25	1.3				
X_{18}	WS13-Obs_c_N	25	1.1				
X_{19}	WS14-Verb_fail_N	25	1.3				
X_{20}	WS15-Verb_succ_N	5	1.1				
X_{21}	WS16-Verb_fail_D	5	1.1				
X_{22}	WS17-Verb_succ_D	5	1.1				
X_{23}	WS18-Call_help_D	5	1.1				
X_{24}	DS1-Call_intub	5	0.3				
X_{25}	DS2-Prep_p_N	5	0.3				
X_{26}	DS3-Prep_eq_N	5	0.6				
X_{27}	DS4-Prep_d_N	50	0.6				
X_{28}	DS5-Pre_oxy_D	50	0.3				
X_{29}	DS6-Prep_team_D	5	0.3				
X_{30}	DS7-Prep_dif_D	15	0.4				
X_{31}	DS8-Give_d_N	15	0.3				
X_{32}	DS9-Give_cr_N	15	0.4				
X_{33}	DS10-E_A_D	25	0.55				
X_{34}	DS11-E_intub_D	25	0.5				
X_{35}	DS12-Mon_p_N	40	0.4				
X_{36}	DS13-Obs_c_N	25	0.4				
X_{37}	DS16-Verb_fail_D	8	1.1				
X_{38}	DS17-Verb_succ_D	8	1.4				
X_{39}	DS18-Call_help_D	5	1.3				
X_{40}	DS19-Mem-Prep_team_D	25	0.5				
X_{41}	DS20-Mem-Prep_dif_D	25	0.5				
X_{42}	DOS5-Pre_oxy_D	45	2				
X_{43}	DOS6-Prep_team_D	5	0.5				
X_{44}	DOS7-Prep_dif_D	25	0.4				

Fig. 11 Aggregation characteristics: role matrix **mcfp** for combination function parameters

X_{45}	DOS10-E_A_D	5	0.5			
X_{46}	DOS11-E_intub_D	5	0.5			
X_{47}	DOS16-Verb_fail_D	5	0.5			
X_{48}	DOS17-Verb_succ_D	5	0.5			
X_{49}	DOS18-Call_help_D	5	0.5			
X_{50}	NS1-Call_intub	5	0.5			
X_{51}	NS2-Prep_p_N	5	0.5			
X_{52}	NS3-Prep_eq_N	5	0.15			
X_{53}	NS4-Prep_d_N	8	0.1			
X_{54}	NS5-Pre_oxy_D	5	0.3			
X_{55}	NS6-Prep_team_D	5	0.3			
X_{56}	NS7-Prep_dif_D	5	0.3			
X_{57}	NS8-Give_d_N	5	0.3			
X_{58}	NS9-Give_cr_N	5	0.3			
X_{59}	NS10-E_A_D	15	0.5			
X_{60}	NS11-E_intub_D	15	0.55			
X_{61}	NS12-Mon_p_N	15	0.3			
X_{62}	NS13-Obs_c_N	15	0.55			
X_{63}	NS14-Verb_fail_N	15	1.3			
X_{64}	NS15-Verb_succ_N	5	0.5			
X_{65}	NS18-Call_help_D	5	1.3			
X_{66}	NS19-Mem_Prep_p_N	25	0.8			
X_{67}	NS20-Mem_Prep_dr_N	25	0.5			
X_{68}	NS21-Mem_Give_dr_N	25	0.5			
X_{69}	NOS2-Prep_p_N	5	0.2			
X_{70}	NOS3-Prep_eq_N	25	1.8			
X_{71}	NOS4-Prep_dr_N	25	0.5			
X_{72}	NOS8-Give_d_N	5	1.1			
X_{73}	NOS9-Give_cr_N	25	0.3			
X_{74}	NOS12-Mon_p_N	5	0.5			
X_{75}	NOS13-Obs_c_N	5	0.5			
X_{76}	NOS14-Verb_fail_N	5	0.5			
X_{77}	NOS15-Verb_succ_N	5	0.5			
X_{78}	W1-**LW**_Prep_p_Nurse→Prep_eq_Nurse			X_{87}		
X_{79}	W2-**IW**_Prep_p_Nurse→Prep_eq_Nurse	5	0.5			
X_{80}	W3-**RW**_Prep_p_Nurse→Prep_eq_Nurse	5	1.2			
X_{81}	W4-**LW**_Prep_p_N,D→Prep_eq_N,D			X_{88}		
X_{82}	W5-**IW**_Prep_p_N,D→Prep_eq_N,D	5	0.5			
X_{83}	W6-**RW**_Prep_p_N,D→Prep_eq_N,D	5	0.5			
X_{84}	W7-**LW**_Verb_fail_D→Call_help_D			X_{89}		
X_{85}	W8-**IW**_Verb_fail_D→Call_help_D	5	0.5			
X_{86}	W9-**RW**_Verb_fail_D→Call_help_D	5	0.5			
X_{87}	M1-**M_LW**_Prep_p_Nurse→Prep_eq_Nurse				1	0
X_{88}	M2-**M_LW**_Prep_p_N,D→Prep_eq_N,D				1	0
X_{89}	M3-**M_LW**_Verb_fail_D→Call_help_D				1	0

Fig. 11 (continued)

Fig. 12 Initial values list **iv**

iv	initial values	1
X_1		
X_2		
X_3		
X_4		
X_5	WS0-Context	0.5
X_6	WS1-Call_intub	0
X_7	WS2-Prep_p_N	0
X_8	WS3-Prep_eq_N	0
X_9	WS4-Prep_d_N	0
X_{10}	WS5-Pre_oxy_D	0
X_{11}	WS6-Prep_team_D	0
X_{12}	WS7-Prep_dif_D	0
X_{13}	WS8-Give_d_N	0
X_{14}	WS9-Give_cr_N	0
X_{15}	WS10-E_A_D	0
X_{16}	WS11-E_intub_D	0
X_{17}	WS12-Mon_p_N	0
X_{18}	WS13-Obs_c_N	0
X_{19}	WS14-Verb_fail_N	0
X_{20}	WS15-Verb_succ_N	0
X_{21}	WS16-Verb_fail_D	0
X_{22}	WS17-Verb_succ_D	0
X_{23}	WS18-Call_help_D	0
X_{24}	DS1-Call_intub	0
X_{25}	DS2-Prep_p_N	0
X_{26}	DS3-Prep_eq_N	0
X_{27}	DS4-Prep_d_N	0
X_{28}	DS5-Pre_oxy_D	0
X_{29}	DS6-Prep_team_D	0
X_{30}	DS7-Prep_dif_D	0
X_{31}	DS8-Give_d_N	0
X_{32}	DS9-Give_cr_N	0
X_{33}	DS10-E_A_D	0
X_{34}	DS11-E_intub_D	0
X_{35}	DS12-Mon_p_N	0
X_{36}	DS13-Obs_c_N	0
X_{37}	DS16-Verb_fail_D	0
X_{38}	DS17-Verb_succ_D	0
X_{39}	DS18-Call_help_D	0
X_{40}	DS19-Mem-Prep_team_D	0
X_{41}	DS20-Mem-Prep_dif_D	0
X_{42}	DOS5-Pre_oxy_D	0
X_{43}	DOS6-Prep_team_D	0
X_{44}	DOS7-Prep_dif_D	0
X_{45}	DOS10-E_A_D	0
X_{46}	DOS11-E_intub_D	0
X_{47}	DOS16-Verb_fail_D	0
X_{48}	DOS17-Verb_succ_D	0
X_{49}	DOS18-Call_help_D	0

Fig. 12 (continued)

X_{50}	NS1-Call_intub	0
X_{51}	NS2-Prep_p_N	0
X_{52}	NS3-Prep_eq_N	0
X_{53}	NS4-Prep_d_N	0
X_{54}	NS5-Pre_oxy_D	0
X_{55}	NS6-Prep_team_D	0
X_{56}	NS7-Prep_dif_D	0
X_{57}	NS8-Give_d_N	0
X_{58}	NS9-Give_cr_N	0
X_{59}	NS10-E_A_D	0
X_{60}	NS11-E_intub_D	0
X_{61}	NS12-Mon_p_N	0
X_{62}	NS13-Obs_c_N	0
X_{63}	NS14-Verb_fail_N	0
X_{64}	NS15-Verb_succ_N	0
X_{65}	NS18-Call_help_D	0
X_{66}	NS19-Mem_Prep_p_N	0
X_{67}	NS20-Mem_Prep_dr_N	0
X_{68}	NS21-Mem_Give_dr_N	0
X_{69}	NOS2-Prep_p_N	0
X_{70}	NOS3-Prep_eq_N	0
X_{71}	NOS4-Prep_dr_N	0
X_{72}	NOS8-Give_d_N	0
X_{73}	NOS9-Give_cr_N	0
X_{74}	NOS12-Mon_p_N	0
X_{75}	NOS13-Obs_c_N	0
X_{76}	NOS14-Verb_fail_N	0
X_{77}	NOS15-Verb_succ_N	0
X_{78}	W1-**LW**_Prep_p_Nurse→Prep_eq_Nurse	0.1
X_{79}	W2-**IW**_Prep_p_Nurse→Prep_eq_Nurse	0
X_{80}	W3-**RW**_Prep_p_Nurse→Prep_eq_Nurse	0
X_{81}	W4-**LW**_Prep_p_N,D→Prep_eq_N,D	0.1
X_{82}	W5-**IW**_Prep_p_N,D→Prep_eq_N,D	1
X_{83}	W6-**RW**_Prep_p_N,D→Prep_eq_N,D	0
X_{84}	W7-**LW**_Verb_fail_D→Call_help_D	0.1
X_{85}	W8-**IW**_Verb_fail_D→Call_help_D	1
X_{86}	W9-**RW**_Verb_fail_D→Call_help_D	0
X_{87}	M1-**M_LW**_Prep_p_Nurse→Prep_eq_Nurse	0.8
X_{88}	M2-**M_LW**_Prep_p_N,D→Prep_eq_N,D	0.9
X_{89}	M3-**M_LW**_Verb_fail_D→Call_help_D	1

References

Abraham, W.C., Bear, M.F.: Metaplasticity: the plasticity of synaptic plasticity. Trends Neurosci. **19**(4), 126–130 (1996)

Bazerman, M.H., Giuliano, T., Appelman, A.: Escalation of commitment in individual and group decision making. Organ. Behav. Hum. Perform. **33**, 141–152 (1984)

Bhalwankar, R., Treur, J.: Modeling the development of internal mental models by an adaptive network model. In: Proceedings of the 11th Annual International Conference on Brain-Inspired Cognitive Architectures for AI, BICA*AI'20. Procedia Computer Science, vol. 190, pp. 90–101. Elsevier (2021a)

Bhalwankar, R., Treur, J.: A second-order adaptive network model for learner-controlled mental model learning processes. In: Benito R.M., Cherifi C., Cherifi H., Moro E., Rocha L.M., Sales-Pardo M. (eds.) Proceedings of the 9th International Conference on Complex Networks and their Applications. Studies in Computational Intelligence, vol. 944, pp. 245–259. Springer Nature Switzerland AG (2021b)

Burthscher, M.J., Kolbe, M., Wacker, J.: Interaction of team mental models and monitoring behaviors predict team performance in simulated anesthesia inductions. J. Exp. Psychol. Appl. **17**(3), 257–269 (2011)

Burtscher, M.J., Manser, T.: Team mental models and their potential to improve teamwork and safety: a review and implications for future research in healthcare. Saf. Sci. **50**(5), 1344–1354 (2012)

Craik, K.J.W.: The Nature of Explanation. University Press, Cambridge, MA (1943)

Canbaloglu, G., Treur, J., Wiewiora, A. eds.: Computational modeling of multilevel organisational learning and its control using self-modeling network models. Springer Nature (2023)

DeChurch, L.A., Mesmer-Magnus, J.R.: Measuring shared team mental models: a meta-analysis. Group Dyn. Theory Res. Practice **14**(1), 1–14 (2010)

Dionne, S.D., Sayama, H., Hao, C., Bush, B.J.: The role of leadership in shared mental model convergence and team performance improvement: an agent-based computational model. Leadersh. q. **21**, 1035–1049 (2010)

Doyle, J.K., Ford, D.N.: Mental models concepts for system dynamics research. Syst. Dyn. Rev. **14**(1), 3–29 (1998)

Fischhof, B., Johnson, S.: Organisational Decision Making. Cambridge University Press, Cambridge (1997)

Furlough, C.S., Gillan, D.J.: Mental models: structural differences and the role of experience. J. Cogn. Eng. Decis. Making **12**(4), 269–287 (2018)

Garcia, R.: Stress, metaplasticity, and antidepressants. Curr. Mol. Med. **2**, 629–638 (2002)

Gentner, D., Stevens, A.L.: Mental Models. Erlbaum, Hillsdale NJ (1983)

Hebb, D.O.: The Organization of Behavior: A Neuropsychological Theory. Wiley, New York (1949)

Higgs, A., McGrath, B.A., Goddard, C., Rangasami, J., Suntharalingam, G., Gale, R., Cook, T.M.: Guidelines for the management of tracheal intubation of critically ill adults. Br. J. Anaesth. **120**(2), 323–352 (2018)

Isenberg, D.J.: Group polarisation: a critical review and meta-analysis. J. Pers. Soc. Psychol. **50**, 1141–1151 (1986)

Janis, I.L.: Victims of Groupthink. Houghton MiZin, Boston (1972)

Johnson-Laird, P.N.: Mental Models: Towards a Cognitive Science of Language, Inference, and Consciousness. Harvard University Press (1983)

Jones, P.E., Roelofsma, P.H.M.P.: The potential for social contextual and group biases in team decision making: biases, conditions and psychological mechanisms. Ergonomics **43**(8), 1129–1152 (2000)

De Kleer, J., Brown, J.: Assumptions and ambiguities in mechanistic mental models. In: Gentner, D., Stevens, A., (eds.) Mental Models, pp. 155–190. Lawrence Erlbaum Associates, Hillsdale, NJ (1983)

Kleindorfer, P.R., Kunreuther, H.C., Schoemaker, P.J.H.: Decision Sciences: An Integrated Perspective. Cambridge University Press, Cambridge (1993)
Krueger, J.: On the perception of social consensus. In: Zanna, M.P. (ed.) Advances in Experimental Social Psychology, vol. 30, pp. 163–240. Academic Press, New York (1998)
Lamm, H., Myers, D.G.: Group induced polarisation of attitudes and behaviour. In: Berkowitz, L. (ed.) Advances in Experimental Social Psychology, vol. 11, pp. 145–195. Academic Press, New York (1978)
Langan-Fox, J., Code, S., Langfield-Smith, K.: Team mental models: techniques, methods, and analytic approaches. Hum. Factors **42**(2), 242–271 (2000)
Mathieu, J.E., Hefner, T.S., Goodwin, G.F., Salas, E., Cannon-Bowers, J.A.: The influence of shared mental models on team process and performance. J. Appl. Psychol. **85**(2), 273–283 (2000)
Nini, M.: All on the same page: How Team Mental Models (TMM) increase team performance. CQ Net (2019). https://www.ckju.net/en/dossier/team-mental-models-increase-team-performance
Outland, N.B.: A Computational Cognitive Architecture for Exploring Team Mental Models. College of Science and Health Theses and Dissertations, p. 289 (2019). https://via.library.depaul.edu/csh_etd/289
Ross, L., Greene, D., House, P.: The 'false consensus effect': an egocentric bias in social perception and attribution process. J. Exp. Soc. Psychol. **13**, 279–301 (1977)
Scheutz M.: Computational mechanisms for mental models in human-robot interaction. In: Shumaker R. (eds.) Virtual Augmented and Mixed Reality. Designing and Developing Augmented and Virtual Environments. VAMR 2013. Lecture Notes in Computer Science, vol. 8021, pp. 304–312. Springer, Berlin (2013)
Seo, S., Kennedy-Metz, L.R., Zenati, M.A., Shah, J.A., Dias, R.D., Unhelkar, V.V.: Towards an AI Coach to Infer Team Mental Model Allignment in Healthcare. Department of Computer Science, Rice University Houston TX, USA (2021)
Shih, Y.F., Alessi, S.M.: Mental models and transfer of learning in computer programming. Journal of research on computing in education **26**(2), 154–175 (1993)
Staw, B.M.: Knee deep in the big muddy: a study of escalating commitment to a chosen course of action. Organ. Behav. Hum. Perform. **16**, 27–44 (1976)
Todd, J.: Audit of compliance with WHO surgical safety checklist and building a shared mental model in the operating theatre. BJM Leader **2**(1), 32–135 (2018)
Treur, J.: Network-Oriented Modeling: Addressing Complexity of Cognitive, Affective and Social Interactions. Springer Nature, Cham (2016)
Treur, J.: Network-Oriented Modeling for Adaptive Networks: Designing Higher-Order Adaptive Biological, Mental and Social Network Models. Springer Nature, Cham (2020)
Treur, J., Van Ments, L. eds.: Mental models and their dynamics, adaptation, and control: a self-modeling network modeling approach. Springer Nature (2022)
Van Ments, L., Treur, J.: A higher-order adaptive network model to simulate development of and recovery from PTSD. In: Proceedings of the 21th International Conference on Computational Science, ICCS'21. Lecture Notes in Computer Science, vol. 12743, pp. 154–166. Springer Nature, Cham (2021a)
Van Ments, L., Treur, J., Roelofsma, P.H.M.P.: A temporal-causal network model for the relation between religion and human empathy. In: Proceedings of the 5th International Workshop on Complex Networks and their Applications. Lecture Notes in Computer Science, vol. 693, pp. 55–67. Springer Publications (2016)
Van Ments, L., Treur, J., Roelofsma, P.H.M.P.: Modeling the effect of religion on human empathy based on an adaptive temporal-causal network model. Comput. Soc. Netw. **5**(1) (2018)
Van Ments, L., Treur, J., Klein, J., Roelofsma, P.H.M.P.: A computational network model for shared mental models in hospital operation rooms. In: Proceedings of the 14th International Conference on Brain Informatics, BI'21. Lecture Notes in AI, vol. 12960, pp. 67–78. Springer Nature, Cham (2021a)
Van Ments, L., Treur, J., Klein, J., Roelofsma, P.H.M.P.: A second-order adaptive network model for shared mental models in hospital teamwork. In: Nguyen, N.T., et al. (eds.) Proceedings of

the 13th International Conference on Computational Collective Intelligence, ICCCI'21. Lecture Notes in AI, vol. 12876, pp. 126–140. Springer Nature, Cham (2021b)
Van Ments, L., Treur, J., Roelofsma, P.H.M.P.: How empathic is your god: an adaptive network model for formation and use of a mental god-model and its effect on human empathy. In: Treur, J., Van Ments, L. (eds.) Mental Models and Their Dynamics, Adaptation, and Control: A Self-Modeling Network Modeling Approach, pp. 285–319. Springer Nature, Cham (2022)
Van Ments, L., Treur, J.: Modeling adaptive cooperative and competitive metaphors as mental models for joint decision making. Cogn. Syst. Res. **69**, 67–82 (2021b)
Van Ments, L., Treur, J.: Reflections on dynamics, adaptation and control: a cognitive architecture for mental models. Cogn. Syst. Res. **70**, 1–9 (2021c)
Williams, D.: The Mind as a Predictive Modelling Engine: Generative Models, Structural Similarity, and Mental Representation. Ph.D. Thesis, University of Cambridge, UK (2018)
Wilson, A.: Creating and applying shared mental models in the operating room. J. Perioperative Nurs. **32**(3), 33–36 (2019)

Learning of Safety and Security Management Through Cyberspace: An Adaptive Self-Modeling Network Model for Multilevel Organizational Learning

Gülay Canbaloğlu, Jan Treur, and Peter H. M. P. Roelofsma

Abstract Multilevel organisational learning concerns an interplay of different types of learning at individual, team, and organisational levels. These processes use complex dynamic and adaptive mechanisms. A second-order adaptive network model for this is introduced here and illustrated.

Keywords Multilevel organisational learning · Adaptive network model · Self-model

1 Introduction

Multilevel organisational learning is a complex, dynamic, adaptive, cyclical and non-linear type of learning involving multiple levels and both dependent on individuals and independent of individuals. It is multilevel because the learning of an organisation

G. Canbaloğlu · J. Treur (✉)
Department of Computer Science, Social AI Group, Vrije Universiteit Amsterdam, Amsterdam, The Netherlands
e-mail: j.treur@vu.nl; j.treur@hhs.nl

G. Canbaloğlu
e-mail: gulaycnbl@gmail.com

J. Treur · P. H. M. P. Roelofsma
Risk Management and Cybersecurity Group, Centre of Expertise Cybersecurity, The Hague University of
Applied Sciences, The Hague, The Netherlands
e-mail: p.h.m.p.roelofsma@hhs.nl

P. H. M. P. Roelofsma
Department of Pediatrics, Division Neonatology, Erasmus University Medical Centre, Erasmus MC, University Medical Center-Sophia Children's Hospital, Rotterdam, The Netherlands

P. H. M. P. Roelofsma et al. (eds.), *Using Shared Mental Models and Organisational Learning to Support Safety and Security Through Cyberspace: A Computational Analysis Approach*, Studies in Systems, Decision and Control 570,
https://doi.org/10.1007/978-3-031-72075-8_3

involves learning at the level of individuals, at the level of teams (or groups or projects), and at the level of the organisation via feed forward and feedback pathways:

> Through feed-forward processes, new ideas and actions flow from the individual to the group to the organisation levels. At the same time, what has already been learned feeds back from the organisation to group and individual levels, affecting how people act and think. (Crossan et al. 1999), p. 532.

> There is growing consensus in the literature that the theory of organisational learning should consider individual, team and organisational levels (Wiewiora et al. 2019), p. 94

There is a huge amount of literature on multilevel organisational learning such as (Argyris and Schön 1978; Bogenrieder 2002; Crossan et al. 1999; Fischhof and Johnson 1997; Kim 1993; McShane and Glinow 2010; Stelmaszczyk 2016; Wiewiora et al. 2019; Wiewiora et al. 2020). However, until recently systematic approaches to obtain (adaptive) computational models were not easy to find. In the current chapter, a self-modeling network modeling perspective is used to model the different adaptive, interacting processes of multilevel organisational learning. In contrast to the previous Chap. 6 of this volume (Canbaloğlu et al. 2023b), in the current chapter also the intermediate level of teams or projects is addressed, following Crossan et al. (1999, Wiewiora et al. (2019) whereas in the previous chapter the paper of Kim (1993) was used as a main source of inspiration, where this intermediate level is not addressed.

Computational modeling of multilevel organisational learning provides a more observable formalization of multilevel organisational learning and provides possibilities to perform 'in silico' (simulation) experiments with it. To this end, the self-modeling network modeling approach introduced in Treur (2020) that is explained in some detail in Sect. 3, is used in this current chapter.

First, Sect. 2 presents how literature provides ideas on mental models at individual, team and organisation level and their role in multilevel organisational learning. Then, Sect. 3 explains the characteristics and details of adaptive self-modeling network models and how they can be used to model the different processes concerning dynamics, adaptation and control of mental models. In Sect. 4 the controlled adaptive network model for multilevel organisational learning is introduced. Then in Sect. 5, an example simulation scenario is explained in detail. Section 6 is a Discussion section.

2 Background Literature

The quotes in the introduction section illustrate the perspective adopted here. Mental models are considered a vehicle to model the interplay of learning at individual, team and organisational level. Individual mental models learnt are a basis for formation of shared team mental models; these shared team mental models provide input for the shared mental models at the organisation level. Conversely, these shared mental models at organisation and team level are used to improve shared team mental models

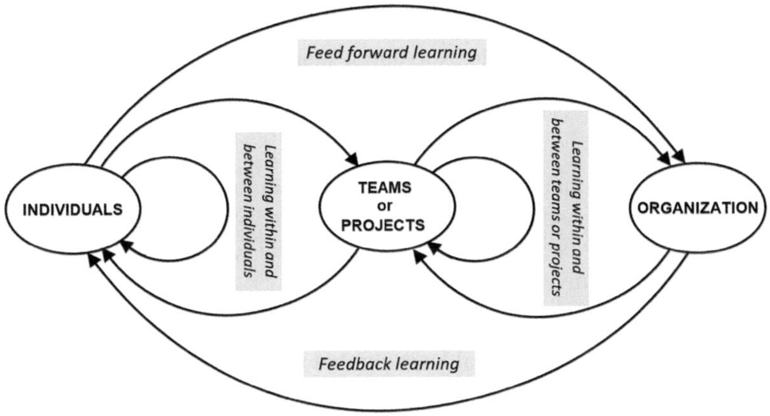

Fig. 1 Multilevel organisational learning: multiple levels and nested cycles (with depth 3) of interactions; see also Wiewiora et al. (2019, 2020)

and individual mental models, respectively. The picture of the different pathways shown in Fig. 1 is based on Fig. 4 of Wiewiora et al. (2019) and Fig. 3 of Wiewiora et al. (2020).

Inspired by this, as a basis for the analysis made here, the considered overall multilevel organisational learning process consists of the following main pathways and interactions; see also Crossan et al. (1999) and Wiewiora et al. (2019):

(a) **Individual level**

   (1) Creating and maintaining individual mental models
   (2) Choosing for a specific context a suitable individual mental model as focus
   (3) Applying a chosen individual mental model for internal simulation
   (4) Improving individual mental models

(b) **From individual level to team level (feed forward learning)**

   (1) Deciding about creation of shared team mental models
   (2) Creating shared team mental models based on developed individual mental models

(c) **From team level to organisation level (feed forward learning)**

   (1) Deciding about creation of shared mental models
   (2) Creating shared mental models based on developed individual mental models

(d) **From organisation level to team level (feedback learning)**

   (1) Deciding about teams to adopt shared organisation mental models
   (2) Teams adopting shared mental models

(e) **From team level to individual level (feedback learning)**

  (1) Deciding about individuals to adopt shared team mental models
  (2) Individuals adopting shared team mental models by learning them

(f) **Individual level**

  (1) Creating and maintaining individual mental models
  (2) Choosing for a specific context a suitable individual mental model as focus
  (3) Applying a chosen individual mental model for internal simulation
  (4) Improving individual mental models

This overview provided useful input to the design of the computational network model for multilevel organisational learning that will be introduced in Sect. 4.

3 The Self-modeling Network Modeling Approach

In this section, the self-modeling modeling approach (Treur 2020) used is explained in some detail. A network model is defined by (where X and Y are nodes or states of the network):

- *Connectivity characteristics*
  Connections from one state X to a state Y with their weights $\boldsymbol{\omega}_{X,Y}$
- *Aggregation characteristics*
  For any state Y, a combination function $\mathbf{c}_Y(..)$ is used to specify the aggregation that is applied to the impacts $\boldsymbol{\omega}_{X,Y}X(t)$ on Y from the incoming connections from states X to Y.
- *Timing characteristics*
  For each state Y a speed factor $\boldsymbol{\eta}_Y$ defines how fast it changes for given causal impact.

Each state or node Y has time t dependent activation values $Y(t)$. The following difference equations are used for simulation; they are based on the network characteristics $\boldsymbol{\omega}_{X,Y}$, $\mathbf{c}_Y(..)$, $\boldsymbol{\eta}_Y$ in a canonical manner:

$$Y(t + \Delta t) = Y(t) + \boldsymbol{\eta}_Y[\mathbf{c}_Y(\boldsymbol{\omega}_{X_1,Y}X_1(t), \ldots, \boldsymbol{\omega}_{X_k,Y}X_k(t)) - Y(t)]\Delta t \quad (1)$$

for each state Y, where X_1 to X_k are the states from which Y receives incoming connections. The dedicated software environment (Treur 2020, Chap. 9), includes a library with currently around 65 basic combination functions. The examples of basic combination functions that are applied in the model introduced here can be found in Table 1.

By using a *self-modeling principle* (also called a *reification principle*), a network-oriented conceptualisation can also be applied to *adaptive* networks; see Treur (2020). Here new states are added to the network (called *self-model states*) representing

Table 1 The combination functions applied in the introduced network model

	Notation	Formula	Parameters
Advanced logistic sum	$\mathbf{alogistic}_{\sigma,\tau}(V_1, \ldots, V_k)$	$\left[\frac{1}{1+e^{-\sigma(V_1+\cdots+V_k-\tau)}} - \frac{1}{1+e^{\sigma\tau})}\right](1+e^{-\sigma\tau})$	Steepness $\sigma > 0$ Excitability threshold τ
Steponce	$\mathbf{steponce}_{\alpha,\beta}(..)$	1 if time t is between α and β, else 0	Start time α End time β
Hebbian learning	$\mathbf{hebb}_{\mu}(V_1, V_2, V_3)$	$V_1 * V_2(1 - V_3) + \mu V_3$	V_1,V_2 activation levels of states X and Y; V_3 activation level of the self-model state $\mathbf{W}_{X,Y}$ Persistence factor μ
Maximum composed with Hebbian learning	$\mathbf{max\text{-}hebb}_{\mu}(V_1, \ldots, V_k)$	$\max\left(\mathbf{hebb}_{\mu}(V_1, V_2, V_3), V_4, \ldots, V_k\right)$	
Scaled maximum	$\mathbf{smax}_{\lambda}(V_1, \ldots, V_k)$	$\max(V_1, \ldots, V_k)/\lambda$	Scaling factor λ

network characteristics. These self-model states are depicted at a next level (called *self-model level* or *reification level*); the original network is at the *base level.*

This is often applied to the weight $\omega_{X,Y}$ of a connection from state X to state Y; this is represented by a self-model state $\mathbf{W}_{X,Y}$. Similarly, any other network characteristic from $\omega_{X,Y}$, $\mathbf{c}_Y(..)$, η_Y can be self-modeled by including self-model states. For example, the speed factor η_Y of a state Y can be represented by a self-model state $\mathbf{H}_Y$.

This self-modeling network construction can be applied iteratively to obtain multiple orders of self-models at multiple (first-order, second-order, ...) self-model levels. For example, a second-order self-model may include a second-order self-model state $\mathbf{H}_{\mathbf{W}_{X,Y}}$ representing the speed factor $\eta_{\mathbf{W}_{X,Y}}$ for the (learning) dynamics of first-order self-model state $\mathbf{W}_{X,Y}$ which in turn represents an adaptative connection weight $\omega_{X,Y}$. Similarly, a persistence factor $\mu_{\mathbf{W}_{X,Y}}$ of such a first-order self-model state $\mathbf{W}_{X,Y}$ used for adaptation (e.g., based on Hebbian learning) can be represented by a second-order self-model state $\mathbf{M}_{\mathbf{W}_{X,Y}}$.

In the current chapter, the self-modeling network perspective is applied to design a second-order adaptive mental network architecture addressing the mental and social processes underlying organisational learning by proper handling of individual mental models and shared mental models. In this self-modeling network architecture, the base level addresses the use of a mental model by internal simulation, the first-order self-model the adaptation of the mental model, and the second-order self-model level models the control over this; see Fig. 2. In this way the three-level cognitive architecture described in Van Ments et al. (2021), Treur and Van Ments (2022) is formalized computationally in the form of a self-modeling network architecture. In Bhalwankar and Treur (2021) it is shown how specific forms of learning and their control can be modeled based on this self-modeling network architecture, in particular learning by observation and learning by instruction and combinations thereof (Yi and Davis 2003; Van Gog et al. 2009). Some of these forms of learning will also be applied in the model for multilevel organisational learning introduced here in Sect. 4.

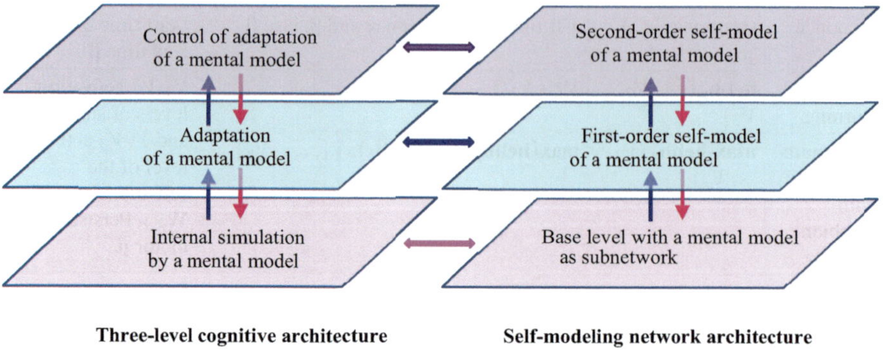

Fig. 2 Computational formalization of the three-level cognitive architecture for mental model handling from Van Ments et al. (2021) by a self-modeling network architecture

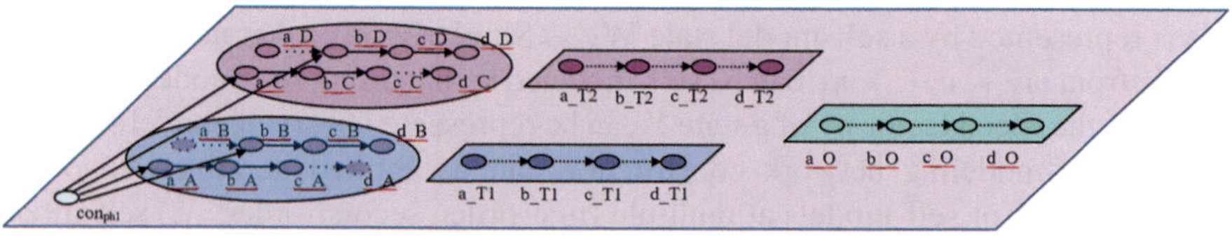

Fig. 3 The connectivity of the base level of the adaptive network model: (1) in the ovals the individual mental models of the team members (A and B for Team T1 and C and D for team T2), (2) in the rectangles the shared team mental models for team T1 (blue-grey) and team T2 (pink-purple) and the shared mental model of the whole organisation O (green)

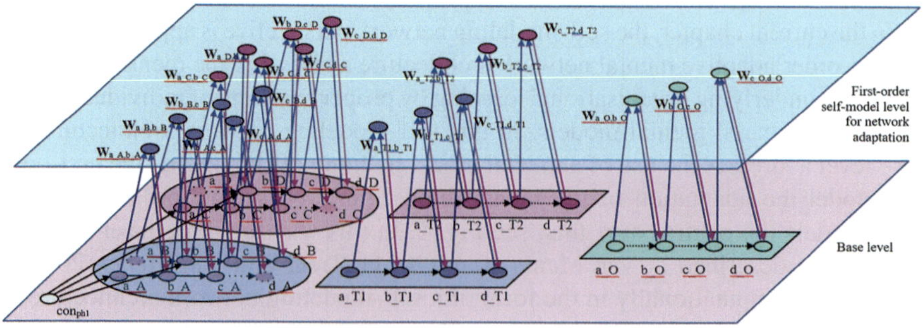

Fig. 4 The connectivity between the base level and first-order self-model level for the adaptation of the mental models by Hebbian learning

4 The Network Model for Multilevel Organisational Learning

In the considered case study concerning tasks *a*, *b*, *c*, and *d*, initially the individual mental models of 4 people are different and based on some strong and some weak connections; they don't use a stronger shared mental model as that does not exist yet. The multilevel organisational learning addressed to improve the situation covers:

1. Individual (Hebbian) learning by persons of their mental models through internal simulation which results in stronger but still incomplete and different mental models. Person A and C's mental models have no connection from task *c* to task *d* and person B and D's mental models have no connection from *a* to *b*.
2. Formation of two shared team mental models for teams T1 (consisting of persons A and B) and T2 (consisting of persons C and D) based on the different individual mental models. A process of unification by aggregation takes place (feed forward learning).
3. Formation of a shared organisation mental model based on the two team mental models. Again, a process of unification by aggregation takes place (feed forward learning).
4. Flow of information and knowledge from organisation mental model to team mental models, e.g., a form of instructional learning (feedback learning).
5. Learning of individual mental models from the shared team mental models, e.g., also a form of instructional learning (feedback learning).
6. Improvements on these individual mental models by individual learning through internal simulation which results in stronger and now complete mental models (by Hebbian learning). Now person A and C's mental models have a connection from task c to task d, and person B and D's mental models have a connection from a to b.

The connectivity of the introduced network model is shown in Figs. 3, 4, 5 and 6; for an overview of the states, see Figs. 7, 8, 9, 10 and 11, and for more details about the connections and how they relate to **(a)** to **(f)** from Sect. 2, see Fig. 12.

The undermost base level of this model has mental model states for individuals, teams and organisation, and also context states for activation of six different phases (like the **(a)** to **(f)** in Sect. 2) at different times. The mental states of persons are connected to each other according to the order of the tasks, and the first ones have a connection from first context state to be able to start to perform internal simulation and learn. As can be seen in Fig. 3, some connections between task states of persons are dashed, which means initially there is no connection (initial lack of knowledge).

Therefore, states where these dashed connections are, are the 'hollow' non-known mental state connections of persons. These states have connections that change over time to enable to observe the improvement of the individual's knowledge with the impact of organisation and team mental models in the fifth phase. The base level mental states relate to the basic tasks and can be considered as the basic ingredients of the mental models representing knowledge on relations between tasks.

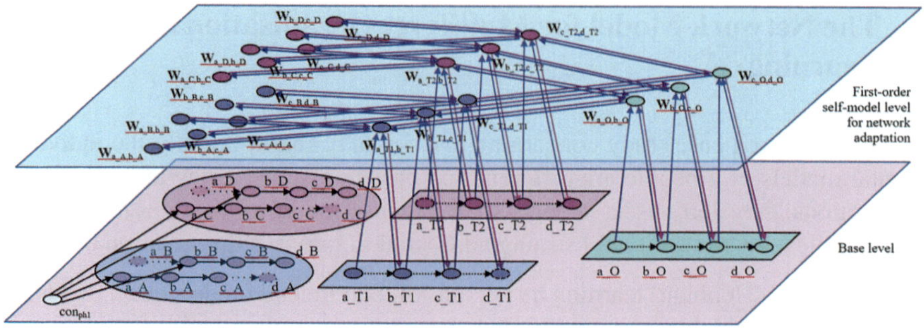

Fig. 5 The connectivity within the first-order self-model level for the adaptation of the mental models by formation of shared team and organisation mental models (links from left to right: feed-forward learning) and by instruction learning of individual mental models and shared team mental models from these shared mental models (links from right to left: feedback learning)

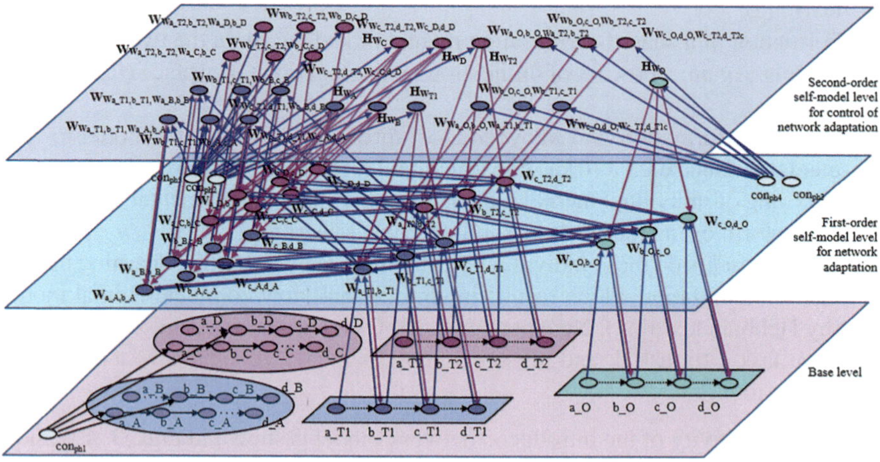

Fig. 6 The connectivity of the second-order adaptive network model for the second-order self-model of the mental models: the interactions between the first-order self-model level and the second-order self-model level: the second-order Hebbian learning for the second-order **W**-states (the $\mathbf{W}_{\mathbf{W}}$-states)

To make the mental models adaptive, first-order self-model states are added in the intermediary (first-order self-model) level (blue plane). These are **W**-states representing adaptive weights for each developed connection between individual, team and organisation mental model states in the base level; see Fig. 4. The (blue) upward and (pink) downward connections between the two levels are used to model individual (Hebbian) learning.

Within the first-order self-model level (blue plane), there are also intralevel **W**-to-**W** connections between first-order **W**-states here to enable feed forward learning (in Phase 2 and Phase 3) and feedback learning (in Phase 4 and Phase 5) (Crossan

Nr	State	Explanation
X_1	a_A	Individual mental model state for person A for task a
X_2	b_A	Individual mental model state for person A for task b
X_3	c_A	Individual mental model state for person A for task c
X_4	d_A	Individual mental model state for person A for task d
X_5	a_B	Individual mental model state for person B for task a
X_6	b_B	Individual mental model state for person B for task b
X_7	c_B	Individual mental model state for person B for task c
X_8	d_B	Individual mental model state for person B for task d
X_9	a_C	Individual mental model state for person C for task a
X_{10}	b_C	Individual mental model state for person C for task b
X_{11}	c_C	Individual mental model state for person C for task c
X_{12}	d_C	Individual mental model state for person C for task d
X_{13}	a_D	Individual mental model state for person D for task a
X_{14}	b_D	Individual mental model state for person D for task b
X_{15}	c_D	Individual mental model state for person D for task c
X_{16}	d_D	Individual mental model state for person D for task d
X_{17}	a_T1	Shared mental model state for team T1 for task a
X_{18}	b_T1	Shared mental model state for team T1 for task b
X_{19}	c_T1	Shared mental model state for team T1 for task c
X_{20}	d_T1	Shared mental model state for team T1 for task d
X_{21}	a_T2	Shared mental model state for team T2 for task a
X_{22}	b_T2	Shared mental model state for team T2 for task b
X_{23}	c_T2	Shared mental model state for team T2 for task c
X_{24}	d_T2	Shared mental model state for team T2 for task d
X_{25}	a_O	Shared mental model state for organization O for task a
X_{26}	b_O	Shared mental model state for organization O for task b
X_{27}	c_O	Shared mental model state for organization O for task c
X_{28}	d_O	Shared mental model state for organization O for task d
X_{29}	con_{ph1}	Context state for Phase 1: individual mental model simulation and learning
X_{30}	con_{ph2}	Context state for Phase 2: creation of shared mental models for teams T1 and T2
X_{31}	con_{ph3}	Context state for Phase 3: creation of a shared mental model for organization O
X_{32}	con_{ph4}	Context state for Phase 4: learning shared team mental models from the shared mental model for organization O
X_{33}	con_{ph5}	Context state for Phase 5: learning individual mental models from the shared mental models for teams T1 and T2
X_{34}	con_{ph6}	Context state for Phase 6: individual mental model simulation and learning

Fig. 7 Base level states of the introduced adaptive network model

et al. 1999). These **W**-to-**W** connections correspond to the arrows for feed forward and feedback learning shown in Fig. 1 (upper part, resp. lower part). Thus, formation of shared team and organisation mental models is performed by this feed forward learning mechanism, and the learning from the shared organisation mental model and the shared team mental model by individuals occurs by the feedback learning mechanism.

To control the adaptivity modeled by the first-order adaptation level, second-order self-model states are added in the uppermost level (purple plane). In first place, there are $\mathbf{W}_{\mathbf{W}}$-states (higher-order **W**-states) for (intralevel) connections between first-order adaptivity level **W**-states, in other words adaptive weight representations of the horizontal connections between adaptive weight representation states in the level below. These control processes (for whether and when to activate feed forward and feedback learning) are not explicitly depicted in Fig. 1 based on Wiewiora et al.

Nr	State	Explanation
X_{35}	$\mathbf{W}_{a\_A,b\_A}$	First-order self-model state for the weight of the connection from a to b within the individual mental model of person A
X_{36}	$\mathbf{W}_{b\_A,c\_A}$	First-order self-model state for the weight of the connection from b to c within the individual mental model of person A
X_{37}	$\mathbf{W}_{c\_A,d\_A}$	First-order self-model state for the weight of the connection from c to d within the individual mental model of person A
X_{38}	$\mathbf{W}_{a\_B,b\_B}$	First-order self-model state for the weight of the connection from a to b within the individual mental model of person B
X_{39}	$\mathbf{W}_{b\_B,c\_B}$	First-order self-model state for the weight of the connection from b to c within the individual mental model of person B
X_{40}	$\mathbf{W}_{c\_B,d\_B}$	First-order self-model state for the weight of the connection from c to d within the individual mental model of person B
X_{41}	$\mathbf{W}_{a\_C,b\_C}$	First-order self-model state for the weight of the connection from a to b within the individual mental model of person C
X_{42}	$\mathbf{W}_{b\_C,c\_C}$	First-order self-model state for the weight of the connection from b to c within the individual mental model of person C
X_{43}	$\mathbf{W}_{c\_C,d\_C}$	First-order self-model state for the weight of the connection from c to d within the individual mental model of person C
X_{44}	$\mathbf{W}_{a\_D,b\_D}$	First-order self-model state for the weight of the connection from a to b within the individual mental model of person D
X_{45}	$\mathbf{W}_{b\_D,c\_D}$	First-order self-model state for the weight of the connection from b to c within the individual mental model of person D
X_{46}	$\mathbf{W}_{c\_D,d\_D}$	First-order self-model state for the weight of the connection from c to d within the individual mental model of person D
X_{47}	$\mathbf{W}_{a\_T1,b\_T1}$	First-order self-model state for the weight of the connection from a to b within the shared mental model of team T1
X_{48}	$\mathbf{W}_{b\_T1,c\_T1}$	First-order self-model state for the weight of the connection from b to c within the shared mental model of team T1
X_{49}	$\mathbf{W}_{c\_T1,d\_T1}$	First-order self-model state for the weight of the connection from c to d within the shared mental model of team T1
X_{50}	$\mathbf{W}_{a\_T2,b\_T2}$	First-order self-model state for the weight of the connection from a to b within the shared mental model of team T2
X_{51}	$\mathbf{W}_{b\_T2,c\_T2}$	First-order self-model state for the weight of the connection from b to c within the shared mental model of team T2
X_{52}	$\mathbf{W}_{c\_T2,d\_T2}$	First-order self-model state for the weight of the connection from c to d within the shared mental model of team T2
X_{53}	$\mathbf{W}_{a\_O,b\_O}$	First-order self-model state for the weight of the connection from a to b within the shared mental model of the organisation O
X_{54}	$\mathbf{W}_{b\_O,c\_O}$	First-order self-model state for the weight of the connection from b to c within the shared mental model of the organisation O
X_{55}	$\mathbf{W}_{c\_O,d\_O}$	First-order self-model state for the weight of the connection from c to d within the shared mental model of the organisation O

Fig. 8 First-order self-model states of the introduced adaptive network model

(2019) and Crossan et al. (1999) but still are crucial for the processes to function well. Additionally, $\mathbf{H_W}$-states for adaptation speeds of connection weight representations in the first-order adaptation level, and $\mathbf{M_W}$-states for persistence of adaptation are placed on the second-order self-model level. This provides the speed and persistence control of the adaptation.

For a full specification of the network model, see the Appendix section. In Figs. 7, 8, 9, 10 and 11 an overview is given of all states of the model. In Table 12 an overview is given of the different types of connections.

Nr	State	Explanation
X_{56}	$\mathbf{W}_{\mathbf{W}_{a\_T1,b\_T1},\mathbf{W}_{a\_A,b\_A}}$	Second-order self-model state for the weight of the connection from shared team mental model connection weight self-model state $\mathbf{W}_{a\_T1,b\_T1}$ to individual mental model connection weight self-model state $\mathbf{W}_{a\_A,b\_A}$ for instructional (feedback) learning from the shared mental model of team T1
X_{57}	$\mathbf{W}_{\mathbf{W}_{b\_T1,c\_T1},\mathbf{W}_{b\_A,c\_A}}$	Second-order self-model state for the weight of the connection from shared team mental model connection weight self-model state $\mathbf{W}_{b\_T1,c\_T1}$ to individual mental model connection weight self-model state $\mathbf{W}_{b\_A,c\_A}$ for instructional (feedback) learning from the shared mental model of team T1
X_{58}	$\mathbf{W}_{\mathbf{W}_{c\_T1,d\_T1},\mathbf{W}_{c\_A,d\_A}}$	Second-order self-model state for the weight of the connection from shared team mental model connection weight self-model state $\mathbf{W}_{c\_T1,d\_T1}$ to individual mental model connection weight self-model state $\mathbf{W}_{c\_A,d\_A}$ for instructional (feedback) learning from the shared mental model of team T1
X_{59}	$\mathbf{W}_{\mathbf{W}_{a\_T1,b\_T1},\mathbf{W}_{a\_B,b\_B}}$	Second-order self-model state for the weight of the connection from shared team mental model connection weight self-model state $\mathbf{W}_{a\_T1,b\_T1}$ to individual mental model connection weight self-model state $\mathbf{W}_{a\_B,b\_B}$ for instructional (feedback) learning from the shared mental model of team T1
X_{60}	$\mathbf{W}_{\mathbf{W}_{b\_T1,c\_T1},\mathbf{W}_{b\_B,c\_B}}$	Second-order self-model state for the weight of the connection from shared team mental model connection weight self-model state $\mathbf{W}_{b\_T1,c\_T1}$ to individual mental model connection weight self-model state $\mathbf{W}_{b\_B,c\_B}$ for instructional (feedback) learning from the shared mental model of team T1
X_{61}	$\mathbf{W}_{\mathbf{W}_{c\_T1,d\_T1},\mathbf{W}_{c\_B,d\_B}}$	Second-order self-model state for the weight of the connection from shared team mental model connection weight self-model state $\mathbf{W}_{c\_T1,d\_T1}$ to individual mental model connection weight self-model state $\mathbf{W}_{c\_B,d\_B}$ for instructional (feedback) learning from the shared mental model of team T1
X_{62}	$\mathbf{W}_{\mathbf{W}_{a\_T2,b\_T2},\mathbf{W}_{a\_C,b\_C}}$	Second-order self-model state for the weight of the connection from shared team mental model connection weight self-model state $\mathbf{W}_{a\_T2,b\_T2}$ to individual mental model connection weight self-model state $\mathbf{W}_{a\_C,b\_C}$ for instructional (feedback) learning from the shared mental model of team T2
X_{63}	$\mathbf{W}_{\mathbf{W}_{b\_T2,c\_T2},\mathbf{W}_{b\_C,c\_C}}$	Second-order self-model state for the weight of the connection from shared team mental model connection weight self-model state $\mathbf{W}_{b\_T2,c\_T2}$ to individual mental model connection weight self-model state $\mathbf{W}_{b\_C,c\_C}$ for instructional (feedback) learning of the shared mental model of team T2
X_{64}	$\mathbf{W}_{\mathbf{W}_{c\_T2,d\_T2},\mathbf{W}_{c\_C,d\_C}}$	Second-order self-model state for the weight of the connection from shared team mental model connection weight self-model state $\mathbf{W}_{c\_T2,d\_T2}$ to individual mental model connection weight self-model state $\mathbf{W}_{c\_C,d\_C}$ for instructional (feedback) learning from the shared mental model of team T2
X_{65}	$\mathbf{W}_{\mathbf{W}_{a\_T2,b\_T2},\mathbf{W}_{a\_D,b\_D}}$	Second-order self-model state for the weight of the connection from shared team mental model connection weight self-model state $\mathbf{W}_{a\_T2,b\_T2}$ to individual mental model connection weight self-model state $\mathbf{W}_{a\_D,b\_D}$ for instructional (feedback) learning from the shared mental model of team T2
X_{66}	$\mathbf{W}_{\mathbf{W}_{b\_T2,c\_T2},\mathbf{W}_{b\_D,c\_D}}$	Second-order self-model state for the weight of the connection from shared team mental model connection weight self-model state $\mathbf{W}_{b\_T2,c\_T2}$ to individual mental model connection weight self-model state $\mathbf{W}_{b\_D,c\_D}$ for instructional (feedback) learning from the shared mental model of team T2
X_{67}	$\mathbf{W}_{\mathbf{W}_{c\_T2,d\_T2},\mathbf{W}_{c\_D,d\_D}}$	Second-order self-model state for the weight of the connection from shared team mental model connection weight self-model state $\mathbf{W}_{c\_T2,d\_T2}$ to individual mental model connection weight self-model state $\mathbf{W}_{c\_D,d\_D}$ for instructional (feedback) learning from the shared mental model of team T2

Fig. 9 Second-order self-model states of the introduced adaptive network model: the higher-order **W**-states for feedback learning from shared team mental model to individual mental models

In summary, first- and second-order self-model states are used to bring multi-order adaptivity to the network model. The first-order adaptation level provides adaptivity of the base level and the second-order one controls this adaptivity. In the first-order self-model level, **W**-states for all the weights of the connections between the base level states are placed. In the first place, these are the adaptive weights of the base level

Nr	State	Explanation
X_{68}	$\mathbf{W}_{\mathbf{W}_{a\_O,b\_O},\mathbf{W}_{a\_T1,b\_T1}}$	Second-order self-model state for the weight of the connection from shared organisation mental model connection weight self-model state $\mathbf{W}_{a\_O,b\_O}$ to shared team mental model connection weight self-model state $\mathbf{W}_{a\_T1,b\_T1}$ for instructional (feedback) learning from the shared organisation mental model
X_{69}	$\mathbf{W}_{\mathbf{W}_{b\_O,c\_O},\mathbf{W}_{b\_T1,c\_T1}}$	Second-order self-model state for the weight of the connection from shared organisation mental model connection weight self-model state $\mathbf{W}_{b\_O,c\_O}$ to shared team mental model connection weight self-model state $\mathbf{W}_{b\_T1,c\_T1}$ for instructional (feedback) learning from the shared organisation mental model
X_{70}	$\mathbf{W}_{\mathbf{W}_{c\_O,d\_O},\mathbf{W}_{c\_T1,d\_T1}}$	Second-order self-model state for the weight of the connection from shared organisation mental model connection weight self-model state $\mathbf{W}_{c\_O,d\_O}$ to shared team mental model connection weight self-model state $\mathbf{W}_{c\_T1,d\_T1}$ for instructional (feedback) learning from the shared organisation mental model
X_{71}	$\mathbf{W}_{\mathbf{W}_{a\_O,b\_O},\mathbf{W}_{a\_T2,b\_T2}}$	Second-order self-model state for the weight of the connection from shared organisation mental model connection weight self-model state $\mathbf{W}_{a\_O,b\_O}$ to shared team mental model connection weight self-model state $\mathbf{W}_{a\_T2,b\_T2}$ for instructional (feedback) learning from the shared organisation mental model
X_{72}	$\mathbf{W}_{\mathbf{W}_{b\_O,c\_O},\mathbf{W}_{b\_T2,c\_T2}}$	Second-order self-model state for the weight of the connection from shared organisation mental model connection weight self-model state $\mathbf{W}_{b\_O,c\_O}$ to shared team mental model connection weight self-model state $\mathbf{W}_{b\_T2,c\_T2}$ for instructional (feedback) learning from the shared organisation mental model
X_{73}	$\mathbf{W}_{\mathbf{W}_{c\_O,d\_O},\mathbf{W}_{c\_T2,d\_T2}}$	Second-order self-model state for the weight of the connection from shared organisation mental model connection weight self-model state $\mathbf{W}_{c\_O,d\_O}$ to shared team mental model connection weight self-model state $\mathbf{W}_{c\_T2,d\_T2}$ for instructional (feedback) learning from the shared organisation mental model

Fig. 10 Second-order self-model states of the introduced adaptive network model: the higher-order **W**-states for feedback learning from shared organisation mental model to shared team mental models

individual mental state connections of persons. In addition, there are **W**-states of the developed shared organisation mental model states. At this first-order adaptation level there are (intralevel) connections from all the **W**-states (two for this case) that specify the weight of a connection between the same tasks for all people (two for this case) to the **W**-states representing the weights of the connections of the shared organisation model (for the formation of the shared organisation mental model) and vice versa (for the learning of the shared organisation mental model by the individuals). At the second-order self-model level, there are $\mathbf{W}_\mathbf{W}$-states specifying the weights of the intralevel connections between the **W**-states to the individual ones (to initiate and control the learning of the shared organisation mental model by the individuals), $\mathbf{H}_\mathbf{W}$-states for adaptation speeds of connection weights in the first-order adaptation level, and $\mathbf{M}_\mathbf{W}$-states for persistence of adaptation. This provides the speed and persistence control of the adaptation.

5 Example Simulation Scenario

In this scenario, for reasons of presentation a multi-phase approach is applied to get a clear picture of the progress of multilevel organisational learning via teams. In general, the model can also process all phases simultaneously. It is possible to see the feed forward flow of the development of shared team mental models from individual

Nr	State	Explanation
X_{74}	$\mathbf{H}_{\mathbf{W}_A}$	Second-order self-model state for the adaptation speed of all individual mental model connection weight self-model states $\mathbf{W}_{x\_A,y\_A}$ for formation or revision of the individual mental model of person A
X_{75}	$\mathbf{H}_{\mathbf{W}_B}$	Second-order self-model state for the adaptation speed of all individual mental model connection weight self-model states $\mathbf{W}_{x\_B,y\_B}$ for formation or revision of the individual mental model of person B
X_{76}	$\mathbf{H}_{\mathbf{W}_C}$	Second-order self-model state for the adaptation speed of all individual mental model connection weight self-model states $\mathbf{W}_{x\_C,y\_C}$ for formation or revision of the individual mental model of person C
X_{77}	$\mathbf{H}_{\mathbf{W}_D}$	Second-order self-model state for the adaptation speed of all individual mental model connection weight self-model states $\mathbf{W}_{x\_D,y\_D}$ for formation or revision of the individual mental model of person D
X_{78}	$\mathbf{H}_{\mathbf{W}_{T1}}$	Second-order self-model state for the adaptation speed of shared mental model connection weight self-model state $\mathbf{W}_{x\_T1,y\_T1}$ for formation or revision of the shared mental model of team T1
X_{79}	$\mathbf{H}_{\mathbf{W}_{T2}}$	Second-order self-model state for the adaptation speed of all shared mental model connection weight self-model states $\mathbf{W}_{x\_T2,y\_T2}$ for formation or revision of the shared mental model of team T2
X_{80}	$\mathbf{H}_{\mathbf{W}_O}$	Second-order self-model state for the adaptation speed of all shared organisation mental model connection weight self-model states $\mathbf{W}_{x\_O,y\_O}$ for formation or revision of the shared organisation mental model
X_{81}	$\mathbf{M}_{\mathbf{W}_{a\_A,b\_A}}$	Second-order self-model state for persistence of adaptation of individual mental model connection weight self-model state $\mathbf{W}_{a\_A,b\_A}$ of person A
X_{82}	$\mathbf{M}_{\mathbf{W}_{b\_A,c\_A}}$	Second-order self-model state for persistence of adaptation of individual mental model connection weight self-model state $\mathbf{W}_{b\_A,c\_A}$ of person A
X_{83}	$\mathbf{M}_{\mathbf{W}_{c\_A,d\_A}}$	Second-order self-model state for persistence of adaptation of individual mental model connection weight self-model state $\mathbf{W}_{c\_A,d\_A}$ of person A
X_{84}	$\mathbf{M}_{\mathbf{W}_{a\_B,b\_B}}$	Second-order self-model state for persistence of adaptation of individual mental model connection weight self-model state $\mathbf{W}_{a\_B,b\_B}$ of person B
X_{85}	$\mathbf{M}_{\mathbf{W}_{b\_B,c\_B}}$	Second-order self-model state for persistence of adaptation of individual mental model connection weight self-model state $\mathbf{W}_{b\_B,c\_B}$ of person B
X_{86}	$\mathbf{M}_{\mathbf{W}_{c\_B,d\_B}}$	Second-order self-model state for persistence of adaptation of individual mental model connection weight self-model state $\mathbf{W}_{c\_B,d\_B}$ of person B
X_{87}	$\mathbf{M}_{\mathbf{W}_{a\_C,b\_C}}$	Second-order self-model state for persistence of adaptation of individual mental model connection weight self-model state $\mathbf{W}_{a\_C,b\_C}$ of person C
X_{88}	$\mathbf{M}_{\mathbf{W}_{b\_C,c\_C}}$	Second-order self-model state for persistence of adaptation of individual mental model connection weight self-model state $\mathbf{W}_{b\_C,c\_C}$ of person C
X_{89}	$\mathbf{M}_{\mathbf{W}_{c\_C,d\_C}}$	Second-order self-model state for persistence of adaptation of individual mental model connection weight self-model state $\mathbf{W}_{c\_C,d\_C}$ of person C
X_{90}	$\mathbf{M}_{\mathbf{W}_{a\_D,b\_D}}$	Second-order self-model state for persistence of adaptation of individual mental model connection weight self-model state $\mathbf{W}_{a\_D,b\_D}$ of person D
X_{91}	$\mathbf{M}_{\mathbf{W}_{b\_D,c\_D}}$	Second-order self-model state for persistence of adaptation of individual mental model connection weight self-model state $\mathbf{W}_{b\_D,c\_D}$ of person D
X_{92}	$\mathbf{M}_{\mathbf{W}_{c\_D,d\_D}}$	Second-order self-model state for persistence of adaptation of individual mental model connection weight self-model state $\mathbf{W}_{c\_D,d\_D}$ of person D

Fig. 11 Second-order self-model states of the introduced adaptive network model: the $\mathbf{H}_{\mathbf{W}}$-states for adaptive learning speed and $\mathbf{M}_{\mathbf{W}}$-states for adaptive persistence of the learning

mental models first, formation of the shared organisation mental model originating from teams' mental models then, and finally by the feedback flow the impact of these shared mental models on teams and individuals. In practice, and also in the model, these phases also can overlap or take place entirely simultaneously. The considered six phases are as follows; see Figs. 13, 14 and 15:

Intralevel connections		
$x\_Z \to y\_Z$	Connection from x to y in individual mental model of person Z: **(a)** from Sect. 2.3.	
$x\_O \to y\_O$	Connection from x to y in shared mental model of organisation O: **(a)** from Sect. 2.3.	
$con_p \to x\_Z$	Connection from context state con_p for phase $p \in \{ph1, ph6\}$ to activate mental model state x of person Z: **(f)** from Sect. 3.2.	
$\mathbf{W}_{x\_Z,y\_Z} \to \mathbf{W}_{x\_T,y\_T}$	Connection for person Z's contribution from the weight of the connection from x to y in the individual mental model of Z to the weight of the connection from x to y in the shared team mental model of team T: **(b)** from Sect. 3.2.	
$\mathbf{W}_{x\_T,y\_T} \to \mathbf{W}_{x\_O,y\_O}$	Connection for team T's contribution from the weight of the connection from x to y in the shared team mental model of Z to the weight of the connection from x to y in the shared organisation mental model of O: **(c)** from Sect. 3.2.	
$\mathbf{W}_{x\_O,y\_O} \to \mathbf{W}_{x\_T,y\_T}$	Connection for O's contribution from the weight of the connection from x to y in the shared organisation mental model of O to the weight of the connection from x to y in the shared team mental model of team T: **(d)** from Sect. 3.2.	
$\mathbf{W}_{x\_T,y\_T} \to \mathbf{W}_{x\_Z,y\_Z}$	Connection for team T's contribution from the weight of the connection from x to y in the shared team mental model of T to the weight of the connection from x to y in the individual mental model of person Z: **(e)** from Sect. 3.2.	
$\mathbf{W}_{x\_Z,y\_Z} \to \mathbf{W}_{x\_Z,y\_Z}$	Persistence connection for Z's mental model connections: **(a)** from Sect. 3.2.	
Interlevel connections		
$x\_Z \to \mathbf{W}_{x\_Z,y\_Z}$	Connection for individual Hebbian learning from state x in person Z's individual mental model to self-model state $\mathbf{W}_{x\_A,y\_A}$ for Z's individual mental model: **(a)** from Sect. 3.2.	Upward from base level to first self-model level
$y\_Z \to \mathbf{W}_{x\_Z,y\_Z}$	Connection for individual Hebbian learning from state y in person Z's individual mental model to self-model state $\mathbf{W}_{x\_A,y\_A}$ for Z's individual mental model: **(a)** from Sect. 3.2.	
$x\_T \to \mathbf{W}_{x\_T,y\_T}$	Connection for individual Hebbian learning from state x in team T's shared mental model to self-model state $\mathbf{W}_{x\_T,y\_T}$ for T's shared mental model: **(b)** from Sect. 3.2.	
$y\_T \to \mathbf{W}_{x\_T,y\_T}$	Connection for individual Hebbian learning from state y in team T's shared mental model to self-model state $\mathbf{W}_{x\_T,y\_T}$ for T's shared mental model: **(b)** from Sect. 3.2.	
$x\_O \to \mathbf{W}_{x\_O,y\_O}$	Connection for Hebbian learning from state x in O's shared mental model to self-model state $\mathbf{W}_{x\_A,y\_A}$ for O's shared mental model: **(c)** from Sect. 3.2.	
$y\_O \to \mathbf{W}_{x\_O,y\_O}$	Connection for Hebbian learning from state y in O's shared mental model to self-model state $\mathbf{W}_{x\_A,y\_A}$ for O's shared mental model: **(c)** from Sect. 3.2.	
$\mathbf{W}_{x\_Z,y\_Z} \to y\_Z$	Connection for the effect of self-model state $\mathbf{W}_{x\_Z,y\_Z}$ for person Z's individual mental model on state y in Z's individual mental model: **(a)** from Sect. 3.2.	Downward from first-order self-model level to base level
$\mathbf{W}_{x\_T,y\_T} \to y\_T$	Connection for the effect of self-model state $\mathbf{W}_{x\_T,y\_T}$ for team T's shared mental model on state y in T's shared mental model: **(b)** from Sect. 3.2.	
$\mathbf{W}_{x\_O,y\_O} \to y\_O$	Connection for the effect of self-model state $\mathbf{W}_{x\_O,y\_O}$ for O's shared mental model on state y in O's shared mental model: **(c)** from Sect. 3.2.	
$con_{ph2} \to \mathbf{H}_{\mathbf{W}_{x\_T,y\_T}}$	Connection from the context state for Phase 2 to second-order self-model state $\mathbf{H}_{\mathbf{W}_{a\_T,b\_T}}$ representing the adaptation speed of first-order self-model state $\mathbf{W}_{x\_T,y\_T}$ for the weight of the connection from x to y in the shared mental model of team T in order to trigger this adaptation speed for shared mental model formation: **(b)** from Sect. 3.2.	Upward from base level to second-order self-model level
$con_{ph3} \to \mathbf{H}_{\mathbf{W}_{x\_O,y\_O}}$	Connection from the context state for Phase 2 to second-order self-model state $\mathbf{H}_{\mathbf{W}_{a\_O,b\_O}}$ representing the adaptation speed of first-order self-model state $\mathbf{W}_{x\_O,y\_O}$ for the weight of the connection from x to y in the shared organisation mental model of O in order to trigger this adaptation speed for shared mental model formation: **(c)** from Sect. 3.2.	
$con_{ph5} \to \mathbf{W}_{\mathbf{W}_{x\_T,y\_T},\mathbf{W}_{x\_Z,y\_Z}}$	Connection from the context state for Phase 3 to second-order self-model state $\mathbf{W}_{\mathbf{W}_{x\_T,y\_T},\mathbf{W}_{x\_Z,y\_Z}}$ representing the weight of the connection from first-order self-model state $\mathbf{W}_{x\_T,y\_T}$ for the weight of the connection from x to y in the shared mental model of team T to first-order self-model state $\mathbf{W}_{x\_Z,y\_Z}$ for the weight of the connection from x to y in the individual mental model of person Z in order to activate this connection for instructional learning of Z from the shared mental model: **(e)** from Sect. 3.2.	

Fig. 12 Types of connections in the adaptive network model and how they relate to **(a)** to **(f)** identified in Sect. 2. For the example scenario, x and y are states from $\{a, b, c, d\}$, T is a team from $\{T1, T2\}$ and Z is a person from $\{A, B, C, D\}$

$\text{con}_{ph4} \rightarrow \mathbf{W}_{\mathbf{W}_{x\_O,y\_O},\mathbf{W}_{x\_Z,y\_Z}}$	Connection from the context state for Phase 3 to second-order self-model state $\mathbf{W}_{\mathbf{W}_{x\_O,y\_O},\mathbf{W}_{x\_Z,y\_Z}}$ representing the weight of the connection from first-order self-model state $\mathbf{W}_{x\_O,y\_O}$ for the weight of the connection from *x* to *y* in the shared mental model of O to first-order self-model state $\mathbf{W}_{x\_Z,y\_Z}$ for the weight of the connection from *x* to *y* in the individual mental model of person *Z* in order to activate this connection for instructional learning of *Z* from the shared mental model: **(d)** from Sect. 3.2.	
$\mathbf{H}_{\mathbf{W}_{x\_O,y\_O}} \rightarrow \mathbf{W}_{x\_O,y\_O}$	Effectuation of control of the adaptation of O's shared mental model connection weight $\mathbf{W}_{x\_O,y\_O}$ for shared mental model formation based on *Z*'s individual mental model: **(b)**, **(c)** from Sect. 3.2.	Downward from second- to first-order self-model level
$\mathbf{W}_{\mathbf{W}_{x\_O,y\_O},\mathbf{W}_{x\_Z,y\_Z}} \rightarrow \mathbf{W}_{x\_Z,y\_Z}$	Effectuation of control of the adaptation of person *Z*'s individual mental model connection weight $\mathbf{W}_{x\_Z,y\_Z}$ for instructional learning of *Z*'s individual mental model from O's shared mental model: **(d)** from Sect. 3.2.	

Fig. 12 (continued)

- **Phase 1: Individual mental model usage and learning**
  This relates to **(a)** in Sect. 2. Different individual mental models by four different persons are constructed and strengthened here. The knowledge of people for the tasks, initially, is not same. Thus, the learning levels are different as can be seen in the first phase between time 25 and 200 in the simulation graph in Fig. 11 (overall) and 12 (the **W**-states). For example, activation levels of first three base states for tasks *a* to *c* of person A from Team 1 and person C from Team 2 (a_A to c_A and a_C to c_C) increase while the activation levels of states for task *d* (d_A and d_C) remain at zero indicating that they do not have knowledge on this task. A similar lack of knowledge is observed for the other persons B from Team 1 and D from Team 2, for task *a* this time. Therefore, the activation levels of their states a_B and a_D remain at zero in this phase, while others get increased (b_B to d_B and b_D to d_D). After this first individual learning phase, forgetting takes place for all persons because they do not have perfect persistence factors self-model **M**-state values (values < 1, meaning imperfection). Increased **W**-states during phase 1, start to slightly decrease after phase 1 at different rates representing the differences between persons concerning forgetting speed.
- **Phase 2: Shared team mental model formation (feed forward learning)**
  This relates to **(b)** in Sect. 2. Formation of two shared team mental models happens in this phase. The collaboration of the individuals creates the aggregation of their mental models as part of feed forward organisational learning (in this case team learning). The **W**-states of the teams ($\mathbf{W}_{a\_T1,b\_T1}$ to $\mathbf{W}_{c\_T1,d\_T1}$ and $\mathbf{W}_{a\_T2,b\_T2}$ to $\mathbf{W}_{c\_T2,d\_T2}$) increase at different rates in Phase 2 between time 250 and 300 in Figs. 11 and 12. Team 1 becomes better at the connection $c \rightarrow d$, and Team 2 becomes better at connection $a \rightarrow b$ because the teams have different persons. Then, these shared mental models are maintained by the two teams.
- **Phase 3: Shared organisation mental model formation (feed forward learning)**
  This relates to **(c)** in Sect. 2. A shared organisation mental model is formed in this phase from the unification and aggregation of the two shared team mental models. The values of shared organisation mental model **W**-states ($\mathbf{W}_{a\_O,b\_O}$ to $\mathbf{W}_{c\_O,d\_O}$) increase here between time 350 and 400.

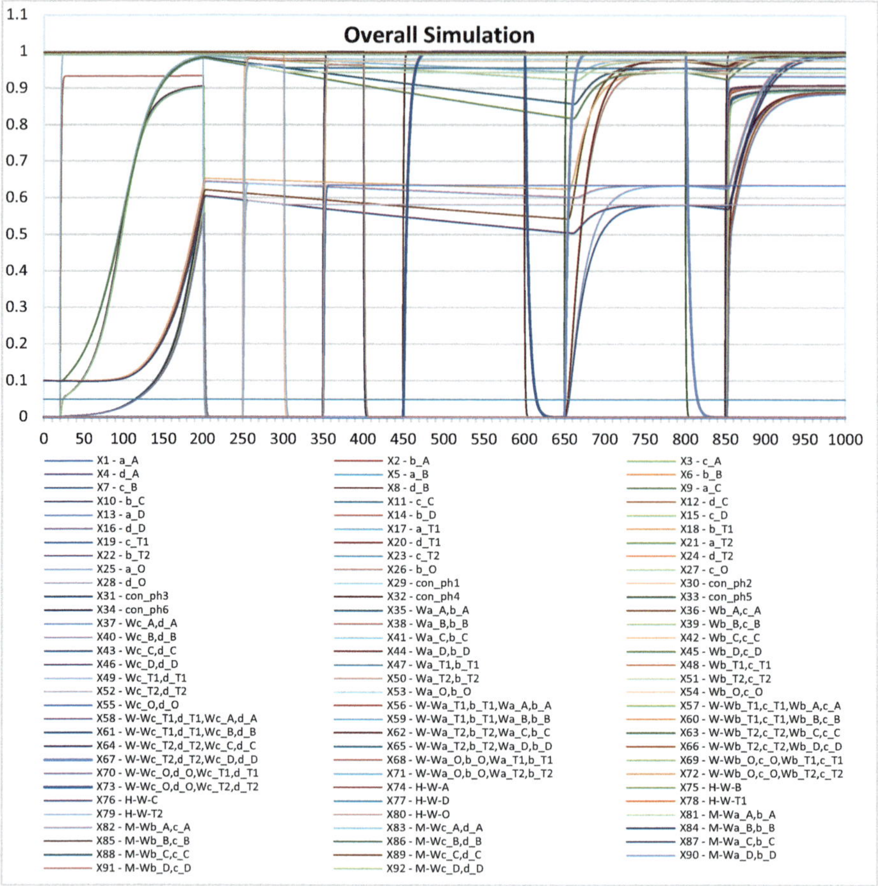

Fig. 13 Simulation graph showing all states

- **Phase 4: Feedback learning of the shared team mental model from the shared organisation mental model**
  This relates to **(d)** in Sect. 2. Knowledge from the shared organisation mental model is received by the team mental models as a form of (instructional) feedback learning here in this phase. The (higher-order adaptive) connections from organisation **W**-states to teams **W**-states (X_{68} to X_{73}) become activated, and the teams start to get stronger connections about tasks.
- **Phase 5: Feedback learning of the individual mental models from the shared team mental models**
  This relates to **(e)** in Sect. 2. Improved knowledge from shared team mental models is received by individuals as a form of (instructional) feedback learning in this phase. Higher-order adaptive weight states for connections from teams **W**-states to individual **W**-states (X_{56} to X_{67}) are activated. This provides the learning

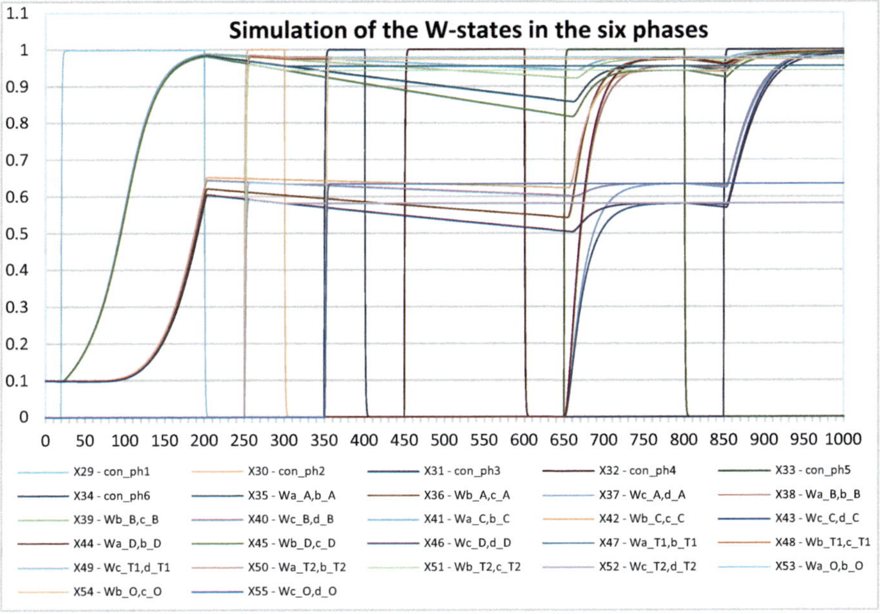

Fig. 14 Simulation graph for the **W**-states

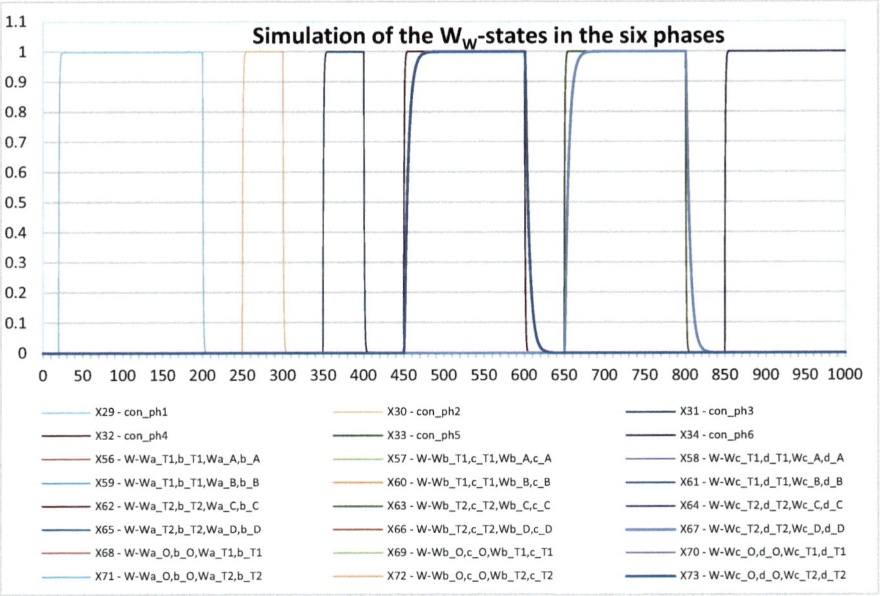

Fig. 15 Simulation graph for the $\mathbf{W_W}$-states

of individual mental models and gives persons the chance of improving their unknown connections in the next phase. For instance, the person A starts to learn about the task *d* that it does not know in the beginning by the help of its team. In Fig. 6, the **W**-states of persons make jumps in this Phase 5 between time 650 and 800.

- **Phase 6: Individual mental model usage and learning**
  This relates to **(f)** in Sect. 2. Persons start to further improve their knowledge and skills (their mental models) already strengthened in Phase 5 by Hebbian learning (Hebb 1949). Person A's knowledge on task *d* (state d_A) becomes nonzero now (obtained via shared team mental model) and similar improvements are observed for other persons and their 'hollow' unknown states.

6 Discussion

The material in this chapter is based on Canbaloğlu et al. (2023a). Within mainstream organisational learning literature such as Crossan et al. (1999), Wiewiora et al. (2019), mental models at individual, team and organisation levels and the interplay of them are considered a vehicle for organisational learning. This is called multilevel organisational learning. Based on developed individual mental models, by socalled feed forward learning the formation of shared team mental models can take place and based on them, a shared mental model for the level of the organisation as a whole (see also Fig. 1 adopted from the mentioned literature). Once these shared mental models have been formed, they can be adopted by individuals within the organisation, indicated as feedback learning. This involves several mechanisms of different types that by their cyclical interaction together can be considered to form the basis of multilevel organisational learning. These mechanisms have been formalized in a computational manner here and brought together in an adaptive self-modeling network architecture. The model was illustrated by a relatively simple but realistic case study. For the sake of presentation, in the case study scenario the different types of mechanisms have been controlled in such a manner that they are sequentially over time. This is not inherent in the designed computational network model: these processes can equally well work simultaneously. The two lowest levels of the three-level network model describe Fig. 1 very well, especially the intralevel connections within the middle level directly correspond to the arrows in Fig. 1. However, the necessary control of these processes is left out of consideration in Fig. 1 but is fully addressed here by the highest (third) level.

One of the extension possibilities concerns the type of aggregation used for the process of shared mental model formation. In the current model this has been based on the maximal knowledge about a specific mental model connection. But other forms of aggregation can equally well be applied, for example weighted averages. Another possible extension is to make states used for the control adaptive in a context-sensitive manner, such as the second-order self-model **H**- and **M**-states for the individuals, which for the sake of simplicity were kept constant in the current example scenario.

Appendix: Role Matrices

In Figs. 16, 17, 18, 19 and 20, the different role matrices are shown that provide a full specification of the network characteristics defining the adaptive network model in a standardised table format. Here in each role matrix, each state has its row where it is listed which are the impacts on it from that role.

Role matrices for connectivity characteristics

The connectivity characteristics are specified by role matrices **mb** and **mcw** shown in Figs. 16 and 17. Role matrix **mb** lists the other states (at the same or lower level) from which the state gets its incoming connections, whereas in role matrix **mcw** the connection weights are listed for these connections.

Nonadaptive connection weights are indicated in **mcw** (in Fig. 17) by a number (in a green shaded cell), but adaptive connection weights are indicated by a reference to the (self-model) state representing the adaptive value (in a peach-red shaded cell). This can be seen at the base level for states X_2 to X_4 (with self-model states X_{35} to X_{36}), states X_6 to X_8 (with self-model states X_{38} to X_{40}), X_{10} to X_{12} (with self-model states X_{41} to X_{43}), X_{14} to X_{16} (with self-model states X_{44} to X_{46}), X_{18} to X_{20} (with self-model states X_{47} to X_{49}), X_{22} to X_{24} (with self-model states X_{50} to X_{52}), and X_{26} to X_{28} (with self-model states X_{53} to X_{55}). Similarly, at the first-order self-model level this can be seen for states X_{35} to X_{52} (with self-model states X_{56} to X_{73}). This specifies that all these 39 connections are indeed adaptive with weights represented by states X_{35} to X_{73}.

Role matrices for aggregation characteristics

The network characteristics for aggregation are defined by the selection of combination functions from the library and values for their parameters. In role matrix **mcfw** it is specified by weights which state uses which combination function; see Fig. 18.

In role matrix **mcfp** (see Fig. 19) it is indicated what the parameter values are for the chosen combination functions. Some of them are adaptive, as can be seen in the rows from X_{35} to X_{46} (the persistence factors μ represented by the self-model states X_{81} to X_{92}). This specifies that these 12 persistence factors are indeed adaptive with weights represented by states X_{81} to X_{92}.

Role matrices for timing characteristics

In Fig. 20, the role matrix **ms** for speed factors is shown, which lists all speed factors. Next to it, the list of initial values can be found. Also, for **ms** some entries are adaptive: the speed factors of X_{35} to X_{55} are represented by (second-order) self-model states X_{74} to X_{80}. This specifies that all these 21 speed factors are indeed adaptive with weights represented by states X_{74} to X_{80}.

Fig. 16 Role matrices for the connectivity: **mb** for base connectivity

mb	**base connectivity**	1	2	3	4	5	6	7	8	9
X_{1}	a_A	X_{29}	X_{34}							
X_{2}	b_A	X_{1}								
X_{3}	c_A	X_{2}								
X_{4}	d_A	X_{3}								
X_{5}	a_B	X_{34}								
X_{6}	b_B	X_{5}	X_{29}							
X_{7}	c_B	X_{6}								
X_{8}	d_B	X_{7}								
X_{9}	a_C	X_{29}	X_{34}							
X_{10}	b_C	X_{9}								
X_{11}	c_C	X_{10}								
X_{12}	d_C	X_{11}								
X_{13}	a_D	X_{34}								
X_{14}	b_D	X_{13}	X_{29}							
X_{15}	c_D	X_{14}								
X_{16}	d_D	X_{15}								
X_{17}	a_T1									
X_{18}	b_T1	X_{17}								
X_{19}	c_T1	X_{18}								
X_{20}	d_T1	X_{19}								
X_{21}	a_T2									
X_{22}	b_T2	X_{21}								
X_{23}	c_T2	X_{22}								
X_{24}	d_T2	X_{23}								
X_{25}	a_O									
X_{26}	b_O	X_{25}								
X_{27}	c_O	X_{26}								
X_{28}	d_O	X_{27}								
X_{29}	con_{ph1}	X_{29}								
X_{30}	con_{ph2}	X_{30}								
X_{31}	con_{ph3}	X_{31}								
X_{32}	con_{ph4}	X_{32}								
X_{33}	con_{ph5}	X_{33}								
X_{34}	con_{ph6}	X_{34}								

mb	**base connectivity**	1	2	3	4	5	6	7	8	9
X_{35}	$\mathbf{W}_{a\_A,b\_A}$	X_{1}	X_{2}	X_{35}	X_{47}					
X_{36}	$\mathbf{W}_{b\_A,c\_A}$	X_{2}	X_{3}	X_{36}	X_{48}					
X_{37}	$\mathbf{W}_{c\_A,d\_A}$	X_{3}	X_{4}	X_{37}	X_{49}					
X_{38}	$\mathbf{W}_{a\_B,b\_B}$	X_{5}	X_{6}	X_{38}	X_{47}					
X_{39}	$\mathbf{W}_{b\_B,c\_B}$	X_{6}	X_{7}	X_{39}	X_{48}					
X_{40}	$\mathbf{W}_{c\_B,d\_B}$	X_{7}	X_{8}	X_{40}	X_{49}					
X_{41}	$\mathbf{W}_{a\_C,b\_C}$	X_{9}	X_{10}	X_{41}	X_{50}					
X_{42}	$\mathbf{W}_{b\_C,c\_C}$	X_{10}	X_{11}	X_{42}	X_{51}					
X_{43}	$\mathbf{W}_{c\_C,d\_C}$	X_{11}	X_{12}	X_{43}	X_{52}					
X_{44}	$\mathbf{W}_{a\_D,b\_D}$	X_{13}	X_{14}	X_{44}	X_{50}					
X_{45}	$\mathbf{W}_{b\_D,c\_D}$	X_{14}	X_{15}	X_{45}	X_{51}					
X_{46}	$\mathbf{W}_{c\_D,d\_D}$	X_{15}	X_{16}	X_{46}	X_{52}					
X_{47}	$\mathbf{W}_{a\_T1,b\_T1}$	X_{17}	X_{18}	X_{35}	X_{38}	X_{47}	X_{53}			
X_{48}	$\mathbf{W}_{b\_T1,c\_T1}$	X_{18}	X_{19}	X_{36}	X_{39}	X_{48}	X_{54}			
X_{49}	$\mathbf{W}_{c\_T1,d\_T1}$	X_{19}	X_{20}	X_{37}	X_{40}	X_{49}	X_{55}			
X_{50}	$\mathbf{W}_{a\_T2,b\_T2}$	X_{21}	X_{22}	X_{41}	X_{44}	X_{50}	X_{53}			
X_{51}	$\mathbf{W}_{b\_T2,c\_T2}$	X_{22}	X_{23}	X_{42}	X_{45}	X_{51}	X_{54}			
X_{52}	$\mathbf{W}_{c\_T2,d\_T2}$	X_{23}	X_{24}	X_{43}	X_{46}	X_{52}	X_{55}			
X_{53}	$\mathbf{W}_{a\_O,b\_O}$	X_{25}	X_{26}	X_{47}	X_{50}	X_{53}				
X_{54}	$\mathbf{W}_{b\_O,c\_O}$	X_{26}	X_{27}	X_{48}	X_{51}	X_{54}				
X_{55}	$\mathbf{W}_{c\_O,d\_O}$	X_{27}	X_{28}	X_{49}	X_{52}	X_{55}				

mb	base connectivity	1	2	3	4	5	6	7	8	9
X_{56}	$\mathbf{W}w_{a\_T1,b\_T1},w_{a\_A,b\_A}$	X_{35}	X_{47}	X_{33}						
X_{57}	$\mathbf{W}w_{b\_T1,c\_T1},w_{b\_A,c\_A}$	X_{36}	X_{48}	X_{33}						
X_{58}	$\mathbf{W}w_{c\_T1,d\_T1},w_{c\_A,d\_A}$	X_{37}	X_{49}	X_{33}						
X_{59}	$\mathbf{W}w_{a\_T1,b\_T1},w_{a\_B,b\_B}$	X_{38}	X_{47}	X_{33}						
X_{60}	$\mathbf{W}w_{b\_T1,c\_T1},w_{b\_B,c\_B}$	X_{39}	X_{48}	X_{33}						
X_{61}	$\mathbf{W}w_{c\_T1,d\_T1},w_{c\_B,d\_B}$	X_{40}	X_{49}	X_{33}						
X_{62}	$\mathbf{W}w_{a\_T2,b\_T2},w_{a\_C,b\_C}$	X_{41}	X_{50}	X_{33}						
X_{63}	$\mathbf{W}w_{b\_T2,c\_T2},w_{b\_C,c\_C}$	X_{42}	X_{51}	X_{33}						
X_{64}	$\mathbf{W}w_{c\_T2,d\_T2},w_{c\_C,d\_C}$	X_{43}	X_{52}	X_{33}						
X_{65}	$\mathbf{W}w_{a\_T2,b\_T2},w_{a\_D,b\_D}$	X_{44}	X_{50}	X_{33}						
X_{66}	$\mathbf{W}w_{b\_T2,c\_T2},w_{b\_D,c\_D}$	X_{45}	X_{51}	X_{33}						
X_{67}	$\mathbf{W}w_{c\_T2,d\_T2},w_{c\_D,d\_D}$	X_{46}	X_{52}	X_{33}						
X_{68}	$\mathbf{W}w_{a\_O,b\_O},w_{a\_T1,b\_T1}$	X_{32}								
X_{69}	$\mathbf{W}w_{b\_O,c\_O},w_{b\_T1,c\_T1}$	X_{32}								
X_{70}	$\mathbf{W}w_{c\_O,d\_O},w_{c\_T1,d\_T1}$	X_{32}								
X_{71}	$\mathbf{W}w_{a\_O,b\_O},w_{a\_T2,b\_T2}$	X_{32}								
X_{72}	$\mathbf{W}w_{b\_O,c\_O},w_{b\_T2,c\_T2}$	X_{32}								
X_{73}	$\mathbf{W}w_{c\_O,d\_O},w_{c\_T2,d\_T2}$	X_{32}								
X_{74}	$\mathbf{H}w_{A}$	X_{30}	X_{1}	X_{2}	X_{3}	X_{4}	X_{35}	X_{36}	X_{37}	X_{74}
X_{75}	$\mathbf{H}w_{B}$	X_{30}	X_{5}	X_{6}	X_{7}	X_{8}	X_{38}	X_{39}	X_{40}	X_{75}
X_{76}	$\mathbf{H}w_{C}$	X_{30}	X_{9}	X_{10}	X_{11}	X_{12}	X_{41}	X_{42}	X_{43}	X_{76}
X_{77}	$\mathbf{H}w_{D}$	X_{30}	X_{13}	X_{14}	X_{15}	X_{16}	X_{44}	X_{45}	X_{46}	X_{77}
X_{78}	$\mathbf{H}w_{T1}$	X_{30}								
X_{79}	$\mathbf{H}w_{T2}$	X_{30}								
X_{80}	$\mathbf{H}w_{O}$	X_{31}								
X_{81}	$\mathbf{M}w_{a\_A,b\_A}$	X_{1}	X_{2}	X_{35}	X_{81}	X_{29}	X_{30}	X_{31}	X_{32}	X_{33}
X_{82}	$\mathbf{M}w_{b\_A,c\_A}$	X_{2}	X_{3}	X_{36}	X_{82}	X_{29}	X_{30}	X_{31}	X_{32}	X_{33}
X_{83}	$\mathbf{M}w_{c\_A,d\_A}$	X_{3}	X_{4}	X_{37}	X_{83}	X_{29}	X_{30}	X_{31}	X_{32}	X_{33}
X_{84}	$\mathbf{M}w_{a\_B,b\_B}$	X_{5}	X_{6}	X_{38}	X_{84}	X_{29}	X_{30}	X_{31}	X_{32}	X_{33}
X_{85}	$\mathbf{M}w_{b\_B,c\_B}$	X_{6}	X_{7}	X_{39}	X_{85}	X_{29}	X_{30}	X_{31}	X_{32}	X_{33}
X_{86}	$\mathbf{M}w_{c\_B,d\_B}$	X_{7}	X_{8}	X_{40}	X_{86}	X_{29}	X_{30}	X_{31}	X_{32}	X_{33}
X_{87}	$\mathbf{M}w_{a\_C,b\_C}$	X_{8}	X_{9}	X_{41}	X_{87}	X_{29}	X_{30}	X_{31}	X_{32}	X_{33}
X_{88}	$\mathbf{M}w_{b\_C,c\_C}$	X_{9}	X_{10}	X_{42}	X_{88}	X_{29}	X_{30}	X_{31}	X_{32}	X_{33}
X_{89}	$\mathbf{M}w_{c\_C,d\_C}$	X_{10}	X_{11}	X_{43}	X_{89}	X_{29}	X_{30}	X_{31}	X_{32}	X_{33}
X_{90}	$\mathbf{M}w_{a\_D,b\_D}$	X_{11}	X_{12}	X_{44}	X_{90}	X_{29}	X_{30}	X_{31}	X_{32}	X_{33}
X_{91}	$\mathbf{M}w_{b\_D,c\_D}$	X_{12}	X_{13}	X_{45}	X_{91}	X_{29}	X_{30}	X_{31}	X_{32}	X_{33}
X_{92}	$\mathbf{M}w_{c\_D,d\_D}$	X_{13}	X_{14}	X_{46}	X_{92}	X_{29}	X_{30}	X_{31}	X_{32}	X_{33}

Fig. 16 (continued)

Fig. 17 Role matrices for the connectivity: **mcw** for connection weights

mcw	connection weights	1	2	3	4	5	6	7	8	9
X_1	a_A	1	1							
X_2	b_A	X_{35}								
X_3	c_A	X_{36}								
X_4	d_A	X_{37}								
X_5	a_B	1								
X_6	b_B	X_{38}	1							
X_7	c_B	X_{39}								
X_8	d_B	X_{40}								
X_9	a_C	1	1							
X_{10}	b_C	X_{41}								
X_{11}	c_C	X_{42}								
X_{12}	d_C	X_{43}								
X_{13}	a_D	1								
X_{14}	b_D	X_{44}	1							
X_{15}	c_D	X_{45}								
X_{16}	d_D	X_{46}								
X_{17}	a_T1									
X_{18}	b_T1	X_{47}								
X_{19}	c_T1	X_{48}								
X_{20}	d_T1	X_{49}								
X_{21}	a_T2									
X_{22}	b_T2	X_{50}								
X_{23}	c_T2	X_{51}								
X_{24}	d_T2	X_{52}								
X_{25}	a_O									
X_{26}	b_O	X_{53}								
X_{27}	c_O	X_{54}								
X_{28}	d_O	X_{55}								
X_{29}	con_{ph1}	1								
X_{30}	con_{ph2}	1								
X_{31}	con_{ph3}	1								
X_{32}	con_{ph4}	1								
X_{33}	con_{ph5}	1								
X_{34}	con_{ph6}	1								

mcw	connection weights	1	2	3	4	5	6	7	8	9
X_{35}	$\mathbf{W}_{a\_A,b\_A}$	1	1	1	X_{56}					
X_{36}	$\mathbf{W}_{b\_A,c\_A}$	1	1	1	X_{57}					
X_{37}	$\mathbf{W}_{c\_A,d\_A}$	1	1	1	X_{58}					
X_{38}	$\mathbf{W}_{a\_B,b\_B}$	1	1	1	X_{59}					
X_{39}	$\mathbf{W}_{b\_B,c\_B}$	1	1	1	X_{60}					
X_{40}	$\mathbf{W}_{c\_B,d\_B}$	1	1	1	X_{61}					
X_{41}	$\mathbf{W}_{a\_C,b\_C}$	1	1	1	X_{62}					
X_{42}	$\mathbf{W}_{b\_C,c\_C}$	1	1	1	X_{63}					
X_{43}	$\mathbf{W}_{c\_C,d\_C}$	1	1	1	X_{64}					
X_{44}	$\mathbf{W}_{a\_D,b\_D}$	1	1	1	X_{65}					
X_{45}	$\mathbf{W}_{b\_D,c\_D}$	1	1	1	X_{66}					
X_{46}	$\mathbf{W}_{c\_D,d\_D}$	1	1	1	X_{67}					
X_{47}	$\mathbf{W}_{a\_T1,b\_T1}$	1	1	X_{68}	1					
X_{48}	$\mathbf{W}_{b\_T1,c\_T1}$	1	1	X_{69}	1					
X_{49}	$\mathbf{W}_{c\_T1,d\_T1}$	1	1	X_{70}	1					
X_{50}	$\mathbf{W}_{a\_T2,b\_T2}$	1	1	X_{71}	1					
X_{51}	$\mathbf{W}_{b\_T2,c\_T2}$	1	1	X_{72}	1					
X_{52}	$\mathbf{W}_{c\_T2,d\_T2}$	1	1	X_{73}	1					
X_{53}	$\mathbf{W}_{a\_O,b\_O}$	1	1	1	1					
X_{54}	$\mathbf{W}_{b\_O,c\_O}$	1	1	1	1					
X_{55}	$\mathbf{W}_{c\_O,d\_O}$	1	1	1	1					

mcw	connection weights	1	2	3	4	5	6	7	8	9
X_{56}	$\mathbf{W}w_{a\_T1,b\_T1},w_{a\_A,b\_A}$	1	1	1						
X_{57}	$\mathbf{W}w_{b\_T1,c\_T1},w_{b\_A,c\_A}$	1	1	1						
X_{58}	$\mathbf{W}w_{c\_T1,d\_T1},w_{c\_A,d\_A}$	1	1	1						
X_{59}	$\mathbf{W}w_{a\_T1,b\_T1},w_{a\_B,b\_B}$	1	1	1						
X_{60}	$\mathbf{W}w_{b\_T1,c\_T1},w_{b\_B,c\_B}$	1	1	1						
X_{61}	$\mathbf{W}w_{c\_T1,d\_T1},w_{c\_B,d\_B}$	1	1	1						
X_{62}	$\mathbf{W}w_{a\_T2,b\_T2},w_{a\_C,b\_C}$	1	1	1						
X_{63}	$\mathbf{W}w_{b\_T2,c\_T2},w_{b\_C,c\_C}$	1	1	1						
X_{64}	$\mathbf{W}w_{c\_T2,d\_T2},w_{c\_C,d\_C}$	1	1	1						
X_{65}	$\mathbf{W}w_{a\_T2,b\_T2},w_{a\_D,b\_D}$	1	1	1						
X_{66}	$\mathbf{W}w_{b\_T2,c\_T2},w_{b\_D,c\_D}$	1	1	1						
X_{67}	$\mathbf{W}w_{c\_T2,d\_T2},w_{c\_D,d\_D}$	1	1	1						
X_{68}	$\mathbf{W}w_{a\_O,b\_O},w_{a\_T1,b\_T1}$	1								
X_{69}	$\mathbf{W}w_{b\_O,c\_O},w_{b\_T1,c\_T1}$	1								
X_{70}	$\mathbf{W}w_{c\_O,d\_O},w_{c\_T1,d\_T1}$	1								
X_{71}	$\mathbf{W}w_{a\_O,b\_O},w_{a\_T2,b\_T2}$	1								
X_{72}	$\mathbf{W}w_{b\_O,c\_O},w_{b\_T2,c\_T2}$	1								
X_{73}	$\mathbf{W}w_{c\_O,d\_O},w_{c\_T2,d\_T2}$	1								
X_{74}	$\mathbf{H}w_{A}$	1	1	1	1	1	-0.1	-0.1	-0.1	1
X_{75}	$\mathbf{H}w_{B}$	1	1	1	1	1	-0.1	-0.1	-0.1	1
X_{76}	$\mathbf{H}w_{C}$	1	1	1	1	1	-0.1	-0.1	-0.1	1
X_{77}	$\mathbf{H}w_{D}$	1	1	1	1	1	-0.1	-0.1	-0.1	1
X_{78}	$\mathbf{H}w_{T1}$	1								
X_{79}	$\mathbf{H}w_{T2}$	1								
X_{80}	$\mathbf{H}w_{O}$	1								
X_{81}	$\mathbf{M}w_{a\_A,b\_A}$	1	1	1	1	-1	-1	-1	-1	-1
X_{82}	$\mathbf{M}w_{b\_A,c\_A}$	1	1	1	1	-1	-1	-1	-1	-1
X_{83}	$\mathbf{M}w_{c\_A,d\_A}$	1	1	1	1	-1	-1	-1	-1	-1
X_{84}	$\mathbf{M}w_{a\_B,b\_B}$	1	1	1	1	-1	-1	-1	-1	-1
X_{85}	$\mathbf{M}w_{b\_B,c\_B}$	1	1	1	1	-1	-1	-1	-1	-1
X_{86}	$\mathbf{M}w_{c\_B,d\_B}$	1	1	1	1	-1	-1	-1	-1	-1
X_{87}	$\mathbf{M}w_{a\_C,b\_C}$	1	1	1	1	-1	-1	-1	-1	-1
X_{88}	$\mathbf{M}w_{b\_C,c\_C}$	1	1	1	1	-1	-1	-1	-1	-1
X_{89}	$\mathbf{M}w_{c\_C,d\_C}$	1	1	1	1	-1	-1	-1	-1	-1
X_{90}	$\mathbf{M}w_{a\_D,b\_D}$	1	1	1	1	-1	-1	-1	-1	-1
X_{91}	$\mathbf{M}w_{b\_D,c\_D}$	1	1	1	1	-1	-1	-1	-1	-1
X_{92}	$\mathbf{M}w_{c\_D,d\_D}$	1	1	1	1	-1	-1	-1	-1	-1

Fig. 17 (continued)

Fig. 18 Role matrices for the aggregation characteristics: **mcfw** for combination function weights

mcfw combination function weights	1 alogistic	2 steponce	3 maxhebb	4 smax
X_1 a_A	1			
X_2 b_A	1			
X_3 c_A	1			
X_4 d_A	1			
X_5 a_B	1			
X_6 b_B	1			
X_7 c_B	1			
X_8 d_B	1			
X_9 a_C	1			
X_{10} b_C	1			
X_{11} c_C	1			
X_{12} d_C	1			
X_{13} a_D	1			
X_{14} b_D	1			
X_{15} c_D	1			
X_{16} d_D	1			
X_{17} a_T1	1			
X_{18} b_T1	1			
X_{19} c_T1	1			
X_{20} d_T1	1			
X_{21} a_T2	1			
X_{22} b_T2	1			
X_{23} c_T2	1			
X_{24} d_T2	1			
X_{25} a_O	1			
X_{26} b_O	1			
X_{27} c_O	1			
X_{28} d_O	1			
X_{29} con_{ph1}		1		
X_{30} con_{ph2}		1		
X_{31} con_{ph3}		1		
X_{32} con_{ph4}		1		
X_{33} con_{ph5}		1		
X_{34} con_{ph6}		1		

mcfw combination function weights	1 alogistic	2 steponce	3 maxhebb	4 smax
X_{35} $\mathbf{W}_{a\_A,b\_A}$			1	
X_{36} $\mathbf{W}_{b\_A,c\_A}$			1	
X_{37} $\mathbf{W}_{c\_A,d\_A}$			1	
X_{38} $\mathbf{W}_{a\_B,b\_B}$			1	
X_{39} $\mathbf{W}_{b\_B,c\_B}$			1	
X_{40} $\mathbf{W}_{c\_B,d\_B}$			1	
X_{41} $\mathbf{W}_{a\_C,b\_C}$			1	
X_{42} $\mathbf{W}_{b\_C,c\_C}$			1	
X_{43} $\mathbf{W}_{c\_C,d\_C}$			1	
X_{44} $\mathbf{W}_{a\_D,b\_D}$			1	
X_{45} $\mathbf{W}_{b\_D,c\_D}$			1	
X_{46} $\mathbf{W}_{c\_D,d\_D}$			1	
X_{47} $\mathbf{W}_{a\_T1,b\_T1}$				1
X_{48} $\mathbf{W}_{b\_T1,c\_T1}$				1
X_{49} $\mathbf{W}_{c\_T1,d\_T1}$				1
X_{50} $\mathbf{W}_{a\_T2,b\_T2}$				1
X_{51} $\mathbf{W}_{b\_T2,c\_T2}$				1
X_{52} $\mathbf{W}_{c\_T2,d\_T2}$				1
X_{53} $\mathbf{W}_{a\_O,b\_O}$				1
X_{54} $\mathbf{W}_{b\_O,c\_O}$				1
X_{55} $\mathbf{W}_{c\_O,d\_O}$				1

mcfw	combination function weights	1 alogistic	2 steponce	3 maxhebb	4 smax
X_{56}	$\mathbf{W}w_{a\_T1,b\_T1},w_{a\_A,b\_A}$				1
X_{57}	$\mathbf{W}w_{b\_T1,c\_T1},w_{b\_A,c\_A}$				1
X_{58}	$\mathbf{W}w_{c\_T1,d\_T1},w_{c\_A,d\_A}$				1
X_{59}	$\mathbf{W}w_{a\_T1,b\_T1},w_{a\_B,b\_B}$				1
X_{60}	$\mathbf{W}w_{b\_T1,c\_T1},w_{b\_B,c\_B}$				1
X_{61}	$\mathbf{W}w_{c\_T1,d\_T1},w_{c\_B,d\_B}$				1
X_{62}	$\mathbf{W}w_{a\_T2,b\_T2},w_{a\_C,b\_C}$				1
X_{63}	$\mathbf{W}w_{b\_T2,c\_T2},w_{b\_C,c\_C}$				1
X_{64}	$\mathbf{W}w_{c\_T2,d\_T2},w_{c\_C,d\_C}$				1
X_{65}	$\mathbf{W}w_{a\_T2,b\_T2},w_{a\_D,b\_D}$				1
X_{66}	$\mathbf{W}w_{b\_T2,c\_T2},w_{b\_D,c\_D}$				1
X_{67}	$\mathbf{W}w_{c\_T2,d\_T2},w_{c\_D,d\_D}$				1
X_{68}	$\mathbf{W}w_{a\_O,b\_O},w_{a\_T1,b\_T1}$				1
X_{69}	$\mathbf{W}w_{b\_O,c\_O},w_{b\_T1,c\_T1}$				1
X_{70}	$\mathbf{W}w_{c\_O,d\_O},w_{c\_T1,d\_T1}$				1
X_{71}	$\mathbf{W}w_{a\_O,b\_O},w_{a\_T2,b\_T2}$				1
X_{72}	$\mathbf{W}w_{b\_O,c\_O},w_{b\_T2,c\_T2}$				1
X_{73}	$\mathbf{W}w_{c\_O,d\_O},w_{c\_T2,d\_T2}$				1
X_{74}	$\mathbf{H}w_{A}$	1			
X_{75}	$\mathbf{H}w_{B}$	1			
X_{76}	$\mathbf{H}w_{C}$	1			
X_{77}	$\mathbf{H}w_{D}$	1			
X_{78}	$\mathbf{H}w_{T1}$	1			
X_{79}	$\mathbf{H}w_{T2}$	1			
X_{80}	$\mathbf{H}w_{O}$	1			
X_{81}	$\mathbf{M}w_{a\_A,b\_A}$	1			
X_{82}	$\mathbf{M}w_{b\_A,c\_A}$	1			
X_{83}	$\mathbf{M}w_{c\_A,d\_A}$	1			
X_{84}	$\mathbf{M}w_{a\_B,b\_B}$	1			
X_{85}	$\mathbf{M}w_{b\_B,c\_B}$	1			
X_{86}	$\mathbf{M}w_{c\_B,d\_B}$	1			
X_{87}	$\mathbf{M}w_{a\_C,b\_C}$	1			
X_{88}	$\mathbf{M}w_{b\_C,c\_C}$	1			
X_{89}	$\mathbf{M}w_{c\_C,d\_C}$	1			
X_{90}	$\mathbf{M}w_{a\_D,b\_D}$	1			
X_{91}	$\mathbf{M}w_{b\_D,c\_D}$	1			
X_{92}	$\mathbf{M}w_{c\_D,d\_D}$	1			

Fig. 18 (continued)

mcfp	combination function parameters	1 alogistic 1 σ	1 alogistic 2 τ	2 steponce 1 α	2 steponce 2 β	3 maxhebb 1 μ	3 maxhebb 2	4 smax 1 λ	4 smax 2
X_1	a_A	5	0.3						
X_2	b_A	5	0.3						
X_3	c_A	5	0.3						
X_4	d_A	5	0.3						
X_5	a_B	5	0.3						
X_6	b_B	5	0.3						
X_7	c_B	5	0.3						
X_8	d_B	5	0.3						
X_9	a_C	5	0.3						
X_{10}	b_C	5	0.3						
X_{11}	c_C	5	0.3						
X_{12}	d_C	5	0.3						
X_{13}	a_D	5	0.3						
X_{14}	b_D	5	0.3						
X_{15}	c_D	5	0.3						
X_{16}	d_D	5	0.3						
X_{17}	a_T1	5	0.3						
X_{18}	b_T1	5	0.3						
X_{19}	c_T1	5	0.3						
X_{20}	d_T1	5	0.3						
X_{21}	a_T2	5	0.3						
X_{22}	b_T2	5	0.3						
X_{23}	c_T2	5	0.3						
X_{24}	d_T2	5	0.3						
X_{25}	a_O	5	0.3						
X_{26}	b_O	5	0.3						
X_{27}	c_O	5	0.3						
X_{28}	d_O	5	0.3						
X_{29}	con_{ph1}			20	200				
X_{30}	con_{ph2}			250	300				
X_{31}	con_{ph3}			350	400				
X_{32}	con_{ph4}			450	600				
X_{33}	con_{ph5}			650	800				
X_{34}	con_{ph6}			850	1000				

mcfp	combination function parameters	1 alogistic 1 σ	1 alogistic 2 τ	2 steponce 1 α	2 steponce 2 β	3 maxhebb 1 μ	3 maxhebb 2	4 smax 1 λ	4 smax 2
X_{35}	W_{a_A,b_A}					X_{81}			
X_{36}	W_{b_A,c_A}					X_{82}			
X_{37}	W_{c_A,d_A}					X_{83}			
X_{38}	W_{a_B,b_B}					X_{84}			
X_{39}	W_{b_B,c_B}					X_{85}			
X_{40}	W_{c_B,d_B}					X_{86}			
X_{41}	W_{a_C,b_C}					X_{87}			
X_{42}	W_{b_C,c_C}					X_{88}			
X_{43}	W_{c_C,d_C}					X_{89}			
X_{44}	W_{a_D,b_D}					X_{90}			
X_{45}	W_{b_D,c_D}					X_{91}			
X_{46}	W_{c_D,d_D}					X_{92}			
X_{47}	W_{a_T1,b_T1}							1	
X_{48}	W_{b_T1,c_T1}							1	
X_{49}	W_{c_T1,d_T1}							1	
X_{50}	W_{a_T2,b_T2}							1	
X_{51}	W_{b_T2,c_T2}							1	
X_{52}	W_{c_T2,d_T2}							1	

Fig. 19 Role matrices for the aggregation characteristics: **mcfp** for combination function parameters

mcfp combination function parameters	1 alogistic		2 steponce		3 maxhebb		4 smax	
	1	2	1	2	1	2	1	2
	σ	τ	α	β	μ		λ	
X_{56} $\mathbf{W}_{w_{a\_T1,b\_T1},w_{a\_A,b\_A}}$							1	
X_{57} $\mathbf{W}_{w_{b\_T1,c\_T1},w_{b\_A,c\_A}}$							1	
X_{58} $\mathbf{W}_{w_{c\_T1,d\_T1},w_{c\_A,d\_A}}$							1	
X_{59} $\mathbf{W}_{w_{a\_T1,b\_T1},w_{a\_B,b\_B}}$							1	
X_{60} $\mathbf{W}_{w_{b\_T1,c\_T1},w_{b\_B,c\_B}}$							1	
X_{61} $\mathbf{W}_{w_{c\_T1,d\_T1},w_{c\_B,d\_B}}$							1	
X_{62} $\mathbf{W}_{w_{a\_T2,b\_T2},w_{a\_C,b\_C}}$							1	
X_{63} $\mathbf{W}_{w_{b\_T2,c\_T2},w_{b\_C,c\_C}}$							1	
X_{64} $\mathbf{W}_{w_{c\_T2,d\_T2},w_{c\_C,d\_C}}$							1	
X_{65} $\mathbf{W}_{w_{a\_T2,b\_T2},w_{a\_D,b\_D}}$							1	
X_{66} $\mathbf{W}_{w_{b\_T2,c\_T2},w_{b\_D,c\_D}}$							1	
X_{67} $\mathbf{W}_{w_{c\_T2,d\_T2},w_{c\_D,d\_D}}$							1	
X_{68} $\mathbf{W}_{w_{a\_O,b\_O},w_{a\_T1,b\_T1}}$							1	
X_{69} $\mathbf{W}_{w_{b\_O,c\_O},w_{b\_T1,c\_T1}}$							1	
X_{70} $\mathbf{W}_{w_{c\_O,d\_O},w_{c\_T1,d\_T1}}$							1	
X_{71} $\mathbf{W}_{w_{a\_O,b\_O},w_{a\_T2,b\_T2}}$							1	
X_{72} $\mathbf{W}_{w_{b\_O,c\_O},w_{b\_T2,c\_T2}}$							1	
X_{73} $\mathbf{W}_{w_{c\_O,d\_O},w_{c\_T2,d\_T2}}$							1	
X_{74} $\mathbf{H}_{w_A}$	5	0.3						
X_{75} $\mathbf{H}_{w_B}$	5	0.3						
X_{76} $\mathbf{H}_{w_C}$	5	0.3						
X_{77} $\mathbf{H}_{w_D}$	5	0.3						
X_{78} $\mathbf{H}_{w_{T1}}$	5	0.3						
X_{79} $\mathbf{H}_{w_{T2}}$	5	0.3						
X_{80} $\mathbf{H}_{w_O}$	5	0.3						
X_{81} $\mathbf{M}_{w_{a\_A,b\_A}}$	5	2						
X_{82} $\mathbf{M}_{w_{b\_A,c\_A}}$	5	2						
X_{83} $\mathbf{M}_{w_{c\_A,d\_A}}$	5	2						
X_{84} $\mathbf{M}_{w_{a\_B,b\_B}}$	5	2						
X_{85} $\mathbf{M}_{w_{b\_B,c\_B}}$	5	2						
X_{86} $\mathbf{M}_{w_{c\_B,d\_B}}$	5	2						
X_{87} $\mathbf{M}_{w_{a\_C,b\_C}}$	5	2						
X_{88} $\mathbf{M}_{w_{b\_C,c\_C}}$	5	2						
X_{89} $\mathbf{M}_{w_{c\_C,d\_C}}$	5	2						
X_{90} $\mathbf{M}_{w_{a\_D,b\_D}}$	5	2						
X_{91} $\mathbf{M}_{w_{b\_D,c\_D}}$	5	2						
X_{92} $\mathbf{M}_{w_{c\_D,d\_D}}$	5	2						

Fig. 19 (continued)

Fig. 20 Role matrix **ms** for the timing characteristics (speed factors) and initial values **iv**

ms	speed factors	1	initial values	1
X_1	a_A	1		0
X_2	b_A	1		0
X_3	c_A	1		0
X_4	d_A	1		0
X_5	a_B	1		0
X_6	b_B	1		0
X_7	c_B	1		0
X_8	d_B	1		0
X_9	a_C	1		0
X_{10}	b_C	1		0
X_{11}	c_C	1		0
X_{12}	d_C	1		0
X_{13}	a_D	1		0
X_{14}	b_D	1		0
X_{15}	c_D	1		0
X_{16}	d_D	1		0
X_{17}	a_T1	1		0
X_{18}	b_T1	1		0
X_{19}	c_T1	1		0
X_{20}	d_T1	1		0
X_{21}	a_T2	1		0
X_{22}	b_T2	1		0
X_{23}	c_T2	1		0
X_{24}	d_T2	1		0
X_{25}	a_O	1		0
X_{26}	b_O	1		0
X_{27}	c_O	1		0
X_{28}	d_O	1		0
X_{29}	con_{ph1}	1		0
X_{30}	con_{ph2}	1		0
X_{31}	con_{ph3}	1		0
X_{32}	con_{ph4}	1		0
X_{33}	con_{ph5}	1		0
X_{34}	con_{ph6}	1		0

ms	speed factors	1	initial values	1
X_{35}	$\mathbf{W}_{a\_A,b\_A}$	X_{74}		0.1
X_{36}	$\mathbf{W}_{b\_A,c\_A}$	X_{74}		0.1
X_{37}	$\mathbf{W}_{c\_A,d\_A}$	X_{74}		0
X_{38}	$\mathbf{W}_{a\_B,b\_B}$	X_{75}		0
X_{39}	$\mathbf{W}_{b\_B,c\_B}$	X_{75}		0.1
X_{40}	$\mathbf{W}_{c\_B,d\_B}$	X_{75}		0.1
X_{41}	$\mathbf{W}_{a\_C,b\_C}$	X_{76}		0.1
X_{42}	$\mathbf{W}_{b\_C,c\_C}$	X_{76}		0.1
X_{43}	$\mathbf{W}_{c\_C,d\_C}$	X_{76}		0
X_{44}	$\mathbf{W}_{a\_D,b\_D}$	X_{77}		0
X_{45}	$\mathbf{W}_{b\_D,c\_D}$	X_{77}		0.1
X_{46}	$\mathbf{W}_{c\_D,d\_D}$	X_{77}		0.1
X_{47}	$\mathbf{W}_{a\_T1,b\_T1}$	X_{78}		0
X_{48}	$\mathbf{W}_{b\_T1,c\_T1}$	X_{78}		0
X_{49}	$\mathbf{W}_{c\_T1,d\_T1}$	X_{78}		0
X_{50}	$\mathbf{W}_{a\_T2,b\_T2}$	X_{79}		0
X_{51}	$\mathbf{W}_{b\_T2,c\_T2}$	X_{79}		0
X_{52}	$\mathbf{W}_{c\_T2,d\_T2}$	X_{79}		0
X_{53}	$\mathbf{W}_{a\_O,b\_O}$	X_{80}		0
X_{54}	$\mathbf{W}_{b\_O,c\_O}$	X_{80}		0
X_{55}	$\mathbf{W}_{c\_O,d\_O}$	X_{80}		0

Fig. 20 (continued)

ms	speed factors	1	initial values	1
X_{56}	$\mathbf{W}_{w_{a\_T1,b\_T1},w_{a\_A,b\_A}}$	0.2		0.1
X_{57}	$\mathbf{W}_{w_{b\_T1,c\_T1},w_{b\_A,c\_A}}$	0.2		0.1
X_{58}	$\mathbf{W}_{w_{c\_T1,d\_T1},w_{c\_A,d\_A}}$	0.2		0.1
X_{59}	$\mathbf{W}_{w_{a\_T1,b\_T1},w_{a\_B,b\_B}}$	0.2		0.1
X_{60}	$\mathbf{W}_{w_{b\_T1,c\_T1},w_{b\_B,c\_B}}$	0.2		0.1
X_{61}	$\mathbf{W}_{w_{c\_T1,d\_T1},w_{c\_B,d\_B}}$	0.2		0.1
X_{62}	$\mathbf{W}_{w_{a\_T2,b\_T2},w_{a\_C,b\_C}}$	0.2		0.1
X_{63}	$\mathbf{W}_{w_{b\_T2,c\_T2},w_{b\_C,c\_C}}$	0.2		0.1
X_{64}	$\mathbf{W}_{w_{c\_T2,d\_T2},w_{c\_C,d\_C}}$	0.2		0.1
X_{65}	$\mathbf{W}_{w_{a\_T2,b\_T2},w_{a\_D,b\_D}}$	0.2		0.1
X_{66}	$\mathbf{W}_{w_{b\_T2,c\_T2},w_{b\_D,c\_D}}$	0.2		0.1
X_{67}	$\mathbf{W}_{w_{c\_T2,d\_T2},w_{c\_D,d\_D}}$	0.2		0.1
X_{68}	$\mathbf{W}_{w_{a\_O,b\_O},w_{a\_T1,b\_T1}}$	0.2		0.1
X_{69}	$\mathbf{W}_{w_{b\_O,c\_O},w_{b\_T1,c\_T1}}$	0.2		0.1
X_{70}	$\mathbf{W}_{w_{c\_O,d\_O},w_{c\_T1,d\_T1}}$	0.2		0.1
X_{71}	$\mathbf{W}_{w_{a\_O,b\_O},w_{a\_T2,b\_T2}}$	0.2		0.1
X_{72}	$\mathbf{W}_{w_{b\_O,c\_O},w_{b\_T2,c\_T2}}$	0.2		0.1
X_{73}	$\mathbf{W}_{w_{c\_O,d\_O},w_{c\_T2,d\_T2}}$	0.2		0.1
X_{74}	$\mathbf{H}_{w_A}$	0		0.05
X_{75}	$\mathbf{H}_{w_B}$	0		0.05
X_{76}	$\mathbf{H}_{w_C}$	0		0.05
X_{77}	$\mathbf{H}_{w_D}$	0		0.05
X_{78}	$\mathbf{H}_{w_{T1}}$	0.9		0
X_{79}	$\mathbf{H}_{w_{T2}}$	0.9		0
X_{80}	$\mathbf{H}_{w_O}$	0.9		0
X_{81}	$\mathbf{M}_{w_{a\_A,b\_A}}$	0		0.994
X_{82}	$\mathbf{M}_{w_{b\_A,c\_A}}$	0		0.994
X_{83}	$\mathbf{M}_{w_{c\_A,d\_A}}$	0		0.994
X_{84}	$\mathbf{M}_{w_{a\_B,b\_B}}$	0		0.997
X_{85}	$\mathbf{M}_{w_{b\_B,c\_B}}$	0		0.997
X_{86}	$\mathbf{M}_{w_{c\_B,d\_B}}$	0		0.997
X_{87}	$\mathbf{M}_{w_{a\_C,b\_C}}$	0		0.998
X_{88}	$\mathbf{M}_{w_{b\_C,c\_C}}$	0		0.998
X_{89}	$\mathbf{M}_{w_{c\_C,d\_C}}$	0		0.998
X_{90}	$\mathbf{M}_{w_{a\_D,b\_D}}$	0		0.992
X_{91}	$\mathbf{M}_{w_{b\_D,c\_D}}$	0		0.992
X_{92}	$\mathbf{M}_{w_{c\_D,d\_D}}$	0		0.992

References

Argyris, C., Schön, D.A.: Organizational Learning: A Theory of Action Perspective. Addison-Wesley, Reading, MA (1978)

Bhalwankar, R., Treur, J.: Modeling learner-controlled mental model learning processes by a second-order adaptive network model. PLoS ONE **16**(8), e0255503 (2021)

Bogenrieder, I.: Social architecture as a prerequisite for organizational learning. Manag. Learn. **33**(2), 197–216 (2002)

Canbaloğlu, G., Treur, J., Roelofsma, P.H.M.P.: An adaptive self-modeling network model for multilevel organisational learning. In: Proceedings of the 7th International Congress on Information and Communication Technology, ICICT'22, vol. 2. Lecture Notes in Networks and Systems, vol. 448, pp. 179–191. Springer Nature, Cham (2023a)

Canbaloğlu, G., Treur, J., Wiewiora, A. (eds.).: Computational Modeling of Multilevel Organisational Learning and Its Control Using Self-Modeling Network Models. Springer Nature, Cham (2023b)

Crossan, M.M., Lane, H.W., White, R.E.: An organizational learning framework: from intuition to institution. Acad. Manag. Rev. **24**, 522–537 (1999)

Fischhof, B., Johnson, S.: Organisational decision making. Cambridge University Press (1997)

Hebb, D.O.: The Organization of Behavior: A Neuropsychological Theory. Wiley, New York (1949)

Kim, D.H.: The link between individual and organizational learning. Sloan Manag. Rev. Fall **1993**, 37–50 (1993)

McShane, S.L., Glinow, M.A. von: Organizational Behavior. McGraw-Hill (2010)

Stelmaszczyk, M.: Relationship between individual and organizational learning: mediating role of team learning. J. Econ. Manag. **26**(4), 1732–1947 (2016). https://doi.org/10.22367/jem.2016.26.06

Treur, J.: Network-Oriented Modeling for Adaptive Networks: Designing Higher-Order Adaptive Biological, Mental and Social Network Models. Springer Nature, Cham (2020)

Treur, J., Van Ments, L. (eds.).: Mental Models and their Dynamics, Adaptation, and Control: a Self-Modeling Network Modeling Approach. Springer Nature, Cham (2022)

Van Gog, T., Paas, F., Marcus, N., Ayres, P., Sweller, J.: The mirror neuron system and observational learning: implications for the effectiveness of dynamic visualizations. Educ. Psychol. Rev. **21**(1), 21–30 (2009)

Van Ments, L., Treur, J., Klein, J., Roelofsma, P.H.M.P.: A second-order adaptive network model for shared mental models in hospital teamwork. In: Proceedings of ICCCI'21. Lecture Notes in AI, vol. 12876, pp. 126–140. Springer Nature, Cham (2021)

Wiewiora, A., Smidt, M., Chang, A.: The 'How' of multilevel learning dynamics: a systematic literature review exploring how mechanisms bridge learning between individuals, teams/projects and the organisation. Eur. Manag. Review **16**, 93–115 (2019)

Wiewiora, A., Chang, A., Smidt, M.: Individual, project and organisational learning flows within a global project-based organisation: exploring what, how and who. Int. J. Project Manag. **38**, 201–214 (2020)

Yi, M.Y., Davis, F.D.: Developing and validating an observational learning model of computer software training and skill acquisition. Infor. Syst. Res. **14**(2), 146–169 (2003)

Supporting Safety and Security in Medical Teamwork in the Neonatal Domain Through Cyberspace

Ensuring safety and security in cyberspace is a critical concern across various domains, especially in high-stakes environments such as healthcare. The third part of this volume delves into the application of shared mental models, organizational learning, and AI-driven support systems to enhance medical teamwork in the neonatal domain. This part presents four chapters that collectively illustrate how adaptive network models and AI coaches can significantly improve decision-making, risk management, and adherence to medical protocols, ultimately leading to better patient outcomes.

The first chapter, "Supporting Medical Team Decision in the Neonatal Domain: How Safety and Security Risk Management Through Cyberspace can be Supported by Adaptive Agent Network Models with Internal Mental Models", by Xu et al., introduces the use of adaptive agent-based network models to support medical teams dealing with high-risk neonatal cases. This chapter emphasizes the importance of internal mental models and the role of an AI coach in enhancing team performance.

The second chapter, "Increasing Safety and Security Through Cyberspace by an Adaptive Network Model for AI-Assisted Risk Management of Neonatal Respiratory Distress", by Mokadem et al., extends this discussion to the management of neonatal respiratory distress. It showcases how an AI coach can assist medical teams by monitoring and supporting their actions, thereby improving safety and security in clinical settings.

In the third chapter, "Supporting Risk Management Through Cyberspace: An Adaptive Network Model Simulating AI Coach Effects by Inducing Adherence to Guidelines in Neonatal Medical Protocols", by Chawla et al., explore the AI coach's role in ensuring adherence to treatment guidelines for neonatal hypoglycemia and hyperbilirubinemia. This chapter highlights the AI coach's ability to intervene when deviations occur, thereby maintaining high standards of safe and secure care.

The final chapter, "Improving Risk Management Through Cyberspace: Optimizing Neonatal Respiratory Support through Network Oriented Modeling", by Sebahi et al., focuses on optimizing neonatal respiratory support using network-oriented modeling. It presents a comprehensive framework that integrates AI-driven interventions to address the complexities of neonatal care.

What Will be Addressed in This Part

Supporting Medical Team Decision in the Neonatal Domain

First, Chapter "Supporting Medical Team Decision in the Neonatal Domain: How Safety and Security Risk Management Through Cyberspace can be Supported by Adaptive Agent Network Models with Internal Mental Models" introduces a second-order adaptive agent-based network model designed to support medical teams in critical neonatal care scenarios. By simulating the internal mental models of team members, the model provides a detailed representation of how shared understanding and real-time interventions can enhance decision-making processes.

In high-stakes environments such as neonatal care, where decisions must be made swiftly and accurately, the integration of an AI coach can be transformative. The chapter illustrates this with the example of a newborn baby in danger, demonstrating how the AI coach supports the team by simulating possible outcomes and suggesting optimal actions. This approach not only improves the immediate decision-making process but also contributes to long-term organizational learning by identifying and addressing gaps in team knowledge and coordination.

The adaptive nature of the model allows it to learn from past experiences, continuously improving its recommendations and interventions. This capability is crucial for managing the dynamic and often unpredictable nature of neonatal care. By incorporating internal mental models, the AI coach can tailor its support to the specific needs and mental states of individual team members, ensuring that interventions are both timely and effective.

AI-Assisted Risk Management of Neonatal Respiratory Distress

Chapter "Increasing Safety and Security Through Cyberspace by an Adaptive Network Model for AI-Assisted Risk Management of Neonatal Respiratory Distress" builds on the previous one by focusing specifically on neonatal respiratory distress management. It presents an adaptive network model that integrates an AI coach to monitor and support hospital teams in managing respiratory distress in neonates. The chapter highlights the practical application of this model through various simulated scenarios.

Respiratory distress in neonates requires prompt and precise interventions, making it an ideal context for demonstrating the benefits of AI-assisted risk management. The AI coach acts as an additional team member, providing real-time monitoring and guidance to ensure that medical procedures are executed correctly. By simulating different scenarios, the model shows how the AI coach can help the team navigate complex situations, anticipate potential complications, and respond effectively to unexpected challenges.

The chapter emphasizes the importance of organizational learning in enhancing safety and security. The AI coach not only supports individual team members but also contributes to the overall learning process by capturing and analyzing data from each scenario. This continuous feedback loop enables the team to refine their practices, improve their coordination, and enhance their overall performance over time.

AI Coach Effects by Inducing Adherence to Guidelines in Neonatal Medical Protocols

In Chapter "Supporting Risk Management Through Cyberspace: An Adaptive Network Model Simulating AI Coach Effects by Inducing Adherence to Guidelines in Neonatal Medical Protocols" the focus shifts to the AI coach's role in ensuring adherence to neonatal treatment guidelines for conditions such as hypoglycemia and hyperbilirubinemia. The second-order adaptive agent-based network model is used to simulate three different scenarios: perfect adherence to guidelines, imperfect adherence, and imperfect adherence with AI coach intervention.

Adherence to medical protocols is critical for patient safety and effective treatment outcomes. The chapter demonstrates how deviations from these protocols can lead to suboptimal care and increased risks for neonates. By integrating an AI coach, the model provides timely interventions to correct deviations and reinforce adherence to guidelines. This capability is particularly important in high-pressure environments where human errors are more likely to occur.

Through detailed simulations, the chapter illustrates the internal dynamics of team members' mental models and how the AI coach influences these dynamics. By continuously monitoring the team's actions and providing targeted feedback, the AI coach helps ensure that medical protocols are followed accurately, thereby enhancing the safety and security of neonatal care.

Optimizing Neonatal Respiratory Support Through Network Oriented Modeling

Finally, Chapter "Improving Risk Management Through Cyberspace: Optimizing Neonatal Respiratory Support through Network Oriented Modeling" presents a network-oriented modeling approach to optimizing neonatal respiratory support. It introduces a comprehensive framework that integrates various states, including world states, doctor states, and AI coach states, to provide a holistic view of the interactions and dependencies involved in neonatal respiratory care.

Neonatal respiratory support is a complex and critical aspect of neonatal care, requiring careful coordination and precise interventions. The chapter highlights the challenges faced in this domain and the potential of AI-driven network modeling to address these challenges. By simulating multiple scenarios, the model investigates the effects of different interventions and identifies optimal strategies for managing respiratory support.

A significant contribution of this chapter is the introduction of an AI coach as a real-time intervention mechanism. The AI coach fills in knowledge gaps, provides guidance during critical moments, and helps maintain adherence to best practices. The simulations demonstrate how the AI coach can improve patient outcomes by ensuring that medical procedures are performed correctly and efficiently.

The chapter concludes by summarizing key findings and discussing potential avenues for future research. It underscores the importance of integrating AI-driven network modeling into neonatal care to enhance safety and security, improve patient outcomes, and streamline healthcare protocols.

Conclusion

The chapters in this part collectively underscore the transformative potential of shared mental models, organizational learning, and AI-driven support systems in enhancing safety and security in neonatal care. By leveraging advanced computational techniques and adaptive network models, these chapters demonstrate how AI coaches can significantly improve decision-making, cyber risk management, and adherence to medical protocols. The insights and methodologies presented in these chapters provide a valuable framework for integrating technology into healthcare, ensuring that safety and security remain at the forefront of medical practice in the digital age.

Supporting Medical Team Decision in the Neonatal Domain: How Safety and Security Risk Management Through Cyberspace Can Be Supported by Adaptive Agent Network Models with Internal Mental Models

Yida Xu, Fakhra Jabeen, Jan Treur, H. Rob Taal, and Peter H. M. P. Roelofsma

Abstract In this chapter, it is shown how second-order adaptive agent-based network models can be used to model a medical team supported by a virtual AI Coach. It is illustrated for the case of a newborn baby in danger. The design of these computational agent models is based on an adaptive self-modeling network modeling approach. It also addresses how the AI Coach can play a central role in organizational learning. The agent models enable representations and processing of all actors' internal mental models and internal simulation of these mental models, adaptive changes of these mental models (learning and forgetting), and the interaction between the actors and the world.

Y. Xu · F. Jabeen
Safety and Security Science Section, Delft University of Technology, Delft, The Netherlands
e-mail: xuyd19@midea.com

F. Jabeen
e-mail: fakhra.jabeen@ru.nl

J. Treur (✉)
Social AI Group, Vrije Universiteit Amsterdam, Amsterdam, The Netherlands
e-mail: j.treur@vu.nl; j.treur@hhs.nl

J. Treur · P. H. M. P. Roelofsma
Risk Management and Cybersecurity Group, Centre of Expertise Cybersecurity, The Hague University of Applied Sciences, The Hague, The Netherlands
e-mail: p.h.m.p.roelofsma@hhs.nl

H. Rob Taal · P. H. M. P. Roelofsma
Department of Pediatrics, Division Neonatology, Erasmus MC, University Medical Center-Sophia Children's Hospital, Rotterdam, The Netherlands
e-mail: h.taal@erasmusmc.nl

P. H. M. P. Roelofsma et al. (eds.), *Using Shared Mental Models and Organisational Learning to Support Safety and Security Through Cyberspace: A Computational Analysis Approach*, Studies in Systems, Decision and Control 570,
https://doi.org/10.1007/978-3-031-72075-8_4

Keywords Adaptive agent model · Self-modeling network · Virtual AI coach · Mental model · Medical team · Neonatology

1 Introduction

A *medical error* indicates that an expected action has not been completed in a medical treatment (Garrouste-Orgeas et al. 2012; Rasmussen 1990; Emanuel et al. 2009; Donaldson 2002). *Patient safety* is defined as 'the prevention of harm to patients' (Aspden et al. 2004; National Quality Forum 2004; Kirwan et al. 2013). *Safety culture* is defined as 'the product of individual and group perceptions, attitudes, values, competencies, and patterns of behavior that decide the commitment to, and the style and proficiency of, an organization's health and safety management' (Canbaloğlu et al. 2023, 2025; Craik 1943; Kirwan et al. 2013; Lee et al. 2019; Tangirala and Ramanujam 2008). The terminology for safety has evolved from a focus on error to patient safety to adopting a safety culture with adequate shared internal mental models. That evolution reflects a process of shifting responsibility from individuals to the whole organization. A mental model means the intelligence that imitates relation structures of external processes (Craik 1943; Williams 2018; Mayer 1989; Piaget 1954; Benbassat 2014; Hogan and Pressley 1997; Van Ments and Treur 2021; Treur and Van Ments 2022). A shared mental model of a team means a large or complete overlap of the team members' individual mental models (Jonker et al. 2010; Maynard and Gilson 2014; Burtscher and Manser 2012; Waller et al. 2004; McComb and Simpson 2014; Mathieu et al. 2000; Edwards 2005). A useful method to analyse shared mental models of a team is adaptive network-oriented modeling, which refers to modeling complex processes by adaptive networks (Treur 2016, 2018, 2020a, b). The safety culture of a healthcare team can be reflected in a change from a blame culture to a just culture. In a blame culture, ineffective communication among team members and silence about hidden mistakes can cause serious problems (Kirwan et al. 2013; Lee et al. 2019; Tangirala and Ramanujam 2008). Organizational learning reflects adaptive dynamics of organizational knowledge. From the perspective of organizational learning, in just culture, an organization is able to identify, report, and investigate incidents and to take appropriate learning actions that increase the quality of patient care and reduce the probability of recurrence of errors (Fiol and Lyles 1985; Schulz 2002; Levitt and March 1988; Bohn 1995; Haunschild and Sullivan 2002; Sitkin 1996; Argote and Miron-Spektor 2011; Glynn et al. 1994; Kogut and Zander 1992; Nonaka and Krogh 2009; Kim 1993). AI technology has the potential to help a medical team deploy just safety culture. But research on how AI can involve in the workflow of existing medical teams is limited. In addition, research on computational formalisation of organizational learning is also limited.

Research Question

The above descriptions indicate the problem that this research addresses: *How can AI contribute to the application of just safety culture for medical treatment*? This study chooses these points to explore how an AI Coach improves the efficiency of team communication, how such AI leverages organizational learning, and how this AI points out defects and prevents errors. From the perspective of results, these converge into how AI can help improve the performance of the medical team. The application scenario of this research is a case of neonatal resuscitation where a doctor and a nurse give ventilation to a newborn baby in danger. That leads to the main research question of this research: *How to use an adaptive network-oriented modeling method to analyse the AI-coach's contribution in a medical team's ventilation operation to save baby in danger?*

2 Research Method: Adaptive Self-modeling Networks

Network-oriented modeling refers to modeling complex processes by network structures. These processes include processes in physical, mental, and social reality. The approach provides a powerful tool to present relevant states (nodes) and causal relations (connections) between them. By using a temporal dimension and the principle of self-modeling, the approach shows not only dynamics of states, but also of adaptive changes of connections and of other network structure characteristics over time (Treur 2018, 2020a, b). Network nodes X have state values varying over time t represented by $X(t)$. Characteristics that define a network are *connectivity characteristics* $\boldsymbol{\omega}_{X,Y}$ (connection weights), *aggregation characteristics* $\mathbf{c}_Y$ (combination functions), and *timing characteristics* $\boldsymbol{\eta}_Y$ (speed factors). When state Y is influenced by several states X_i, the aggregated impact of connections from these states X_i to Y is represented by a specified combination function $\mathbf{c}_Y$. There is a library with multiple forms of combination functions (Treur 2020b). For the purpose of simulation, the following numerical form incorporating the mentioned network characteristics is used (where state Y gets its incoming connections from the X_i):

$$Y(t+\Delta t) = Y(t) + \boldsymbol{\eta}_Y\left[\mathbf{c}_Y\left(\boldsymbol{\omega}_{X_1,Y}X_1(t), \ldots, \boldsymbol{\omega}_{X_k,Y}X_k(t)\right) - Y(t)\right]\Delta t \qquad (1)$$

In the software environment described in Treur (2020b, Chap. 9), a computational network engine based on generic Eq. (1) processes all network states based on their network characteristics. As network characteristics can be adaptive (also indicated by *plasticity*), first-order self-modeling is applied to represent these adaptive network characteristics. For instance, a first-order self-model **W**-state $\mathbf{W}_{X,Y}$ and **T**-state $\mathbf{T}_Y$ are applied to express an adaptive connection weight $\boldsymbol{\omega}_{X,Y}$ and an adaptive parameter of excitability threshold value $\boldsymbol{\tau}_Y$. Such first-order adaptation can be controlled by second-order adaptation (also called *metaplasticity*). For instance, the persistence factor $\boldsymbol{\mu}_{\mathbf{W}_{X,Y}}$ for a first-order self-model **W**-state $\mathbf{W}_{X,Y}$ for Hebbian learning can be adaptive (e.g., depending on stress) and be represented by a second-order self-model $\mathbf{M_W}$-state $\mathbf{M}_{\mathbf{W}_{X,Y}}$. Similarly, an adaption speed factor $\boldsymbol{\eta}_{\mathbf{T}_Y}$ of a first-order self-model **T**-state $\mathbf{T}_Y$ can be adaptive and be represented by a second-order self-model $\mathbf{H_T}$-state $\mathbf{H}_{\mathbf{T}_Y}$ (Treur 2020b).

In Treur (2021), Hendrikse et al. (2023), it is shown that any smooth dynamical system has a canonical representation as a temporal-causal network and any smooth (multi-order) adaptive dynamical system has a canonical representation as a (multi-order) self-modeling temporal-causal network. Therefore, compared to adaptive dynamical systems in general, the network-oriented modeling approach used here does not introduce any fundamental limitations concerning what it can model. This has also been confirmed by applications to case studies in practice, e.g., for mental models in Treur and Van Ments (2022) and for organizational learning in Canbaloğlu et al. (2023).

Mental Models

A model is an imitation of a process and has a similar relational structure to that process. A mental model means the intelligence that imitates relation structures of external processes. For an agent using an internal mental model, the agent gets the ability to understand and predict the corresponding external process. In organisms, these representations and predictions are based on neural structures. These neural structures only imitate the (causal) relation structure of external process, instead of material changes of the external world (Craik 1943; Williams 2018; Mayer 1989; Piaget 1954; Benbassat 2014; Hogan and Pressley 1997; Van Ments and Treur 2021; Treur and Van Ments 2022).

A shared mental model is the overlap or convergence among how members represent their team and task in their mental models. In other words, shared mental models represent the agreement among members (Jonker et al. 2010). A good shared mental model is necessary to a team's good performance, as it not only enables team members to understand, explain, and predict team behaviors and but also enable team members understand how to cooperate to complete a task according to some form of protocol (Burtscher and Manser 2012; Waller et al. 2004). A shared mental model for teamwork means team members' agreement on how they will cooperate on a given task. A shared mental model for the task refers to the agreement among team members on the structure of the task and their roles in it McComb and Simpson (2014), Mathieu et al. (2000). *Similarity*, *accuracy* and *dynamics* are three important properties of a shared mental model (Mathieu et al. 2000). The former one reflects the extent of how team members' mental models are consistent with others. The middle one reflects the degree of how team members' mental models correctly resemble corresponding real-world process. The latter one means the shared mental model can vary over time.

Organizational Learning

Organizational learning is defined as a change of organizational knowledge occurring as a function of experience and exchange between individuals (Fiol and Lyles 1985). *Experience* is what happens when an organization performs a task. *Knowledge* is defined as justified personal belief. The knowledge can be expressed in multiple ways including dynamics of behaviors, routines and cognitions (McComb and Simpson 2014; Quality and Forum 2004; Nonaka and Krogh 2009; Piaget 1954; Rasmussen 1990; Schulz 2002; Levitt and March 1988; Bohn 1995; Haunschild and Sullivan 2002; Sitkin 1996). Organizational learning only occurs when the change of individual knowledge is embedded into organizational knowledge and is available for other people within the organization to access. Creating, retaining and transferring knowledge can be considered to be three sub-processes of organizational learning (Rasmussen 1990; Roelofsma and Treur 2025; Schulz 2002; Sitkin 1996; Argote and Miron-Spektor 2011; Glynn et al. 1994). Currently, research on computational formalization of organizational learning is rare. Kim (1993) and Canbaloğlu et al. (2021) mention the following mechanisms of organizational learning; see also (Roelofsma and Treur 2025; Canbaloğlu et al. 2025), this volume. To begin with, individuals generate their mental models. Then, the organization obtains knowledge from individuals. Based on this, the organization processes this knowledge and forms an organizational mental model (feed forward learning). Finally, individuals update their knowledge from organizations (feedback learning). These mechanisms are useful for computational analysis of organizational learning as shown in Canbaloğlu et al. (2021, 2025).

Case Description

A teaching video from University of Nottingham about giving ventilation to newborn baby in danger provides a case of the application scenario (Treur 2020a). In this video, the medical team operates two times of pure ventilation and one time of the combination of compression and ventilation and finally save the baby. This process will be abstracted to build our adaptive agent-based network models. The sequence of protocol without organizational learning is as follows:

1. The nurse (N) prepares the baby on the operating table.
2. The team prepares for pure ventilation.
3. The nurse (N) gives the first time of pure ventilation for 30 s.
4. The nurse (N) helps assess the baby. The doctor reports the baby's condition. The team decides to give a combination of compression and ventilation.
5. The team prepares for the combination of compression and ventilation.
6. The team gives the combination of compression and ventilation for 30 s.
7. The nurse (N) helps assess the baby. The doctor (D) reports the baby's condition. The team decides to give the second time pure ventilation.
8. The nurse (N) gives the second time of pure ventilation for 30 s.
9. The nurse (N) helps assess the baby and the doctor (D) reports the baby is out of danger.

Proper operation of ventilation requires good preparation. That means that in mental models a connection from preparation to the action of ventilation is important. Hebbian learning, learning from instruction and organizational learning all can be introduced here. In addition, mistakes of team members can be modeled with adequate application scenarios for the AI-coach's intervention.

3 Adaptive Agent Network Model for the AI's Engagement Without Organizational Learning

In this first considered condition, the nurse has imperfect knowledge in preparing pure ventilation. But when she just starts giving the first time of pure ventilation and is at the early phase of making a mechanical frequency related error, the AI-coach detects her action through sensors. To prevent further damage, the AI-coach warns her for she is about to fail to give correct pure ventilation and tells her the right way to do that. Learning from the AI-coach, the nurse is able to do the first time of pure ventilation correctly. In this condition there is no introduction of organizational learning. The agent network model is structured (depicted in 3D) according to three planes for base level, first-order self-model level, and second-order self-model level; see Fig. 1.

Base Level Overview

The base plane of the introduced agent network model, addresses from top to bottom:

- context states prefixed with 'Con'
- the nurse's mental model states prefixed with 'NS'
- the nurse's action ownership states prefixed with 'NOS'
- the world states prefixed with 'WS'
- the doctor's action ownership states prefixed with 'DOS'
- the doctor's mental model states prefixed with 'DS'
- the baby's condition states prefixed with 'BS'
- the AI-coach's action ownership states with prefix 'AOS'
- the AI-coach's mental model states prefixed with 'AS'.

The world states express the steps that take place in the world's reality for the followed protocol condition. The world states are depicted by light blue nodes and their connections in the middle part of the base plane. The shared mental model concerns individual models for the nurse and the doctor. The nurse's mental model is depicted by light green nodes and their connections, while the doctor's mental model is depicted by light-yellow nodes and their connections. Individual mental models reflect modeling of a real application scenario taking into account individual minds. The shared mental model reflects the team's consensus of mental models for the real application scenario. Recall that mental models specify who does what and in what order.

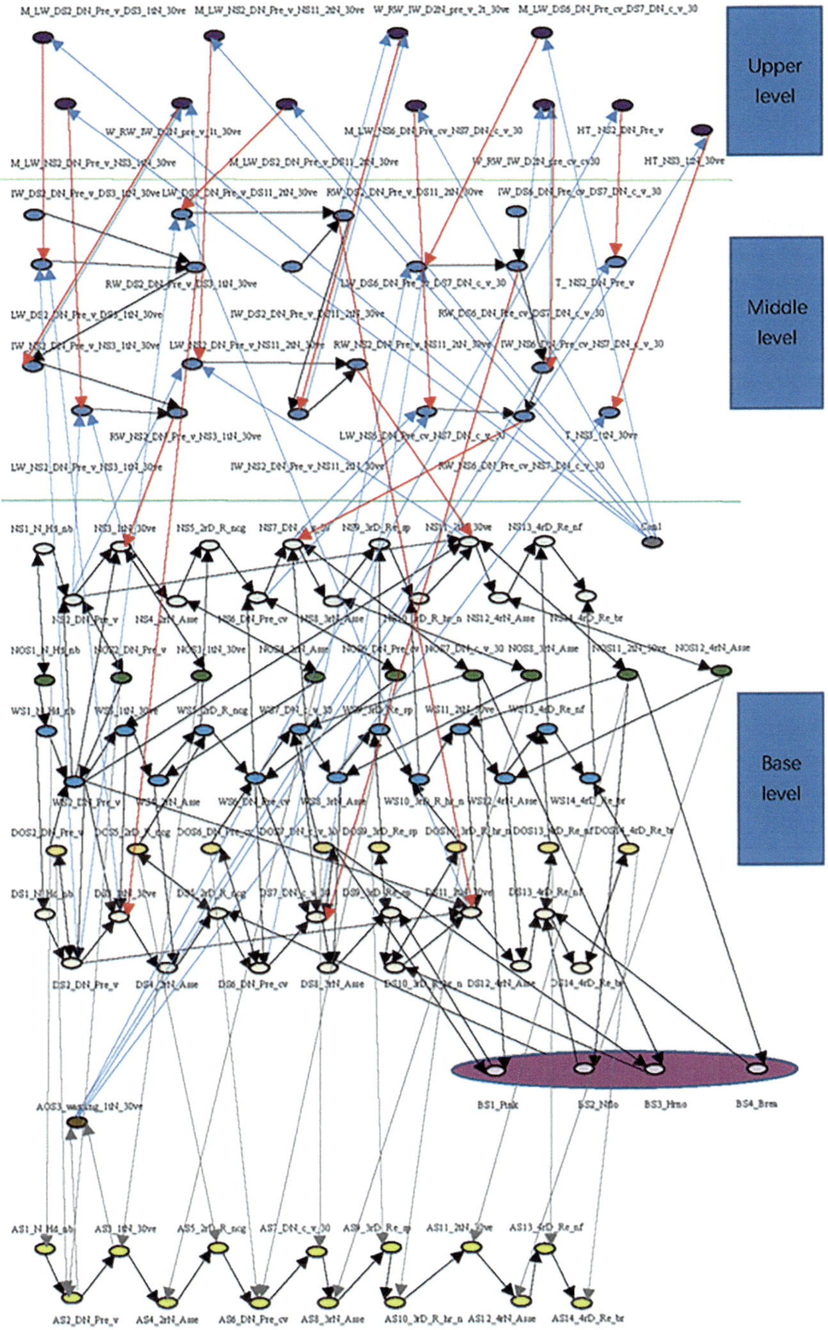

Fig. 1 The adaptive network model for the AI-coach's engagement without organizational learning

The application scenario consists of 14 steps, and each step has a corresponding mental model state for both team members and a corresponding world state. These 14 steps can be classified into 3 ventilation actions. Each ventilation action involves preparations, giving ventilation to the baby, and assessing the baby after ventilation. An overview of base level states is presented in Table 1 in Appendix. In Table 1, the state prefix indicates the type of state. The state names mark the steps that make up the application scenario. The state names contain information of 'who does what' and 'on which time or round'. The explanation describes the content of these steps.

Within mental models, mental states are connected indicating the order in which the corresponding world states are to occur in the application scenario in the real world. For example, mental model states of giving ventilations are positively influenced by incoming connections from mental model states of preparation of ventilations, e.g., *NS11_2tN_30ve* has an incoming connection from *NS2_ DN_Pre_v*. That means, if the nurse doesn't activate well the preparation state for pure ventilation in her mental model, she will not successfully activate the mental model state of giving the second time of ventilation. In addition, mental model states are also influenced by observations. When one team member influences the world states, the other member(s) will adjust his/her mental model states according to the effects of observation in the real world of events involving change of world states and the baby's condition states. Moreover, each mental model state is connected to corresponding team members' action ownership states. That means in particular that the AI-coach is monitoring the team members' actions.

Action ownership states are mental states but not mental model states themselves. Instead, they are the mediators through which mental model states influence world states. These mediators get input from mental model states and then become input of world states. This process expresses how the nurse's and the doctor's mental model determines execution of actions, then these actions change the real world. In addition, actions will give feedback information to the mental model states; this is expressed by mental model states that are influenced by incoming connections from corresponding action ownership states.

World states are connected according to the sequence in reality. In addition, to model the influence from actions, world states have connections from action ownership states. Besides, preparation of ventilations in real world will influence team members' mental model states of giving ventilation as well. The baby's condition is influenced by the actions of ventilation. The team's assessments also rely on observing the baby's condition. As a result, in the network model, the action ownerships states of three ventilations have connections towards the relevant baby's condition states. And the baby's condition states have connections toward the doctor's reporting mental model states.

The base level by itself can be used to model a nonadaptive agent. However, for adaptivity the other levels are included.

Middle Level and Upper-Level Overview

The middle level consists of states that represent the adaptation of the agents' internal mental models. This is accomplished through the principle of (first-order) self-modeling by self-model states $\mathbf{W}_{X,Y}$ applied for connection weights $\boldsymbol{\omega}_{X,Y}$ of the mental models. These self-model states and their dynamics highlight a mental model's learning process, which can be an integration of instructional learning based on communication and internal hebbian learning. The states for instructional learning are prefixed by **IW**, the states for internal hebbian learning are prefixed by **LW**, and their integrated resulting learning states are prefixed by **RW**. For the nurse, **IW**-states have an incoming connection from the doctor's **RW**-states. This represents that the nurse is learning from doctor's instructional communication. For the doctor, **IW**-states are with stable values assumed to have been obtained by instructional learning in the past. For Hebbian learning, a persistence factor represents the fraction that persists each time unit. For example, $\boldsymbol{\mu} = 0.95$ means that 5 percent of a learnt value is forgotten each time unit. The persistence factor $\boldsymbol{\mu}$ can be adapted as well, as will be discussed for the upper level. For both team members, all their **LW**-states have second-order adaptive persistence factors which are influenced by the context state (concerning stress) in the base level. From these learning states, the **RW**-states have downward connections to base level states to effectuate their role as representing connection weights. In addition, **T**-states are first-order self-model states of threshold values of the nurse's mental model states for preparing and giving first time of pure ventilation. An overview of the states in the middle level is presented at Table 2 in Appendix.

States in the upper-level control the adaptation of mental models. Within neuroscience, such control (over plasticity) is also called metaplasticity. In the upper level, second-order self-model **M**-states represent the persistence factors $\boldsymbol{\mu}$ of the hebbian learning states (**LW**-states). These **M**-states have an upward connection from the context state representing the stress level. This means that the persistence factors $\boldsymbol{\mu}$ of the Hebbian learning states (**LW**-states) in middle level are context-sensitive: they depend on the circumstances. In this application scenario, a highly stressful context is assumed to bring damage to learning efficiency: learnt effects are easily forgotten. That means, the value of the context state determines the values of the **M**-states. In the middle level, it is also adaptive how the total learning effect of the doctor transfers into the nurse's instructional learning effect. More specifically, the weights of the connections from the doctor's **RW**-states to the nurse's **IW**-states all are represented by second-order self-model states in the upper level. They all have a (second-order) Hebbian learning effect too. That means, knowledge is impaired as it is delivered from doctor to nurse. The scale of impairing is controlled by the persistence factor $\boldsymbol{\mu}$ of $\mathbf{W}_{\mathbf{RW},\mathbf{IW}}$-states in the upper level. $\mathbf{H}_{\mathbf{T}}$-states are second-order self-model states for adaptation speeds of **T**-states. An overview of the upper-level states is presented at Table 4 in Appendix. The combination functions used in this network model are presented in Table 3 in Appendix.

4 Simulation for the Example Scenarios

If the agent network model for this condition is valid, expected results should occur in simulations. When the nurse lacks perfect knowledge on pure ventilation, the nurse's mental model states for her preparing and giving pure ventilation will have low values. The threshold for these two states should stay at high levels. After the AI-coach's action ownership state for warning the nurse start rising, second-order self-model $\mathbf{H_T}$-states for adaptation speed of first-order self-model **T**-states for these threshold values should start to rise as well. After that, the first-order self-model **T**-states for these threshold values should decrease because of the influence from the AI-coach's action ownership state for warning. That will lead to increasing values of the nurse's mental model states for preparing and giving a first time of pure ventilation, and that marks that the nurse's imperfect knowledge has been fixed.

The network model was specified using the standard table format called role matrices for the network characteristics which can also be input to the dedicated software environment for simulation. The example simulation presented here was run from 0 to 50 with step size $\Delta t = 0.5$. Simulation results of state activation based on the current network model and its characteristics will be discussed below. The contextual stress level is set to be relatively high, so that the M-states have values no more than 0.95. Due to page size limitation and for clear analysis, the results of the same simulation at the indicated time points have been split into (1) the graph of states that embodies the function of the AI-coach, (2) the graph of the world states, the baby states and the context state.

Figure 2 reflects the AI-coach's function of monitoring. In Fig. 2, the dashed lines, the dashed and the dotted lines, the solid lines, and the dotted lines represent the nurse's action ownership states, the doctor's action ownership states, the AI-coach's mental model states and the AI-coach's action ownership states, respectively. In the AI-coach's mental model, except *AS2_DN_Pre_v* and *AS3_1tN_30ve*, all other AI-coach's mental model states get activated right after team members' corresponding action ownership states. This reflects the AI-coach is monitoring the team's actions. For the AI-coach's mental model state for the team preparing pure ventilation, it rises to 0.5 first when the doctor's action ownership state for preparing pure ventilation gets activated. Since the nurse's action ownership state for that preparation is not activated, the AI-coach's warning function is activated. The AI-coach's mental model state for the nurse giving first time of pure ventilation automatically reaches activation from as a result of its own database. Then, the AI-coach is able to activate the state for warning nurse to give correct pure ventilation.

Figure 3 presents the warning process. The rise of the AI-coach's mental model states for the team preparing pure ventilation and the nurse giving pure ventilation make that the AI-coach's mental model state for warning the nurse giving correct pure ventilation gets activated. That leads second-order self-model $\mathbf{H_T}$-states for (adaptation) speed of the corresponding first-order self-model **T**-states for threshold

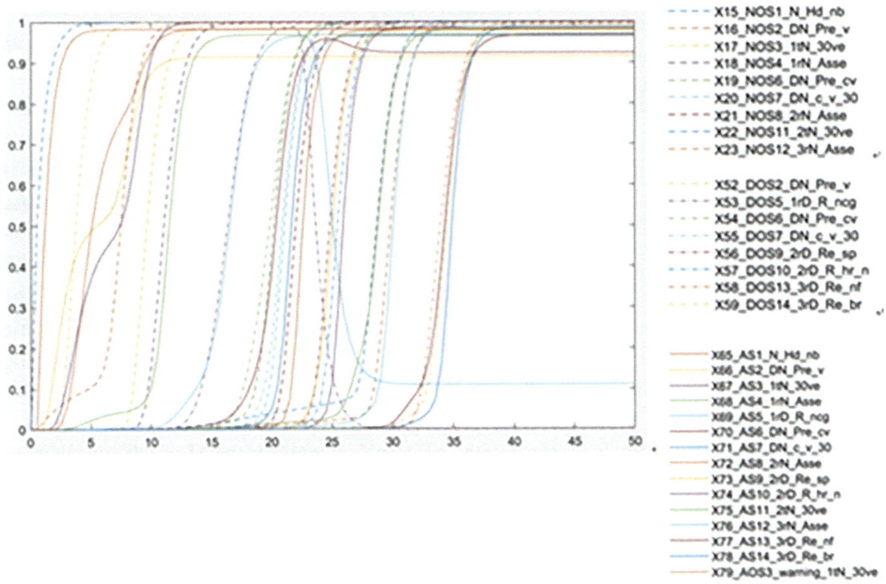

Fig. 2 Simulation results that show the AI-coach's monitoring in the AI-coach's engagement without organizational learning

values to increase. Since the first-order self-model **T**-states for threshold values has negative connections from the AI-coach's action ownership state for warning the nurse giving pure ventilation, the rise of latter one lead the former ones to decrease. With lower threshold value, the nurse's mental model states for the team preparing the pure ventilation and for giving first time of pure ventilation get activated at around $t = 7.5$ and $t = 10$. Thus, the nurse is well prepared and successfully giving first time of pure ventilation with the help of the AI-coach. Activations of the remaining states will be introduced in latter subsections.

Figure 4 shows the world states, the context state and the baby states over time. In Fig. 4, the solid lines, the dashed and dotted line, and the dashed lines reflect values of the world states, the context state and the baby states over time, respectively. The world state for nurse's holding the new-born baby on the table is activated at around $t = 2$. That means the team starts saving the baby. The world state for the team preparing for pure ventilation gets activated at around $t = 9$. That means, after the AI-coach warns the nurse, she is finally well-prepared so that the team are finally well- prepared as well. Then, the world state for the nurse giving the first time of pure ventilation gets activated at around $t = 11.5$, that means with the warning of the AI-coach, she successfully gives a pure ventilation. In the partial mistake condition, that state is never activated, so that the AI-coach makes a good contribution. In the

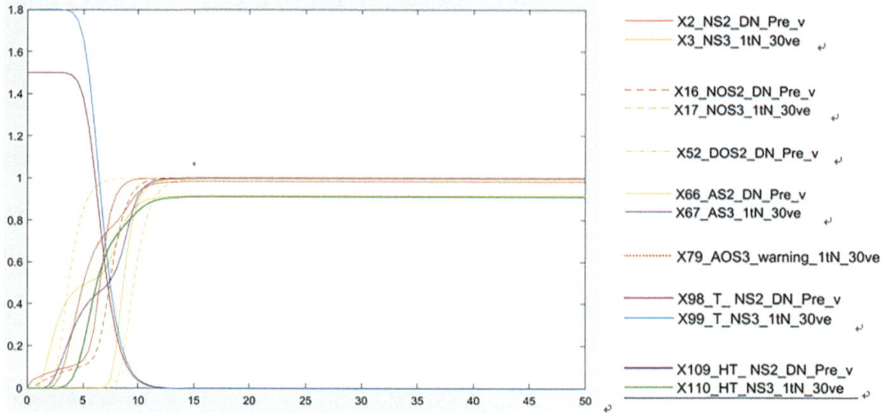

Fig. 3 Simulation results that show effects of the AI-coach's Warning in the AI-coach's engagement without organizational learning

first round of assessment, the world state of the nurse helping that assessment and for the doctor reporting the baby has no change reach activation at around $t = 12.5$ and $t = 17$. The baby still needs more treatment. Thus, the world states for the team preparing and giving a combination of compression and ventilation get activated at around $t = 20.5$ and $t = 21$. After that, the baby partially recovers. The baby states for the baby is slightly pink and has normal heart rate are active at around $t = 25$. So, the world state for the nurse helping the second round of assessment and for the doctor reporting the baby is slightly pink and with normal heart rate reach activation at around $t = 24$, $t = 26$, and $t = 27$. To have the baby completely safe, the world state for the nurse giving the second time of pure ventilation gets activated at around 30.5. Almost at the same time, the baby is not floppy anymore and start breathing. The baby states for these reach activation at around $t = 30.5$. In the third round of assessment, the world state for the nurse helping the assessment and for the doctor reporting the baby is not floppy and starts breathing get activated at around $t = 31$, $t = 35.5$ and $t = 37$.

5 Adaptive Agent Network Model for the AI's Engagement Covering Organizational Learning

In this condition, at the beginning the nurse has imperfect knowledge in preparing pure ventilation and the doctor has imperfect knowledge in preparing the combination of compression and ventilation. The AI-coach can help them avoid mistakes by applying organizational learning to get their knowledge exchanged. In the team, the knowledge of the doctor and nurse is complementary. They each mastered the knowledge that the other lacks. The AI-coach can obtain both their knowledge by

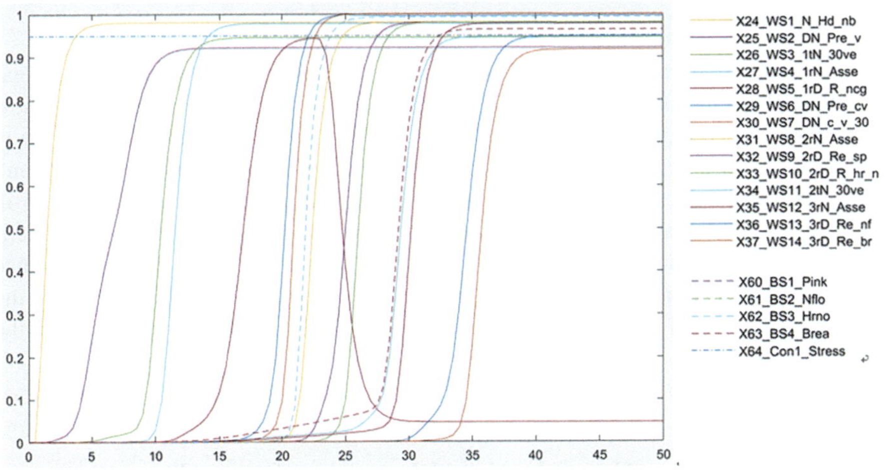

Fig. 4 Simulation results for the world states, baby states and the context state in the AI-coach's engagement without organizational learning

feed forward learning and send improved knowledge back through feedback learning. The process of how the AI-coach help the team correct their imperfect knowledge is presented in Fig. 5. In this process, initially the AI-coach obtains knowledge from the doctor, so that the AI-coach has an imperfect mental model, the same as the doctor. Since the AI-coach masters the knowledge that the nurse lacks, the nurse learns from the AI-coach to correct her imperfect knowledge. As a result, the nurse gets a perfect mental model. Then, by another feed forward learning process the AI-coach learns from the nurse and achieves a perfect mental model as well. Finally, by another feedback learning process the doctor learns from the AI-coach and obtain her missing knowledge. That means all team members are well prepared in the end. For tables with overviews of the different states, see Appendix.

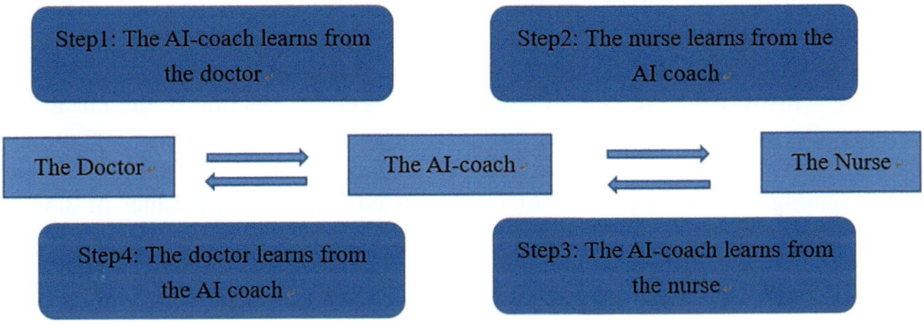

Fig. 5 How the AI-coach applies organizational learning

Network Overview

Figure 6 shows the agent network model for this condition. In the network, organizational learning is reflected at the middle level and the upper level. Three new first-order self-model states reflect the AI-coach's mastery of knowledge. Newly added second-order self-model states control the organizational learning effects, and they themselves are affected by the four newly added context states. In Table 5 in Appendix, the first context state is always active while the rest of them will be activated at a specific time determined by the new context states (set by the steponce function). This steponce function has a parameter $\boldsymbol{\alpha}$ for start time. Before $t = \boldsymbol{\alpha}$, the value of the function is 0. After $t = \boldsymbol{\alpha}$, the value of the function is 1, up to time $\boldsymbol{\beta}$, the indicated end time.

6 Simulation for Organisational Learning Example Scenarios

In this agent network model, in order to highlight the effects of organizational learning, when team members lack perfect knowledge, the corresponding self-learning effects and instructional learning effects of this knowledge is set to be 0. That means, for clear results the choice is made that team members obtain their missing knowledge only through feedback learning from the AI-coach and not by individual learning.

Like before, the example simulation presented here was run from 0 to 50 with stepsize $\Delta t = 0.5$. Below the simulation results of state activation based on current network characteristics are discussed. The contextual stress level is set to be extremely low, so that the context state has a value of 1, and that will lead $\mathbf{M_W}$-states have an upper limit value 1.

Figure 7 shows the adaptation states over time and reflects how the AI coach applies organizational learning principles: feed forward learning and feedback learning, like in Treur (2018). In Fig. 7, second-order self-model $\mathbf{W_{WW}}$-states for the weight of the connection from the doctor's total learning effects **RW** to the nurse's instructional learning effects **IW** all stay under 0.1. That's to eliminate the nurse's instructional learning effects to highlight the function of organizational learning. Second-order self-model $\mathbf{M_W}$-states for persistence factors for all self-learning effects all get activated at around $t = 2$. Since first-order self-model **W**-states corresponding to team members' missing knowledge have speed values of 0, these self-learning effects is also eliminated to highlight the contribution of organizational learning.

The doctor's first-order self-model **W**-states for total weight of the connection from preparation of pure ventilation to giving 2 times of pure ventilation both get activated at around $t = 2$. But the doctor's first-order self-model **W**-state for total weight of the connection from preparation of the combination of compression and ventilation to giving that ventilation is stable at 0 before $t = 25$. This means the

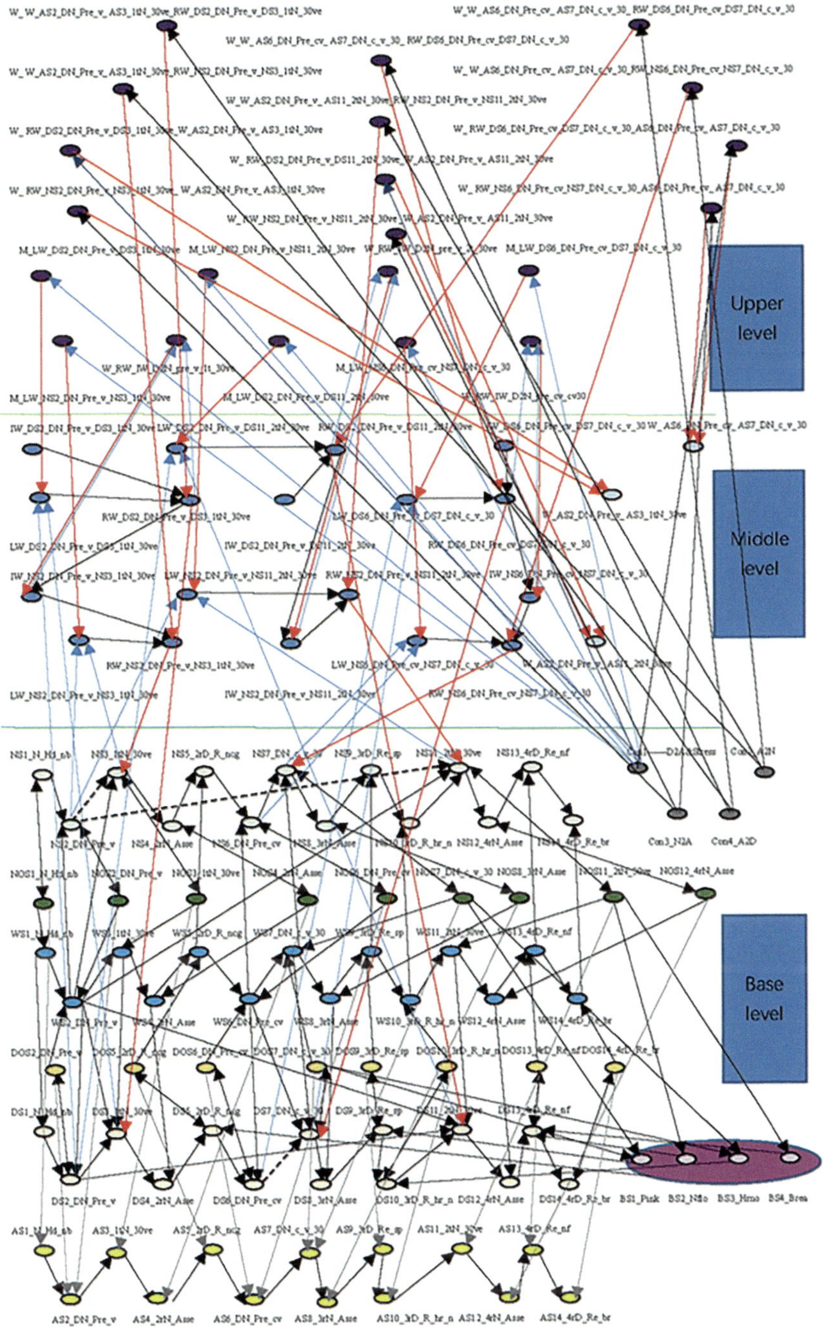

Fig. 6 Adaptive network model for the AI-coach covering its engagement with organizational learning

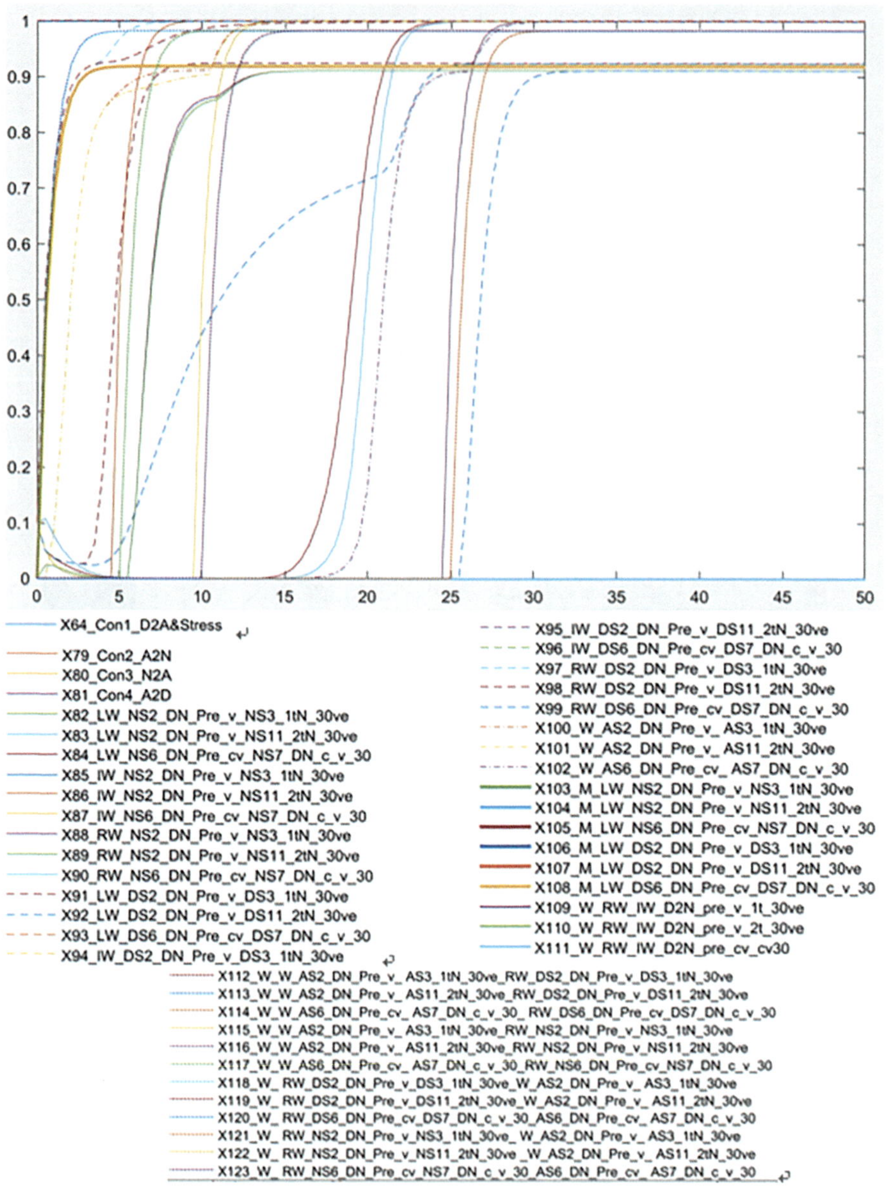

Fig. 7 Simulation results for adaptation states in the AI-coach's engagement with organizational learning

doctor is well prepared for pure ventilation but lacks perfect knowledge for the combination of compression and ventilation. The context state that controls the feed forward learning from the doctor to the AI-coach is always active. So, the AI-coach can copy the doctor's total learning effects at the beginning. The AI-coach's first-order self-model **W**-states for total weight of the connection from preparation of pure ventilation to giving 2 times of pure ventilation both get activated at around $t = 3.5$. But the AI-coach's first-order self-model **W**-state for total weight of the connection from preparation of the combination of compression and ventilation to giving that ventilation is stable at 0 before $t = 15$. The nurse's first-order self-model **W**-states for total weight of the connection from preparation of pure ventilation to giving 2 times of pure ventilation are both stable at 0 before $t = 5$. But the nurse's first-order self-model **W**-states for total weight of the connection from preparation of the combination of compression and ventilation to giving that ventilation reaches activation at around $t = 21$. This means the nurse has imperfect knowledge for pure ventilation but she is well-prepared for the combination of compression and ventilation.

The doctor's first-order self-model **W**-states for total weight of the connection from preparation of pure ventilation to giving 2 times of pure ventilation both get activated at around $t = 2$. But the doctor's first-order self-model **W**-states for total weight of the connection from preparation of the combination of compression and ventilation to giving that ventilation is stable at 0 before $t = 25$. This means the doctor is well prepared for pure ventilation but lacks perfect knowledge for the combination of compression and ventilation.

The context state that triggers the feed forward learning from doctor to the AI-coach is always active. So, the AI-coach copies the doctor's total learning effects at the beginning. The AI-coach's first-order self-model **W**-states for total weight of the connection from preparation of pure ventilation to giving 2 times of pure ventilation both get activated at around $t = 3.5$. But the AI-coach's first-order self-model W-states for total weight of the connection from preparation of the combination of compression and ventilation to giving that ventilation is stable at 0 before $t = 15$.

The nurse's first-order self-model **W**-states for total weight of the connection from preparation of pure ventilation to giving 2 times of pure ventilation are both stable at 0 before t = 5. But the nurse's first-order self-model **W**-states for total weight of the connection from preparation of the combination of compression and ventilation to giving that ventilation reaches activation at around $t = 21$. This means the nurse has imperfect knowledge for pure ventilation but she is well-prepared for the combination of compression and ventilation.

At $t = 5$, the context state that triggers the AI-coach giving feedback learning to the nurse gets activated. So, the nurse starts to gain her missing knowledge from the AI-coach. Her first-order self-model **W**-states for total weight of the connection from preparation of pure ventilation to giving 2 times of pure ventilation both reach activation at around $t = 8$. That means the nurse gets perfect knowledge.

At $t = 10$, the context state that triggers the AI-coach getting feed forward learning from the nurse reaches activation. Thus, the AI-coach starts learning from the nurse to make its knowledge perfect as well. Its first-order self-model **W**-states for total weight of the connection from preparation of the combination of compression and ventilation to giving that ventilation reaches activation at around $t = 22.5$. It's right after the nurse's corresponding state's activation. That means the AI-coach gets perfect knowledge as well.

At $t = 25$, the context state that triggers the AI-coach giving feedback learning to the nurse reaches activation. That means the doctor finally is able to get her missing knowledge from the AI-coach. Her first-order self-model **W**-states for total weight of the connection from preparation of the combination of compression and ventilation to giving that ventilation become active at around $t = 28$. The doctor also has perfect knowledge in the end. The adaptation states and the context states clearly show the process of organizational learning. The AI-coach promotes communicational efficiency between both team members. These results confirm the validity of the network.

Figure 8 reflects the AI-coach's monitoring function. In this graph, the AI-coach's mental model states for individual actions all get activated right after activation of corresponding action ownership states. The AI-coach's mental model states for teamwork actions get activated slightly earlier than action ownership states, that's because both team members' action ownership states have influence of the AI-coach's mental model states. Simultaneous influence superposition makes the AI-coach's mental model states reach activation faster than team members' action ownership states.

Here the AI-coach's mental model states will be described. The AI-coach's mental model state for the nurse holding the newborn baby on the table is active at around $t = 2.5$. After observing first time of pure ventilation, The AI-coach's mental model states for the team preparing and the nurse giving that ventilation get activated at around $t = 4.5$ and $t = 10$. The 5.5 gap is because of the organizational learning. By monitoring the first round of assessment, the AI-coach's mental model states for the nurse helping that assessment and for the doctor reporting the baby has no change reach activation at around get activated at around $t = 13.5$ and $t = 15$, respectively. The AI-coach knows new treatment will come. By observing the team complete preparing and giving the combination of compression and ventilation, the AI-coach's mental model states for these reach activation at around $t = 17$ and $t = 22$, respectively. The time gap is because of organizational learning as well. By monitoring the second round of assessment, the AI-coach's mental model states for the nurse helping the second round of assessment and for the doctor reporting the baby is slightly pink and with normal heart rate reach activation at around $t = 24$, $t = 25$, and $t = 26$, respectively. Knowing the baby has change, the AI-coach's mental model state for the doctor reporting no change become not active at $t = 23$. These are followed with the nurse giving a second time of pure ventilation. The AI-coach's mental model state for that reach activation at around $t = 26.5$. By observing the third round of assessment, the AI-coach's mental model state for the nurse helping the assessment and for the doctor reporting the baby is not floppy and starts breathing get activated at around $t = 27.5$, $t = 28.5$ and $t = 29$, respectively.

Figure 9 shows the world states. The world state for the nurse holding the newborn baby on table is active at around $t = 2.5$. After observing first time of pure ventilation,

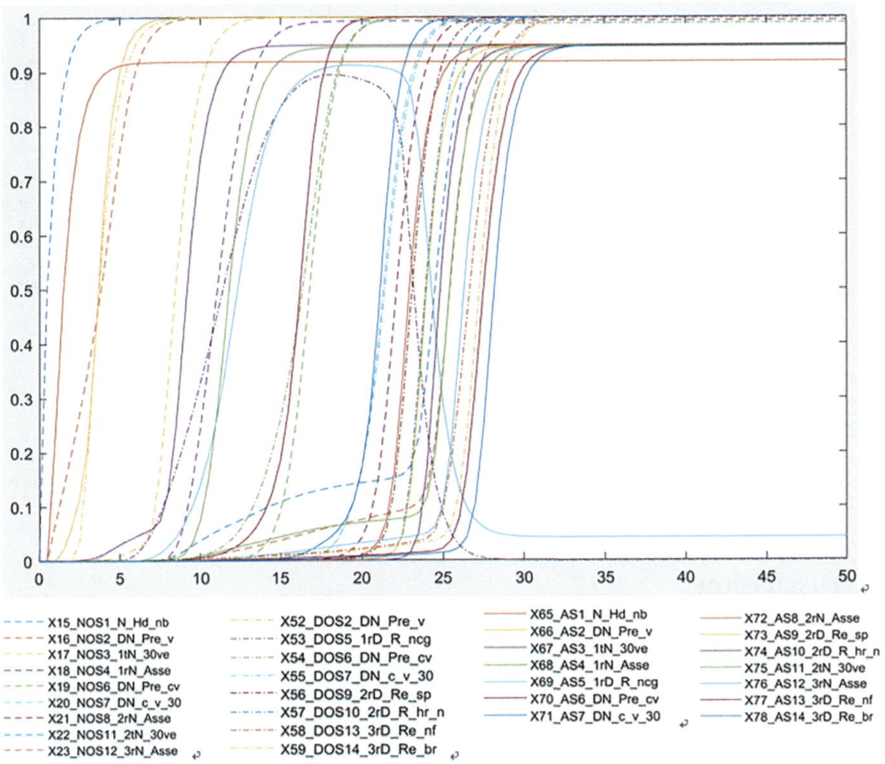

Fig. 8 Simulation results shows effects of the AI-coach's monitoring in the AI-coach's engagement with organizational learning

The world states for the team preparing and the nurse giving that ventilation get activated at around $t = 5$ and $t = 9.5$, respectively. Right after, the world states for the nurse helping that assessment and for the doctor reporting the baby has no change reach activation at around get activated at around $t = 11$ and $t = 16.5$, respectively. That means the baby needs new treatment. So the team complete preparing and giving the combination of compression and ventilation, the world states for these reach activation at around $t = 17$ and $t = 21.5$, respectively. That follows with the second round of assessment. The world state for the nurse helping the second round of assessment and for the doctor reporting the baby is slightly pink and with normal heart rate reach activation at around $t = 23$, $t = 24.5$, and $t = 25.5$, respectively. At the same time, the AI-coach's mental model state for the doctor reporting no change becomes not active at $t = 22$. These follows with the activation of the world state for the nurse giving second time of pure ventilation and for she is helping the third round of assessment get activated at around $t = 26$ and $t = 26.5$, respectively. Right after, the world state for the doctor reporting the baby is not floppy and starts breathing get activated at around $t = 28$, $t = 30$, respectively.

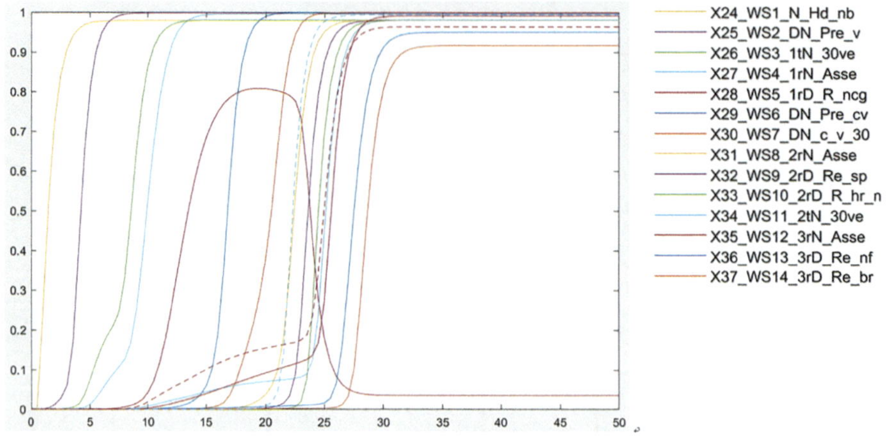

Fig. 9 Simulation results for the world states in the AI-coach's engagement with organizational learning

7 Discussion

In this chapter, based on material from Xu et al. (2022), two second-order adaptive computational agent network models show processes for two conditions of a medical team of a doctor and a nurse giving ventilation to a newborn baby in danger by involving a shared mental model. The design of these computational models is based on an adaptive network-oriented modeling approach and are simulated in the dedicated software environment described in Treur (2020a, b, Chap. 9). The computational mechanisms for organizational learning were introduced in Canbaloğlu et al. (2021), see also Canbaloğlu et al. (2025). In these network models, actors include the nurse and the doctor, and sometimes the AI and the organization. Actors change the world by action ownership states. In turn, the world states and action ownerships states can change an actor's mental model as well. All network models enable representations and processing of all actors' internal simulation of their mental models, adaptive changes of these mental models (learning and forgetting), all actors' actions in the world, the dynamics of the interaction between actors and the world. To show the influence of stress on the forgetting effect of an internal hebbian learning process, a context state representing the context stress was introduced. In these network models, learning effects consists of internal hebbian learning effects, instructional learning effects and organizational learning effects, all represented by self-model states and their dynamics and control. That shows how these learning effects influence the relation structure of mental models. Simulation experiments show how the AI-coach promotes the formation of shared mental models for correct ventilation.

The condition of the AI-coach's engagement without organizational learning corresponds to a partial mistake condition. In that condition, the AI-coach keeps monitoring team members' actions. When the AI-coach notices the nurse's tendency to make an error, it warns her to avoid making that error. The simulation results illustrate the world states for that. In the simulation, the direct result of the AI-coach's intervention is the rise of second-order self-model $\mathbf{H}_T$-self-model states and decrease of first-order self-model **T**-states. That enables the activation of the nurse's mental model states for preparing and giving first time of pure ventilation which she fails in partial mistake condition. The condition of the AI-coach's engagement without organizational learning provides two directions for the development of medical AI, those are, monitoring and warning. The involvement of the AI-coach will eliminate mistakes before they are formed. In this way, the AI-coach can play a role of a speaker without fear of authority and punishment and with effective communication ability. The AI-coach is well aware of the team members' respective responsibilities and knowledge and urges them to take responsibility, acquire the right knowledge and make the right actions. The condition of the AI-coach's engagement with organizational learning corresponds to a more extreme mistake condition. Compared to the situation without organizational learning, the AI-coach in this condition addresses more on promoting knowledge exchange between team members to prevent errors. In this condition, the AI-coach interacts with team members by feed forward learning and feedback learning from organisational learning (Canbaloğlu et al. 2023), obtains knowledge from both team members, and transfers correct knowledge back to both team members. In this way, both team members correct their imperfect knowledge. In simulation results, these are reflected on the adaptation level (first-order self-model) and the control of adaptation level (second-order self-model). The AI-coach's first-order self-model **W**-states for its learning effects and team members' first-order self-model **W**-states for their learning effects influence each other. The transmission direction of the influence is controlled by second-order self-model $\mathbf{W}_{WW}$-states that control the first-order adaptation level. In this condition, the team members' tendency of making mistakes doesn't even appear because the team members' imperfect knowledge is corrected in time so that their actions are guided by correct knowledge. In addition to the monitoring function discussed already above, this condition shows that medical AI developers can improve knowledge exchange between team members as a direction of development. In this direction, an AI-coach enables the organization to identify, report, and investigate incidents and to take appropriate actions that increase reduce the probability of recurrence of errors and in turn improve performance of members within this organization.

These adaptive agent network models visually present how to analyse the AI-coach's contribution in a medical team's ventilation operation to save a baby in danger. The AI-coach's engagement shows how it can prevent errors by monitoring, warning, and promoting knowledge transferring. The storyline of these network models proves that this research achieves its research objective of exploring how AI can help improve the performance of a medical team. Since the AI-coach monitors team members' actions, speak up on time to prevent mistakes, promotes the team's ability to organizational learning, the AI-coach helps overcome the shortcomings of a blame culture, reluctance to speak up, ineffective communication, and inability to detect and learn from potential mistakes early. In other words, such an AI-coach promotes just safety culture within an organization.

Future study for extensions can address several directions. Field experiments can be designed to test the performance of the AI-coach in the real world. In addition, application scenarios can be changed. The medical scene is massive, there are always urgent cases where it is intolerable to waste time because of mistakes. Besides, medical teams can be bigger. That will result in a more complex teamwork process.

Appendix

See Tables 1, 2, 3, 4, 5 and 6.

Table 1 Overview of baselevel states' names and their explanation of adaptive network model for the AI-coach's engagement without organizational learning

State name	Explanation
N_Hd_nb	Nurse holds the new born baby on the table and prepares for intubation
DN_Pre_v	Doctor and nurse together prepare for pure ventilation
1tN_30ve	Nurse gives baby 30 s of pure ventilation for the first time
1rN_Asse	Nurse helps assess the baby's condition at the first round of assessment
1rD_R_ncg	Doctor reports the baby has no change at the first round of assessment
DN_Pre_cv	Doctor and nurse together prepare for the combination of compression and ventilation
DN_c_v_30	Doctor and nurse together give baby 30 s of the combination of compression and ventilation
2rN_Asse	Nurse helps assess the baby's condition at second assessment round
2rD_Re_sp	Doctor reports the baby is slightly pink at second assessment round
2rD_R_hr_n	Doctor reports the baby has normal heart rate at second assessment round
2tN_30ve	Nurse gives baby 30 s of pure ventilation for the second time
3rN_Asse	Nurse helps assess the baby's condition at third assessment round
3rD_Re_nf	Doctor reports the baby is not floppy at third assessment round
3rD_Re_br	Doctor reports the baby is breathing at third assessment round
Con1	Contextual stress factor
Pink	Baby is pink
Nflo	Baby is not floppy
Hrno	Baby has normal heart rate
Brea	Baby is breathing

Table 2 Overview of middle-level states' names and their explanation of adaptive network model for the AI-coach's engagement without organizational learning

State name	Explanation
LW_NS2_ DN_Pre_v_ NS3_1tN_ 30ve	First-order self-model state for the nurse's weight of the connection from 'the state for preparing pure ventilation' to 'the state for giving pure ventilation for 30 s for the first time' as learnt by Hebbian learning
LW_NS2_ DN_Pre_v_ NS11_2tN_ 30ve	First-order self-model state for the nurse's weight of the connection from 'the state for preparing pure ventilation' to 'the state for giving pure ventilation for 30 s for the second time' as learnt by Hebbian learning
LW_NS6_ DN_Pre_cv_ NS7_DN_c_ v_30	First-order self-model state for the nurse's weight of the connection from 'the state for preparing the combination of compression and ventilation' to 'the state for giving the combination of compression and ventilation for 30 s' as learnt by Hebbian learning
IW_NS2_ DN_Pre_v_ NS3_1tN_ 30ve	First-order self-model state for the nurse's weight of the connection from 'the state for preparing pure ventilation' to 'the state for giving pure ventilation for 30 s for the first time' as learnt from instruction by the doctor
IW_NS2_ DN_Pre_v_ NS11_2tN_ 30ve	First-order self-model state for the nurse's weight of the connection from 'the state for preparing pure ventilation' to 'the state for giving pure ventilation for 30 s for the second time' as learnt from instruction by the doctor
IW_NS6_ DN_Pre_cv_ NS7_DN_c_ v_30	First-order self-model state for the nurse's weight of the connection from 'the state for preparing the combination of compression and ventilation' to 'the state for giving the combination of compression and ventilation for 30 s' as learnt from instruction by the doctor
RW_NS2_ DN_Pre_v_ NS3_1tN_ 30ve	First-order self-model state for the nurse's overall weight of the connection from 'the state for preparing pure ventilation' to 'the state for giving pure ventilation for 30 s for the first time'
RW_NS2_ DN_Pre_v_ NS11_2tN_ 30ve	First-order self-model state for the nurse's overall weight of the connection from 'the state for preparing pure ventilation' to 'the state for giving pure ventilation for 30 s for the second time'
RW_NS6_ DN_Pre_cv_ NS7_DN_c_ v_30	First-order self-model state for the nurse's overall weight of the connection from 'the state for preparing the combination of compression and ventilation' to 'the state for giving the combination of compression and ventilation for 30 s'

(continued)

Table 2 (continued)

State name	Explanation
LW_DS2_ DN_Pre_v_ DS3_1tN_ 30ve	First-order self-model state for the doctor's weight of the connection from 'the state for preparing pure ventilation' to 'the state for giving pure ventilation for 30 s for the first time' as learnt by Hebbian learning
LW_DS2_ DN_Pre_v_ DS11_2tN_ 30ve	First-order self-model state for the doctor's weight of the connection from 'the state for preparing pure ventilation' to 'the state for giving pure ventilation for 30 s for the second time' as learnt by Hebbian learning
LW_DS6_ DN_Pre_cv_ DS7_DN_c_ v_30	First-order self-model state for the doctor's weight of the connection from 'the state for preparing the combination of compression and ventilation' to 'the state for giving the combination of compression and ventilation for 30 s' as learnt by Hebbian learning
IW_DS2_ DN_Pre_v_ DS3_1tN_ 30ve	First-order self-model state for the doctor's weight of the connection from 'the state for preparing pure ventilation' to 'the state for giving pure ventilation for 30 s for the first time' as known to the doctor
IW_DS2_ DN_Pre_v_ DS11_2tN_ 30ve	First-order self-model state for the doctor's weight of the connection from 'the state for preparing pure ventilation' to 'the state for giving pure ventilation for 30 s for the second time' as known to the doctor
IW_DS6_ DN_Pre_cv_ DS7_DN_c_ v_30	First-order self-model state for the doctor's weight of the connection from 'the state for preparing the combination of compression and ventilation' to 'the state for giving the combination of compression and ventilation for 30 s' as known to the doctor
RW_DS2_ DN_Pre_v_ DS3_1tN_ 30ve	First-order self-model state for the doctor's overall weight of the connection from 'the state for preparing pure ventilation' to 'the state for giving pure ventilation for 30 s for the first time'
RW_DS2_ DN_Pre_v_ DS11_2tN_ 30ve	First-order self-model state for the doctor's overall weight of the connection from 'the state for preparing pure ventilation' to 'the state for giving pure ventilation for 30 s for the second time'
RW_DS6_ DN_Pre_cv_ DS7_DN_c_ v_30	First-order self-model state for the doctor's overall weight of the connection from 'the state for preparing the combination of compression and ventilation' to 'the state for giving the combination of compression and ventilation for 30 s'
T_ NS2_ DN_Pre_v	First-order self-model state for threshold value of the nurse's mental model state for the nurse preparing pure ventilation
T_NS3_1tN_ 30ve	First-order self-model state for threshold value of the nurse's mental model state for the nurse giving first time of pure ventilation

Table 3 Combination functions

Function	Notation	Formula	Parameters
Advanced logistic sum	$\mathbf{alogistic}_{\sigma,\tau}(V_1, \ldots, V_k)$	$[\frac{1}{1+e^{-\sigma(V_1+\cdots+V_k-\tau)}} - \frac{1}{1+e^{\sigma\tau}}](1+e^{-\sigma\tau})$	Steepness σ Threshold τ
Hebbian Learning	$\mathbf{hebb}_{\mu}(V_1, V_2, V_3)$	$V_1V_2(1-V_3)+\mu V_3$	Persistence factor μ
Identity	$\mathbf{id}(V)$	V	

Table 4 Overview of upper-level states' names and their explanation of adaptive network model for the AI-Coach's engagement without organizational learning

M_LW_NS2_DN_Pre_v_NS3_1tN_30ve	Second-order self-model state for the persistence factor of the Nurse's weight of the connection from 'the state for preparing pure ventilation' to 'the state for giving pure ventilation for 30 s for the first time' as learnt by Hebbian learning
M_LW_NS2_DN_Pre_v_NS11_2tN_30ve	Second-order self-model state for the persistence factor of the Nurse's weight of the connection from 'the state for preparing pure ventilation' to 'the state for giving pure ventilation for 30 s for the second time' as learnt by Hebbian learning
M_LW_NS6_DN_Pre_cv_NS7_DN_c_v_30	Second-order self-model state for the persistence factor of the Nurse's weight of the connection from 'the state for preparing the combination of compression and ventilation' to 'the state for giving the combination of compression and ventilation for 30 s' as learnt by Hebbian learning
M_LW_DS2_DN_Pre_v_DS3_1tN_30ve	Second-order self-model state for the persistence factor of the Doctor's weight of the connection from 'the state for preparing pure ventilation' to 'the state for giving pure ventilation for 30 s for the first time' as learnt by Hebbian learning
M_LW_DS2_DN_Pre_v_DS11_2tN_30ve	Second-order self-model state for the persistence factor of the Doctor's weight of the connection from 'the state for preparing pure ventilation' to 'the state for giving pure ventilation for 30 s for the second time' as learnt by Hebbian learning

(continued)

Table 4 (continued)

M_LW_NS2_DN_Pre_v_NS3_1tN_30ve	Second-order self-model state for the persistence factor of the Nurse's weight of the connection from 'the state for preparing pure ventilation' to 'the state for giving pure ventilation for 30 s for the first time' as learnt by Hebbian learning
M_LW_DS6_DN_Pre_cv_DS7_DN_c_v_30	Second-order self-model state for the persistence factor of the Doctor's weight of the connection from 'the state for preparing the combination of compression and ventilation' to 'the state for giving the combination of compression and ventilation for 30 s' as learnt by Hebbian learning
W_RW_IW_D2N_pre_v_1t_30ve	Second-order self-model state for the nurse's weight of connection from 'the first-order self-model state for the doctor's total learning effect' to 'the first-order self-model state for nurse's instructional learning effect' on 'preparation of ventilation to giving first time of pure ventilation' as learnt by Hebbian learning
W_RW_IW_D2N_pre_v_2t_30ve	Second-order self-model state for the nurse's weight of connection from 'the self-model state for the doctor's total learning effect' to 'the first-order self-model state for nurse's instructional learning effect' on 'preparation of ventilation to giving second time of pure ventilation' as learnt by Hebbian learning
W_RW_IW_D2N_pre_cv_cv30	Second-order self-model state for the nurse's weight of connection from 'the first-order self-model state for the doctor's total learning effect' to 'the first-order self-model state for nurse's instructional learning effect' on 'preparation of the combination of compression and ventilation to giving the combination of compression and ventilation' as learnt by Hebbian learning

Table 5 An overview of new added context states of adaptive network model for the AI-Coach's engagement without organizational learning

State name	Explanation
Con1_D2A&Stress	The context state that control context stress and the start of feedforward learning from the doctor to the AI-coach
Con2_A2N	The context state that control the start of feedback learning from the AI-coach to the nurse
Con3_N2A	The context state that control the start of feedforward learning from the nurse to the AI-coach
Con4_A2D	The context state that control the start of feedback learning from the AI-coach to the doctor

Table 6 An Overview of new added adaption states of adaptive network model for the AI-Coach's engagement without organizational learning

State name	W_AS2_DN_Pre_v_ AS3_1tN_30ve
Explanation	First-order self-model state for the AI-coach's weight of the connection from 'the state for the team preparing pure ventilation' to 'the state for the nurse giving first time of pure ventilation for 30 s' as learnt by Hebbian learning
State Name	W_AS2_DN_Pre_v_ AS11_2tN_30ve
Explanation	First-order self-model state for the AI-coach's weight of the connection from 'the state for the team preparing pure ventilation' to 'the state for the nurse giving second time of pure ventilation for 30 s' as learnt by Hebbian learning
State Name	W_AS6_DN_Pre_cv_ AS7_DN_c_v_30
Explanation	First-order self-model state for the AI-coach's weight of the connection from 'the state for the team preparing the combination of compression and ventilation' to 'the state for the team giving the combination of compression and ventilation for 30 s' as learnt by Hebbian learning
State Name	W_W_AS2_DN_Pre_v_ AS3_1tN_30ve_RW_DS2_DN_Pre_v_DS3_1tN_30ve
Explanation	Second order self-model state for feedback learning effects from the AI-coach to the doctor for the weight of the connection from the team preparing pure ventilation to the nurse giving first time of pure ventilation
State Name	W_W_AS2_DN_Pre_v_ AS11_2tN_30ve_RW_DS2_DN_Pre_v_DS11_2tN_ 30ve
Explanation	Second order self-model state for feedback learning effects from the AI-coach to the doctor for the weight of the connection from the team preparing pure ventilation to the nurse giving second time of pure ventilation
State Name	W_W_AS6_DN_Pre_cv_AS7_DN_c_v_30_RW_DS6_DN_Pre_cv_DS7_DN_ c_v_30
Explanation	Second order self-model state for feedback learning effects from the AI-coach to the doctor for the weight of the connection from the team preparing the combination of compression and ventilation' to 'the state for the team giving the combination of compression
State Name	W_W_AS2_DN_Pre_v_ AS3_1tN_30ve_RW_NS2_DN_Pre_v_NS3_1tN_30ve
Explanation	Second order self-model state for feedback learning effects from the AI-coach to the nurse for the weight of the connection from the team preparing pure ventilation to the nurse giving first time of pure ventilation
State Name	W_W_AS2_DN_Pre_v_ AS11_2tN_30ve_RW_NS2_DN_Pre_v_NS11_2tN_ 30ve
Explanation	Second order self-model state for feedback learning effects from the AI-coach to the nurse for the weight of the connection from the team preparing pure ventilation to the nurse giving second time of pure ventilation
State Name	W_W_AS6_DN_Pre_cv_AS7_DN_c_v_30_RW_NS6_DN_Pre_cv_NS7_DN_ c_v_30
Explanation	Second order self-model state for feedback learning effects from the AI-coach to the nurse for the weight of the connection from the team preparing the combination of compression and ventilation' to 'the state for the team giving the combination of compression
State Name	W_ RW_DS2_DN_Pre_v_DS3_1tN_30ve_W_AS2_DN_Pre_v_ AS3_1tN_30ve

(continued)

Table 6 (continued)

State name	W_AS2_DN_Pre_v_ AS3_1tN_30ve
Explanation	Second order self-model state for feedforward learning effects from the doctor to the AI-coach for the weight of the connection from the team preparing pure ventilation to the nurse giving first time of pure ventilation
State Name	W_ RW_DS2_DN_Pre_v_DS11_2tN_30ve_W_AS2_DN_Pre_v_ AS11_2tN_ 30ve
Explanation	Second order self-model state for feedforward learning effects from the doctor to the AI-coach for the weight of the connection from the team preparing pure ventilation to the nurse giving second time of pure ventilation
State Name	W_ RW_DS6_DN_Pre_cv_DS7_DN_c_v_30_AS6_DN_Pre_cv_ AS7_DN_c_ v_30
Explanation	Second order self-model state for feedforward learning effects from the doctor to the AI-coach for the weight of the connection from the team preparing the combination of compression and ventilation' to 'the state for the team giving the combination of compression
State Name	W_ RW_NS2_DN_Pre_v_NS3_1tN_30ve_ W_AS2_DN_Pre_v_ AS3_1tN_30ve
Explanation	Second order self-model state for feedforward learning effects from the nurse to the AI-coach for the weight of the connection from the team preparing pure ventilation to the nurse giving first time of pure ventilation
State Name	W_ RW_NS2_DN_Pre_v_NS11_2tN_30ve _W_AS2_DN_Pre_v_ AS11_2tN_ 30ve
Explanation	Second order self-model state for feedforward learning effects from the nurse to the AI-coach for the weight of the connection from the team preparing pure ventilation to the nurse giving second time of pure ventilation
State Name	W_ RW_NS6_DN_Pre_cv_NS7_DN_c_v_30_AS6_DN_Pre_cv_ AS7_DN_c_ v_30
Explanation	Second order self-model state for feedforward learning effects from the nurse to the AI-coach for the weight of the connection from the team preparing the combination of compression and ventilation' to 'the state for the team giving the combination of compression

References

Argote, L., Miron-Spektor, E.: Organizational learning: From experience to knowledge. Organ. Sci. **22**(5), 1123–1137 (2011)

Aspden, P., Corrigan, J.M., Wolcott, J., Erickson, S.M.: Patient safety reporting systems and applications. In: Patient Safety: Achieving a New Standard for Care. National Academies Press (US) (2004)

Benbassat, J.: Role modeling in medical education: the importance of a reflective imitation. Acad. Med. **89**(4), 550 (2014)

Bohn, R.E.: Noise and learning in semiconductor manufacturing. Manag. Sci. **41**(1), 31–42 (1995)

Burtscher, M.J., Manser, T.: Team mental models and their potential to improve teamwork and safety: a review and implications for future research in healthcare. Saf. Sci. **50**(5), 1344–1354 (2012)

Canbaloğlu, G., Treur, J., Roelofsma, P.H.M.P.: Computational modeling of organisational learning by self-modeling networks. Cogn. Syst. Res. **73**, 51–64 (2021)

Canbaloğlu, G., Treur, J., Wiewiora, A. (eds.): Computational Modeling of Multilevel Organisational Learning and Its Control Using Self-modeling Network Models. Springer Nature, Cham (2023)

Canbaloğlu, G., Treur, J., Roelofsma, P.H.M.P.: Learning of safety and security management through cyberspace: an adaptive self-modeling network model for multilevel organizational learning. In: Roelofsma, P.H.M.P., Jabeen, F., Taal, H.R., Treur, J. (eds.) Using Shared Mental Models and Organisational Learning to Support Safety and Security through Cyberspace: A Computational Analysis Approach. Springer Nature, Cham, Switzerland (this volume) (2025)

Craik, K.J.W.: The nature of explanation. CUP Archive **445** (1943)

Donaldson, L.: An organisation with a memory. Clin. Med. **2**(5), 452 (2002)

Edwards, W.H.: Patient safety in the neonatal intensive care unit. Clin. Perinatol. **32**(1), 97–106 (2005)

Emanuel, L., Berwick, D., Conway, J., Combes, J., Hatlie, M., Leape, L., et al.: What exactly is patient safety? J. Med. Regul. **95**(1), 13–24 (2009)

Fiol, C.M., Lyles, M.A.: Organizational learning. Acad. Manag. Rev. **10**(4), 803–813 (1985)

Garrouste-Orgeas, M., Philippart, F., Bruel, C., Max, A., Lau, N., Misset, B.: Overview of medical errors and adverse events. Ann. Intensive Care **2**(1), 1–9 (2012)

Glynn, M.A., Lant, T.K., Milliken, F.J.: Mapping learning processes in organizations: a multi-level framework linking learning and organizing. Adv. Manag. Cogn. Organ. Inf. Process. **5**(2), 43–83 (1994)

Haunschild, P.R., Sullivan, B.N.: Learning from complexity: effects of prior accidents and incidents on airlines' learning. Adm. Sci. Q. **47**(4), 609–643 (2002)

Health E-Learning and Media Team (HELM).: Neonatal Resuscitation Demonstration [Video]. YouTube (2020). https://www.youtube.com/watch?v=mz88FSxbmtI&t=431s

Hendrikse, S.C.F., Treur, J., Koole, S.L.: Modeling emerging interpersonal synchrony and its related adaptive short-term affiliation and long-term bonding: a second-order multi-adaptive neural agent model. Int. J. Neural Syst. **33**(7), 2350038, 1–41 (2023). https://doi.org/10.1142/S0129065723500387

Hogan, K.E., Pressley, M.E.: Scaffolding Student Learning: Instructional Approaches and Issues. Brookline Books (1997)

Jonker, C.M., Riemsdijk, M., Vermeulen, B.: Shared mental models. In: International Workshop on Coordination, Organizations, Institutions, and Norms in Agent Systems, pp. 132–151. Springer, Berlin (2010)

Kim, D.H.: A framework and methodology for linking individual and organizational learning: Applications in TQM and product development (Doctoral dissertation, Massachusetts Institute of Technology) (1993)

Kirwan, M., Matthews, A., Scott, P.A.: The impact of the work environment of nurses on patient safety outcomes: a multi-level modelling approach. Int. J. Nurs. Stud. **50**(2), 253–263 (2013)

Kogut, B., Zander, U.: Knowledge of the firm, combinative capabilities, and the replication of technology. Organ. Sci. **3**(3), 383–397 (1992)

Lee, S.E., Scott, L.D., Dahinten, V.S., Vincent, C., Lopez, K.D., Park, C.G.: Safety culture, patient safety, and quality of care outcomes: a literature review. West. J. Nurs. Res. **41**(2), 279–304 (2019)

Levitt, B., March, J.G.: Organizational learning. Ann. Rev. Soc. 319–340 (1988)

Mathieu, J.E., Heffner, T.S., Goodwin, G.F., Salas, E., Cannon-Bowers, J.A.: The influence of shared mental models on team process and performance. J. Appl. Psychol. **85**(2), 273 (2000)

Mayer, R.E.: Models for understanding. Rev. Educ. Res. **59**(1), 43–64 (1989)

Maynard, M.T., Gilson, L.L.: The role of shared mental model development in understanding virtual team effectiveness. Group Org. Manag. **39**(1), 3–32 (2014)

McComb, S., Simpson, V.: The concept of shared mental models in healthcare collaboration. J. Adv. Nurs. **70**(7), 1479–1488 (2014)

National Quality Forum.: National voluntary consensus standards for nursing-sensitive care: An initial performance measure set. National Quality Forum, Washington, DC (2004)

Nonaka, I., Von Krogh, G.: Perspective—Tacit knowledge and knowledge conversion: Controversy and advancement in organizational knowledge creation theory. Organ. Sci. **20**(3), 635–652 (2009)

Piaget, J.: The Construction of Reality in the Child. Ballantine, New York (1954)

Rasmussen, J.: Human error and the problem of causality in analysis of accidents. Philos. Trans. R. Soc. Lond. B, Biol. Sci. **327**(1241), 449–462 (1990)

Roelofsma, P.H.M.P., Treur, J.: Computational analysis to support safety and security through cyberspace using shared mental models and organisational learning. In: Roelofsma, P.H.M.P., Jabeen, F., Taal, H.R., Treur, J. (eds.) Using Shared Mental Models and Organisational Learning to Support Safety and Security through Cyberspace: A Computational Analysis Approach. Springer Nature, Cham, Switzerland (this volume) (2025)

Schulz, M.: Organizational Learning'in Baum, JAC (2002)

Sitkin, S.: Learning through failure: the strategy of small losses. In: Cohen, T.M., Sproull Teoksessa, L. (eds.) Organizational Learning, pp. 541–577 (1996)

Tangirala, S., Ramanujam, R.: Employee silence on critical work issues: The cross level effects of procedural justice climate. Pers. Psychol. **61**(1), 37–68 (2008)

Treur, J.: Network-Oriented Modeling. Springer International Publishing Switzerland (2016)

Treur, J.: Network reification as a unified approach to represent network adaptation principles within a network. In: Proc. of the 7th International Conference on Theory and Practice of Natural Computing, TPNC 2018. Lecture Notes in Computer Science, vol. 11324, pp. 344–358. Springer, Cham (2018)

Treur, J.: Modeling higher order adaptivity of a network by multilevel network reification. Netw. Sci. **8**(S1), S110–S144 (2020a)

Treur, J.: Network-Oriented Modeling for Adaptive Networks: Designing Higher-Order Adaptive Biological, Mental and Social Network Models. Springer Nature, Cham (2020b)

Treur, J.: On the dynamics and adaptivity of mental processes: relating adaptive dynamical systems and self-modeling network models by mathematical analysis. Cogn. Syst. Res. **70**, 93–100 (2021)

Treur, J., Van Ments, L. (eds.).: Mental Models and Their Dynamics, Adaptation, and Control: A Self-Modeling Network Modeling Approach. Springer Nature, Cham (2022)

Van Ments, L., Treur, J.: Reflections on dynamics, adaptation and control: a cognitive architecture for mental models. Cogn. Syst. Res. **70**, 1–9 (2021)

Waller, M.J., Gupta, N., Giambatista, R.C.: Effects of adaptive behaviors and shared mental models on control crew performance. Manag. Sci. **50**(11), 1534–1544 (2004)

Williams, D.G.: The mind as a predictive modelling engine: generative models, structural similarity, and mental representation (Doctoral dissertation, University of Cambridge) (2018)

Xu, Y., Jabeen, F., Treur, J., Taal, H.R., Roelofsma, P.H.M.P.: Adaptive agent network models with internal mental models supporting patient safety. In: Proceedings of the 15th International Conference on Social Computing and Networking, SocialCom'22. IEEE Computer Society Press (2022)

Increasing Safety and Security Through Cyberspace by an Adaptive Network Model for AI-Assisted Risk Management of Neonatal Respiratory Distress

Nisrine Mokadem, Fakhra Jabeen, Jan Treur, H. Rob Taal, and Peter H. M. P. Roelofsma

Abstract This chapter presents the use of second-order adaptive network models of hospital teams consisting of doctors and nurses, interacting together. A variety of scenarios are modelled and simulated, in relation with respiratory distress of a neonate, along with the integration of an AI-Coach for monitoring and support of such teams and of organizational learning. The research highlights the benefits of introducing a virtual AI-Coach in a hospital setting. The practical application setting revolves around a medical team responsible for managing neonates with respiratory distress. In this setting an AI-Coach act as an additional team member, to ensure correct execution of medical procedure. Through simulation experiments, the adaptive network models demonstrate that the AI-Coach not only aids in maintaining correct medical procedure execution but also facilitates organizational learning, leading to significant improvements in procedure adherence and error reduction during neonatal care.

N. Mokadem · J. Treur (✉)
Social AI Group, Department of Computer Science, Vrije Universiteit Amsterdam, Amsterdam, The Netherlands
e-mail: j.treur@vu.nl; j.treur@hhs.nl

N. Mokadem
e-mail: n.mokadem@student.vu.nl

F. Jabeen
Safety and Security Science Section, Delft University of Technology, Delft, The Netherlands
e-mail: fakhra.jabeen@ru.nl

J. Treur · P. H. M. P. Roelofsma
Risk Management and Cybersecurity Group, Centre of Expertise Cybersecurity, The Hague University of Applied Sciences, The Hague, The Netherlands
e-mail: p.h.m.p.roelofsma@hhs.nl

H. Rob Taal · P. H. M. P. Roelofsma
Department of Neonatal and Pediatric Intensive Care, Division Neonatology, Erasmus University Medical Center, Rotterdam, The Netherlands
e-mail: h.taal@erasmusmc.nl

P. H. M. P. Roelofsma et al. (eds.), *Using Shared Mental Models and Organisational Learning to Support Safety and Security Through Cyberspace: A Computational Analysis Approach*, Studies in Systems, Decision and Control 570,
https://doi.org/10.1007/978-3-031-72075-8_5

Keywords Adaptive network · AI-Coach · Respiratory distress · Organizational learning · Hospital teams

1 Introduction

Organizations face the ongoing challenge of continual development and learning not only at personal but also at the team level. They encompass strategic management objectives which aim at creating better working conditions and structures that enhance employees' willingness and ability to learn, thereby improving organizational processes of adjustment and adaptation. Safety culture in healthcare needs perception, competency, and attitude of an individual and whole organization to take responsibility of their actions. However, actions without adequate training or the implementation of new processes may introduce the possibility of malfunction or errors and necessitate adjustments and modifications (Keith and Frese 2011).

One approach that holds promise in this regard is 'organizational learning'. Organizational learning encompasses the process by which an organization acquires, interprets, and applies knowledge to improve its performance and adapt to changing circumstances. It involves capturing lessons learned, sharing information, fostering a learning culture, and implementing effective feedback mechanisms within the organization (Saadat and Saadat 2016). By leveraging the principles of organizational learning, teams can promote their performance by enhancing their shared mental models. Through continuous learning, teams can identify and rectify deficiencies in their mental models, fostering a shared understanding of the tasks, roles, and interdependencies (Jonker et al. 2011).

This paper show cases the application of the adaptive network-oriented modelling approach from Treur (2020), coupled with the cognitive architecture for usage, adaptation and control of mental models described in Van Ments and Treur (2021), Treur and Van Ments (2022), see also Van Ments et al. (2025), this volume. This study primarily aims to develop and analyze adaptive network models, homing in on the decision-making and mental processes of medical staff in neonatal respiratory distress scenarios. This is accomplished by modeling the mental processes of the medical staff on the floor, which are second-order adaptive and incorporating an AI-Coach to monitor and support these processes.

The integration of an AI-Coach in healthcare is a crucial advancement with significant implications, contributing to the wider context of AI support in the health area, e.g., Abdi et al. (2021), Catania (2021), Jayachitra et al. (2023), Kavitha et al. (2023), Sitterding et al. (2019). It marks a critical step in improving patient care quality and safety, especially in critical environments like neonatal intensive care units. By enhancing medical team decision-making and reducing errors, an AI-Coach not only betters patient outcomes but also aids healthcare professionals in their vital roles. This research offers valuable insights into integrating technology in healthcare, emphasizing patient safety and staff efficiency.

Specifically, it illustrates how an AI-Coach can enhance the knowledge within the mental processes of medical team members. Uniquely, this study applies an adaptive network-oriented model to the intricate context of neonatal respiratory distress. It allows for a comprehensive analysis of how an AI-coach can effectively drive organizational learning in such specialized scenarios. Through simulation experiments, the models illustrate the dual role of the AI-Coach in both ensuring accurate medical procedure execution and fostering organizational learning. This facilitation by the AI-Coach leads to notable enhancements in adherence to procedures and a reduction of errors within neonatal care, underpinning the importance of technological support in critical healthcare settings. In Sect. 2, a comprehensive introduction and background are provided, encompassing topics such as respiratory distress, mental models, and organizational learning. Section 3 outlines the general approach used in designing the adaptive network models. The subsequent section, Sect. 4, presents the various scenarios to be examined in the study. Section 5 details the second-order models for each scenario, elucidating their design principles. Thereafter, Sect. 6 presents the simulations conducted, and the paper concludes with a conclusion and discussion in Sect. 7.

2 Background Literature

The following section provides a comprehensive overview of the background literature. It begins by elucidating the concept of respiratory distress in neonates, including its clinical manifestations and observable signs. Additionally, we delve into the significance of mental models, shared mental models, and organizational learning in this context.

2.1 Respiratory Distress in Neonates

Section 2.1 presents the critical aspects of neonatal respiratory distress, including its prevalence, causes, clinical signs, and diagnostic approaches. This overview establishes a context for the advanced interventions explored in the following sections.

2.1.1 Introduction to Respiratory Distress in Neonates

Respiratory distress in neonates refers to a condition where newborn experience difficulty in breathing or inadequate oxygenation. It is a significant concern in clinical practice as it can be a marker of underlying respiratory disorders or other medical conditions. Prompt recognition and appropriate management are crucial to prevent complications and ensure optimal outcomes for neonates. Common respiratory disorders observed include Respiratory Distress Syndrome (RDS), Transient Tachypnea of the Newborn (TTN), and Meconium Aspiration Syndrome (MAS) (Swarnkar and Swarnkar 2015).

2.1.2 Understanding the Varied Prevalence and Factors Influencing Respiratory Distress in Neonates

The prevalence of respiratory distress among neonates can vary depending on various factors such as gestational age, birth weight, and underlying medical conditions such as sepsis and congenital malformations. Premature infants, in particular, are more susceptible to respiratory distress due to prematurity. Studies have shown that respiratory distress syndrome affects approximately 50–60% of infants born before 28 weeks of gestation, while the prevalence decreases with increasing gestational age (Sweet et al. 2017). As mentioned before, specific respiratory disorders commonly observed in neonates include RDS, TTN, and MAS. RDS is one of the most prevalent respiratory disorders among premature infants, especially those born before 32 weeks of gestation (Saboute et al. 2015). TTN typically affects near-term and full-term infants, with an incidence of around 1–2% of all deliveries. MAS, although less common, can occur when the baby inhaled meconium-stained amniotic fluid during or before birth.

2.1.3 Recognizing and Assessing Clinical Manifestations of Respiratory Distress in Neonates: Diagnostic Approaches and Indicators

Clinical manifestations of respiratory distress in neonates can vary depending on the underlying cause and severity of the condition. Common signs include tachypnea, which is characterized by rapid breathing with an increased respiratory rate above the normal range for the infant's gestational age. Retractions, observed as inward movement of the chest wall during inspiration, indicate increased work of breathing. Nasal flaring, where the nostrils dilate during inspiration, is another common manifestation. Grunting, a sound produced by the infant during expiration, signifies an attempt to maintain lung volume and improve oxygenation. Cyanosis, a bluish discoloration of the skin, lips, or mucous membranes, may also occur due to inadequate oxygenation (Machogu and Gaston 2021).

To assess and confirm respiratory distress in neonates, healthcare professionals employ diagnostic approaches such as physical examination, pulse oximetry, chest X-rays, blood gas analysis, and other investigations (Bass et al. 2015). Bed-side lung ultrasound is another diagnostic technique that is now increasingly used. Physical examination involves a thorough assessment of the infant's respiratory effort, chest movement, breath sounds, and signs of distress. Pulse oximetry, a non-invasive method, measures the oxygen saturation level in the infant's blood, providing information about oxygenation adequacy. Chest X-rays are used to identify abnormalities like lung consolidations, fluid accumulation, or signs of RDS (Copetti et al. 2008). Blood gas analysis helps evaluate the acid–base status, oxygenation, and ventilation in neonates with respiratory distress. Additionally, SpO2 (oxygen saturation) and FiO2 (fraction of inspired oxygen) are evaluated to examine the oxygenation. Depending on the clinical presentation and suspected underlying cause, additional tests such as blood tests, respiratory viral panels, or echocardiography may be conducted.

2.2 *Mental Models and Associated Mental Processes*

Understanding the complexity of the brain poses a significant challenge due to the presence of cyclic patterns, recursion, and the absence of linear causality. As aptly stated by Scherer (2009), the conventional notion of linear causality cannot be applied to these mechanisms, primarily due to their constant recursion. This feature aligns with the characteristics typically found in self-organizing systems, where a simple and unidirectional sense of causality is absent. Also Nigg (2017, 2023) emphasizes that often several kinds of self-regulation play an important role in mental processes, which are typically cyclic processes without linear causality.

These phenomena observed within the brain share similarities with network structures, further supporting the idea that networks can serve as a suitable framework for describing such complex mental processes. Within their complex mental processes, humans often make use of mental models (Craik 1943). According to Kim (1997), mental models serve as representations of an individual's perspective on the world. They provide the necessary context for understanding and interpreting new information. In essence, mental models act as cognitive frameworks through which individuals make sense of their experiences. Similarly, Jonker et al. (2011) describe mental models as internal representations that enable humans to interact with the world. These representations, known as mental models, consist of knowledge about physical systems and their behaviors. Mental models encompass an understanding of the system's structure, overall behavior, and the impact of disturbances on the system. By possessing these mental models, individuals can effectively engage with and navigate the systems they encounter. For more information and reviews about mental models, see Van Ments et al. (2021), Van Ments and Treur (2021), Treur and Van Ments (2022) or Van Ments et al. (2025), this volume.

By combining these perspectives, it becomes clear that mental models play a crucial role in shaping our understanding of the world. They serve as internal constructs that enable individuals to navigate their surroundings and make sense of new information based on pre-existing knowledge and cognitive frameworks. To get the idea, mental models are like maps in our minds, connecting ideas like streets on a city map. They guide us in understanding new information, much like using a map to navigate a new place and to make informed decisions based on this information.

Mental models have proven valuable not only in elucidating human interactions with complex physical systems that require comprehension and control but also in the realm of teamwork. In the study conducted by Cannon-Bowers et al. (1993), the concept of a shared mental model is defined as knowledge structures collectively held by team members, see also further work in Stout et al. (2017). These structures enable accurate explanations and expectations regarding the task, facilitating coordinated actions and adaptive behavior in response to task demands and the presence of other team members. By this, a shared mental model adds valuable insights to the understanding of the tasks described by it and the impact on expert team decision making.

2.3 *Organizational Learning*

Importance of seeking consultation and developing effective procedures through adaption is highly emphasized to effectively respond to the organizational challenges in order to optimize their performance (Wilden et al. 2018).

Organizational learning encompasses a multidisciplinary approach, drawing upon various fields such as information processing, organizational culture, and strategy implementation, resulting in a diverse range of conceptualizations (Crossan et al. 1999, Wiewiora et al. 2019, 2020). According to Garcia Moralez et al. (2012), organizational learning refers to an organization's capacity to preserve and enhance its performance through the utilization of previous experiences. This involves transitioning from tacit and explicit knowledge to the sharing and application of knowledge within the organization. For example, in a healthcare setting, when a hospital faces a surge in patient admissions during flu season, it learns from past experiences to adjust staffing and resources efficiently. In a similar vein, Saadat et al. defines organizational learning as an ongoing process of identifying errors and mistakes and subsequently resolving and rectifying them. This continuous process is facilitated by acquiring knowledge and improving performance over time (Hosseini et al. 2023; Kucharska and Bedford 2020; Kucharska and Rebelo 2022; Saadat and Saadat 2016), see also Hosseini et al. (2025), this volume. Consider an example case where a healthcare facility identifies a recurring issue in patient data entry errors, leading to billing discrepancies. Through organizational learning, the facility not only addresses the immediate errors but also implements training programs and improved data entry protocols to prevent future mistakes. Altman and Iles (1998) identified four significant theoretical advancements that contribute to the formation of the concept of organizational learning:

1. Strategic Management: This theoretical progression focuses on the internal processes and procedures within an organization. It recognizes the importance of aligning organizational goals and strategies with the external environment. Strategic management plays a vital role in shaping the understanding and implementation of organizational learning.

2. Systems Theory: This framework has played a crucial role in the development of the concept of organizational learning. Systems theory provides a holistic perspective that examines the interconnections and interdependencies within an organization. It helps to conceptualize how learning progress over the time, and how it influences the overall functioning and effectiveness of the organization.
3. Social Learning: Social learning theory encompasses a set of concepts that delve into learning at various levels within the organization. It emphasizes the role of interactions, social networks, and knowledge sharing in the learning process. Social learning theory recognizes that learning is not limited to individuals but also flourish through social interactions and collective experiences among a team.
4. Theoretical Approaches: This theoretical progression pays attention to organizational backgrounds, considering factors such as organizational culture and structure. It provides a fourth theoretical foundation for understanding organizational learning. By examining the organizational context, this approach sheds light on how factors like culture, structure, and processes influence the learning capabilities of the organization.

For the integration of these different perspectives on organizational learning mental models can be used as a crucial factor. Individual learning of mental models serves as the foundation for the formation of shared mental models within the organization, thereby facilitating organizational learning (Canbaloğlu et al. 2023a, b), see also Canbaloğlu et al. (2025), this volume. By embracing and actively cultivating these mental models, organizations can develop a mechanism for continuous learning and improvement at all levels.

3 The Network-Oriented Modeling Approach

3.1 Network-Oriented Modelling: Capturing Dynamics and Interconnected Causal Processes in Complex Systems

In the field of modelling complex systems, a technique has emerged known as network-oriented modelling. This approach captures the intricate interplay of interconnected causal processes with temporal nature. Building upon the fundamental understanding of the direct relationship between dynamics and causal relations, network-oriented modelling integrates the dynamic aspects into the network structure which leads to a more comprehensive and accurate representation of the underlying phenomena (Treur 2016). Thus, one key aspect of network-oriented modelling is its ability to model dynamic, temporal-causal networks. These networks operate on a continuous time dimension, enabling the modelling of cyclic causal networks and timing of causal effects. Consequently, this approach opens up possibilities for modelling a wide range of networks. In this modelling framework, the nodes within the networks represent state variables, while the connections between them represent causal relations that exert influence. This interconnectedness allows for a holistic depiction of the system's behavior, capturing the intricate cause-and-effect relationships between different components (Treur 2020).

3.2 Modelling Mental Processes and Adaptive Networks

A temporal-causal network model includes various characteristics represented by causal states connected to each other. For example, states X and Y, are interconnected through weighted connections $\boldsymbol{\omega}_{X,Y}$ in a model. These connections enable the transmission of impacts $\boldsymbol{\omega}_{X,Y}X(t)$ from state X to Y. To aggregate the impact of all states connected to Y, a combination function $\mathbf{c}_Y(\ldots)$ is applied to state Y, determining how the impacts from incoming connections from states X to Y are aggregated. Table 1 provides an overview of the specific combination functions utilized in this chapter. Furthermore, each state Y possesses a speed factor $\boldsymbol{\eta}_Y$, which governs the rate at which it undergoes change in response to a causal impact. Together, these characteristics—connectivity, aggregation, and timing—form the foundation of the temporal-causal network model. The difference equation used to compute the change of all states from time t to a later time $t + \Delta t$ is:

$$Y(t + \Delta t) = Y(t) + \boldsymbol{\eta}_Y\left[\mathbf{c}_Y\left(\boldsymbol{\omega}_{X_1,Y}X_1(t), \ldots, \boldsymbol{\omega}_{X_k,Y}X_k(t)\right) - Y(t)\right]\Delta t \quad (1)$$

In Eq. (1), at the current time t, we have a state value $Y(t)$. This represents the current value of state Y. The equation determines how this state $Y(t)$ will change in the future, specifically at time $t + \Delta t$. To calculate this change, the concept of a speed factor is introduced, denoted as $\boldsymbol{\eta}_Y$, which characterizes how rapidly state Y responds to impacts from its environment. A higher $\boldsymbol{\eta}_Y$ implies quicker change. In addition to the speed factor, the influences from other states are considered, denoted as X_1 to X_k, which are interconnected toward state Y. These influences are modulated by weighted connections with weights $\boldsymbol{\omega}_{X_1,Y}$ to $\boldsymbol{\omega}_{X_k,Y}$. These weights modulate the strength of the impact that each state X has on Y by multiplying each state value $X_i(t)$ by $\boldsymbol{\omega}_{X_i,Y}$, thus obtaining impact $\boldsymbol{\omega}_{X_i,Y}X_i(t)$. Importantly, the current state values of these influencing states at time t, which are X_1 (t) and X_k (t), need to be known at this point. These states represent where X_1 and X_k are at the present moment. The next step involves a combination function $\mathbf{c}_Y(\ldots)$, which takes into account all the incoming influences from states X_1 to X_k. This function aggregates the impacts $\boldsymbol{\omega}_{X_i,Y}X_i(t)$ and determines how they collectively affect the change in state Y. Finally, the aggregated impact is multiplied by a small time interval Δt and incorporates this to obtain the future state value $Y(t + \Delta t)$ following state $Y(t)$: this calculation results in the prediction of this future state value $Y(t + \Delta t)$. These calculations were computed by the network engine implemented within the dedicated software environment in MATLAB and described in detail in Treur (2020), Chap. 9.

Table 1 Combination functions used in the network models

Function	Notation	Formula	Parameters
Alogistic–advanced logistic sum	$\mathbf{alogistic}_{\sigma,\tau}(V_1, \ldots, V_k)$	$\left[\frac{1}{1+e^{-\sigma(V_1+\cdots+V_k-\tau)}} - \frac{1}{1+e^{\sigma\tau}}\right](1 + e^{-\sigma\tau})$	steepness σ threshold τ
Steponce–one period of activation	$\mathbf{steponce}_{\alpha,\beta}(V_{1,\ldots,}V_k)$	time t $1 \text{if} \alpha \leq t \leq \beta \text{else} 0$	start time α end time β
Stepmod–step modulo	$\mathbf{stepmod}_{\rho,\delta}(V_1, \ldots, V_k)$	time$t$ $0 \text{ if } \bmod (t, \rho) < \delta \text{ else } 1$	repeated time duration ρ tipping point δ
Scalemap–mapping activation scale	$\mathbf{scalemap}_{\lambda,\upsilon}(V)$	$\lambda + (\upsilon - \lambda)V$	lower bound λ upper bound υ
Max with Hebbian Learning	$\mathbf{maxhebb}_{\mu}(V_1, \ldots, V_k)$	$\max(\mathbf{hebb}_{\mu}(V_1, V_2, V_3), V_4, \ldots, V_k)$ Where $\mathbf{hebb}_{\mu}(V_1, V_2, V_3) = V_1V_2(1 - V_3) + \mu V_3$	persistence factor μ

In summary, this difference equation serves as a valuable tool for both simulation and analysis within temporal-causal networks as it captures the dynamics and behaviors exhibited by the interconnected states within the network.

A self-modelling network, also known as a reified network, offers a powerful framework for conceptualizing adaptive networks using mathematically defined functions and relations. By introducing self-model states, it becomes possible to represent and capture the adaptive network characteristics of a network model. These self-model states are visualized at a higher level referred to as the self-model level or reification level, while the original network structure remains at the base level. For example, the weight $\boldsymbol{\omega}_{X,Y}$ of a connection between two states, X and Y, can be represented at the self-model level by introducing a self-model state named $\mathbf{W}_{X,Y}$. By changing the activation values of such self-model states over time, the related network characteristic becomes adaptive. Similarly, other network characteristics can also be made adaptive by incorporating corresponding self-model states. The process of network reification not only enhances the representation of adaptive networks but also generates a temporal-causal network model itself. This enables the iterative application of self-modelling network construction, resulting in the creation of multiple orders of self-models at different self-model levels. Each subsequent level builds upon the previous one, further enriching the modelling capabilities and capturing the complex dynamics and adaptability within network systems (Treur 2020). This allows modeling of higher-order adaptivity.

In Treur (2021), Hendrikse et al. (2023), it is shown that any smooth dynamical system has a canonical representation as a temporal-causal network and any smooth (multi-order) adaptive dynamical system has a canonical representation as a (multi-order) self-modeling temporal-causal network. Therefore, compared to adaptive dynamical systems in general, the network-oriented modeling approach used here does not introduce any fundamental limitations concerning what it can model. This has also been confirmed by applications to case studies in practice, e.g., for mental models in Treur and Van Ments (2022) and for organizational learning in Canbaloğlu et al. (2023a, b).

To illustrate self-modeling networks, consider their application in healthcare, specifically in neonatal care. In a neonatal intensive care unit, optimizing oxygen delivery to premature infants facing respiratory distress is critical. A self-modeling network can be used to achieve this. At the base level, the network represents components like sensors, oxygen sources, and monitors, with connection weights based on protocols. At the self-model level, we introduce self-model states for connection weights like

$$\mathbf{W}_{\text{Sensor, ControlUnit}}$$

which adapt connection behaviors. By adjusting the values of these self-model states, the control system can autonomously optimize oxygen delivery in real-time based on individual patient needs. This adaptivity enhances healthcare systems like the neonatal intensive care unit, ensuring optimal care for infants.

4 Description of the Scenarios

In this paper, three distinct scenarios for simulation experiments will be presented. While each scenario presents a unique storyline, the central focus remains on the collaborative efforts of a doctor and a nurse in a hospital setting. Together, they form a cohesive team, utilizing their specialized knowledge, skills, and dedication to ensure the best possible care for neonates facing respiratory distress. The scenarios are designed to provide a narrative that illustrates the sequence of medical procedures in neonatal care, offering a storyline that brings the context to life. It is important to note that these scenarios are merely illustrative and serve as representations rather than being definitive.

4.1 The Roles and Responsibilities of the Nurse and Doctor in Managing Respiratory Distress: Scenario 1

Scenario 1 has two subvariants, namely 1a and 1b, which intricately illustrate the separate roles and responsibilities assumed by both the nurse and the doctor. This division is critical in delineating and emphasizing the distinct responsibilities involved in providing care for neonates experiencing respiratory distress in a hospital setting. In both scenarios 1a and 1b, the team members share a perfect team mental model. As a result, there are no differing individual instances of the team mental model, as all team members have the same understanding of their roles, responsibilities, and goals.

4.1.1 Nurse-Led Assessment and Intervention: Scenario 1a

A premature baby of 31 weeks of gestational age at birth is delivered with the assistance of a doctor. After birth there was a delay in the clearance of the fetal lung fluid. The nurse assumes responsibility for assessing the baby's respiratory vital signs. The nurse's duties include:

- monitor respiratory rate
- measuring oxygen saturation level
- observing for grunting
- observing for chest retractions
- observing for cyanosis

Upon completing the assessment and obtaining good results, the nurse permits the baby to bond with their mother by kangaroo care which involves skin-to-skin contact. However, if the assessment yields poor results, the nurse will contact the doctor for advice. In such cases, the nurse may provide:

- supplemental oxygen;

- respiratory support;
- transfers neonate to a NICU (a premature baby is always transferred to a NICU).

4.1.2 Nurse-Led Intervention with Subsequent Doctor's Intervention: Scenario 1b

In the previous Scenario 1a, the nurse assesses the baby's respiratory vital signs and seeks the doctor's advice if respiratory distress is observed. In this case, the nurse administers supplemental oxygen without consulting the doctor. When the baby does not improve and enters a critical state a few hours later, the nurse contacts the doctor for immediate intervention. This makes the family of the baby emotionally distressed. The doctor's duties include:

- prepare medical procedure
- take X-ray and other evaluations
- providing treatment
- give advice.

4.2 Collaborative Adaptation and Learning in Nurse-Led Intervention and Impact of a Fatigued Doctor: Scenario 2

Scenario 2 is a variant of Scenario 1. This implies that Scenario 2 exhibits certain similarities with Scenario 1, while also introducing notable variations.

The nurse independently administers supplemental oxygen without consulting the doctor. The baby's condition does not improve and enters a critical state. Subsequently, the nurse seeks help, and the doctor's responsibilities encompass preparing medical procedures, conducting evaluations such as X-rays, providing suitable treatment, and offering guidance and advice to address the critical situation. However, in this scenario, the doctor experiences symptoms of fatigue, affecting his ability to perform optimally. Moreover, the team members do not possess a flawless team mental model. Consequently, individual instances of the team mental model differ, as team members hold varying understandings of their roles, responsibilities, and objectives. This situation fosters a learning environment for both the doctor and nurse. They observe real-world events, which impact their mental models, leading to an exchange of knowledge and an evolution of their understanding.

4.3 Collaborative Adaptation and Learning in Nurse-Led Intervention with Subsequent Doctor's Intervention and the Role of an AI-Coach: Scenario 3

Scenario 3 is a variant of Scenario 2, introducing additional factors such as workplace stress and the presence of an underlying congenital or genetic condition in the baby. Despite the doctor's illness and the critical state of the baby, the doctor's responsibilities remain unchanged.

A key difference is the introduction of a virtual AI-Coach in Scenario 3. This AI-Coach serves as a monitoring system during the medical procedure, signalling any deviations from the expected course. With its comprehensive mental model and vast knowledge, the AI-Coach provides guidance and support to both the doctor and the nurse. It continuously monitors the situation and activates its decision and communication features in the event of a mistake occurring in the real-world setting.

5 The Adaptive Network Models

Every scenario is addressed its own network model, which invariably comprises a base level and for Scenarios 2 and 3 additional levels of reification. In the 3D format used to depict these models, the oval shapes represent the states, while the arrows signify the influences of one state on another. Each section will provide a comprehensive explanation of these 3D pictures. The detailed exploration of these models is critical to achieving the research objectives, as it allows for the simulation and analysis of the dynamic interactions within medical teams and the impact of an AI-Coach on decision-making and learning processes in a neonatal care context.

5.1 The Roles and Responsibilities of the Nurse and Doctor in Managing Respiratory Distress: The Computational Models for Scenario 1

Scenario 1a and 1b solely comprise a base level as there is no requirement for additional levels of reification. This is due to the absence of any learning needs in both scenarios, as all team members possess flawless mental models, signifying perfect knowledge. Given the numerous similarities between the two models, it is justified to explain them together within a single section.

5.1.1 Base Level Scenario 1a and 1b: Overview

Within the base planes, the world states indicating the actual steps in the world for the Scenarios 1a and 1b are depicted in Figs. 1 and 2 respectively by the green nodes with their connections in the middle area of the base plane. A few of these world states have a lighter green node because they are not directly influenced by the other world states. The actor is indicated within a world state name by D for doctor or N for nurse.

In both cases the base levels consist of two mental models, one of the nurse and one for the doctor. These mental models reflect the structure of the medical procedure explained in Sects. 4.1.1 and 4.1.2. The nodes of the mental model of the doctor are depicted by the orange ovals and the nodes of the nurse by pink ovals.

In order for the doctor and the nurse to determine their actions in the world, they rely on their individual mental models which in turn activate action ownership states. The doctor's action ownership states are visually depicted in a light orange colour, while the nurse's states are represented in light pink. These states incorporate input from the respective mental models and enable a form of mediation from the mental model to the real world by initiating the execution of the designated actions. Therefore, these states act as decision states to perform certain action by both of the actors in the real world, resulting from a shared mental model. By mutually observing each others' actions or by accompanying verbal comments about them, the nurse and doctor interact.

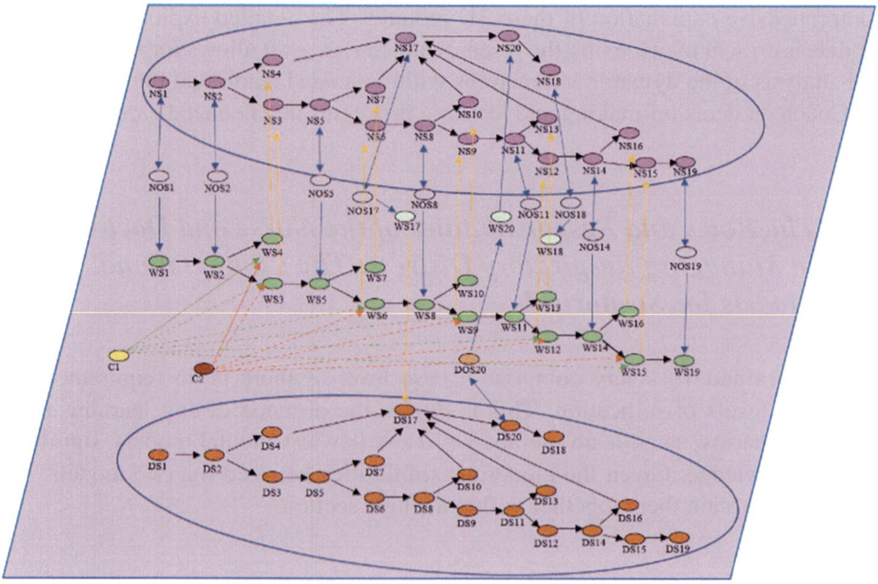

Fig. 1 Computational model for Scenario 1a

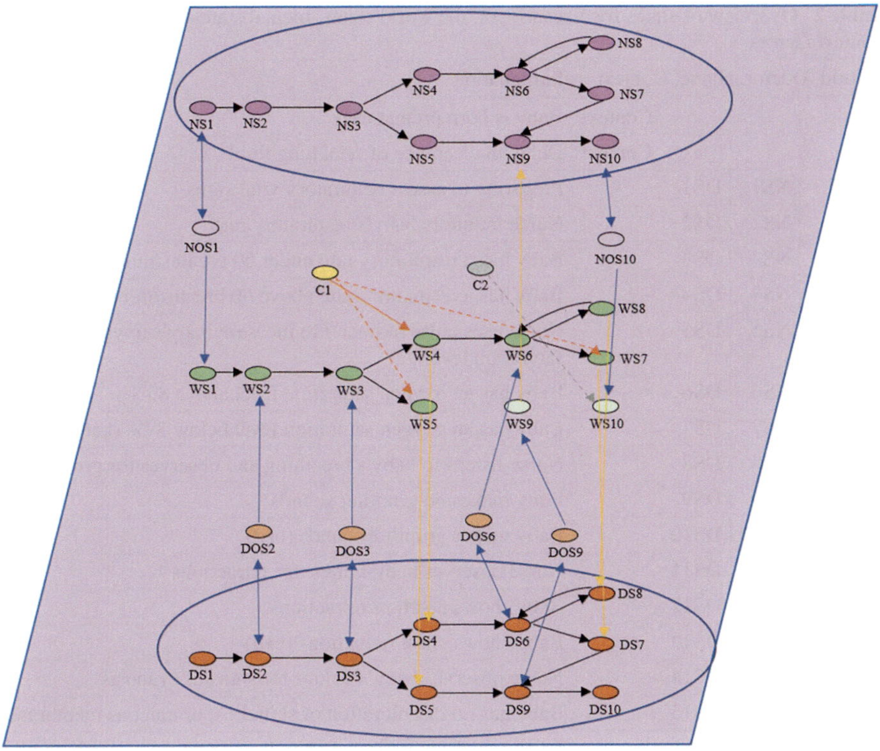

Fig. 2 Computational model for Scenario 1b

In Fig. 1, two contextual factors are depicted as nodes, with one in yellow and the other in dark orange. The yellow node signifies that the baby is born prematurely, while the dark orange node indicates that there was a delay in the clearance of fetal lung fluid. Within Fig. 2, the two contextual factors can be observed through a yellow node, symbolizing the critical state of the baby, and a grey shade, indicating emotional distress experienced by the family.

In the case of the model for Scenario 1a, Table 2 provides an overview encompassing the world states, the mental model states for both the doctor and nurse, their ownership states, and contextual factors, each accompanied by an explanation. A similar overview of the model of Scenario 1b can be found in Table 3.

Black arrows are used to denote the influence of one world state on another. Blue lines indicate the connection between ownership states and world states. The yellow lines represent observation arrows, indicating that the nurse and doctor observe events occurring in the world and activate the corresponding states in their mental models. Observation lines are selectively included from particular world states to the mental states only when those specific world states indicate observations, such as "Baby has a respiratory rate under 60 breaths per minute". Solid arrows originating from the

Table 2 Overview of states for Scenario 1a: the world states, mental states, ownership states and context factors

World, Doctor, Nurse, Context				Explanation
C1			Context	Baby is born prematurely
C2			Context	Delay in clearance of fetal lung fluid
WS1	NS1	DS1		Preparing to assess respiratory vital signs
WS2	NS2	DS2		Nurse monitors baby's respiratory rate
WS3	NS3	DS3		Baby has a respiratory rate under 60 breaths/minute
WS4	NS4	DS4		Baby has a respiratory rate above 60 breaths/minute (bad)
WS5	NS5	DS5		Nurse uses pulse oximeter to measure baby's oxygen saturation level
WS6	NS6	DS6		Baby has an oxygen saturation level above 89%
WS7	NS7	DS7		Baby has an oxygen saturation level below 89% (bad)
WS8	NS8	DS8		Nurse listens to baby's breathing and observes for grunting
WS9	NS9	DS9		Baby makes no grunting sounds
WS10	NS10	DS10		Baby makes grunting sounds (bad)
WS11	NS11	DS11		Nurse observes baby's chest for retractions
WS12	NS12	DS12		Baby shows no chest retractions
WS13	NS13	DS13		Baby shows chest retractions (bad)
WS14	NS14	DS14		Nurse observes baby's colour for signs of cyanosis
WS15	NS15	DS15		Baby has no discoloration of skin, lips, or mucous membranes
WS16	NS16	DS16		Baby has discoloration of skin, lips, or mucous membranes (bad)
WS17	NS17	DS17		Nurse calls for doctor's help
WS18	NS18	DS18		Nurse provides supplemental oxygen, respiratory support, or transfers the baby to a NICU
WS19	NS19	DS19		Nurse allows the baby to bond with the mother
WS20	NS20	DS20		Doctor gives advice
NOS1				Ownership state for the action of preparing to assess respiratory vital signs
NOS2				Ownership state for the action of nurse monitors baby's respiratory rate
NOS5				Ownership state for the action of nurse uses pulse oximeter to measure baby's oxygen saturation level
NOS8				Ownership state for the action of nurse listens to baby's breathing and observes for grunting
NOS11				Ownership state for the action of nurse observes baby's chest for retractions
NOS14				Ownership state for the action of nurse observes baby's colour for signs of cyanosis
NOS17				Ownership state for the action nurse calls for doctor's help

(continued)

Table 2 (continued)

World, Doctor, Nurse, Context				Explanation
NOS18				Ownership state for the action of nurse provides supplemental oxygen, respiratory support, or transfers the baby to a NICU
NOS19				Ownership state for the action of nurse allows the baby to bond with the mother
DOS20				Ownership state for the action of doctor gives advice

Table 3 Overview of states for Scenario 1b: the world states, mental states, ownership states and context factors

World, Doctor, Nurse, Context				Explanation
C1			Context	Baby is in critical state
C2			Context	Family is emotionally distressed
WS1	NS1	DS1		Nurse alarms doctor
WS2	NS2	DS2		Doctor preparing necessary equipment and supplies for a medical procedure
WS3	NS3	DS3		Doctor takes an X-ray and other medical evaluations
WS4	NS4	DS4		X-ray and other evaluations show lung expansion, airway obstruction, pneumothorax, infiltrates, or/and cardiomegaly
WS5	NS5	DS5		X-ray and other evaluations do not show anything abnormal
WS6	NS6	DS6		Doctor provides oxygen therapy, medications, re-expands lung, surgery or other treatments
WS7	NS7	DS7		Treatment works
WS8	NS8	DS8		Treatment does not work
WS9	NS9	DS9		Doctor advises to let neonate rest
WS10	NS10	DS10		Nurse informs family about doctor's advice
NOS1				Ownership state for the action of nurse alarms doctor
NOS10				Ownership state for the action of nurse informs family about doctor's advice
DOS2				Ownership state for the action of doctor preparing necessary equipment and supplies for a medical procedure
DOS3				Ownership state for the action of doctor takes an X-ray and other medical evaluations
DOS6				Ownership state for the action of doctor provides oxygen therapy, medications, re-expands lung, surgery or other treatments
DOS9				Ownership state for the action of doctor advises to let neonate rest

context factors signify the strengthening of specific world states, while dotted arrows indicate the weakening of world states.

The **alogistic** combination function stands for advanced logistic sum and includes a steepness and threshold parameter. The steepness parameter in the **alogistic** function determines how quickly the output changes in response to changes in the (weighted) sum of inputs. The threshold parameter in the **alogistic** function defines the input value at which the output starts to change significantly. Inputs below the threshold have minimal impact on the output, while inputs above the threshold cause the output to respond more significantly. The threshold essentially sets a boundary or activation point that triggers the transition in the output. On the other hand, the **steponce** combination function operates by encompassing a specific period of activation. It takes one period of activation from **α** to **β**, in which **α** stands for the start time and **β** for the end time. The combination functions utilized in the network model are the alogistic function for all the world states, mental model states, and ownership states. For the context factors, the **stepmod** function is used (see Table 1). This function takes parameters for a repeated time duration and a tipping point. The tipping point represents the critical point at which the output of the stepmod function changes its value taking into account the repeated time duration specified by its parameter.

5.2 *Collaborative Adaptation and Learning in Nurse-Led Intervention and Impact of a Fatigued Doctor: The Computational Model for Scenario 2*

Scenario 2 diverges from the foundation established in Scenario 1, presenting itself as an extensive variant rather than a direct extension. The model for Scenario 2 introduces two adaptation levels (for first-order and second-order adaptation) on top of the base level. Each of these levels will be addressed in a separate section, providing a comprehensive explanation of their intricacies and details. In Scenario 2, the central concept revolves around different periods. The upcoming section will delve into the role of breaks and how they come into play. This is where the factors pertaining to periods and contextual states will be thoroughly explained to provide a more comprehensive understanding.

5.2.1 Base Level for Scenario 2: Overview

Once more, the world states are depicted by green nodes with their connections in the middle area of the base plane (see Fig. 3). As usual, the lighter green nodes indicate that these nodes are not directly influenced by the other world states. In this case, there are again two mental models for the two actors, and these reflect the structure of the medical procedure explained in Sect. 4.1.3. The nodes of the mental model of the doctor are depicted by the orange nodes and the nodes of the nurse by pink nodes.

Also in this case, the doctor and nurse have certain actions that they have to do in the world. To determine at specific time points their actions, they again rely on their individual mental models and use action ownership states to initiate actions accordingly. The doctor's action ownership states are visually depicted in a light orange colour, while the nurse's states are represented in light pink.

In the model for Scenario 2, in total three contextual factors are at play. The first one, represented by the colour yellow, signifies that the baby is in a critical condition, causing both the nurse and doctor to experience stress. The second factor, depicted as a brown–red node, indicates that the doctor is personally unwell. As shown in Fig. 3, this particular context factor strengthens WS6. While it may initially seem

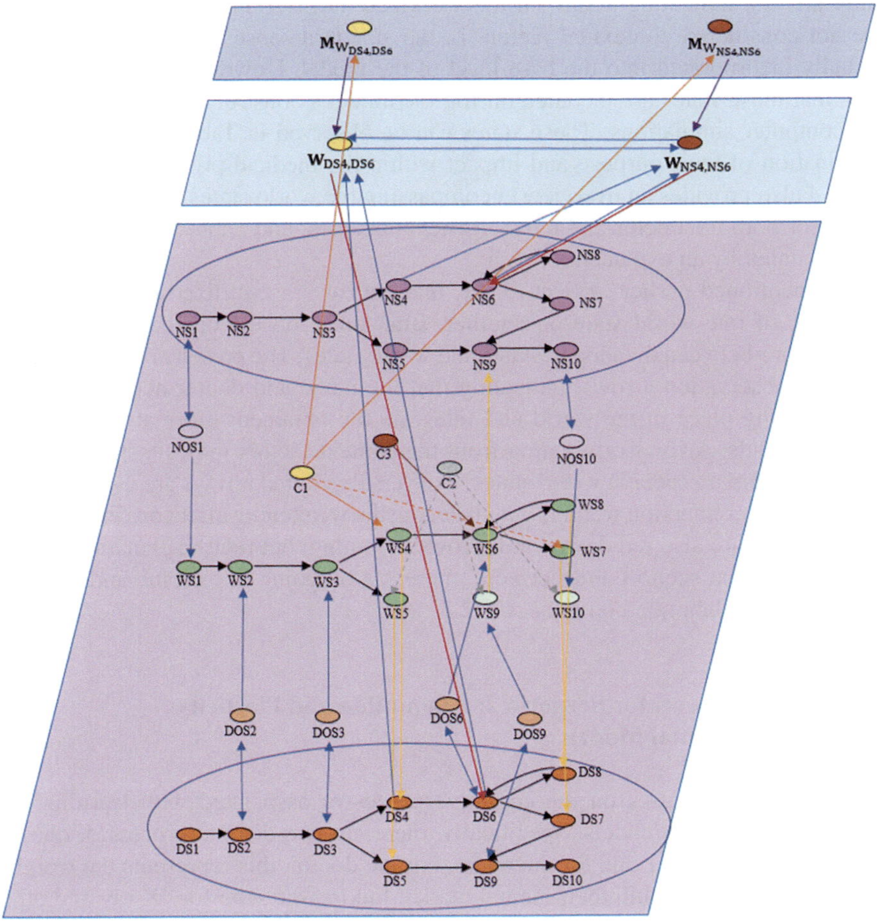

Fig. 3 Computational model for Scenario 2

illogical for it to strengthen the action of "Doctor gives treatment," there is a logical explanation behind it. Despite being sick, the doctor is making an effort to stay focused and provide the necessary treatment. Lastly, the third factor (represented by grey node) represents the premature birth of the baby, which diminishes the likelihood of favourable results in X-ray and other evaluations. Consequently, this also slightly weakens the doctor's ability to provide advice and the nurse's ability to communicate that advice to the family.

In addition to the three context factors mentioned earlier, there are three more states that are not included in the base level of the model. These states are part of the medical procedure and serve to represent different periods within the care process. They are intentionally designed to create pauses or breaks in the workflow. For instance, there is a state that weakens every world state during a specific time interval when there is no baby requiring immediate attention. Conversely, after a certain period, a new baby arrives, indicating a shift in the workload and responsibilities. These states are not considered contextual factors in the standard sense and therefore are not visually incorporated into the base level of the model. However, it is important to note that these states are included during the model's construction and used within the computer simulations. These states can be observed in Table 4, along with an explanation of their purpose and impact within the medical procedure. Moreover, Table 4 also provides an overview encompassing the world states, the mental model states for both the doctor and nurse, ownership states, and contextual factors, each accompanied by an explanation.

As mentioned earlier in Sect. 5.1.1, black arrows are utilized to illustrate the influence of one world state on another. Blue lines, on the other hand, establish connections between ownership states and world states. The presence of yellow lines signifies observation arrows, indicating that the nurse and doctor actively observe events taking place in the world and integrate the obtained information into their mental models. Arrows originating from the context factors indicate strengthening effects of them on specific world states. Conversely, dotted arrows are used to represent negative connection weights which denote the weakening effect on certain world states. There are also red downward arrows incoming, but their explanation will be provided in next section. Furthermore, there are outgoing lines (blue and orange), which will be elaborated on in Sect. 5.2.3.

5.2.2 Middle Level for Scenario 2: Adaptation and Plasticity of the Mental Models

The blue plane focuses on the cognitive processes associated with learning and forgetting of mental models. Specifically, there are two distinct processes, one for the doctor and another one for the nurse. For the doctor, this represents the weights of the connections within their mental model, linking the state for "X-ray and other evaluations show something abnormal" to the state for "Doctor provides treatment", i.e., self-model state $\mathbf{W}_{DS4,DS6}$. From this **W**-state a downward adaptation effectuation link points to the state "Doctor provides treatment" within the doctor's mental model

Table 4 Overview of the states for Scenario 2: the world states, mental states, ownership states, context factors, periods, and first- and second-order adaptation states (for the complete table, see the appendix)

World, Doctor, Nurse, and Context				Explanation	Level
C1			Context	Baby is in critical state	Base level
C2			Context	Doctor is feeling sick himself	
C3			Context	Baby is born prematurely	
P1			Context	Period in which there is no critical ill baby to take care of	
P2			Period	Doctor takes a break	
P3			Period	Doctor is back doing their job	
WS1	NS1	DS1		Nurse alarms doctor	
......	......	......			
WS10	NS10	DS10		Nurse informs family about doctor's advice	
NOS1				Ownership state for the action of nurse alarms doctor	
......	......	......			
DOS9				Ownership state for the action of doctor advises to let neonate rest	
$\mathbf{W}_{DS4,DS6}$				First-order self-model state for the doctor's weight of the connection from DS4 to DS6	First-order adaptation level
$\mathbf{W}_{NS4,NS6}$				First-order self-model state for the nurse's weight of the connection from NS4 to NS6	
$\mathbf{MW}_{DS4,\ DS6}$				Second-order self-model state for the persistence factor of the doctor's weight of the connection from DS4 to DS6	Second-order adaptation level
$\mathbf{MW}_{NS4,\ NS6}$				Second-order self-model state for the persistence factor of the nurse's weight of the connection from NS4 to NS6	

which is indicated by a red arrow. The nurse undergoes a similar process with their respective mental model and connections. Furthermore, the mental models of both the doctor and nurse mutually influence each other as a form of social learning: they learn from one another by communicating, modeled by an interplay of their respective **W**-states.

The other type of learning covered is individual learning modeled by Hebbian Learning (see Table 1). The Hebbian learning combination function is commonly used to update on the fly the connection weights between neurons in neural networks when both neurons are often activated together. To integrate the social learning and the individual Hebbian learning processes, the $\mathbf{maxhebb}_{\mu}$ combination function is used which combines the regular Hebbian function with the max function. The $\mathbf{maxhebb}_{\mu}$ function selects the highest value of the incoming connections including the value by Hebbian learning. This approach ensures that the influence of the most significant connection is preserved while minimizing the impact of less relevant connections.

Like the standard Hebbian learning, the $\mathbf{maxhebb}_\mu$ function incorporates a persistence factor μ, which quantifies the fraction of the learned value that remains over each unit of time. By including this factor, Hebbian learning accounts for the phenomenon of extinction or forgetting. The persistence factor μ determines the extent to which the learned value is retained or forgotten over time. A value of μ between 0 and 1 represents the fraction of the learned value that persists per time unit. A higher value of μ indicates a slower rate of forgetting, meaning that the learned associations will persist for a longer duration. On the other hand, a lower value of μ accelerates the rate of forgetting, leading to a more rapid extinction of the learned associations.

5.2.3 Upper Level for Scenario 2: Meta-Plasticity of the Mental Models

The adaptivity of the first-order self-model within the network is captured by incorporating a second-order self-model, represented by the $\mathbf{M_W}$-state, in the upper-level plane. The presence of the $\mathbf{M_W}$-state within the upper-level plane effectively captures the adaptive nature of the persistence factor μ. It enables dynamic adjustments and responses to changing conditions or contexts, ensuring that the network can adaptively modify the extent of persistence as needed. In the given scenario, it is assumed that when the baby is in a critical state, causing stress for both the doctor and nurse, the value of the $\mathbf{M_W}$-state decreases. This modelling approach captures the concept of forgetting under stressful circumstances. More precisely, the relationship described involves upward connections from the context state C1 in the base level to the $\mathbf{M_W}$-states, as well as downward connections from the $\mathbf{M_W}$-states to the **W**-states. These $\mathbf{M_W}$-states are modelled with the help of the combination function scalemap (see Table 1) which maps activation scale [0, 1] to scale $[\lambda, \upsilon]$ with λ as lower bound and υ as upper bound. The scalemap function enables flexible manipulation and control of the $\mathbf{M_W}$-state activation levels, allowing them to effectively contribute to the model's behaviour and outcomes. By applying this function, the activation values of the $\mathbf{M_W}$-states can be effectively scaled to match the desired range of values which in this case is from 1.0 to 0.5.

5.3 *Collaborative Adaptation and Learning in Nurse-Led Intervention with Subsequent Doctor's Intervention and the Role of an AI-Coach: The Computational Model for Scenario 3*

While Scenario 3 is a variant of Scenario 2, it introduces numerous significant changes that will be extensively examined in the following sections. In order to provide a clearer explanation of the models, each section focuses on different aspects of the overall model. As a result, multiple figures depicting the same model are presented throughout the various sections.

5.3.1 Base Level for Scenario 3: Overview

In the base plane (see Fig. 4), the world states are once again represented by green nodes, with their corresponding connections situated in the middle region of the base plane. As usual, the lighter green nodes indicate that they are not directly influenced by other world states. For an overview of the most important states, see Table 5.

In this particular scenario, an additional mental model has been introduced at the base level. Alongside the mental models of the doctor and nurse, which reflect the structure of the medical procedure they follow, there is now a mental model of the virtual AI-Coach as well. The nodes of the AI-Coach's mental model are depicted as yellow nodes.

However, there are some distinctions in this case. While the doctor and nurse have specific actions to perform in the world, relying on their individual models mental and utilizing action ownership states to initiate their actions, the AI-Coach does not possess such action ownership states. This is because the AI-Coach does not directly interact with the physical world or execute real actions. Instead, its primary function revolves around observing each world state and initiating communication to the others. These observations by the AI-Coach are crucial for the monitoring and communication functions performed by it. Further elaboration on this function will be provided in the subsequent section, focusing on the middle level of the model.

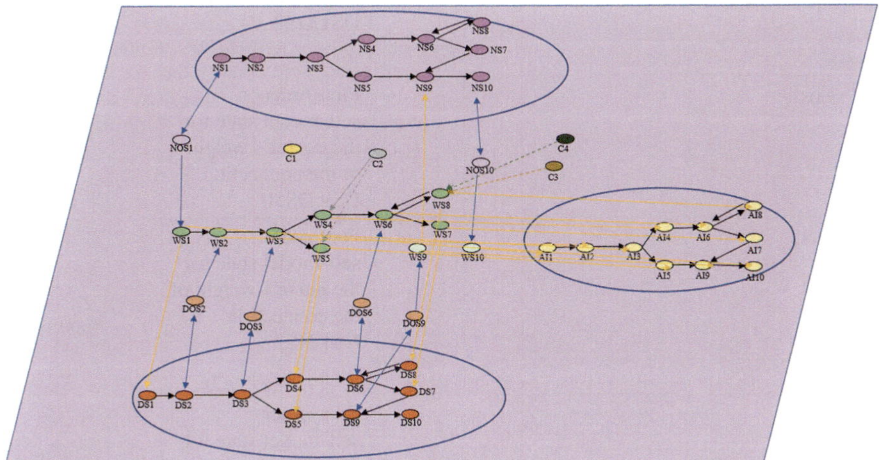

Fig. 4 Base Level for Scenario 3: World states, Mental model states, Ownership states, Context states

Table 5 Overview the states for Scenario 3: the world states, mental states, ownership states, context factors, and first- and second-order states (for the complete table, see the appendix)

World, Doctor, Nurse, AI-Coach, Context					Explanation	Level
C1				Context	Experiences of stress	Base level
C2				Context	Baby is in critical state	
C3				Context	Underlying congenital or genetic condition	
C4				Context	Doctor is feeling sick himself	
WS1	NS1	DS1	AI1		Nurse alarms doctor	
……	……	……	….			
WS10	NS10	DS10	AI10		Nurse informs family about doctor's advice	
NOS1					Ownership state for the action of nurse alarms doctor	
……	……	……	……			
DOS9					Ownership state for the action of doctor advises to let neonate rest	
$\mathbf{W}_{DS1,DS2}$					First-order self-model state for the doctor's weight of the connection DS1,DS2	First-order adaptation level
….	….	….	….			
$\mathbf{W}_{DS9,DS10}$					First-order self-model state for the doctor's weight of the connection DS9,DS10	
$\mathbf{W}_{NS1,NS2}$					First-order self-model state for the nurse's weight of the connection NS1,NS2	
….	….	….	….			
$\mathbf{W}_{NS9,NS10}$					First-order self-model state for the doctor's weight of the connection NS9,NS10	

(continued)

Table 5 (continued)

World, Doctor, Nurse, AI-Coach, Context					Explanation	Level
$\mathbf{W}_{AI1,AI2}$					First-order self-model state for the nurse's weight of the connection AI1,AI2	
….	….	….	….			
$\mathbf{W}_{AI9,AI10}$					First-order self-model state for the doctor's weight of the connection between AI9,AI10	
Monitor WS2					Monitor feature of the AI-Coach to monitor doctor preparing necessary equipment and supplies for medical procedure	
….	….	….	….			
Monitor WS10					Monitor feature of the AI-Coach to monitor nurse informing family about doctor's advice	
Decision WS2					Decision feature of the AI-Coach to decide to signal about doctor preparing necessary equipment and supplies for medical procedure	
….	….	….	….			
Decision WS10					Decision feature of the AI-Coach to decide to signal about nurse informing family about doctor's advice	

(continued)

Table 5 (continued)

World, Doctor, Nurse, AI-Coach, Context					Explanation	Level
CS2					Communication feature of the AI-Coach to signal/communicate about doctor preparing necessary equipment and supplies for medical procedure	Second-order adaptation level
....	....	....	....			
CS10					Communication feature of the AI-Coach to signal/communicate about nurse informing family about doctor's advice	
$\mathbf{W}_{\mathbf{W}_{AI1,AI2},\mathbf{W}_{DS1,DS2}}$					Second-order self-model state representing connection weight for $\mathbf{W}_{DS1,DS2}$	
....	....	....	....			
$\mathbf{W}_{\mathbf{W}_{AI7,AI9},\mathbf{W}_{DS7,DS9}}$					Second-order self-model state representing connection weight $\mathbf{W}_{DS7,DS9}$	
$\mathbf{W}_{\mathbf{W}_{AI9,AI10},\mathbf{W}_{NS9,NS10}}$					Second-order self-model state representing connection weight for $\mathbf{W}_{NS9,NS10}$	

Within this network model, four contextual factors come into play. The first factor, represented by the colour yellow, signifies the presence of stress experiences. The second factor, depicted as a grey node, indicates that the baby is in a critical state. The dark green colour represents the presence of an underlying congenital or genetic condition in the baby, while the brown factor denotes that the doctor is feeling ill themselves.

In the base level, the colours of the arrows remained the same. Black arrows depict the influence of one world state on another, while blue lines connect ownership states to world states. Yellow lines represent observation arrows, indicating active observation by the nurse, doctor, and this time also by the AI-Coach. Solid arrows from context factors strengthen specific world states, while dotted arrows weaken certain world states.

The **alogistic** function is employed once more for the states at the base level, encompassing world states, mental model states, and ownership states. Context factors, on the other hand, make use of the **stepmod** function (see Table 1).

5.3.2 Middle Level for Scenario 3: Adaptation and Plasticity of the Mental Models

The middle level, denoted by the colour blue, exhibits significant differences compared to Scenario 2. At this level, numerous crucial processes of the AI-Coach take place. As previously mentioned, the AI-Coach possesses comprehensive knowledge and by observing the actions and states in the world is at each point in time is already aware of all relevant information. More specifically, it can observe each step of the medical procedure in the world, and this information is conveyed to the monitor states, represented by dark red nodes. Subsequently, this information is fed into the AI-Coach's decision-making states for performing signalling actions through communication, indicated by lighter red nodes. However, these decision states are suppressed by the world states if the medical procedure is progressing as intended, thereby eliminating the need for the AI-Coach to perform any signalling actions (see Fig. 5).

Nevertheless, if the AI-Coach's decision-making state becomes partly activated, it initiates its communication action, which resides in the upper level of the model (see Sect. 5.3.3). The communication states serve the purpose of signalling the correct **W**-state of either the doctor or nurse, depending on which step deviates from the expected course and who is responsible for that particular step. For instance, if the AI-Coach detects that WS6 has not occurred, it will decide to communicate this discrepancy to the doctor, as the doctor is responsible for administering the treatment. This signalling mechanism assists the team members in the process by helping to prevent mistakes.

Additionally, team members can enhance their accuracy by leveraging organizational learning. The AI-Coach, assumed to have perfect knowledge, establishes connections between the **W**-states of itself and the corresponding **W**-states of the doctor and nurse. This enables the transfer of knowledge, aiding the team members in their decision-making processes and reducing the likelihood of errors.

In this specific scenario, Hebbian Learning is not employed. Instead, the **alogistic** function is utilized for each **W**-state. The **alogistic** function's characteristic of being suitable for modelling states that require a smooth transition between different values makes it an appropriate choice. Therefore, the **alogistic** function is chosen to ensure the appropriate handling of the **W**-states, as well as the monitor, decision, and communication states, maintaining consistency throughout the system.

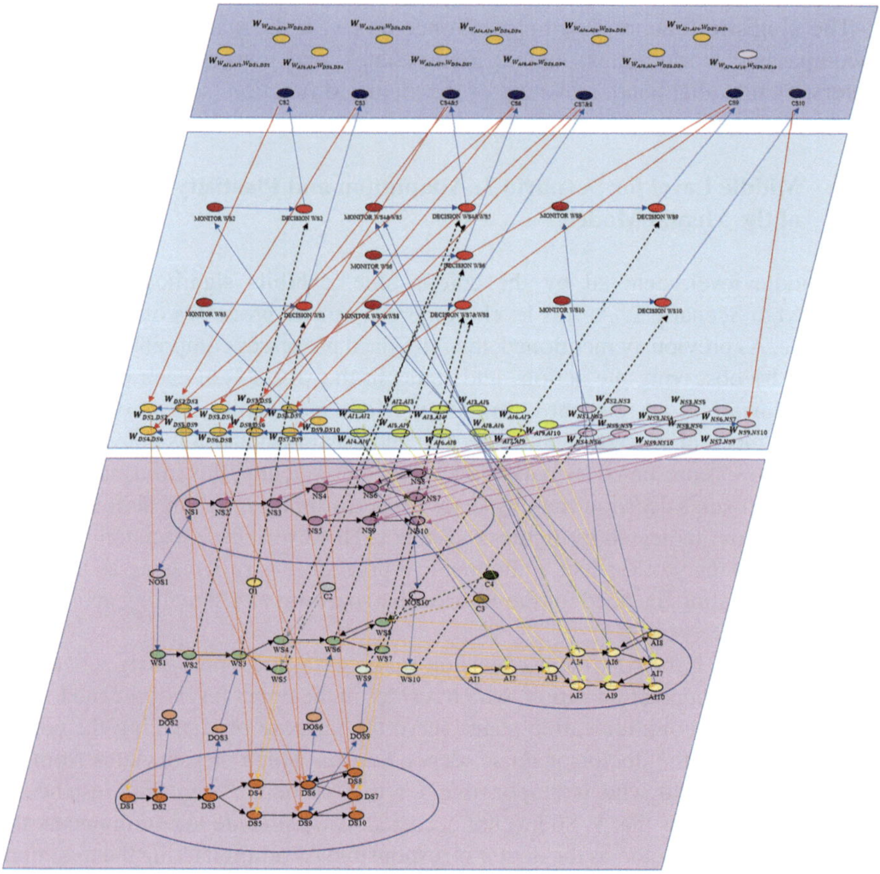

Fig. 5 Middle level for Scenario 3: **W**-states, monitor states, and decision states

5.3.3 Upper Level Scenario 3: Meta-Plasticity of the Mental Models

The upper level, represented by the colour purple, contains second-order self-model states that determine the weight of connections from the AI Coach's **W**-states to individual mental model connection weight self-model states in the middle level (see Fig. 6). These higher order **W**-states are distinct from the $\mathbf{M_W}$-states utilized in the model of scenario 2. The $\mathbf{W_{WW}}$-states, in particular, specify the weights of connections from the AI Coach's **W**-states to the individual **W**-states of doctor or nurse and vice versa, enabling the initiation and control of shared organizational mental model learning by individuals. As previously mentioned, the communication states are situated in the upper level. This arrangement is necessary because the influence of the communication states needs to be adaptation of the **W**-states in the

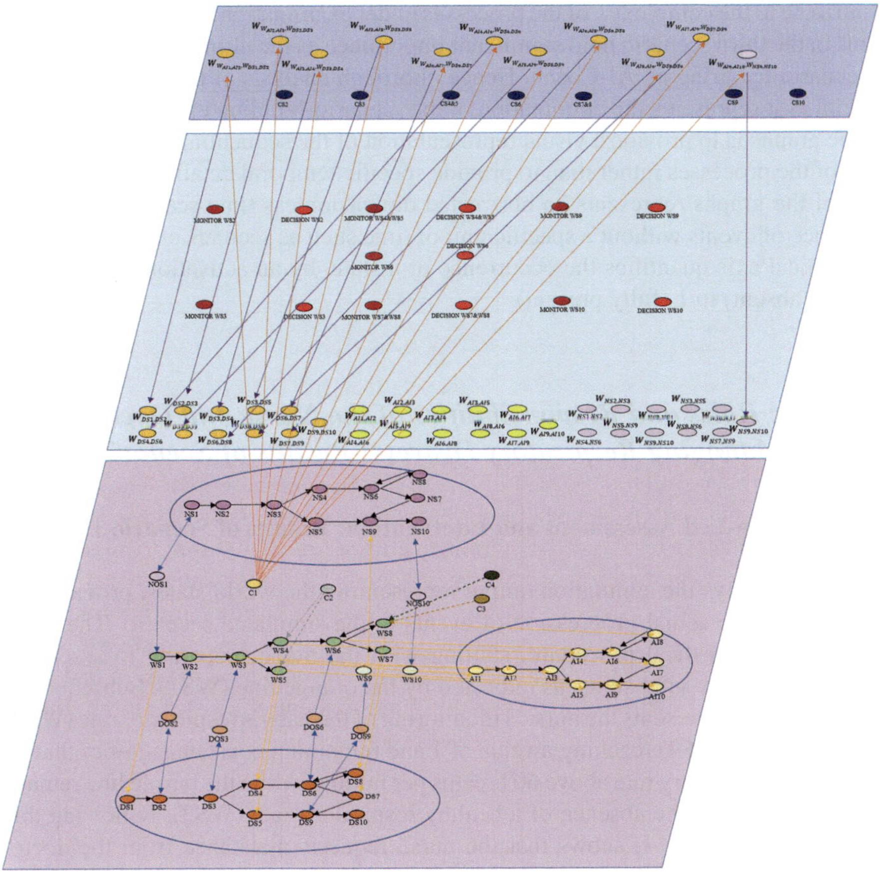

Fig. 6 Upper-level for Scenario 3: $\mathbf{W_{WW}}$-states and communication states

middle level. The communication states employ the **alogistic** function as their combination function, while the $\mathbf{W_{WW}}$-states are modelled using the **scalemap** function (see Table 1).

6 Simulation Experiments

Each scenario has its own set of simulation results. Scenario 1a and 1b will be discussed together in one section, while the remaining scenarios will each have their dedicated sections. In Scenario 1 and 2, the analysis primarily focuses on the world states. However, for Scenario 3, additional aspects such as mental states, ownership states, and W-states are analyzed. These elements are of particular interest as they contribute to the complexity of the processes under examination. Notably, the simulations in the third scenario hold significant importance as the ultimate goal is to create a scenario featuring an AI-Coach. Hence, thorough analysis of these simulations is crucial as it represents the culmination of the scenario-building efforts. The purpose of the graphs is to provide a visual representation of the sequential order or progression of the processes rather than to provide specific temporal details. The horizontal axis of the graphs represents an abstracted dimensionless time scale, denoting the sequence of events without a specific unit of time such as seconds or minutes, while the vertical axis quantifies the occurrence of a state, by an activation level ranging from 0 (absent) to 1 (fully present).

6.1 The Roles and Responsibilities of the Nurse and Doctor in Managing Respiratory Distress: Results of Scenario 1

6.1.1 Nurse-Led Assessment and Intervention: Results of Scenario 1a

Figure 7 displays the simulation output representing the world states, providing an overview of the actual progression of events in the simulated scenario. The figure illustrates the medical procedure, starting with the nurse preparing to assess the baby's respiratory vital signs, as indicated by the orange line (WS1). Subsequently, the yellow line represents the nurse's monitoring of the baby's respiratory rate (WS2). The green line (WS4) reaching a value of 1 and remaining constant signifies that the baby has a respiratory rate above 60 breaths per minute, while the purple line remains at zero, indicating the absence of a healthy respiratory rate (WS3). Following this, the purple line (WS17) shows that the nurse requests assistance from the doctor, triggered by the state where the baby has a respiratory rate that is too high. The nurse promptly notified the doctor as a high respiratory rate in the baby is undesirable and requires attention. The doctor promptly provides advice, represented by the dark green line (WS20). Additionally, the light blue line (WS5) depicts the nurse using a pulse oximeter to measure the baby's oxygen saturation level. The dark blue line at 0 indicates that the oxygen saturation level is below 89% (WS7), while the absence of a red line above 89% suggests a healthy saturation level (WS6). Continuing further, the light cyan line (WS8) represents the nurse's assessment of the baby's breathing, particularly for grunting sounds. The yellow line (WS9) at 0 indicates the absence of grunting sounds, while a non-zero value on the yellow line would indicate the presence of grunting. This is corroborated by the pink line (WS10) reaching a level of 0.35, indicating minimal grunting sounds. However, considering the earlier observations of poor oxygen levels and respiratory distress, the presence of even minimal grunting supports the notion that the baby is experiencing respiratory distress. The medical procedure proceeds with the nurse monitoring for chest retractions, as depicted by the light grey line (WS11). The bordeaux red line (WS13) suggests that the baby is indeed experiencing chest retractions, while the

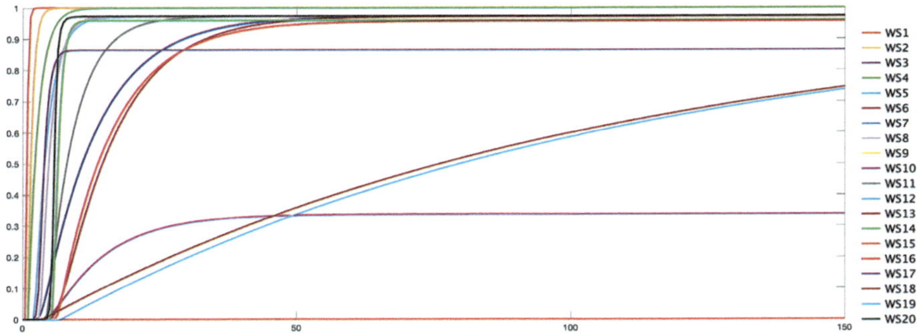

Fig. 7 Results Scenario 1a: the world states

light blue line (WS12) at 0 indicates the absence of chest retractions. Next, the nurse observes the baby's colour for signs of cyanosis, represented by the neon green line (WS14) in the graph. The red line (WS16) indicates the presence of skin, lip, or mucous membrane discoloration, suggesting the baby is experiencing cyanosis. Finally, based on the observations made, the nurse may provide supplemental oxygen, respiratory support, or transfer the baby to a Neonatal Care Intensive Unit, as shown by the brown line (WS18). The last light blue line signifies that the baby is allowed to bond with the mother (WS19).

6.1.2 Nurse-Led Intervention with Subsequent Doctor's Intervention: Results of Scenario 1b

Figure 8 reveals the presence of two context factors: the baby being in a critical state from $t = 1$ to 50 and the family experiencing emotional distress from $t = 1$ to 80. The initial light blue line (WS1) indicates the nurse alarming the doctor. Subsequently, the bordeaux red line (WS2) represents the doctor preparing the necessary equipment and supplies for a medical procedure, followed by the dark blue line (WS3) indicating that the doctor performs an X-ray and other evaluations. The orange line (WS4) demonstrates that the X-ray and other evaluations indeed reveal lung expansion, airway obstruction, pneumothorax, infiltrates, and/or cardiomegaly. It is worth noting a slight decrease at t = 50, as the period of the baby being in a critical state has ended, making it slightly less likely for the X-ray and other evaluations to indicate an abnormality. Afterward, the doctor provides oxygen therapy, medications, re-expands the lung, performs surgery, or administers other treatments (WS6). From $t = 50$ onward, the yellow line (WS5) illustrates a rapid increase in the state "X-ray and other evaluations do not show anything abnormal". This indicates that during the period when the baby was in a critical state, we observed abnormalities in the X-ray and other evaluations. However, after treatment was administered, the state "X-ray and other evaluations do not show anything abnormal" starts increasing because the treatment was effective, as evidenced by the green line (WS7). The light purple line

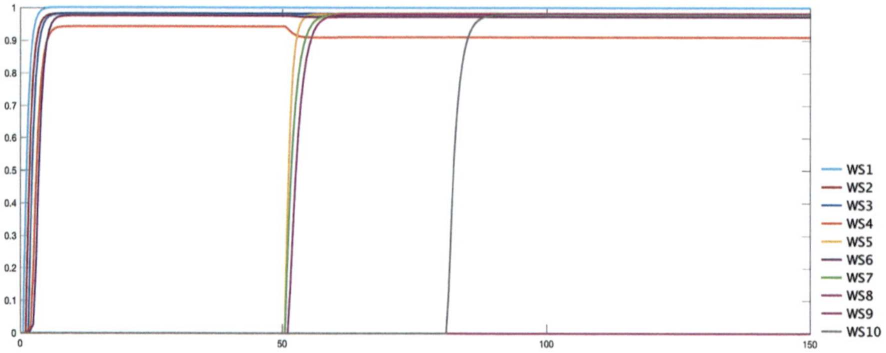

Fig. 8 Results Scenario 1b: the world states

(WS8) remains constant at 0, indicating that the treatment is not ineffective. The pink line (WS9) signifies the doctor providing advice, while the nurse only informs the family about this at $t = 80$, as indicated by the grey line (WS10). This timing is chosen because it aligns with the point when the family's emotional distress subsides, making it an appropriate moment for the nurse to communicate with them.

6.2 *Collaborative Adaptation and Learning in Nurse-Led Intervention and Impact of a Fatigued Doctor: Results of Scenario 2*

This scenario incorporates multiple distinct periods occurring between $t = 0$ and $t = 1000$. It is important to note that the world steps progress rapidly, causing them to have slightly different time values but appearing on the same timeline in the graph (Fig. 9).

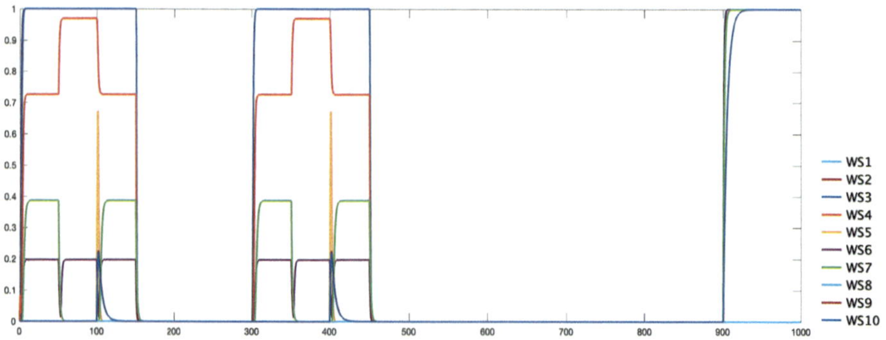

Fig. 9 Results Scenario 2: the world states

The first period takes place between $t = 0$ and $t = 150$. During this time, the nurse contacts the doctor for assistance, as depicted by the light blue line (WS1). The doctor promptly prepares the necessary equipment and supplies for a medical procedure, indicated by the bordeaux red line (WS2). Following this preparation, the doctor conducts an X-ray and other medical evaluations, as shown by the dark blue line (WS3). These evaluations reveal an abnormality (orange line, WS4), leading the doctor to proceed with treatment (purple line, WS6). However, due to the doctor being sick himself, errors occur during the administration of the treatment, resulting in even worse X-ray and evaluation results, as indicated by the increasing orange line. In an attempt to address the situation, the doctor switches to a different treatment (referred to as treatment 2). Unfortunately, this alternative treatment does not appear to be effective, and the doctor makes mistakes during the process, as evidenced by the green line (WS7). Consequently, the doctor reverts back to treatment 1 (purple line), which proves successful this time. The improved X-ray and evaluation results are depicted by the yellow line (WS5), while the decrease in the orange line signifies a positive outcome. Furthermore, the doctor advises the nurse to ensure that the neonate gets plenty of rest indicated by the red line (WS9). The nurse promptly conveys this information to the baby's family, depicted by the dark blue line (WS10).

Between $t = 150$ and $t = 300$, there are no critically ill babies requiring immediate attention. As a result, both the doctor and nurse get a break from their duties. However, between $t = 300$ and $t = 450$ the results show that the nurse reaches out to the doctor once more to help with a new baby. The doctor proceeds to prepare the required equipment and follow the usual steps, but unfortunately, he repeats the same mistakes as before. It's apparent that he is not feeling well, and this is reflected in his incorrect actions. The results show that he has already failed to administer the appropriate treatment to two babies due to his condition. Between $t = 450$ and $t = 900$, the doctor takes a significant break to recuperate and overcome his illness, with the aim of performing his duties more effectively once he returns to work.

After $t = 900$, the figure shows that once more, the nurse reaches out to the doctor for assistance with baby 3. The doctor follows the same steps as before, but this time, the results indicate that he has performed his duties accurately. He administers the treatment correctly on the first attempt, which proves to be effective, and he avoids any further errors. The break he took before did him good.

6.3 Collaborative Adaptation and Learning in Nurse-Led Intervention with Subsequent Doctor's Intervention and the Role of an AI-Coach: Results of Scenario 3

As mentioned earlier, Scenario 3 incorporates a virtual AI-Coach. This coach can monitor and signal when mistakes are made during the medical procedure. However, the coach can also provide the doctor and nurse with its perfect knowledge, facilitating organizational learning. Hence, there will be a differentiation between two

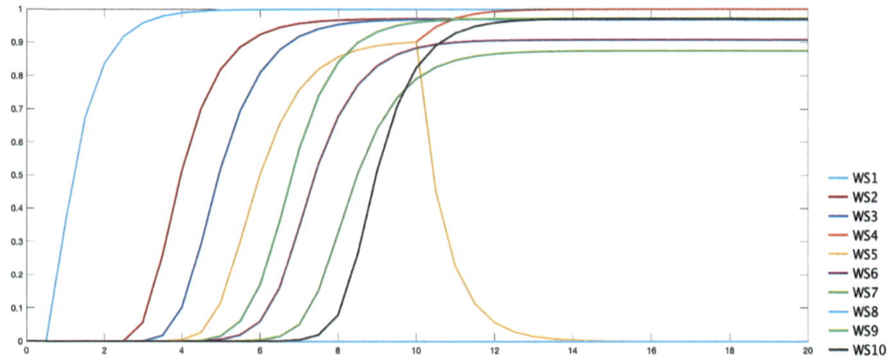

Fig. 10 Results Scenario 3: the world states during case 1

cases: one in which organizational learning plays a big role and another in which the communication feature is important.

6.3.1 Case 1: Organizational Learning

At the first-order adaptation level, specific **W**-states of the doctor and the nurse receive inputs from the **W**-states generated by the AI-Coach. This arrangement facilitates a comprehensive learning and recalling experience for the doctor and nurse, as the AI-Coach possesses a flawless mental model and imparts knowledge to them. In both cases, the $\mathbf{W}_{\mathbf{WW}}$-states won't be further explored in detail since these states remain constant at 1 with no speed and exhibit no variation.

The World States in Case 1

As Fig. 10 depicts, the usual medical procedure happens by the nurse alarming the doctor (light blue line, WS1). This is followed by a bordeaux line (WS2) which means the doctor is preparing necessary equipment and supplies for the medical procedure. The dark blue line (WS3) shows that the doctor takes an X-ray and other evaluations, and the orange line (WS4) shows that the evaluations show that there indeed is something abnormal. While the yellow line (WS5) first started growing too because it seemed that there was nothing wrong, but eventually quickly went to zero because both states cannot happen at the same time; if the X-ray and other evaluations show there is something abnormal, it cannot show that there is nothing wrong. After this, the doctor already advises to let the neonate rest as much as possible because rest is needed for recovery (WS9). After this, the purple line (WS6) shows that the doctor is giving treatment with the green line (WS7) indicating that the treatment is working. In the end, the dark green line (WS10) shows that the nurse informs the family when the medical procedure is done.

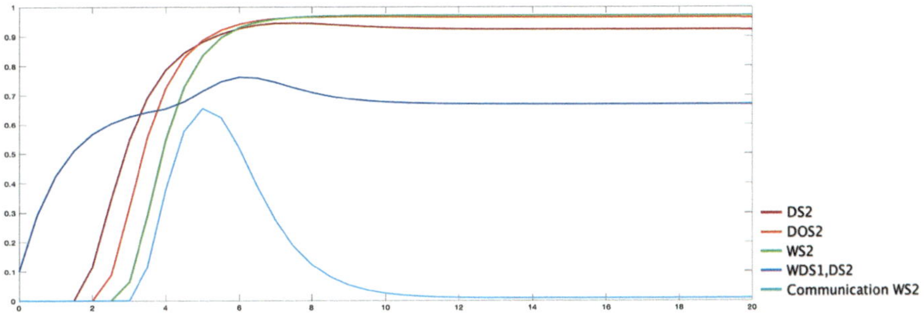

Fig. 11 Results Scenario 3: example WS2 during case 1

Example WS2 in Case 1: Learning State, Mental State, Ownership State, World State, and Communication State

For instance, consider the graph (Fig. 11) below that illustrates the process behind WS2, which involves the doctor preparing necessary equipment and supplies for a medical procedure. The doctor's learning curve, represented by the blue line for adaptation state $\mathbf{W}_{DS1,DS2}$, experiences an immediate rise as the knowledge provided by the AI-Coach is assimilated. The orange line represents DOS2, the ownership state triggered by the mental model state DS2 (bordeaux red line), followed by the neon green line (WS2), which signifies the actual occurrences in the real world. Furthermore, there is a temporary spike in learning as the AI-Coach briefly communicates (depicted by the light blue line, Communication WS2), but this interaction quickly diminishes because the need for continuous signalling from the AI-Coach diminishes as WS2 actively unfolds, resulting in a swift cessation of communication. The doctor's learning journey reaches a sustained level of approximately 0.7 instead of reaching a perfect score of 1. This is because even with the invaluable knowledge provided by the AI-Coach, it is impossible for the doctor to retain and recall every piece of information; in this case, the doctor retains around 70% of what the AI-Coach imparts, resulting in the learning curve plateauing at this level. Therefore, in this particular scenario, the need for frequent communication by the AI-Coach is not significant, thanks to the presence of organizational learning that ensures the seamless advancement of medical procedures within the hospital.

This behaviour is applicable to all other states in this particular case. The learning states consistently increases, ensuring the correct progression of all subsequent states and rendering the communication function of the AI-Coach unnecessary.

Understanding the Learning States of the Doctor in Case 1: A Comprehensive Exploration

Within the model, there are three key actors: the doctor, the nurse, and the AI-Coach. Each of these actors exhibits a distinct process in their learning states. Consideration should be given to the fact that the learning states of the doctor are here taken as an

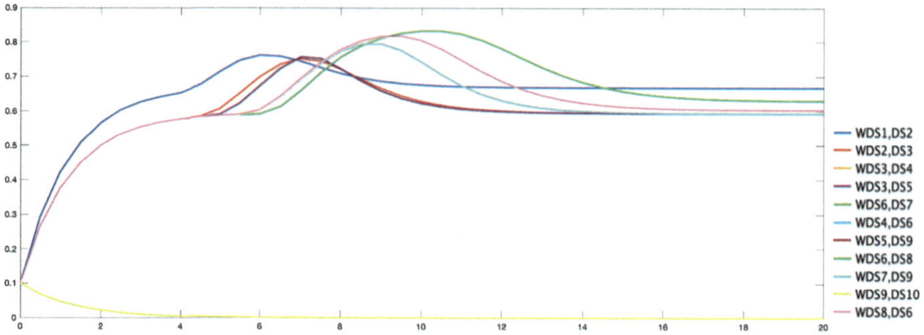

Fig. 12 Case 1 Results Scenario 3: learning done by doctor

example, and a similar exploration of learning states could have been conducted for the nurse, for instance.

Figure 12 illustrates the learning states of the doctor. At the beginning, all W-states demonstrate a comparable pattern of learning, characterized by a gradual increase, followed by a noticeable spike, and then a slight decline indicating some degree of forgetting. However, the learning process is consistently supported by the AI-Coach's impeccable knowledge, preventing the knowledge states from ever reaching zero. As a result, a stable level of knowledge is maintained, ensuring a continuous learning experience throughout the process. However, it is worth noting that WDS9,DS10 follows a different trajectory. This learning state decreases to 0 since it is not triggered by the communication state of the AI-Coach. This is due to the fact that the doctor is not involved in the final step of the medical procedure, which involves informing the family, a responsibility undertaken by the nurse.

Case 2: AI-Coach's Monitoring and Communication Functions

In the previous case, at the initial reification level, the **W**-states of the doctor and nurse are influenced by the **W**-states generated by the AI-Coach. However, in cases where these links are not present, the AI-Coach can still contribute to the improvement of the nurse and doctor's knowledge in a different manner. This is achieved through the AI-Coach's monitoring and communication feature, which enables it to identify and address mistakes.

The World States in Case 2

In this case, the process of the world states is significantly different from that of case 1, see Fig. 13. It is important to present these world states too to provide a comprehensive view of the entire story, ensuring that no crucial details are overlooked.

As usual, the nurse initially alarms the doctor (light blue line, WS1), prompting the doctor to prepare the necessary equipment and supplies (bordeaux line, WS2).

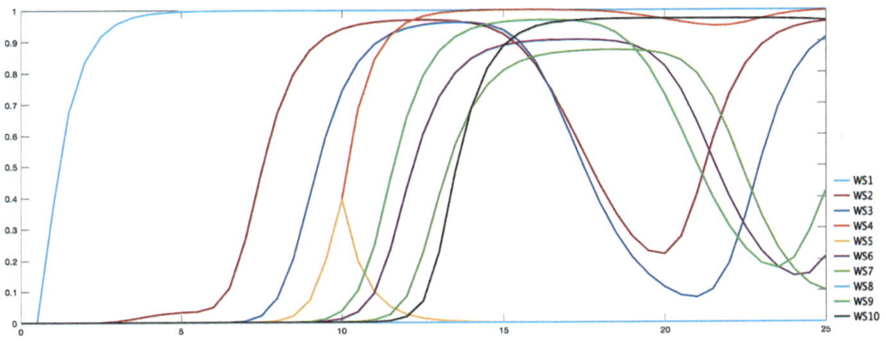

Fig. 13 Results Scenario 3: the world states during Case 2

Subsequently, the doctor performs X-ray and other evaluations (dark blue line, WS3), which confirm the presence of an abnormality (orange line, WS4). The yellow line (WS5) briefly suggests no issue, but it quickly returns to zero as the abnormal and normal X-ray and other evaluations results cannot coexist simultaneously. The doctor advises providing ample rest for the neonate's recovery (WS9). Treatment is then administered (purple line, WS6) and proves effective (green line, WS7). Finally, the nurse informs the family when the medical procedure is completed (dark green line, WS10).

However, when compared to case 1, it is evident that in this particular scenario the word states experience a decline instead of remaining constant. This decline signifies the occurrence of mistakes during the medical procedure, which were uncommon in the case where organizational learning was present. However, it is important to note that these world states never fully reach zero, except for cases where it is necessary for them to be zero, such as when "X-ray and other evaluations show nothing abnormal." The world states never reach zero because the communication feature of the AI-Coach promptly signals the mistakes, causing the lines to increase again, indicating that the doctor/nurse is aware of their error.

Example WS2 in Case 2: Learning State, Mental State, Ownership State, World State, and Communication State

As previously mentioned, in the initial reification level, certain **W**-states of the doctor and nurse receive inputs from the **W**-states generated by the AI-Coach, resulting in optimal knowledge for both. However, if these inputs were hindered and the connections between the **W**-states of the AI-Coach and the nurse/doctor were disrupted, a different outcome would occur. In such a scenario, the 'communication' feature provided by the AI-Coach becomes valuable.

The following graph (see Fig. 14) illustrates the same WS2 state but with a delay in the learning state receiving input from the AI-Coach's learning state. Consequently, the learning first decreases and when the AI-Coach detects forgetfulness, it signals

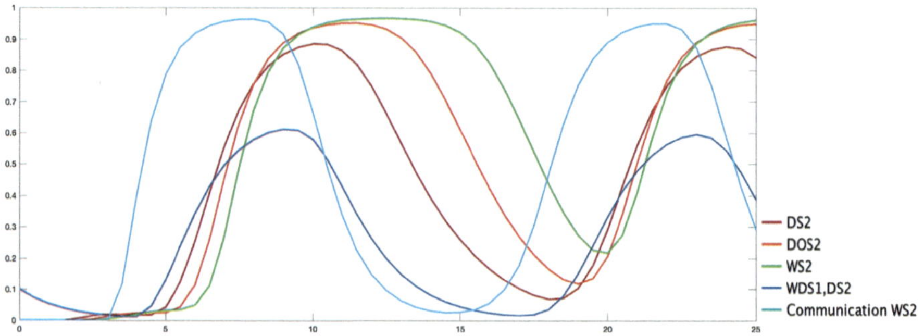

Fig. 14 Results Scenario 3: example WS2 during Case 2

accordingly. The graph depicts a dark blue line for adaptation state $\mathbf{W}_{DS1,DS2}$ representing the learning process. It is evident that the doctor initially begins to forget, prompting the AI-Coach to initiate communication to indicate that WS2 should be taking place (light blue line, Communication WS2). Subsequently, the orange line representing DOS2 gets triggered by the mental model state DS2 which is indicated by the bordeaux red line. Eventually, the occurrence of the neon line indicates the manifestation of WS2 in the real world. As a result, the need for the AI-Coach to communicate and signal the necessity of WS2 diminishes. Subsequently, there is a subsequent occurrence of forgetfulness, resulting in a decline in the mental model state, action state, and world state. This prompts the AI-Coach to signal once more, indicating the need for further learning. As a response, the doctor resumes the learning process, striving to enhance their understanding and proficiency. Thus, it becomes apparent that the communication feature is particularly useful when the doctor and nurse lack perfect knowledge and experience forgetfulness.

Understanding the Learning States of the Doctor in Case 2: A Comprehensive Exploration

Figure 15 illustrates the learning states of the doctor. Initially, all **W**-states exhibit a similar pattern of forgetting, followed by subsequent learning, and eventually a big decrease indicating forgetting again.

This learning process is facilitated by the AI-Coach's signalling when something goes wrong. Thus, when the big decrease in learning happens, the AI-Coach signals this and helps the doctor in increasing their knowledge again. However, it is worth noting that $\mathbf{W}_{DS9,\ DS10}$ follows a different trajectory. This learning state decreases to 0 since it is not triggered by the communication state of the AI-Coach. This is due to the fact that the doctor is not involved in the final step of the medical procedure, which involves informing the family, a responsibility undertaken by the nurse.

In this particular case, the example focuses on the doctor; however, it is important to note that similar processes of learning and forgetting occur with the nurse as well. Both the doctor and the nurse experience the same patterns of learning and forgetting

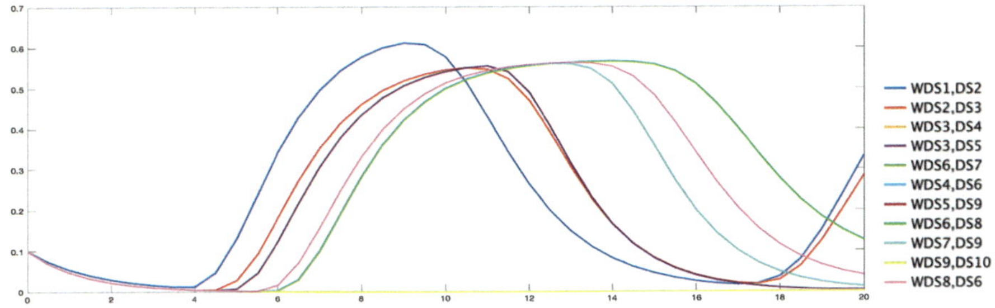

Fig. 15 Case 2 Results Scenario 3: learning by doctor

within their respective roles. Additionally, it is worth highlighting that the learning states of the AI-Coach consistently maintain a constant line at 1. This is due to the AI-Coach possessing perfect knowledge from the outset and not requiring any further learning itself.

7 Conclusion and Discussion

This study introduces adaptive computational network models that examine the processes involved in the execution of a medical procedure related to respiratory distress in a hospital setting. The material comes from Mokadem et al. (2024). The development of these models incorporates the adaptive network-oriented modelling approach (Treur 2020) in combination with the specialized software written in MATLAB. The full specifications by role matrices of all models can be found as Linked Data at https://www.researchgate.net/publication/371986708

Initially, the study presents models that include the mental models of a doctor and a nurse. Subsequently, it explores the integration of a mental model and the utilization of a virtual AI-Coach. This AI-Coach possesses advanced capabilities, employing a monitoring system to assess whether it should provide alerts or not. The initial simulation results highlight the progression of the first two scenarios without the presence of the AI-Coach. However, subsequent simulations demonstrate that the AI-Coach can provide significant benefits in two distinct ways. Firstly, by leveraging organizational learning, the AI-Coach educates team members about each step of the medical procedure, effectively reducing the likelihood of errors in the workplace. This approach ensures that team members have a comprehensive understanding of the procedure, minimizing the occurrence of mistakes. Secondly, the monitoring feature of the AI-Coach plays a crucial role. It promptly alerts team members whenever it detects deviations or delays in the procedure. This vigilance helps mitigate forgetfulness among nurses and doctors, enabling them to relearn and operate without making mistakes. It is important to note that organizational learning, as observed in case 1, has a superior effect compared to the communication feature alone. In case 1, learning consistently increased and reached a constant state, indicating the correct execution of the medical procedure. On the other hand, in case 2, learning increased only when the AI-Coach signalled an issue. Furthermore, in case 2, the world states decreased, while in case 1, these word states remained constant, indicating a successful medical procedure. However, despite the effectiveness of organizational learning facilitated by the AI-Coach, the communication feature played a crucial role in averting critical mistakes. When something went wrong, the AI-Coach's communication feature promptly addressed the issue, ensuring timely resolution. In conclusion, while organizational learning via the AI-Coach resulted in better outcomes, the communication feature proved essential in preventing fatal mistakes by swiftly identifying and rectifying issues as they arose.

Xu et al. (2022) demonstrated a model of doctors and nurses assisting a distressed newborn with breathing, see also Xu et al. (2025), this volume. The case they addressed is whether the equipment is used effectively. Also, their findings demonstrated an AI-coach that prevents errors. This is consistent with the findings in this chapter, which show that an AI-coach enables organizations to identify and reduce errors, ultimately improving performance. An adaptive model for communication in labour and delivery rooms, including doctor-nurse communication, was developed by Doornkamp et al. (2023), see also Doornkamp et al. (2024), this volume. This doctor-nurse communication was also done with observation arrows in this chapter. However, one significant difference is that their study encourages a doctor to speak up when problems arise, whereas this chapter focuses on how an AI-coach monitors and warns doctors and nurses about mistakes. Weigl et al. (2022) investigated computational network models depicting communication between fathers and healthcare providers during childbirth, see also Weigl et al. (2025), this volume. They discovered that organizational learning promotes knowledge growth and communication. This is consistent with this chapter, which emphasizes how organizational learning improves the knowledge of the medical team and the overall procedure. However, in this chapter, a distinction was made between two scenarios: one involving organizational learning facilitated by the AI-Coach, and another involving only the AI-Coach's monitoring and warning function. This enabled a thorough examination of the differences in their impact.

The differences between their work and this chapter vary not only in scope but also in models. In the upper level of their work, Xu et al. used $\mathbf{M_W}$-states; Scenario 2 of this paper also used $\mathbf{M_W}$-states but Scenario 3 used $\mathbf{W_{WW}}$-states. This decision was made with the intention of facilitating individual learning of shared organizational mental models, as $\mathbf{W_{WW}}$-states define connections between AI-Coach's $\mathbf{W}$-states and a doctor's or nurse's individual $\mathbf{W}$-states. In order to maintain the strongest connection, this study used the maximum Hebbian function in Scenario 2 as opposed to Xu et al. who used the Hebbian combination function for $\mathbf{W}$-states. In the papers of Doornkamp et al. and Weigl et al., no Hebbian learning was used, whereas here, two different versions of learning are used: Hebbian learning in Scenario 2 and the alogistic function for the $\mathbf{W}$-states in Scenario 3. Furthermore, whereas Doornkamp et al.'s model focuses on fear of repercussions, this paper takes into account a variety of contextual factors. The AI-Coach process differs significantly between the three papers and this paper. In this study, the AI-Coach observes what happens in the world and this information is sent to the monitoring state, which then activates the decision states. If everything goes well, the decision states are suppressed, but if they activate, the upper-level communication states are activated. The chapters by Xu et al. (2025), Doornkamp et al. (2024), Weigl et al. (2025) in this volume do not address this specific process.

The primary limitation of the models is their narrow focus on neonates in respiratory distress, which limits their applicability to other medical conditions and patient populations. Future research should prioritize validating and adapting the model to encompass a wider range of healthcare scenarios, enhancing its versatility and effectiveness across diverse medical contexts. Moreover, expanding the communication capabilities of the AI-Coach beyond signalling the doctor and nurse to include other stakeholders, such as patients and their families, has the potential to enhance the overall functioning of the hospital work floor. By establishing communication channels with these additional stakeholders, the AI-Coach can improve coordination, collaboration, and overall efficiency within the healthcare environment. Additionally, it is essential to explore strategies that promote effective collaboration and trust between healthcare professionals and the AI-Coach. Developing methodologies that position the AI-Coach as a supportive tool, rather than a substitute for human expertise, is crucial. By fostering collaboration and trust, healthcare professionals can effectively leverage the capabilities of the AI-Coach while maintaining their essential roles in decision-making and patient care. Furthermore, future work will extend to validating these models against clinical outcomes from neonatal care, adjusting them based on empirical data to improve their predictive power and practical application. Engaging with healthcare professionals will be essential to this validation process to ensure the models' effectiveness in real-world settings.

References

Abdi, S., Bagherzadeh, J., Gholami, G., Tajbakhsh, M.S.: Using an auxiliary dataset to improve emotion estimation in users' opinions. J. Intell. Inf. Syst. **56**, 581–603 (2021)

Altman, Y., Iles, P.: Learning, leadership, teams: corporate learning and organizational change. J. Manag. Dev. **17**(1), 44–55 (1998)

Bass, C., Sajed, D., Adedipe, A., West, E.: Pulmonary ultrasound and pulse oximetry versus chest radiography and arterial blood gas analysis for the diagnosis of acute respiratory distress syndrome: a pilot study. Critical Care **19**, 282 (2015). https://doi.org/10.1186/s13054-015-0995-5

Canbaloğlu, G., Treur, J., Wiewiora, A.: From conceptual to computational mechanisms for multi-level organisational learning. In: Canbaloğlu, G., Treur, J., Wiewiora, A., (eds.) Computational Modeling of Multilevel Organisational Learning And Its Control Using Self-Modeling Network Models, pp. 5. Springer Nature, Cham (2023b). https://doi.org/10.1007/978-3-031-28735-0_5

Canbaloğlu, G., Treur, J., Roelofsma, P.H.M.P.: Learning of Safety And Security Management Through Cyberspace: An Adaptive Self-Modeling Network Model For Multilevel Organizational Learning. In: Roelofsma, P.H.M.P., Jabeen, F., Taal, H.R., Treur, J., (eds.) Using Shared Mental Models and Organisational Learning to Support Safety and Security through Cyberspace: A Computational Analysis Approach. Springer Nature, Cham, Switzerland (this volume) (2025)

Canbaloğlu, G., Treur, J. Wiewiora, A. (eds.).: Computational Modeling of Multilevel Organisational Learning and Its Control Using Self-Modeling Network Models. Springer Nature, Cham (2023a)

Cannon-Bowers, J.A., Salas, E., Converse, S.A.: Shared mental models in expert team decision-making. In: Castellan, Jr., N.J., (ed.) Individual and Group Decision Making, pp. 221–246. Lawrence Erlbaum (1993)

Catania, L.J.: AI applications in prevalent diseases and disorders. In: Foundations of Artificial Intelligence in Healthcare and Bioscience, pp. 293–444. Elsevier (2021)

Copetti, R., Cattarossi, L., Macagno, F., Violino, M., Furlan, R.: Lung ultrasound in respiratory distress syndrome: a useful tool for early diagnosis. Neonatology **94**(1), 52–59 (2008). https://doi.org/10.1159/000113059

Craik, K. J. W.: The nature of explanation. Cambridge, MA: University Press (1943)

Crossan, M.M., Lane, H.W., White, R.E.: An organizational learning framework: from intuition to institution. Acad. Manag. Rev. **24**(3), 522–537 (1999). https://doi.org/10.2307/259140

Doornkamp, S., Jabeen, F., Treur, J., Taal, H.R., Roelofsma, P.H.M.P.: A controlled adaptive computational network model of a virtual coach supporting speaking up by healthcare professionals to optimize patient safety. Cogn. Syst. Res. **81**, 37–49 (2023). https://doi.org/10.1016/j.cogsys.2023.02.002

Doornkamp, S., Jabeen, F., Treur, J., Taal, H.R., Roelofsma, P.H.M.P.: Optimizing speaking up behavior for safety and security through cyberspace: a controlled adaptive computational network model of a virtual coach. In: Roelofsma, P.H.M.P., Jabeen, F. Taal, H. R. Treur, J. (eds.) Using Shared Mental Models and Organisational Learning to Support Safety and Security through Cyberspace: A Computational Analysis Approach. Springer Nature, Cham, Switzerland (this volume) (2024)

García-Morales, V., Jiménez-Barrionuevo, M., Gutierrez, L.: Transformational leadership influence on organizational performance through organizational learning and innovation. J. Bus. Res. **65**(7), 1040–1050 (2012). https://doi.org/10.1016/j.jbusres.2011.03.005

Hendrikse, S.C.F., Treur.J., Koole. S.L.: Modeling emerging interpersonal synchrony and its related adaptive short-term affiliation and long-term bonding: a second-order multi-adaptive neural agent model. Int. J. Neural Syst. **33**(7), 2350038, 1–41 (2023). https://doi.org/10.1142/S0129065723500387

Hosseini, M., Kucharska, W., Treur, J.: An adaptive network model for a double bias perspective on learning from mistakes within organizations. In Proceedings of the 12th International Conference on Complex Networks and their Applications, ComplexNetworks' 23. Studies in Computational Intelligence, vol. 1144, pp. 91–103. Springer Nature Switzerland AG (2023)

Hosseini, M., Treur, J., Kucharska, W.: Learning from mistakes within organizations: An adaptive network-oriented model for a double bias perspective for safety and security through cyberspace. In: Roelofsma, P.H.M.P., Jabeen, F., Taal, H.R., Treur, J. (eds.) Using Shared Mental Models and Organisational Learning to Support Safety and Security through Cyberspace: A Computational Analysis Approach. Springer Nature, Cham, Switzerland (this volume) (2025)

Jayachitra, S., Prasanth, A., Hariprasath, S., Benazir Begam, R., Madiajagan, M.: AI enabled internet of medical things in smart healthcare. In: AI Models for Blockchain-Based Intelligent Networks in IoT Systems: Concepts, Methodologies, Tools, and Applications, pp. 141–161. Springer International Publishing, Cham (2023)

Jonker, C.M., van Riemsdijk, M.B., Vermeulen, B.: Shared mental models. In: Parsons, S., Wooldridge, M., Müller, J., (eds.) Coordination, Organizations, Institutions, and Norms in Agent Systems VI. Lecture Notes in Computer Science, vol. 6541, pp. 132–151. Springer, Berlin (2011)

Kavitha, M., Roobini, S., Prasanth, A., Sujaritha, M.: Systematic view and impact of artificial intelligence in smart healthcare systems, principles, challenges and applications. Mach. Learn. Artif. Intell. Healthcare Syst. 25–56 (2023)

Keith, N., Frese, M.: Enhancing firm performance and innovativeness through error management culture. In: Anderson, N., Herriot, P. (eds.), Handbook of Research on Professional Misconduct in Management and Public Administration, pp. 137–157. Edward Elgar Publishing (2011). https://doi.org/10.4135/9781483307961.n9

Kim, D.H.: The link between individual and organizational learning. In: Klein, D.A., (ed.) The Strategic Management of Intellectual Capital: Vol. 1st Edition. Routledge (1997). https://doi.org/10.4324/9780080517926

Kucharska, W., Bedford, D.A.: Love your mistakes!—they help you adapt to change. How do knowledge, collaboration and learning cultures foster organizational intelligence? J. Organ. Change Manag. **33**(7), 1329–1354 (2020)

Kucharska, W., Rebelo, T.: Knowledge sharing and knowledge hiding in light of the mistakes acceptance component of learning culture-knowledge culture and human capital implications. Learn. Organ. **29**(6), 635–655 (2022)

Machogu, E., Gaston, B.: Respiratory distress in the newborn with primary ciliary dyskinesia. Children **8**(2), 153 (2021). https://doi.org/10.3390/children8020153

March, J.G..: Special issue: organizational learning: papers in honor of (and by) James G. March. Organ. Sci. **2**(1) (1991). Accessed from https://www.jstor.org/stable/i345200

Mokadem, N., Jabeen, F., Treur, J., Taal, H.R., Roelofsma, P.H.M.P.: An adaptive network model for AI-assisted monitoring and management of neonatal respiratory distress. Cogn. Syst. Res. **86**, 101231 (2024)

Nigg, J.T.: Annual Research Review: On the relations among self-regulation, self-control, executive functioning, effortful control, cognitive control, impulsivity, risk-taking, and inhibition for developmental psychopathology. J. Child Psychol. Psychiatry **58**(4), 361–383 (2017)

Nigg, J.T.: Considerations toward an epigenetic and common pathways theory of mental disorder. J. Psychopathol. Clin. Sci. **132**(3), 297 (2023)

Saadat, V., Saadat, Z.: Organizational learning as a key role of organizational success. Procedia–Soc. Behav. Sci. **230**, 219–225 (2016). https://doi.org/10.1016/j.sbspro.2016.09.028

Saboute, M., Kashaki, M., Bordbar, A., Khalessi, N., Farahani, Z.: The incidence of respiratory distress syndrome among preterm infants admitted to neonatal intensive care unit: a retrospective study. Open J. Pediatrics **5**, 285–289 (2015). https://doi.org/10.4236/ojped.2015.54043

Scherer, K.R.: Emotions are emergent processes: they require a dynamic computational architecture. Philos. Trans. R. Soc. B: Biol. Sci. **364**(1535), 3459–3474 (2009). https://doi.org/10.1098/rstb.2009.0141

Sitterding, M.C., Raab, D.L., Saupe, J.L., Israel, K.J.: Using artificial intelligence and gaming to improve new nurse transition. Nurse Lead. **17**(2), 125–130 (2019)

Stout, R.J., Cannon-Bowers, J.A., Salas, E.: The role of shared mental models in developing team situational awareness: implications for training. In: Situational Awareness, pp. 287–318. Routledge (2017)

Swarnkar, M., Swarnkar, K.: Neonatal respiratory distress in early neonatal period and its outcome. Int. J. Biomed. Adv. Res. **6**, 643–647 (2015)

Sweet, L.R., Keech, C., Klein, N.P., Marshall, H.S., Tagbo, B.N., Quine, D., Kaur, P., Tikhonov, I., Nisar, M.I., Kochhar, S., Muñoz, F.M.: Respiratory distress in the neonate: case definition and guidelines for data collection, analysis, and presentation of maternal immunization safety data. Vaccine **35**(48 Pt A), 6506–6517 (2017). https://doi.org/10.1016/j.vaccine.2017.01.046

Treur, J.: Network-Oriented Modeling: Addressing Complexity of Cognitive, Affective and Social Interactions. Springer Nature, Cham (2016)

Treur, J.: Network-Oriented Modeling for Adaptive Networks: Designing Higher-Order Adaptive Biological, Mental and Social Network Models. Springer Nature, Cham, Switzerland (2020)

Treur, J.: On the dynamics and adaptivity of mental processes: relating adaptive dynamical systems and self-modeling network models by mathematical analysis. Cogn. Syst. Res. **70**, 93–100 (2021)

Treur, J., Van Ments, L. (eds.).: Mental Models and Their Dynamics, Adaptation and Control: A Self-Modeling Network Modeling Approach. Springer Nature, Cham, Switzerland (2022)

Van Ments, L., Treur, J., Klein, J., Roelofsma, P.H.M.P.: A second-order adaptive network model for shared mental models in hospital teamwork. In: Nguyen, N.T., et al. (eds.), Proceedings of the 13th International Conference on Computational Collective Intelligence, ICCCI'21. Lecture Notes in AI, vol. 12876, pp. 126–140. Springer Nature, Cham (2021)

Van Ments, L., Treur, J., Klein, J. Roelofsma, P.H.M.P.: Developing a safety and security AI coach: A second-order adaptive network model of shared mental models in hospital teamwork. In: Roelofsma, P.H.M.P., Jabeen, F., Taal, H.R., Treur, J. (eds.) Using Shared Mental Models and Organisational Learning to Support Safety and Security through Cyberspace: A Computational Analysis Approach. Springer Nature, Cham, Switzerland (this volume) (2025)

Van Ments, L., Treur, J.: Reflections on dynamics, adaptation and control: a cognitive architecture for mental models. Cogn. Syst. Res. **70**, 1–9 (2021)

Weigl, L.M., Jabeen, F., Treur, J., Taal, H.R., Roelofsma, P.H.M.P.: Modeling learning for a better safety culture within an organisation using a virtual AI coach: reducing the risk of postpartum depression by more communication with parents. Cogn. Syst. Res. **80**, 1–36 (2023). https://doi.org/10.1016/j.cogsys.2023.06.001

Weigl, L., Jabeen, F., Treur, J., Taal, H.R., Roelofsma, P.H.M.P.: Learning for a better safety and security culture within an organization: Reducing the risk in communication with AI coaching for security communication through cyberspace. In: Roelofsma, P.H.M.P., Jabeen, F. Taal, H. R. Treur, J. (eds.) Using Shared Mental Models and Organisational Learning to Support Safety and Security through Cyberspace: A Computational Analysis Approach. Springer Nature, Cham, Switzerland (this volume) (2025)

Wiewiora, A., Smidt, M., Chang, A.: The 'How' of multilevel learning dynamics: a systematic literature review exploring how mechanisms bridge learning between individuals, teams/projects and the organization. Eur. Manag. Rev. **16**, 93–115 (2019)

Wiewiora, A., Chang, A., Smidt, M.: Individual, project and organizational learning flows within a global project-based organization: exploring what, how and who. Int. J. Project Manag. **38**, 201–214 (2024)

Wilden, R., Hohberger, J., Devinney, T.M., Lavie, D.: Revisiting James March (1991): whither exploration and exploitation? Strateg. Organ. **16**(4) (2018). https://doi.org/10.1177/147612701876503

Xu, Y., Jabeen, F., Treur, J., Taal, H.R., Roelofsma, P.H M.P.: Adaptive agent network models with internal mental models supporting patient safety. In: Proceedings of the 15th International Conference on Social Computing and Networking, SocialCom'22 IEEE Computer Society (2022). https://www.researchgate.net/publication/363832332

Xu, Y., Jabeen, F., Treur, J., Taal, H.R., Roelofsma, P.H.M.P.: Supporting medical team decision in the neonatal domain: how safety and security risk management through cyberspace can be supported by adaptive agent network models with internal mental models. In: Roelofsma, P.H.M.P., Jabeen, F., Taal, H.R., Treur, J. (eds.) Using Shared Mental Models and Organisational Learning to Support Safety and Security through Cyberspace: A Computational Analysis Approach. Springer Nature, Cham, Switzerland (this volume) (2025)

Supporting Risk Management Through Cyberspace: An Adaptive Network Model Simulating AI Coach Effects by Inducing Adherence to Guidelines in Neonatal Medical Protocols

Raghav Chawla, Fakhra Jabeen, Jan Treur, H. Rob Taal, and Peter H. M. P. Roelofsma

Abstract In this chapter, it is shown how second-order adaptive agent-based network models can be used to support a medical team in healthcare institutions to adhere to specific Neonatal Hypoglycemia and Neonatal Hyperbilirubinemia treatment guidelines through the integration of an Artificial Intelligence (AI) Virtual Coach. The proposed AI Coach is designed to provide timely interventions and correct deviations when lapses in the health care practitioner's internal mental model occur. Through simulating three different scenarios, the internal dynamics of these mental models, adaptive changes of these mental models (learning and forgetting), and the interaction between health care practitioners and the world is shown when: (1) There is perfect adherence to guidelines, (2) There is imperfect adherence to guidelines and (3) There is both perfect and imperfect adherence to guidelines alongside interventions of the AI Coach in the latter case.

R. Chawla · J. Treur (✉)
Social AI Group, Vrije Universiteit Amsterdam, Amsterdam, The Netherlands
e-mail: j.treur@vu.nl; j.treur@hhs.nl

R. Chawla
e-mail: raghavchawlaportfolio@gmail.com

F. Jabeen
Radboud University Nijmegen, Nijmegen, The Netherlands
e-mail: fakhra.jabeen@ru.nl

J. Treur · P. H. M. P. Roelofsma
The Hague University of Applied Sciences, Centre of Expertise Cybersecurity, Risk Management and Cybersecurity Group, Hague, The Netherlands
e-mail: p.h.m.p.roelofsma@hhs.nl

H. Rob Taal · P. H. M. P. Roelofsma
Erasmus MC Rotterdam, Department of Neonatal and Pediatric Intensive Care, Division of Neonatology, Rotterdam, The Netherlands
e-mail: h.taal@erasmusmc.nl

P. H. M. P. Roelofsma et al. (eds.), *Using Shared Mental Models and Organisational Learning to Support Safety and Security Through Cyberspace: A Computational Analysis Approach*, Studies in Systems, Decision and Control 570,
https://doi.org/10.1007/978-3-031-72075-8_6

1 Introduction

Healthcare services provided at any care centre, are based upon the laid-out guidelines of medical protocols. These operating procedures play a critical role in ensuring that healthcare standards are justly maintained and that patients receive the most optimal care. In an ideal world, such guidelines would be followed perfectly without any mistakes committed. Often however, practically due to the nature of human error, deviations from established protocols by medical professionals can lead to suboptimal patient care, medical errors and even compromise the safety of clinical subjects. The complexity of human behavior and cognitive processes makes monitoring, predicting, and enforcing adherence to these protocols challenging. Considering this, the introduction of Artificial Intelligence (AI) offers promising capabilities for mitigating this issue. The research embodied in this chapter explores the potential of a virtual AI Coach to foster adherence to medical protocols, by providing timely intervention in instances of deviation by medical professionals. By drawing upon the modeling approach of self-modeling network modeling (Treur 2020), this paper aims to integrate these principles to construct mental models in a temporal-causal network of a medical health care practitioner (HCP) and an AI Coach (AIC), to simulate and explore the effect of the AI on the intricate interactions that dictate adherence to various medical protocols in the domain of neonatal care.

2 Background Literature

This section delivers an overview of the background literature and concepts, serving as a foundational basis for the experimental simulations that the paper entails.

2.1 Neonatal Hypoglycemia

Glucose is a type of sugar that serves as a vital energy source for the metabolic processes in the brain. Hypoglycemia is the condition of having an abnormally low blood sugar (glucose) concentration in the body (Hawdon 2019). Depending on its intensity and persistence, Hypoglycemia may lead to unfavourable neurological effects which can have potentially long-term implications if not treated on time. Clinically, Neonatal Hypoglycemia is defined when a newborn neonate has a blood sugar level of less than 2.5 mmol/L (45 mg/dL) and a glucose cut-off threshold of above 2.6 mmol/L (47 mg/dL) is considered to be clinically out of risk if glucose levels remain stable for three consecutive monitor readings (Rao 2022; Rasmussen et al. 2017).

The cornerstone of Hypoglycemia prevention is foremostly the identification of at-risk neonates. There are significant challenges associated with clinical detection of this condition when a neonate is born as quite often, the infants show no visible signs of Hypoglycemia and are otherwise healthy neonates (asymptomatic) (Rasmussen et al. 2017). The module protocol guideline followed while creating the network-oriented self-model concerns preventing Neonatal Hypoglycemia in this group of otherwise healthy neonates who are at risk for developing Hypoglycemia within the first 24 h after birth without showing any external symptoms. These neonates adhere to a different set of guidelines for preventing and treating Hypoglycemia when compared to neonates who show visible symptoms of Hypoglycemia such as sweating, feeding difficulties, poor suck, weak or high-pitched cry, tremors, hypothermia, and seizures (Abramowski and Hamdan 2021). Asymptomatic otherwise healthy neonates are likely to develop neonatal hypoglycemia if they are born having major risk factors, including prematurity (gestational age <37 weeks), small for gestational age (SGA, birth weight <10th percentile of neonates), large for gestational age (LGA, birth weight >90th percentile of neonates), and infants born to mothers with diabetes (type 1, type 2) who received insulin or oral blood sugar lowering medication during pregnancy (NVK 2020).

The susceptibility of these neonates to neonatal hypoglycemia arises from a combination of factors. Firstly, there is a limited capacity for endogenous glucose production, which is further compounded by increased glucose consumption and inadequate food intake during the initial days following birth. This precarious metabolic situation, disrupts the normal glucose metabolism adaptation in neonates and hampers their access to alternative fuel sources, thus elevating the risk of potential brain damage (NVK 2020). Timely feeding, preferably breastfeeding, is often helpful in preventing hypoglycemia in high-risk otherwise healthy neonates (NVK 2020). Once diagnosed with Neonatal Hypoglycemia, depending on the severity of glucose concentration levels in the blood, the preventive measures taken would be from a heavy or normal intravenous (IV) glucose dose to raising nutritional intake of carbohydrates for less severe glucose levels (NVK 2020). It is important to note that various types of intravenous (IV) therapies exist. To treat the most severe hypoglycemia cases, a 'bolus' or a push IV dose is given. Here a heavy glucose dose is injected through a syringe in a short span of time. There also exist continuous and drip IV therapies, which either continuously feed glucose intravenously over a period of time or keep giving small doses after certain time intervals respectively (NVK 2020). In the treatment of Hypoglycemia, a high dose and a normal dose of continuous IV therapy are used. The treatment of neonatal hypoglycemia presents several vulnerabilities that can become a cause for healthcare practitioners to deviate from following parts of the medical protocol inaccurately. One of the biggest weaknesses concerns late or missed diagnosis of risk factors for hypoglycemia leading to missed or delayed diagnosis especially for neonates who are asymptomatic, thus exacerbating long-term neurodevelopmental outcomes (McKinlay et al. 2015).

Thus, it is vital to screen every neonate to see if the neonate has any risk factors and observe that these warrant further precaution, enough to start following the medical protocol. The issue of monitoring is also critical in managing neonatal hypoglycemia. Insufficient monitoring and improper blood sampling can lead to unnoticed episodes of hypoglycemia, which may result in sub-optimal treatment approaches (Thornton et al. 2015). Moreover, the management protocols also demonstrate vulnerabilities in treatment. Deviations from treatment guidelines or delays in implementing necessary interventions, such as a bolus glucose dose before intravenous glucose therapy in the case of severe hypoglycemia, can prolong the onset of the condition in the neonate, causing potential harm to the infant (Adamkin 2017). Furthermore, the importance of an appropriate follow-up after glucose levels stabilise outside the risk zone is essential but is often not given the attention it deserves. Without proper post-treatment surveillance, clinicians may fail to identify persistent symptoms or observe a potential relapse into hypoglycemia, thus failing to provide optimal care (Narvey and Mark 2019). These vulnerabilities in neonatal hypoglycemia management emphasize the need for rigorous adherence to guidelines, accurate and frequent monitoring, appropriate treatment, and comprehensive follow-up care.

2.2 Neonatal Hyperbilirubinemia

Hyperbilirubinemia, colloquially known as Jaundice, is a medical condition characterized by a visible yellowish discoloration of the skin, sclera, and mucous membranes due to abnormally high concentrations of serum bilirubin levels combined with a simultaneously low bilirubin conjugation capacity in the liver (Ullah et al. 2016). Bilirubin is a yellow pigmented molecule that is a product resulting from the catabolism of haem-containing compounds such as haemoglobin in the blood. It is a known toxic metabolite and should normally be processed and conjugated by the liver with glucuronic acid into bilirubin glucuronides as part of bile and excreted out of the body. However, this process is hindered in individuals with liver deficiencies as mentioned above. High bilirubin levels cause blood plasma and thus subsequently the skin to develop a yellow tint, being one of the distinct symptoms of Jaundice.

It is important to note that bilirubin exists in the blood in 2 types, namely, indirect bilirubin also known as unconjugated bilirubin which is not water soluble and thus not measurable, and direct bilirubin also known as conjugated bilirubin which is water soluble after being conjugated as indirect bilirubin in the liver. Total bilirubin also known as Serum Bilirubin (SBR) is the total bilirubin in the blood plasma of the body and encompasses the sum of the 2 types described above (Laxmi et al. 2018).

Neonatal Hyperbilirubinemia is a cause of major concern as 9–11% of neonates develop upon birth with 2% experiencing severe Jaundice (Ullah et al. 2016). Severe Hyperbilirubinemia can lead to potentially lethal complications. A notable example of this is Kernicterus, a neurological syndrome resulting from bilirubin toxicity that can inflict permanent damage to the basal ganglia, hippocampus, and cranial nerve nuclei (Johnson and Bhutani 2011). Clinically, Jaundice is diagnosed in neonates when serum bilirubin level is observed to be above 5 mg/dL, whereas severe Jaundice is recognised when serum bilirubin levels reach 20 mg/dL (Slusher et al. 2017). Premature infants are particularly vulnerable to Hyperbilirubinemia, owing to their immature hepatic function, increased bilirubin production, and enhanced enterohepatic circulation (AAP 2004). There are several observed symptoms which point to the developing concentration levels of bilirubin in blood plasma. These include, foremostly visually observable yellow tinge of the skin and eyes, followed by dark urine, pale stools and maternal blood group incompatibility (Gillott 2017). Maternal blood group incompatibility in particular can give rise to various conditions such as rhesus disease, and hemolytic disease which are linked to the occurrence of jaundice in neonates (Gillott 2017).

Current clinical protocol necessitates monitoring of serum bilirubin levels and tracking the infant's jaundice visually (AAP 2004). Depending on the severity of the concentration of total serum bilirubin (TSB), different treatment strategies are used which include various levels of phototherapy and, in severe cases, exchange transfusion. Phototherapy uses light waves to allow for the absorption of light through the skin which converts unconjugated bilirubin into bilirubin photoproducts that are excreted in the stool and urine (Muchowski 2014). Single phototherapy refers to phototherapy with a single light source, however with higher bilirubin levels, double phototherapy is used to cover a larger surface area of the baby's body which will further help bring down the bilirubin levels in the body. Intensive phototherapy implies irradiance in the blue-green spectrum with wavelengths of 460–490 nm and is the most effective in converting unconjugated bilirubin (Muchowski 2014).

Despite its efficacy, phototherapy is not without drawbacks. Among these are potential side effects such as dehydration, due to increased fluid loss through the skin, and skin rashes. In addition, a less common yet critical condition known as bronze baby syndrome may develop in infants with liver disease undergoing phototherapy. This condition is characterized by a greyish-brown discoloration of the skin. Phototherapy also necessitates a physical separation between the infant and the mother, causing an emotional and psychological toll on the parents particularly in a period where parent-infant bonding is critical (Muchowski 2014).

In severe cases of hyperbilirubinemia, exchange transfusion might be required. Exchange transfusion is a complex medical procedure, during which the baby's blood is gradually removed and replaced with fresh donor blood or plasma (Maisels et al. 2009). This procedure aims to rapidly decrease bilirubin levels in the blood and prevent or treat acute bilirubin encephalopathy, a potentially life-threatening condition. Specifically, infants presenting with TSB levels of 25 mg/dL or higher, along with symptoms of jaundice and signs of acute bilirubin encephalopathy, are considered for this intervention (Muchowski 2014).

However, exchange transfusion is not a risk-free procedure. Mortality within the first six hours post-procedure in full-term infants without hemolysis is between three and four per 1,000 infants (Muchowski 2014). Moreover, approximately 5% of infants undergoing exchange transfusion may experience serious complications, including blood-related problems like changes in blood clotting factors, cardiorespiratory instability, metabolic abnormalities such as electrolyte imbalances and complications from central line placement (Muchowski 2014). Thus, the judicious application of these strategies is crucial as unwarranted treatment could expose the neonate to unnecessary risks including Hemolysis and Sepsis.

The guidelines laid out for preventing neonatal hyperbilirubinemia present a variety of vulnerabilities that may compromise the effectiveness of its treatment.

One such vulnerability is the failure to observe jaundice symptoms visually. In particular, the possibility of not realising that the skin of a neonate has tinged into a light shade of yellow (a physical sign of hyperbilirubinemia) is likely, and could lead to an insufficient or late response, potentially causing serious complications (Keren et al. 2009). Ceasing phototherapy prematurely is another vulnerability in the treatment of neonatal hyperbilirubinemia as abrupt or premature discontinuation of phototherapy without proper post-treatment supervision could lead to a rebound increase in bilirubin levels, thereby precipitating a need for re-treatment (Paul and Maisels 2017). It is also equally important to dynamically reduce and increase the treatment intensity by routinely following and checking the serum bilirubin levels (Yetman et al. 1998). It is important not to continue intensive treatment such as exchange transfusion and intensive phototherapy for prolonged times, while automatically reducing the treatment threshold from intensive to a lighter single or double phototherapy alongside frequently checking total serum bilirubin measurements (Yetman et al. 1998). Universally, it is maintained that phototherapy can be stopped once three consecutive measurements have indicated that bilirubin levels are consistently 50 mmol/L below the phototherapy threshold line (Osmond and Northrop 2022). Further on, it is important to transfer the neonate back into maternal care after phototherapy is stopped and not keep the neonate under supervision inside the neonatal intensive care unit (NICU) to ensure the maternal-neonate bond continues to develop (Clinical and Guidelines 2022) and prevent the drawbacks as stated above. Lastly, there is a need to maintain vigilance for the re-emergence of jaundice post-discharge. Therefore, keeping these patients under close supervision after stopping interventions is critical (Yetman et al. 1998). This underscores the pressing need to address potential vulnerabilities in the protocol for managing neonatal jaundice, calling for adherence to guidelines.

2.3 *Potential for an AI Coach in Promoting Medical Protocol Adherence*

Recent integrations of Artificial Intelligence in healthcare have shown considerable potential to play an important role in promoting adherence to healthcare guidelines by medical professionals. By evaluating real-world data, AI systems can produce insights, improve decision-making and optimise health outcomes and patient safety by providing real-time feedback and decision support to healthcare professionals, ensuring that the care provided aligns with established guidelines and best practices. For example, AI-integrated Clinical Decision Support Systems (CDSS) can recognise real-time actions in the operation room and recommend or alert practitioners when a deviation from guidelines is detected. Such systems don't merely act as passive references, but actively guide and support clinical decisions (Wasylewicz and Scheepers-Hoeks 2018). Such proactive and dynamic engagements reinforce adherence to guidelines, ultimately driving improved patient outcomes and system efficiency. This is opposed to older systems such as the Modified Early Warning Score (MEWS), a tool used by hospitals, which calculates the mathematical physiological probability of a patient's likelihood of experiencing clinical decline in a short subsequent time interval of an operation, to prevent delay in intervention and promote more vigilance in doctors (AHA 2023). The implementation of aforementioned AI algorithms-based systems can, however, undoubtedly allow for better support in neonatal medical protocols. The implementation and effect of these AI coaches can be systematically explored by network models, including the simulation of optimal processes, common deviations, and AI interventions which is presented below.

3 Methodology

This section delivers an overview of the network-oriented temporal-causal modelling technique, drawn from the approach outlined by Treur (2016, 2020), which allows the construction of self-models in networks used to run the simulations this chapter revolves around. Additionally, this section details the motives followed to create the adaptive network model.

3.1 Introduction to the Network-Oriented Self-modeling Approach

Network-oriented modeling is a technique that can be used for representing, analysing and understanding numerous complex cognitive, biological and social processes that are temporal in nature (Treur 2016). Through graphically conceptualising a phenomenon, even the space between individual and collective behaviour can be bridged. Adding to this, adaptive reified network models, also known as self-modelling networks, adds in an easy way the concept of adaptation to network models and together create a temporal-causal network which is better suited to model learning and forgetting in a dynamically changing environment.

To elaborate, firstly, a complex real-world process can be represented conceptually as a base-level network which can include connections to different stakeholders in the process with their own subprocesses as well. Other than graphically representing such networks, they can also be represented in table format through role matrices which is the required format needed as input to simulate the networks in the software environment used. By focusing on self-modelling and creating adaptive networks, which adjust their structure based on internal and external feedback, a powerful environment which allows capturing the dynamic nature of complex scenario settings in the real world such as healthcare and mental models used in protocols to be followed can be created (Treur 2020).

Mental models operate as essential structures that shape the behavior of individuals. Kim (1993) elucidates these models as encompassing an individual's worldview, inclusive of both implicit and explicit understandings. These models offer a medium for processing and interpreting new information while prescribing the relevance of previously stored knowledge to a given context. These mental models enable individuals to create internal simulations based on specific circumstances. Furthermore, individuals can acquire novel mental models or adapt and improve existing ones, as seen through Piaget's theory, where accommodation (altering one's cognitive concepts based on experiences) and assimilation (integrating experiences into existing cognitive concepts) are keys for learning (Piaget 1954).

It is possible to add adaptivity to the mental models in the base network by creating a self-model of some of the network characteristics in an added reification (self-modeling) level: the first-order self-modeling or reification level. Reification is defined as "representing something abstract as a material or concrete thing or making something abstract more concrete or real" (Treur 2020) and is visually placed in a plane above the plane of the base level in the conceptual model. Further on, a second-order self-modeling level can be further added to add adaptivity to the first-order self modeling level as well (metaplasticity) (Treur 2020). This allows for control of the adaptation in the model (the extent of learning or forgetting) as a whole and creates a three-level architecture to enable control over network characteristics like adaptation speed, the strength of connections, persistence and other fundamental or advanced network attributes. Additionally, in a situation where there exist different pathways in the base network representing a different variation of the processes, context factors exist to exert an external influence to specify which states get activated through having a strong connection while the other context factors have a null weight on their connections (Treur 2020). In the brain for example, the base-level network could be equated to the neural pathways in the brain which holistically execute a process together by propagation of neuron activations. The first-order self-model could be seen as determining the synaptic transmission between neurons, thereby controlling the strength of the connection. The second-order self-model, would then be seen as modeling the meta-plasticity (plasticity of synaptic plasticity), indicating that the states at this level can modify the extent of (strengthening or weakening) synaptic plasticity (Treur 2020).

This pattern of self-modelling in a network can be recursively applied to make even higher levels of self-models beyond the second-order level. For example, taking several adaptation levels into account that have been formed during evolution (Treur 2024). And up to five orders of adaptation can be identified for epigenetic effects on mental disorders, e.g., Kathusing et al. (2024), Treur (2024). For the experimental simulations conducted during the course of this research, a second-order adaptive self-modeling network is employed for creating simulations of neonatal medical protocols through using internal mental models.

3.2 Design and Specification of Self-modeling Networks

The fundamental concept in the network-oriented modeling approach is the presence of a labelled graph containing states, which are nodes, where a node X can be connected through a link to another state Y where X has a causal impact on Y at a certain time, and takes into account the strength of connections, the speed of causal effects, and the combination of different incoming connections and the way in which their collective aggregated effect on the state Y. In summary, such characteristics are described as follows Treur (2016):

- **Connectivity** characteristics: Connections from a state X to a state Y have a connection weight value, represented by $\mathbf{\omega}_{X,Y}$ which represents the strength of the causal impact of state X on state Y throgh the connection
- **Timing** characteristics: For each state Y, a speed factor represented by $\mathbf{\eta}_Y$ indicates how fast the state is changing upon a causal impact.
- **Aggregation** characteristics: For each state Y, a combination function represented by $\mathbf{c}_Y(..)$ determines mathematically how the causal impacts of other states on state Y is combined.

The above concepts are numerically represented and incorporated within related difference equation which represents timing the causal impact (Treur 2016):

$$Y(t + \Delta t) = Y(t) + \mathbf{\eta}_Y[\mathbf{aggimpact}_Y(t) - Y(t)]\Delta t \quad (1)$$

where $\mathbf{aggimpact}_Y(t) = \left[\mathbf{c}_Y\left(\mathbf{\omega}_{X_1,Y}X_1(t), \ldots, \mathbf{\omega}_{X_k,Y}X_k(t)\right) - Y(t)\right]\Delta t$

This shows how the causal impact on Y is exerted over time gradually, using speed factor $\mathbf{\eta}_Y$. This includes single causal impacts represented by $\mathbf{\omega}_{X_i,Y}X_i(t)$, where at time t each state X_i with connection to state Y has an impact on Y, using weight $\mathbf{\omega}_{X_i,Y}$. Equation (1) collects the aggregated impact of multiple states X_i on Y at time t, which is determined using combination function $\mathbf{c}_Y(..)$ (Treur 2016).

Designing a network model means that at the end the above network characteristics are specified in a systematic manner. A dedicated software environment based on the above network characteristics is utilized to generate the simulation output; see Chap. 9 of Treur (2020). A visual graph schema of a conceptual self-modeling network model (See Fig. 6) was initially built, and subsequently translated into various role matrices to delineate the network features based on their specific roles. The software environment for the computational work was MATLAB, which took the different role matrices as input to generate the graphical simulation output. These input role matrices include both adaptive and non-adaptive types. Table 1 provides a summary of the descriptions of these matrices, and a comprehensive review of all the role matrices used for each scenario can be found in the Appendix.

The software environment employs a numerically indexed library of various combination functions to delineate the aggregation traits of the model. The fundamental combination functions are denoted as $\text{bcf}_i(\text{p, v})$, utilizing the assigned list of parameters 'p' and values 'v', as detailed in Table 2. In this library, every index

Table 1 Overview of role matrices used in a detailed design and in the software environment

Role matrix	Description
iv	Initial values: specifies initial values of all state
mb	Base matrix: specifies incoming causal connections for every state
mcwv	Connection weight value matrix: specifies non-adaptive connection weights $\boldsymbol{\omega}$
mcwa	Connection weight adaptation matrix: specifies adaptive connection weights
mcfwv	Combination functions weight matrix: specifies function **c** used for each state
mcfwa	Combination functions adaptation matrix: specifies adaptative function used for a state
mcfpv	Combination functions parameter value matrix: specifies function parameter values for non-adaptive parameters
mcfpa	Combination functions parameter adaptation matrix: specifies function parameters for adaptive parameters
msv	Speed value matrix: specifies nonadaptive speed factor $\boldsymbol{\eta}$ for states
msa	Speed adaptation matrix: specifies adaptive speed values for states

number *i* corresponds to one of over 65 unique mathematical functions that simulate social contagion, learning, synchrony, and homophily, among other concepts (Treur 2020). Each state *Y* can be assigned only one combination function; however, the complete model may incorporate multiple different functions for different states. For every state *Y*, the weighted mean is computed using the combination function weights that are housed in the **mcfwv** and **mcfwa** matrices with the following calculation:

$$\mathbf{c}_Y\left(\boldsymbol{\pi}_{1,1,Y}, \boldsymbol{\pi}_{2,1,Y}, \ldots, \boldsymbol{\pi}_{1,\mathrm{m},Y}, \boldsymbol{\pi}_{2,\mathrm{m},Y}, V_1, \ldots, V_k\right)$$

$$= \frac{\boldsymbol{\gamma}_{1,\mathrm{Y}}\mathrm{bcf}_1\left(\boldsymbol{\pi}_{1,1,Y}, \boldsymbol{\pi}_{2,1,Y}, V_1, \ldots, V_k\right) + \cdots + \boldsymbol{\gamma}_{\mathrm{m,Y}}\mathrm{bcf}_{\mathrm{m}}\left(\boldsymbol{\pi}_{1,\mathrm{m,Y}}, \boldsymbol{\pi}_{2,\mathrm{m,Y}}, V_1, \ldots, V_k\right)}{\boldsymbol{\gamma}_{1,\mathrm{Y}} + \cdots + \boldsymbol{\gamma}_{\mathrm{m,Y}}}$$

Table 2 Overview of combination functions and their parameters selected to model the simulations in this chapter

	Function	Formula	Parameters
Identity	$\mathbf{id}(V_1,\ldots,V_k)$	V	No parameter
Advanced logistic sum	$\mathbf{alogistic}_{\sigma,\tau}(V_1,\ldots,V_k)$	$[\frac{1}{1+e^{-\sigma(V_1+\cdots+V_k-\tau)}} - \frac{1}{1+e^{\sigma\tau}})](1+e^{-\sigma\tau})$	Steepness σ Excitability threshold τ
Max Hebbian learning composition	$\mathbf{maxhebb}_{\mu}(V_1,\ldots,V_k)$	$\max(V_1V_2\,(1-V_3) + \mu\, V_3,\, V_4,\ldots,\, V_k)$	Persistence factor μ

In Treur (2021), Hendrikse et al. (2023), it is shown that any smooth dynamical system has a canonical representation as a temporal-causal network and any smooth (multi-order) adaptive dynamical system has a canonical representation as a (multi-order) self-modeling temporal-causal network. Therefore, compared to adaptive dynamical systems in general, the network-oriented modeling approach used here does not introduce any fundamental limitations concerning what it can model. This has also been confirmed by applications to case studies in practice, e.g., Treur and Van Ments (2022), Canbaloğlu et al. (2023).

To assimilate the above and implement the self-modelling network in the context of this chapter, a model of the healthcare guidelines has to be made for the real-world process. In addition to the process of the guidelines in the real world, the inclusion of a health care practitioner (doctor) to execute the protocol and showcase the interactions between the process being followed is needed. This inclusion will be done through the creation of an internal mental model of the health care practitioner at the base level, to show their thinking process which acts as the preparation before translating into their actions towards the extent they follow the medical protocol in the real world. The healthcare practitioner will be abbreviated to "HCP" in further explanations and in representing the mental states and the mental model of the healthcare practitioner in the conceptual model. Moreover, this paper aims to model and determine the effectiveness of a virtual Artificial Intelligence Coach in fostering adherence of the HCP in following the medical protocol. To observe this, a separate mental model of the AI coach is included alongside that of the HCP at the base level of the conceptual model. The AI coach will be abbreviated to AIC in further explanations and in representing the mental states and the mental model of the AI coach in the conceptual model. The AIC should become aware of any deviations in the execution of the medical protocol and may provide support to the HCP by communicating to alert the HCP of the deviation so as to correct them. Both the HCP and AIC mental models contain a fully identical rendition of the medical protocol to signify that both entities in principle possess the complete knowledge representation of the medical protocol in order to be able to carry out their tasks.

3.3 Modelling Decisions and Methods

In the context of this chapter, three scenarios were modelled using the self-modelling approach to show how an artificial intelligence coach (AIC) can intervene to encourage adherence to guidelines. This will be addressed for both the neonatal medical protocols for the treatment of Neonatal Hypoglycemia and Neonatal Hyperbilirubenemia. The first scenario simulates the medical protocol being followed by the health care practitioner HCP with a perfect mental model. This means that the HCP does not commit any mistakes while following the protocol and receives input from the real world, comprehends, stores it perfectly, and effectuates the corresponding actions needed to follow the protocol without any deviations, delays or mistakes. It is important to note that due to prior experience and education the doctor knows what steps to take when exposed to contextual stimuli from the real-world states such as specific measurement readings. The second scenario deals with simulating the HCP following the medical protocols with an imperfect mental model. This means that the HCP makes mistakes while following the protocol and does not effectuate the actions in an ideal manner in the real world. The third scenario will in two parts simulate the HCP following the medical protocol with the same perfect and imperfect mental model as Scenario 1 and 2, however an artificial intelligence coach (AIC) with perfect mental model is introduced to intervene and prevent any deviations when required. It is important to note that unlike the doctor the AIC does not have prior experience and education and does not know the steps that will be taken or followed by the doctor when exposed to contextual stimuli from real-world states such as specific measurement readings and thus needs constant inputs from all world-states to become aware of which steps in the protocol, thus which world-states in the model, are being executed and to what extent have they been carried out.

Based on the background literature described in Sect. 2, a conceptual model (see Figs. 1 and 2) for each medical protocol was built to establish the framework upon which the final self-modeling network would be simulated. The protocol for the prevention of neonatal hypoglycemia was modelled conceptually following the medical protocol flowchart provided by the Dutch National Union of Child Medicine (NVK 2020). The protocol for prevention of neonatal hyperbilirubinemia was modelled conceptually following the medical protocol flow chart provided by University Hospital Bristol and Weston (Osmond and Northrop 2022). The concepts discussed in the background literature and in these medical protocols were used in the designed computational network model that was simulated using the software described in Sect. 3.2. The conceptualised model envisages a procedure in line with the established protocol which models different paths to be followed depending on different situations in the real world concerning the physical conditions of the neonate. Sections 4.2 and 4.4 describe the specific pathway that was followed in the medical protocols during the simulation. There are crucial measuring states incorporated within the protocols which branch into various measurement readings and give rise to following a different pathway in the protocol, tailored to treat the specific circumstances corresponding to the measurements. To capture a specific circumstantial situation related to the unique conditions of the neonate, context factors are used to allow for the activation of certain measurements and thus their consequent pathways and make sure that the states exclusively included in the pathways which arise from other measurements are not activated as that would be illogical and contradictory. Setting the context factors' initial value and the incoming connections to the contextual nodes at '1', activates the measurements and their subsequent pathway while setting the value and weights to '0' ensures that those states are not activated. For instance, if a baby does not have small gestational age as a risk factor but has large gestational age (can never have both), then the context factors for small gestational age will be set at '0' and '1' will be set for the context factor of the latter. In the conceptualised model of this chapter, the context factors and thus the pathway taken to treat the most severe conditions and symptoms of the Hypoglycemia and Hyperbilirubinemia has been chosen (nhs and sJ respectively).

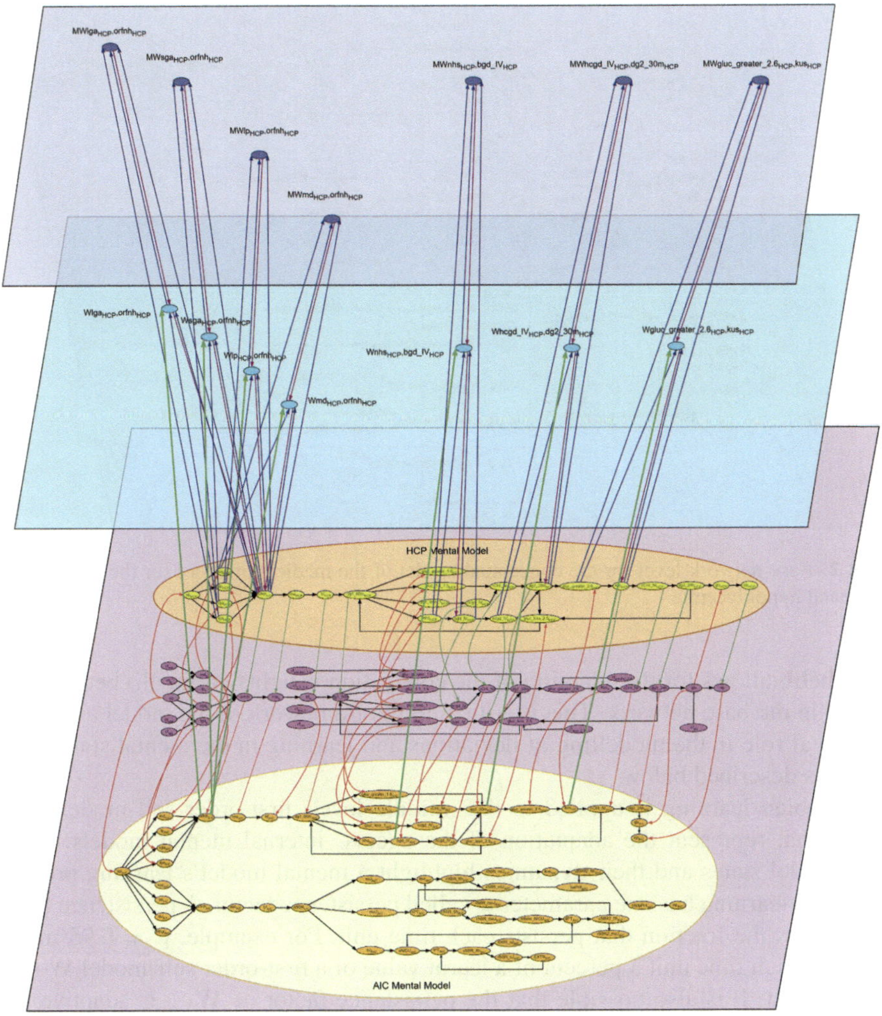

Fig. 1 Final conceptual model depicting medical protocol to treat neonatal hypoglycemia including the presence of hcp and aic mental models

As highlighted above, the combination functions used for aggregations in the states of the conceptual model have been chosen to be **id**, **alogistic** and **maxhebb**. The **alogistic** function is used for states on the base level which have incoming connections to them. For states with no incoming connections such as context factors and the initial state of neonate being born all have combination function set to **id** meaning they need no aggregations done to them. For the first-order adaptation states, the max Hebbian learning composition function was assigned. This combines the max function with an adaptation principle (Hebb 1943) which in a simplified form follows the motto "neurons that fire together, wire together". Specifically, the function

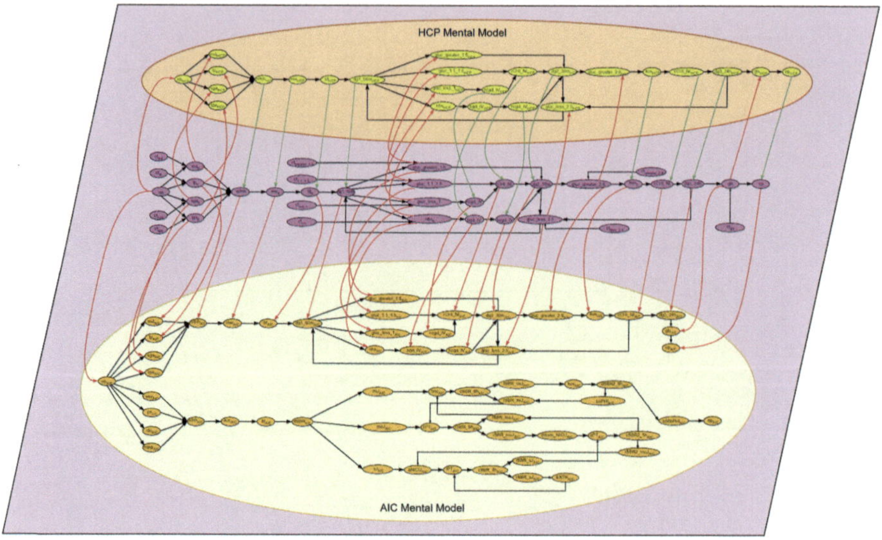

Fig. 2 Base network level for the conceptual model of the medical protocol for the treatment of neonatal hypoglycemia

maxhebb allows for the adaption of the connection weights (strength) between the states in the base network. This is critical to the framework of the model as it plays a pivotal role in the modelling of deviations and learning in the mental state of the HCP as described below.

Hebbian learning is applied to the (middle level) first-order self-model states $\mathbf{W}_{X,Y}$ that represent the adaptation of the agents' internal mental models. These self-model states and their dynamics highlight a mental model's learning process. Hebbian learning has one parameter μ called persistence factor. A persistence factor represents the fraction that persists each time unit. For example, $\mu = 0.95$ means that for each time unit 5 percent of a learnt value of a first-order self-model $\mathbf{W}$-state can be lost. It is also possible that the persistence factor of $\mathbf{W}_{X,Y}$ is adaptive too, represented by a second-order self-model $\mathbf{M}_{\mathbf{W}}$-state $\mathbf{M}_{\mathbf{W}_{X,Y}}$. In the context of this chapter, the second-order $\mathbf{M}_{\mathbf{W}}$-states are influenced by the context state in the base level. They control the adaptation of mental models (metaplasticity), for example stressful circumstances may make the level of $\mathbf{M}_{\mathbf{W}_{X,Y}}$ lower which means more forgetting.

Note that **all** the actions of the HCP (and thus the prerequisite preparation states in the mental model of the HCP due to to which those actions resulted) will be monitored by the AI Coach to ensure that not a single deviation takes place. In the context of this paper to show that the AIC can display behaviour which better induces adherence to guidelines in different medical protocols, specific vulnerabilities for each protocol which were described in the background literature (Sect. 2) have been chosen to be the parts of the protocol where the HCP makes a mistake when it has an imperfect mental model so that AIC has to notice that and intervene.

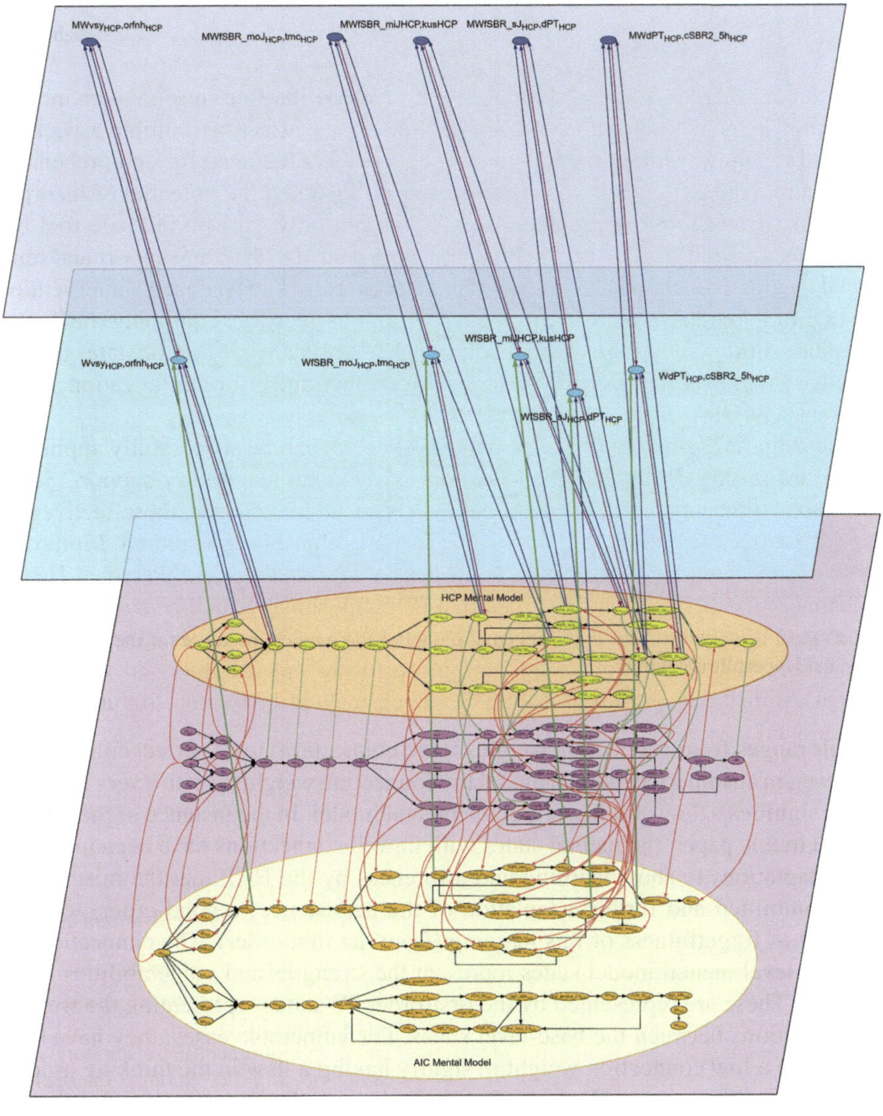

Fig. 3 Final conceptual model depicting medical protocol for treatment of neonatal hyperbilirubinemia including the presence of HCP and AIC mental models

The methodology used to model deviations for simulating Scenario 2 in both medical protocols is as follows. To show that the HCP has not followed a certain step in the medical protocol correctly, there has to be a flaw in their thinking within their mental model which has caused this flaw to happen in the real world. In a perfect mental model, to show that all the connections in the mind are perfect, the weight of the connections is set to 1. Note that any connection in any of the mental

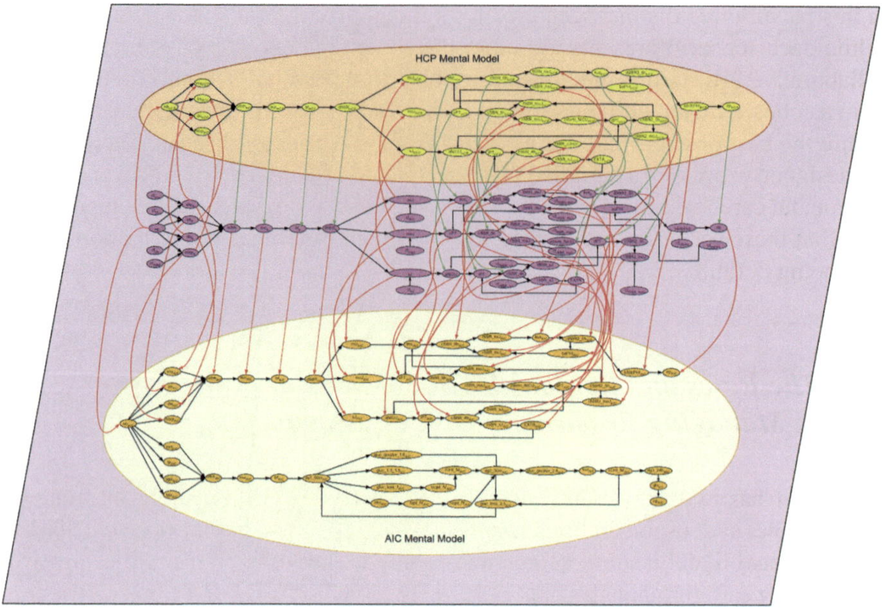

Fig. 4 Base network level for the conceptual model of the medical protocol for the treatment of Neonatal Hyperbilirubinemia

models ranges from connection weights 0 (no connection) to 1 (perfect connection). However, in an imperfect mental model the connection weight is set to a very low level which signifies a flaw or omission in the mental model. In the instance of the models created in this paper, the mental states with flawed connections have been modelled with adaptations to show learning and awareness by the HCP that the mistake has been committed and further adaptation of the persistence (second-order) to show the level of forgetfulness of this learning (from the first order). The connections of the base-level mental model states represent the strengths and vulnerabilities in the protocol. These are represented by the first-order **W**-states representing the weights of connections between the base-level states. For vulnerable cases, they have been set to have a low connection weight to signify having a flaw in the thinking process of the person. Since these **W**-states determine the actual connection weight strength between the states, the situation can change dynamically by learning or forgetting.

The AIC constantly monitors the changes in the actions taken in the real world and recognises whether there has been a shortcoming in the presence and execution of a certain step in a medical protocol corresponding to a state in the mental model. It is assumed here that the monitoring of the real world done by the AIC is carried out perfectly, with the weights of the monitoring connection from the world-states to the AIC mental model always having a value of 1. If there is a state in the world which has to be executed by the HCP and has not been executed ideally, the corresponding state in the AIC mental model signifies to the HCP (and thereby specifically to the HCP mental model) that a mistake has been committed and the specific weak connection should be strengthened to attain a more ideal thought process which translates into performing the protocol with more accuracy and adherence. This communication from AIC to HCP has been modelled by linking an arrow from the vulnerable state in the AIC mental model to the 1st-order self-model **W**-state representing the weak connection for the vulnerable state in the HCP mental model. This arrow has a strong positive weight connection of 1 and increases the strength of the 1st-order **W**-state, translating into a positive adaptation and learning and thus a better functioning of the HCP mental model. Note that the AIC only provides such support when the HCP mental model is flawed, and the AIC detects those flaws: these actions are suppressed when no mistake is committed.

The granularity of the states listed in Table 3 make them ideal for network modelling. Through the application of the various previously tabled combination functions for different linear and non-linear connections a complex simulation can be created for various scenarios that reproduce real-life conditions.

It is also important to detail the starting points of various states. This is modelled by initial values which in simulation graphs signify a value on the vertical axis corresponding to that state's activation level in the simulation. Specifically, while most states start a 0 level, depending on the scenario of the simulation the **W**-states and the $\mathbf{M}_{\mathbf{W}}$-persistence states can start at an initial higher level corresponding to the mental model the HCP has. In an ideal scenario for example, the doctor has an ideal mental model and start with a high level of knowledge which was previously learnt by him and which he remembers and also a strong level of persistence as to not

Table 3 Legend for schematic elements in the adaptive network model

Component	Description
Purple node	Second Order Self-Modelling state **M**, representing persistence μ
Purple level	Second Order reification level
Blue node	First Order Self-Modelling state **W**, representing connection weight ω
Blue level	First Order reification level
Pink node	World States in real world
Pink level	Base Level
Yellow node	HCP mental model state
Beige level	HCP Mental Model
Beige node	AIC mental model state
Yellow level	AIC Mental Model
Red arrow	Incoming connections from real world states to mental model states
Green arrow	Outgoing connection from mental model sates
Blue arrow	Upward connections from base level states to first order self-model state
Pink arrow	Downward connections between reification levels
Black arrow	Connections between real world states

Table 4 Overview of states used in neonatal hypoglycemia conceptual model alongside their corresponding description

State nr	State name	Explanation	Level
X1	nb	Neonate Born: Otherwise healthy neonates (≤ 24 hours old) with a gestational age > 35 weeks and birth weight > 2000 grams	
X2	md_n	Maternal Diabetes concerning mother of neonate	
X3	lp_n	Late Prematurity of Neonate	
X4	sga_n	Neonate has weight Small for Gestational Age	
X5	lga_n	Neonate has weight Large for Gestational Age	
X6	orfnh	Observe risk factors in neonate for neonatal hypoglycemia	
X7	mw_n	Measure weight of the neonate for future reference needed for administering certain amount of glucose to the neonate dependant on the body weight	
X8	bf_n	Begin breastfeeding neonate within 1 hour of birth	
X9	dg1_60m	Determine glucose concentration 60 minutes after birth; after the first feeding. Target value ≥ 2.0 mmol/L	
X10	gluc_greater_1.6	Glucose concentration measured is greater than or equal to 1.6 mmol/L	
X11	gluc_1.1_1.5	Glucose concentration measured is between 1.1 mmol/L and 1.5 mmol/L	
X12	gluc_less_1	Glucose concentration measured is less than or equal to 1 mmol/L	
X13	nhs_n	Neonatal hypoglycemia symptoms shown due to extremely low glucose concentration	
X14	ncgd_IV	Normal continuous glucose dose administered intravenously.	
X15	bgd_IV	bolus (push) glucose dose administered intravenously, in case of severe hypoglycemia	Base Level -
X16	rCHI_NI	Raise Carbohydrate Intake through a Nutritional Intervention	World States
X17	hcgd_IV	High continuous glucose dose administered intravenously, right after the bolus glucose dose in case of severe hypoglycemia.	
X18	dg2_30m	Determine glucose concentration 30 minutes after initial interventions. Target value ≥ 2.6 mmol/L	
X19	gluc_less_2.5	Glucose concentration measured is less than or equal to 2.5 mmol/L	
X20	gluc_greater_2.6	Glucose concentration measured is greater than or equal to 2.6 mmol/L	

(continued)

Table 4 (continued)

X21	kus_n	Keep neonate Under Supervision for 24 hours to further see whether glucose level remains stable. Remove all other interventions	
X22	cCHI_NI	Continue Carbohydrate Intake through a Nutritional Intervention.	
X23	dg3_24h	Determine glucose concentration 24 hours after supervision. Target value ≥ 2.6 mmol/L	
X24	gls	Glucose Level is Stable and above 2.6 mmol/L	
X25	ep	End Protocol	
X26	cf_{md}	Context factor state md_n	Base Level - Context Factor States
X27	cf_{lp}	Context factor state lp_n	
X28	cf_{sga}	Context factor state sga_n	
X29	cf_{lga}	Context factor state lga_n	
X30	$cf_{greater\_1.6}$	Context factor state gluc_greater_1.6	
X31	$cf_{1.1\_1.5}$	Context factor state gluc_1.1_1.5	
X32	cf_{less_1}	Context factor state gluc_less_1	
X33	cf_{nhs}	Context factor state nhs_n	
X34	$cf_{less\_2.5}$	Context factor state gluc_less_2.5	
X35	$cf_{gretaer\_2.6}$	Context factor state gluc_greater_2.6	
X36	cf_{gls}	Context factor state gls_n	
X37	nb_{HCP}	HCP mental model state for registering that Neonate is Born	
X38	md_{HCP}	HCP mental model state for registering that neonates mother has Diabetes	
X39	lp_{HCP}	HCP mental model state for registering that Neonate has had Late Prematurity in pregnancy	
X40	sga_{HCP}	HCP mental model state for registering that Neonate has weight Small for Gestational Age	
X41	lga_{HCP}	HCP mental model state for registering that Neonate has weight Large for Gestational Age	
X42	$orfnh_{HCP}$	HCP mental model state for realising and observing the risk factors for neonatal hypoglycemia in the neonate	
X43	mw_{HCP}	HCP mental model state for Measuring the weight of the neonate	
X44	bf_{HCP}	HCP mental model state to initiate breastfeeding procedure neonate within 1 hour of birth	
X45	$dg1\_60m_{HCP}$	HCP mental model state to Determine glucose concentration 60 minutes after birth; after the first feeding. Target value ≥ 2.0 mmol/L	
X46	$gluc\_greater\_1.6_{HCP}$	HCP mental model state to register that Glucose concentration measured is greater than or equal to 1.6 mmol/L	

(continued)

Table 4 (continued)

X47	$gluc\_1.1\_1.5_{HCP}$	HCP mental model state to register that Glucose concentration measured is between 1.1 mmol/L and 1.5 mmol/L	
X48	$gluc\_less\_1_{HCP}$	HCP mental model state to register that Glucose concentration measured is less than or equal to 1 mmol/L	
X49	nhs_{HCP}	HCP mental model state to register that Neonatal hypoglycemia symptoms shown (due to extremely low glucose concentration)	
X50	$ncgd\_IV_{HCP}$	HCP mental model state to administer Normal continuous glucose dose intravenously.	
X51	bgd_IV_{HCP}	HCP mental model state to administer bolus (push) glucose dose intravenously, in case of severe hypoglycemia	
X52	$rCHI\_NI_{HCP}$	HCP mental model state to effectuate Raising Carbohydrate Intake through a Nutritional Intervention	
X53	$hcgd\_IV_{HCP}$	HCP mental model state to administer a High continuous glucose dose intravenously, in case of severe hypoglycemia.	Base Level - HCP Mental Model
X54	$dg2\_30m_{HCP}$	HCP mental model state to Determine glucose concentration 30 minutes after initial interventions. Target value ≥ 2.6 mmol/L	
X55	$gluc\_less\_2.5_{HCP}$	HCP mental model state to register that Glucose concentration was measured less than or equal to 2.5 mmol/L	
X56	$gluc\_greater\_2.6_{HCP}$	HCP mental model state to register that Glucose concentration was measured greater than or equal to 2.6 mmol/L	
X57	kus_{HCP}	HCP mental model state to effectuate Keep neonate Under Supervision for 24 hours to further see whether glucose level remains stable. All other interventions removed	
X58	$cCHI\_NI_{HCP}$	HCP mental model state to effectuate Continuing Carbohydrate Intake through a Nutritional Intervention.	
X59	$dg3\_24h_{HCP}$	HCP mental model state to Determine glucose concentration 24 hours after supervision. Target value ≥ 2.6 mmol/L	
X60	gls_{HCP}	HCP mental model state to register that Glucose Level is Stable and above 2.6 mmol/L	
X61	ep_{HCP}		

(continued)

Table 4 (continued)

		HCP mental model state to affectuate Ending the Protocol	
X62	**W**md_{HCP},$orfnh_{HCP}$	Reified representation state for connection weight ωmd_{HCP},$orfnh_{HCP}$	First Reification Level
X63	**W**lp_{HCP},$orfnh_{HCP}$	Reified representation state for connection weight ωlp_{HCP},$orfnh_{HCP}$	
X64	**W**sga_{HCP},$orfnh_{HCP}$	Reified representation state for connection weight ωsga_{HCP},$orfnh_{HCP}$	
X65	**W**lga_{HCP},$orfnh_{HCP}$	Reified representation state for connection weight ωlga_{HCP},$orfnh_{HCP}$	
X66	**W**nhs_{HCP},bgd_IV_{HCP}	Reified representation state for connection weight ωnhs_{HCP},bgd IV_{HCP}	
X67	**W**hcgd_IV_{HCP},dg2_$30m_{HCP}$	Reified representation state for connection weight ωhcgd_IV_{HCP},dg2 $30m_{HCP}$	
X68	**W**gluc_greater_2.6_{HCP},kus_{HCP}	Reified representation state for connection weight ωgluc_greater_2.6_{HCP},kus_{HCP}	
X69	**MW**md_{HCP},$orfnh_{HCP}$	Reified representation state for persistence factor μ**W**md_{HCP},$orfnh_{HCP}$ for reified representation state **W**md_{HCP},$orfnh_{HCP}$	Second Reification Level
X70	**MW**lp_{HCP},$orfnh_{HCP}$	Reified representation state for persistence factor μ**W**lp_{HCP},$orfnh_{HCP}$ for reified representation state **W**lp_{HCP},$orfnh_{HCP}$	
X71	**MW**sga_{HCP},$orfnh_{HCP}$	Reified representation state for persistence factor μ**W**sga_{HCP},$orfnh_{HCP}$ for reified representation state **W**sga_{HCP},$orfnh_{HCP}$	
X72		Reified representation state for persistence factor	

(continued)

Table 4 (continued)

	$\mathbf{MW}lga_{HCP},orfnh_{HCP}$	$\boldsymbol{\mu}\mathbf{W}lga_{HCP},orfnh_{HCP}$ for reified representation state $\mathbf{W}lga_{HCP},orfnh_{HCP}$	
X73	$\mathbf{MW}nhs_{HCP},bgd\_IV_{HCP}$	Reified representation state for persistence factor $\boldsymbol{\mu}\mathbf{W}nhs_{HCP},bgd\_IV_{HCP}$ for reified representation state $\mathbf{W}nhs_{HCP},bgd\_IV_{HCP}$	
X74	$\mathbf{MW}hcgd\_IV_{HCP},dg2\_30m_{HCP}$	Reified representation state for persistence factor $\boldsymbol{\mu}\mathbf{W}hcgd\_IV_{HCP},dg2\_30m_{HCP}$ for reified representation state $\mathbf{W}hcgd\_IV_{HCP},dg2\_30m_{HCP}$	
X75	$\mathbf{MW}gluc\_greater\_2.6_{HCP},kus_{HCP}$	Reified representation state for persistence factor $\boldsymbol{\mu}\mathbf{W}gluc\_greater\_2.6_{HCP},kus_{HCP}$ for reified representation state $\mathbf{W}gluc\_greater\_2.6_{HCP},kus_{HCP}$	
X76	nb_{AIC}	AIC mental model state for registering that Neonate is Born	
X77	md_{AIC}	AIC mental model state for registering that neonates mother has Diabetes	
X78	lp_{AIC}	AIC mental model state for registering that Neonate has had Late Prematurity in pregnancy	
X79	sga_{AIC}	AIC mental model state for registering that Neonate has weight Small for Gestational Age	
X80	lga_{AIC}	AIC mental model state for registering that Neonate has weight Large for Gestational Age	
X81	$orfnh_{AIC}$	AIC mental model state for realising and observing the risk factors for neonatal hypoglycemia in the neonate	
X82	mw_{AIC}	AIC mental model state for monitoring that the weight of the neonate has been measured	
X83	bf_{AIC}	AIC mental model state for monitoring that the breastfeeding procedure neonate happens within 1 hour of birth	
X84	$dg1\_60m_{AIC}$	AIC mental model state to monitor that the glucose concentration has been determined 60 minutes after birth; after the first feeding.	Base Level - AIC Mental Model
X85	$gluc\_greater\_1.6_{AIC}$	AIC mental model state to register that Glucose concentration measured is greater than or equal to 1.6 mmol/L	

(continued)

Table 4 (continued)

X86	$gluc\_1.1\_1.5_{AIC}$	AIC mental model state to register that Glucose concentration measured is between 1.1 mmol/L and 1.5 mmol/L
X87	$gluc\_less\_1_{AIC}$	AIC mental model state to register that Glucose concentration measured is less than or equal to 1 mmol/L
X88	nhs_{AIC}	AIC mental model state to register that Neonatal hypoglycemia symptoms shown (due to extremely low glucose concentration)
X89	$ncgd\_IV_{AIC}$	AIC mental model state for monitoring that the Normal continuous glucose dose intravenously has been administered.
X90	bgd_IV_{AIC}	AIC mental model state for monitoring that the bolus (push) glucose dose has been administered intravenously, in case of severe hypoglycemia
X91	$rCHI\_NI_{AIC}$	AIC mental model state for monitoring that the neonate's Carbohydrate Intake has been raised through a Nutritional Intervention
X92	$hcgd\_IV_{AIC}$	AIC mental model state for monitoring that a High continuous glucose dose has been administered intravenously, in case of severe hypoglycemia.
X93	$dg2\_30m_{AIC}$	AIC mental model state for monitoring that glucose concentration 30 minutes after initial interventions has been determined. Target value ≥ 2.6 mmol/L
X94	$gluc\_less\_2.5_{AIC}$	AIC mental model state to register that Glucose concentration was measured less than or equal to 2.5 mmol/L
X95	$gluc\_greater\_2.6_{AIC}$	AIC mental model state to register that Glucose concentration was measured greater than or equal to 2.6 mmol/L
X96	kus_{AIC}	AIC mental model state for monitoring that neonate Kept Under Supervision for 24 hours to further see whether glucose level remains stable. All other interventions removed
X97	$cCHI\_NI_{AIC}$	AIC mental model state for monitoring Continuing Carbohydrate Intake through a Nutritional Intervention for the neonate.
X98	$dg3\_24h_{AIC}$	AIC mental model state for monitoring glucose concentration 24 hours after supervision. Target value ≥ 2.6 mmol/L

(continued)

Table 4 (continued)

X99	gls_{AIC}	AIC mental model state to register that Glucose Level is Stable and above 2.6 mmol/L	
X100	ep_{AIC}	AIC mental model state for monitoring that the Protocol is ended	
X101	$monitor\_md_{orfnh}$	Monitor state for AIC mental model for monitoring world state orfnh with connection from state md	First Reification Level
X102	$monitor\_lp_{orfnh}$	Monitor state for AIC mental model for monitoring world state orfnh with connection from state lp	
X103	$monitor\_sga_{orfnh}$	Monitor state for AIC mental model for monitoring world state orfnh with connection from state sga	
X104	$monitor\_lga_{orfnh}$	Monitor state for AIC mental model for monitoring world state orfnh with connection from state lga	
X105	$monitor_{bgd\_IV}$	Monitor state for AIC mental model for monitoring world state bgd_IV	
X106	$monitor_{dg2\_30m}$	Monitor state for AIC mental model for monitoring world state dg2_30m	
X107	$monitor_{kus}$	Monitor state for AIC mental model for monitoring world state kus	

Table 5 Overview of states used in Neonatal Hyperbilirubinemia conceptual model alongside their corresponding description

State nr	State name	Explanation	Level
X1	nb	Neonate Born	
X2	vsy_n	Neonate Skin is Visually Yellow	
X3	ps_n	Neonate has pale stools	
X4	du_n	Neonate has dark urine	
X5	$mbgi_n$	Neonate has incompatibility with maternal blood type	
X6	orfnh	Observe risk factors in neonate for neonatal hyperbilirubinemia	

(continued)

Table 5 (continued)

X7	auc_n	Assess underlying causes for neonatal hyperbilirubinemia, needed for future purposes if initial treatments do not work.	
X8	bf_n	Begin breastfeeding neonate within 1 hour of birth	
X9	$dSBR_n$	Determine serum bilirubin concentration of neonate in the blood by taking a blood sample	
X10	miJ	Neonate has Mild Jaundice	
X11	moJ	Neonate has Moderate Jaundice	
X12	sJ	Neonate has Severe Jaundice	
X13	tmc_n	Transfer neonate to mother	
X14	sPT	Single Phototherapy done on neonates	
X15	aNICU	Neonate admitted to Neonatal Intensive care unit	Base Level - World States
X16	cSBR_6h	Check Serum Bilirubin levels in Neonate after 6 hours	
X17	cSBR_5h	Check Serum Bilirubin levels in Neonate after 5 hours	
X18	iPT	Intensive Phototherapy done on Neonates	
X19	fSBR_miJ	Serum Bilirubin levels falling after having mild Jaundice	
X20	rSBR_miJ	Serum Bilirubin levels rising after having mild Jaundice	
X21	fSBR_moJ	Serum Bilirubin levels falling after having moderate Jaundice	
X22	rSBR_moJ	Serum Bilirubin levels rising after having moderate Jaundice	
X23	cSBR_4h	Check Serum Bilirubin levels in Neonate after 4 hours	
X24	inform_NICU	Inform Neonatal Intensive care unit	
X25	fSBR_sJ	Serum Bilirubin levels falling after having severe Jaundice	
X26	rSBR_sJ	Serum Bilirubin levels rising after having severe Jaundice	
X27	kus_n	Keep Neonate under supervision	
X28	dPT	Double Phototherapy done on neonates	
X29	EXTR	Neonate undergoes exchange transfusion	

(continued)

Table 5 (continued)

X30	dSBR2_6h	Determine Serum Bilirubin levels in Neonate again after 6 hours	
X31	baPHI	Bilirubin above Phototherapy line	
X32	cSBR2_5h	Check Serum Bilirubin levels in Neonate again after 5 hours	
X33	rSBR2_moJ	Serum Bilirubin levels rising again after having moderate Jaundice	
X34	b50bPHI	Bilirubin 50 mg/l below Phototherapy line	
X35	ep	End Protocol	
X36	nb_{HCP}	HCP mental model state for registering that Neonate is Born	
X37	vsy_{HCP}	HCP mental model state for registering that neonates skin is visually yellow	
X38	ps_{HCP}	HCP mental model state for registering that Neonate has pale stools	
X39	du_{HCP}	HCP mental model state for registering that Neonate has dark urine	
X40	$mbgi_{HCP}$	HCP mental model state for registering that Neonate has incompatibility with maternal blood type	
X41	$orfnh_{HCP}$	HCP mental model state for realising and observing the risk factors for neonatal hyperbilirubinemia	
X42	auc_{HCP}	HCP mental model state for assessing the underlying causes for neonatal hyperbilirubinemia, needed for future purposes if initial treatments do not work.	
X43	bf_{HCP}	HCP mental model state to initiate Neonate with breastfeeding procedure within 1 hour of birth	Base Level - HCP Mental Model
X44	$dSBR_{HCP}$	HCP mental model state to Determine Serum Bilirubin concentration of neonate in the blood by taking a blood sample	
X45	miJ_{HCP}	HCP mental model state to register that Neonate has mild Jaundice	
X46	moJ_{HCP}	HCP mental model state to register that Neonate has moderate Jaundice	

(continued)

Table 5 (continued)

X47	sJ_{HCP}	HCP mental model state to register that Neonate has severe Jaundice
X48	tmc_{HCP}	HCP mental model state to affectuate transferring Neonate to maternal care
X49	sPT_{HCP}	HCP mental model state to administer Neonate with Single Phototherapy
X50	$aNICU_{HCP}$	HCP mental model state to administer Neonate to the Neonatal intensive care unit
X51	$cSBR\_6h_{HCP}$	HCP mental model state to affectuate checking Serum Bilirubin levels after 6 hours
X52	$cSBR\_5h_{HCP}$	HCP mental model state to affectuate checking Serum Bilirubin levels after 5 hours
X53	iPT_{HCP}	HCP mental model state to administer Neonate with Intensive Phototherapy
X54	$fSBR\_miJ_{HCP}$	HCP mental model state to register falling Serum Bilirubin levels in Neonate after having mild Jaundice
X55	$rSBR\_miJ_{HCP}$	HCP mental model state to register rising Serum Bilirubin levels in Neonate after having mild Jaundice
X56	$fSBR\_moJ_{HCP}$	HCP mental model state to register falling Serum Bilirubin levels in Neonate after having moderate Jaundice
X57	$rSBR\_moJ_{HCP}$	HCP mental model state to register rising Serum Bilirubin levels in Neonate after having moderate Jaundice
X58	$cSBR\_4h_{HCP}$	HCP mental model state to affectuate checking Serum Bilirubin levels after 4 hours
X59	$inform\_NICU_{HCP}$	HCP mental model state to affectuate informing Neonatal Intensive Care Unit
X60	$fSBR\_sJ_{HCP}$	HCP mental model state to register falling Serum Bilirubin levels in Neonate after having severe Jaundice
X61	$rSBR\_sJ_{HCP}$	HCP mental model state to register rising Serum Bilirubin levels in Neonate after having severe Jaundice
X62	kus_{HCP}	HCP mental model state to affectuate keeping Neonate under supervision
X63	dPT_{HCP}	HCP mental model state to administer Neonate with Double Phototherapy
X64	$EXTR_{HCP}$	HCP mental model state to administer Neonate with exchange transfusion

(continued)

Table 5 (continued)

X65	$dSBR2\_6h_{HCP}$	HCP mental model state to Determine Serum Bilirubin levels in Neonate again after 6 hours	
X66	$baPHI_{HCP}$	HCP mental model state to register that neonate's Bilirubin levels are above Phototherapy line	
X67	$cSBR2\_5h_{HCP}$	HCP mental model state to Check Serum Bilirubin levels in Neonate again after 5 hours	
X68	$rSBR2\_moJ_{HCP}$	HCP mental model state to register Serum Bilirubin levels rising again after having moderate Jaundice	
X69	$b50bPHI_{HCP}$	HCP mental model state to register that neonate's Bilirubin levels 50 mg/l below Phototherapy line	
X70	ep_{HCP}	HCP mental model state to affectuate Ending the Protocol	
X71	$\mathbf{W}_{vsy_{HCP},orfnh_{HCP}}$	Reified representation state for connection weight $\omega_{vsy_{HCP},orfnh_{HCP}}$	First Reification Level
X72	$\mathbf{W}_{fSBR\_sJ_{HCP},dPT_{HCP}}$	Reified representation state for connection weight $\omega_{fSBR\_sJ_{HCP},dPT_{HCP}}$	
X73	$\mathbf{W}_{dPT_{HCP},cSBR2\_5h_{HCP}}$	Reified representation state for connection weight $\omega_{dPT_{HCP},cSBR2\_5h_{HCP}}$	
X74	$\mathbf{W}_{fSBR\_moJ_{HCP},tmc_{HCP}}$	Reified representation state for connection weight $\omega_{fSBR\_moJ_{HCP},tmc_{HCP}}$	
X75	$\mathbf{W}_{fSBR\_miJ_{HCP},kus_{HCP}}$	Reified representation state for connection weight $\omega_{fSBR\_miJ_{HCP},kus_{HCP}}$	
X76	nb_{AIC}	AIC mental model state for registering that Neonate is Born	
X77	vsy_{AIC}	AIC mental model state for registering that Neonate Skin is Visually Yellow	
X78	ps_{AIC}	AIC mental model state for registering that Neonate has pale stools	
X79	du_{AIC}	AIC mental model state for registering that Neonate has dark urine	
X80	$mbgi_{AIC}$	AIC mental model state for registering that Neonate has incompatibility with maternal blood type	

(continued)

Table 5 (continued)

X81	$orfnh_{AIC}$	AIC mental model state for Observing risk factors in neonate for neonatal hyperbilirubinemia	
X82	auc_{AIC}	AIC mental model state for Assessing underlying causes for neonatal hyperbilirubinemia, needed for future purposes if initial treatments do not work.	
X83	bf_{AIC}	AIC mental model state for monitoring that breast-feeding is done within 1 hour of birth	Base Level - AIC Mental Model
X84	$dSBR_{AIC}$	AIC mental model state for monitoring that serum bilirubin concentration of neonate in the blood is determined by taking a blood sample	
X85	miJ_{AIC}	AIC mental model state for registering that Neonate has Mild Jaundice	
X86	moJ_{AIC}	AIC mental model state for registering that Neonate has Moderate Jaundice	
X87	sJ_{AIC}	AIC mental model state for registering that Neonate Severe Jaundice	
X88	tmc_{AIC}	AIC mental model state for monitoring that Neonate has to be transferred to maternal care	
X89	sPT_{AIC}	AIC mental model state for monitoring that Single Phototherapy has to be done on Neonate	
X90	$aNICU_{AIC}$	AIC mental model state for monitoring a Neonates admission to the Neonatal Intensive care unit	
X91	$cSBR\_6h_{AIC}$	AIC mental model state for monitoring Serum Bilirubin levels check done to Neonate after 6 hours	
X92	$cSBR\_5h_{AIC}$	AIC mental model state for monitoring Serum Bilirubin levels check done to Neonate after 5 hours	
X93	iPT_{AIC}	AIC mental model state for monitoring that Intensive Phototherapy is done on Neonates	
X94	$fSBR\_miJ_{AIC}$	AIC mental model state for monitoring falling Serum Bilirubin levels after having mild Jaundice	
X95	$rSBR\_miJ_{AIC}$	AIC mental model state for monitoring rising Serum Bilirubin levels after having mild Jaundice	
X96	$fSBR\_moJ_{AIC}$	AIC mental model state for monitoring falling Serum Bilirubin levels after having moderate Jaundice	

(continued)

Table 5 (continued)

X97	$rSBR\_moJ_{AIC}$	AIC mental model state for monitoring rising Serum Bilirubin levels after having moderate Jaundice
X98	$cSBR\_4h_{AIC}$	AIC mental model state for monitoring a Serum Bilirubin levels done to Neonate after 4 hours
X99	$inform\_NCU_{AIC}$	AIC mental model state to monitor that the Neonatal Intensive Care Unit is informed of a neonates worsening condition
X100	$fSBR\_sJ_{AIC}$	AIC mental model state to register falling Serum Bilirubin levels in Neonate after having severe Jaundice
X101	$rSBR\_sJ_{AIC}$	AIC mental model state to register rising Serum Bilirubin levels in Neonate after having severe Jaundice
X102	kus_{AIC}	AIC mental model state to monitor that a Neonate is kept under supervision
X103	dPT_{AIC}	AIC mental model state to monitor that Double Phototherapy is done to neonates
X104	$EXTR_{AIC}$	AIC mental model state to monitor that exchange transfusion is carried out on a neonate
X105	$dSBR2\_6h_{AIC}$	AIC mental model state to monitor the determining of Serum Bilirubin levels in Neonate again after 6 hours
X106	$baPHI_{AIC}$	AIC mental model state to register that neonate's Bilirubin levels are above Phototherapy line
X107	$cSBR2\_5h_{AIC}$	AIC mental model state to monitor that the Serum Bilirubin levels in Neonate are checked again after 5 hours
X108	$rSBR2\_moJ_{AIC}$	AIC mental model state to register Serum Bilirubin levels rising again after having moderate Jaundice
X109	$b50bPHI_{AIC}$	AIC mental model state to register that neonate's Bilirubin levels 50 mg/l below Phototherapy line

(continued)

Table 5 (continued)

X110	ep_{AIC}	AIC mental model state for monitoring that the Protocol is ended.	
X111	$\mathbf{MW}vsy_{HCP},orfnh_{HCP}$	Reified representation state for persistence factor $\mathbf{\mu W}vsy_{HCP},orfnh_{HCP}$ for reified representation state $\mathbf{W}vsy_{HCP},orfnh_{HCP}$	Second Reification Level
X112	$\mathbf{MW}fSBR\_sJ_{HCP},dPT_{HCP}$	Reified representation state for persistence factor $\mathbf{\mu W}fSBR\_sJ_{HCP},dPT_{HCP}$ for reified representation state $\mathbf{W}fSBR\_sJ_{HCP},dPT_{HCP}$	
X113	$\mathbf{MW}dPT_{HCP},cSBR2\_5h_{HCP}$	Reified representation state for persistence factor $\mathbf{\mu W}dPT_{HCP},cSBR2\_5h_{HCP}$ for reified representation state $\mathbf{W}dPT_{HCP},cSBR2\_5h_{HCP}$	
X114	$\mathbf{MW}fSBR\_moJ_{HCP},tmc_{HCP}$	Reified representation state for persistence factor $\mathbf{\mu WW}fSBR\_moJ_{HCP},tmc_{HCP}$ for reified representation state $\mathbf{W}fSBR\_moJ_{HCP},tmc_{HCP}$	
X115	$\mathbf{MW}fSBR\_miJ_{HCP},kus_{HCP}$	Reified representation state for persistence factor $\mathbf{\mu W}fSBR\_miJ_{HCP},kus_{HCP}$ for reified representation state $\mathbf{W}fSBR\_miJ_{HCP},kus_{HCP}$	
X116	cf_{vsy}	Context factor state vsy_n	Base Level - Context Factor States
X117	cf_{ps}	Context factor state ps_n	
X118	cf_{du}	Context factor state du_n	
X119	cf_{mbgi}	Context factor state $mbgi_n$	
X120	cf_{miJ}	Context factor state miJ	
X121	cf_{moJ}	Context factor state moJ	
X122	cf_{sJ}	Context factor state sJ	
X123	cf_{fSBR_miJ}	Context factor state fSBR_miJ	

(continued)

Table 5 (continued)

X124	cf_{rSBR_miJ}	Context factor state rSBR_miJ	
X125	cf_{fSBR_moJ}	Context factor state fSBR_moJ	
X126	cf_{rSBR_moJ}	Context factor state rSBR_moJ	
X127	cf_{fSBR_sJ}	Context factor state fSBR_sJ	
X128	cf_{rSBR_sJ}	Context factor state rSBR_sJ	
X129	cf_{rSBR2_moJ}	Context factor state rSBR2_moJ	
X130	cf_{baPHI}	Context factor state baPHI	
X131	$cf_{b50bPHI}$	Context factor state b50bPHI	
X132	$monitor_{orfnh}$	Monitor state for AIC mental model for monitoring world state orfnh	First Reification Level
X133	$monitor_{tmc}$	Monitor state for AIC mental model for monitoring world state orfnh	
X134	$monitor_{kus}$	Monitor state for AIC mental model for monitoring world state orfnh	
X135	$monitor_{dPT}$	Monitor state for AIC mental model for monitoring world state orfnh	
X136	$monitor_{cSBR2\_5h}$	Monitor state for AIC mental model for monitoring world state orfnh	

forget that knowledge. However, in a scenario with a weak mental model, the initial values of the adaptation **W**-states are the opposite. In the models, while the **W**-states of a weak mental model have 0.1 as an initial value, the $\mathbf{M_W}$-states have a null initial value while in a strong mental model scenario the **W**-states have an initial value of 0.9 while the $\mathbf{M_W}$-states have 0.8 as their initial value, respectively. Additionally, to start the simulation the first node where the neonate is born in real life has to have an initial value of 1, to show the occurrence of the event and lastly context factors also have an initial value of 1 or 0 depending on their presence in the real world. Further on, while due to convention this is not conveyed in the graphical picture of the adaptive networks, all **W**-state have a self-connection to themselves to reinforce persistence and be able to improve due to their learning.

Lastly, it is important to touch on the topic of the specific values of parameters which have been chosen throughout all scenarios of the conceptual models. The speed of all states, signifying their rate of change, has been kept at 0.5 to ensure control of a state not changing its value unrealistically relative to other states, while context factor states have a speed of 0, only an initial value of 1 to indicate their presence as it is not a dynamic factor which can change its level of existence in a real-life scenario. Monitoring states have a speed of 0.05 to ensure there is a time delay between a fault taking place in the protocol, and then being detected by the monitoring states after which their signals allow the protocol to be corrected gradually. **W**-states, however, have a speed value of 0.2 to ensure changes to the level of learning happen gradually, reflecting how evolution of thinking in mental models happens in real life and not drastically. Further on steepness and threshold of any present state in the protocol using the **alogistic** function in the model is kept at 5 and 0.5 respectively. Monitoring states are allocated steepness 30 and threshold 0.4 respectively to display after being dormant, a sudden and steep rise in presence the moment they find a fault arising in the protocol. The same reasoning is given to the vulnerable mental model states which have downward effectuating connections incoming from **W**-states and represent where the flaw in the mental model and the mental process of the HCP occurs, further affecting the that corresponding part of the protocol in the real-world states. These mental model states thus need to reflect a similar pattern of sudden change as the monitoring states, which is why they also have a higher steepness and lower threshold value of 10 and 0.3 respectively.

4 Design of the Adaptive Network Models

This section depicts the conceptual models of the self-modelling adaptive networks of the Neonatal Hypoglycemia and neonatal hyperbilirubinemia medical treatment protocols and their corresponding legends for giving an overview of the states shown in the models.

4.1 Neonatal Hypoglycemia Model

To show the base-level network more clearly and without any interference by connections to the second order states from the base-level mental model, the following figure presents the base level mental model independently. Note that zooming into the following figure allows more clarity in observing what the names of each of the states are.

To model the treatment as a real-life situation, a certain path trajectory of the conceptual model of the treatment has to be followed. As stated earlier in, the worst scenario is followed and in the context of the above treatment, this corresponds to the following schema:

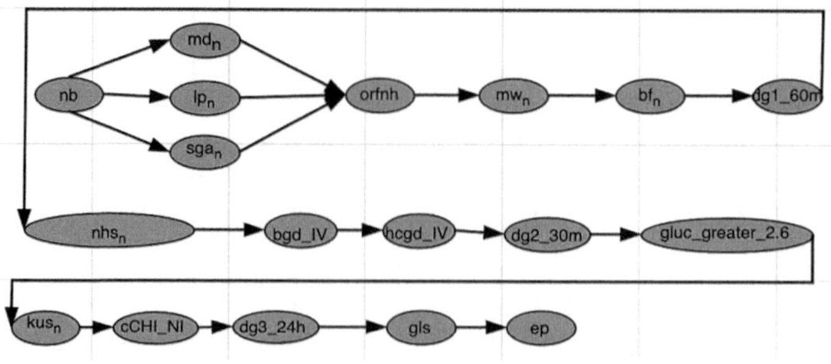

When the neonate is born, it is small for its gestational age, it is born to a mother who has diabetes, and it is born with late prematurity. These are major risk factors for neonatal hypoglycemia and are observed by the doctor. To treat this, the doctor measures weight and gives the baby for breastfeeding to ensure glucose levels stay normal. The glucose level is then determined after 60 min and shown to be extremely low showing symptoms of neonatal hypoglycemia. An intravenous bolus glucose dose is administered as a drastic measure, immediately followed by a high continuous glucose dose. The glucose level is determined the second time and shown to have improved to above 2.6 mmol/L. The neonate is kept under supervision and given carbohydrate food, until the third glucose measurement after 24 h shows glucose levels have stabilised and thus the protocol is ended.

4.2 Legend for Neonatal Hypoglycemia Model

4.3 Neonatal Hyperbilirubinemia Model

To show the base-level network more clearly and without any interference by connections to the second order states from the base-level mental model, the following figure presents the base level mental model independently.

Note that zooming into the following figure allows more clarity in observing what the name of each of the states is.

In order to model the treatment as a real-world situation in the software environment, a certain path trajectory of the conceptual model of the treatment has to be followed. As stated earlier in the report, the worst scenario is followed and in the context of the above treatment, this corresponds to the following schema:

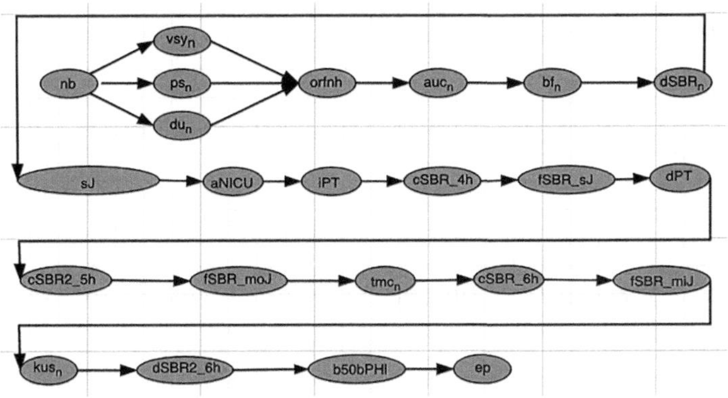

When the neonate is born, it has yellow skin, and discharges pale stool and dark urine. These are major risk factors for neonatal hyperbilirubinemia and are observed by the doctor. To treat this, the doctor assesses the underlying cause for jaundice and gives the baby for breastfeeding. The serum bilirubin concentration is then determined to be extremely high representing severe jaundice. It is then admitted to the neonatal intensive care unit and undergoes intensive phototherapy as a drastic measure. The SBR level is checked after 4 h and is shown to be falling. Thereafter, the neonate undergoes double phototherapy. The SBR level is checked after 5 h and shown to have fallen. The neonate is therefore transferred back to the mother, where SBR level check after 6 h and still shown to be falling. With neonate now below mild jaundice, it is kept under supervision and SBR level checked once again after 6 h. Since it has fallen further to below 50 mg/l of the phototherapy line, the protocol is ended.

4.4 Legend for the Neonatal Hyperbilirubinemia Model

It is important to note that, particularly in Scenario 3 where the AI Coach is present in the simulations, the AI Coach states occupy the same space and state numbers, namely from X76 till X110 for both neonatal medical protocols. This signifies the AI Coach is adaptable and flexible to recognise the symptoms for either of the neonatal medical protocols and choose the path corresponding to those symptoms and thus on the pathway of the correct medical guidelines, if any of them are stored within its mental model. It also shows that the AIC remains the same as choosing a different pathway in its mental model corresponding to a different medical guideline is simply a subsection of the AIC mental model, and that even more medical protocols could be stored in the same AIC mental model without changing any dynamics of how the AIC interacts with the HCP and inducing adherence to guidelines.

In the attached Appendix, tables for the role matrices that are used for the simulation of the scenarios can be found.

5 Results

This section depicts the final output of the simulation charts modelling the self-adaptive networks for the two medical protocols in their corresponding three scenarios.

5.1 *Simulation Outputs for the Neonatal Hypoglycemia Medical Protocol*

Figure 5 shows Scenario 1 of the Neonatal Hypoglycemia treatment where the HCP performs the treatment and follows the corresponding protocol with a perfect mental model without any flaws. It can be observed that each real-world state in the protocol as well as the mental model states are executed rapidly in succession without any steps taken being delayed in their execution (see Scenario 2) or not being executed properly and reaches a level between 0.8 and 1 signifying a high degree of completion. In addition to the protocol being followed properly the learning of HCP can be seen to be rising to full saturation as the treatment progresses. However, note that the **W**-state lines, starting at the 0.9 threshold, have a slight initial decrease in value before recovering and increasing to a full threshold as the treatment progresses and the HCP completes the corresponding protocol (vulnerable) steps in the real world which boosts learning. Note here that a **W**-state which corresponds to the mental model state which has to be followed later on in the protocol has a larger decrease before recovering when compared to a **W**-state for which the corresponding mental model state has to be followed earlier on in the protocol. The $\mathbf{M}_{\mathrm{W}}$-states beginning from 0.8, however, do not decrease in value but only increase to full saturation as the treatment progresses, albeit in a similar fashion to the trend observed while observing the **W**-states, where $\mathbf{M}_{\mathrm{W}}$-states corresponding to mental model states which are followed later on in the protocol increase much more slower when compared to $\mathbf{M}_{\mathrm{W}}$-states for which the corresponding mental model states are followed earlier on in the protocol. The entire protocol is completed by $t = 25$.

Figure 6 shows Scenario 2 of the Neonatal Hypoglycemia treatment where the HCP follows the corresponding protocol with an imperfect mental model having flaws in the connections between mental model states vulnerable to encompass mistakes. It can be witnessed that the **W**- and **M**- states, while following the same trend of an initial decline followed by a recovery, have a very gradual and late rise, and as before, also have a slower rise for the **W**-states for which the mental model states appear later on in the protocol. The corresponding real-world and mental model states to the **W**-states also follow a late and gradual rise. As an effect of these steps not being completely executed and incurring delays in reaching their required threshold levels, firstly the mental model states which are not vulnerable as well as their corresponding world states, that are executed before the vulnerable states in the protocol rise rapidly and reach their final thresholds. However, states executed after

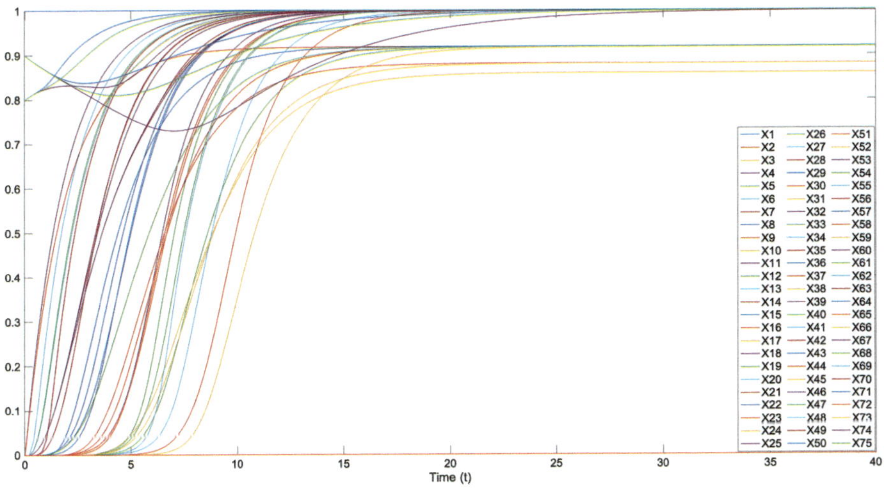

Fig. 5 Simulation results for Neonatal Hypoglycemia treatment Scenario 1

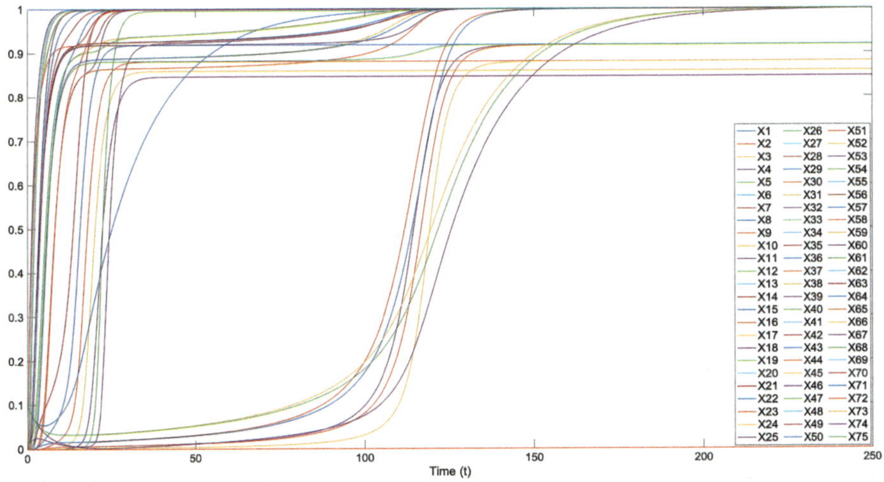

Fig. 6 Simulation results for Neonatal Hypoglycemia treatment Scenario 2

the vulnerable states, which have not been followed properly, do not reach their final thresholds rapidly. Instead they rise to a certain threshold and stay fairly stagnant until the vulnerable mental model states, corresponding **W**- and **M**- states and world states rise rapidly after a delay, signifying that the HCP has realized the flaw and corrects the mistake to complete the step properly. In case of this model, around $t = 100$ is where **W**-states and their corresponding vulnerable mental model and world states rise rapidly to their final thresholds and it is here where states executed after the vulnerable mental model states which were stagnant between 0.8 and 0.9 further increase to level 1 signifying their correct implementation after the previous flaw has been corrected. The entire protocol is completed by $t = 200$.

Figure 7a, b shows Scenario 3a of the Neonatal Hypoglycemia treatment where the HCP performs the treatment and follows the corresponding protocol with a perfect mental model without any flaws with the additional presence of an AI Coach. It can be observed that the model follows the exact trend as scenario 1 of the Neonatal Hypoglycemia treatment. Specifically, it can be seen that the real-world state in the protocol as well as the mental model states are executed rapidly in succession without any steps taken being delayed in their execution or not being executed properly and reaches a threshold level between 0.8 and 1 signifying a high degree of completion. In addition to the protocol being followed properly the learning of HCP, can be seen to be rising to full saturation as the treatment progresses. In this case, possibly due to the additional presence of the AIC, there is no initial decrease in the threshold levels of the **W**-states. The $\mathbf{M_W}$-states beginning from 0.8, also, do not decrease in value but only increase to full saturation as the treatment progresses, albeit in a similar fashion to the trend observed while observing the **W**-states, where $\mathbf{M_W}$-states corresponding to mental model states which are followed later on in the protocol increase afterwards when compared to $\mathbf{M_W}$-states for which the corresponding mental model states are followed earlier on in the protocol. The entire protocol is completed by $t = 25$. It can be further noted that unlike in scenario 2, there is not any state which increases to a certain threshold level and stays stagnant and then further increases to a final threshold level, due to other mental model states and **W**-states having a slow and gradual rise. Upon closer inspection of the model in Fig. 7b, it can also be seen that the monitor states stay dormant due to the dynamic nature of the AIC which does not trigger support signals if the doctor does not commit a mistake and if the protocol is being followed to a high degree of satisfaction.

Figure 8a, b below, depict scenario 3b of the Neonatal Hypoglycemia treatment where the HCP follows the corresponding protocol with an imperfect mental model having flaws in the connections between mental model states vulnerable to mistakes, however there is an added presence of the AI coach which serves to correct those deviations. It can be observed in Fig. 8b that unlike in scenario 2, while there is an initial staggered execution of the vulnerable HCP mental model states and corresponding real-world and **W**- and **M**- states, this deviation is quickly fixed the moment there is a sudden and steep rise in the monitoring states around $t = 5$, which occurs few moments after the protocol has started as the AIC recognizes mistakes committed through incomplete and faulty execution of the protocol as seen in the real-world states by the AIC which then further relays support signals to the mental model of the HCP. When this happens, there is a sudden and rapid rise firstly in persistence, then in the learning in W-states, thereafter a further increase in the correct execution of the mental model state, solidifying the correction of the mistake in the HCP mental model and lastly this translates into an increase in the threshold levels of the corresponding real-world states. The correction of the deviation in the protocol thus happens far quicker than when compared to scenario 2 where the correction happened around $t = 100$, as here the entire protocol is completed by $t = 50$. This also means that unlike in scenario 2, all real-world states rise rapidly to their final

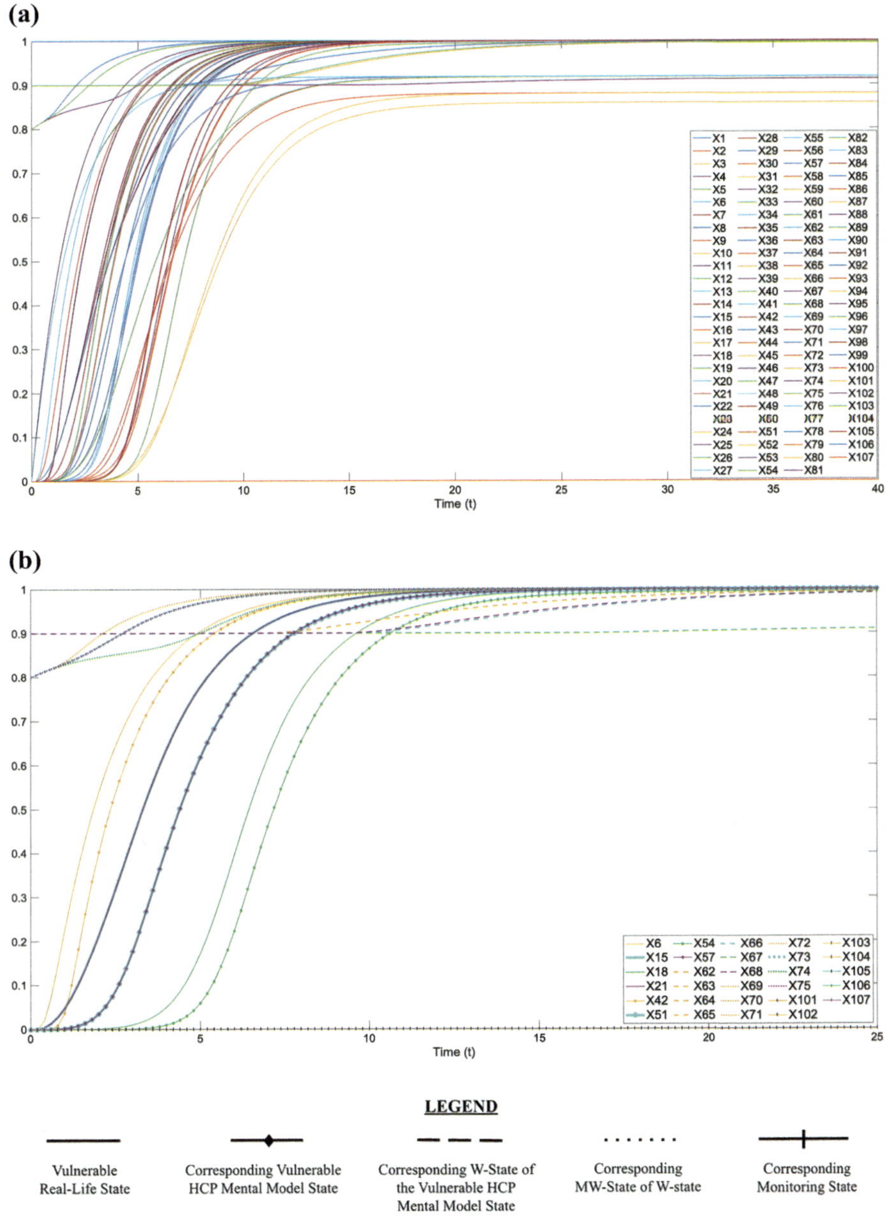

Fig. 7 Simulation results for Neonatal Hypoglycemia treatment Scenario 3a. **b** Magnified chart of the monitor and adaptive states in the reification levels and vulnerable HCP mental model and corresponding real-world states in the base level of the Neonatal Hypoglycemia treatment: Scenario 3a

thresholds, and even states executed after the vulnerable states, which have not been initially followed properly, do not experience any stagnant plateau, but only just have a slight drop in the steepness of the rise before quickly continuing to rise to their final thresholds.

5.2 *Simulation Outputs for the Neonatal Hyperbilirubinemia Medical Protocol*

Figure 9 shows Scenario 1 of the Neonatal Hyperbilirubinemia treatment where the HCP performs the treatment and follows the corresponding protocol with a perfect mental model without any flaws. It can be extrapolated upon close inspection that the states in the observed model follow the exact same pattern and trends as of those in Scenario 1 of the Neonatal Hypoglycemia treatment. Each real-world state in the protocol as well as the mental model states are executed rapidly in succession without any steps taken being delayed in their execution or not being executed properly and reaches a level between 0.8 and 1 signifying a high degree of completion. The learning of HCP can be seen to be rising to full saturation as the treatment progresses with the **W**-states, starting at the 0.9 threshold, have a slight initial decrease in value before recovering and increasing to a full threshold as the treatment progresses. As in Scenario 1 of the Neonatal Hypoglycemia treatment, the **W**-states which are executed later on in the protocol have a larger initial decrease when compared to a **W**-state for which the corresponding mental model state has to be followed earlier on in the protocol. The $\mathbf{M}_{\mathbf{W}}$-states beginning from 0.8, however, do not decrease in value but only increase to a full threshold level as the treatment progresses, however with $\mathbf{M}_{\mathbf{W}}$-states followed later on in the protocol increasing more gradually when compared to $\mathbf{M}_{\mathbf{W}}$-states which are followed earlier on in the protocol. Protocol is completed by $t = 25$.

Figure 10 shows Scenario 2 of the Neonatal Hyperbilirubinemia treatment where the HCP follows the corresponding protocol with an imperfect mental model having flaws in the connections between mental model states vulnerable to mistakes. Keeping in line with the same trends as seen in Scenario 2 of the Neonatal Hypoglycemia treatment, the **W**- and **M**- states, along with their corresponding mental model states and real-world states have a very gradual and late rise, and as before, also have a slower rise for the **W**-states for which the mental model states appear later on in the protocol. Further on, as described before an effect of these steps not being completely executed and incurring delays in reaching their required threshold levels leads to states executed after the vulnerable states, rise to a certain threshold and stay fairly stagnant until the vulnerable mental model states, corresponding **W**- and **M**- states and world states rise rapidly after a delay, signifying that the HCP has realized the flaw and corrects the mistake to complete the step properly. In case of this model, at $t = 100$ and then again at $t = 150$ is where **W**-states start rising rapidly to their final thresholds. The moment the rapid rise of the mental model and

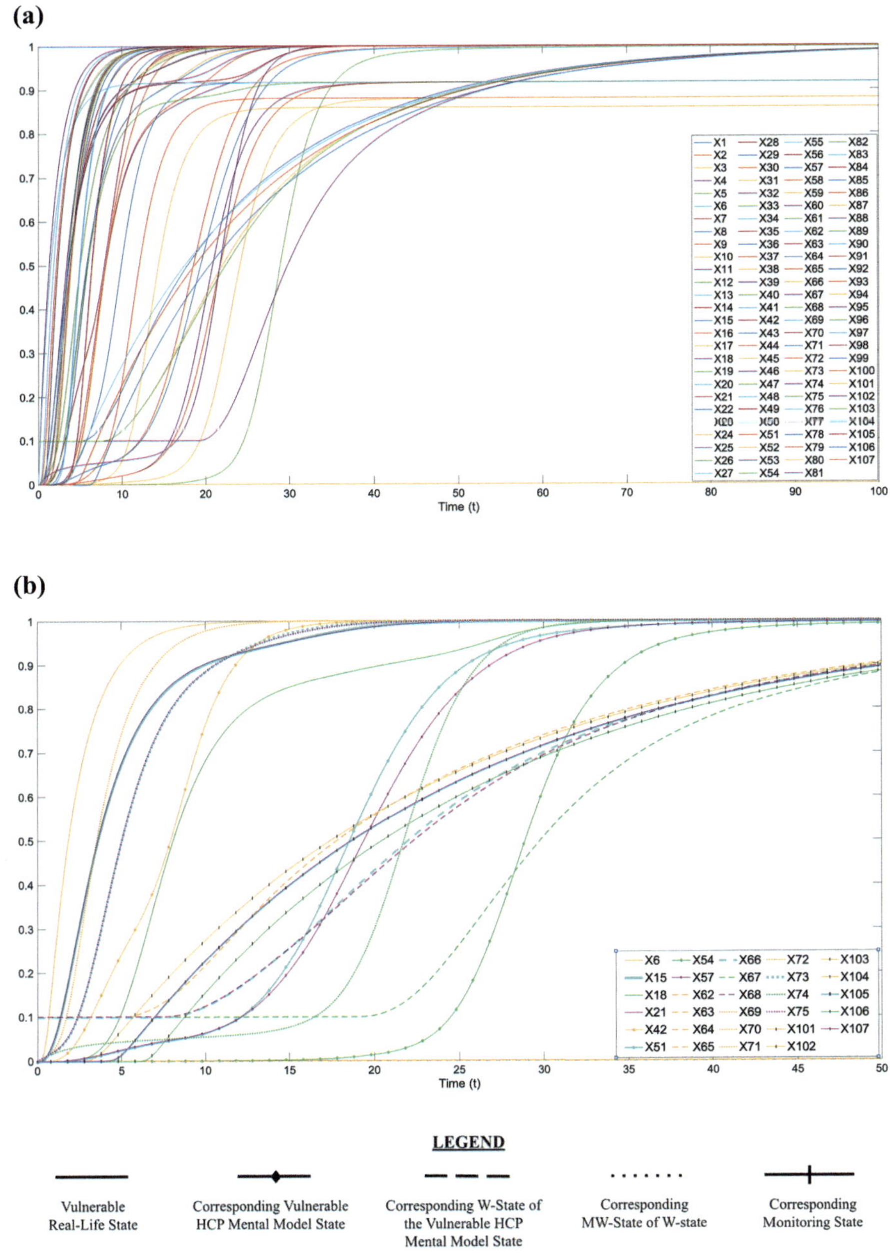

Fig. 8 Simulation results for Neonatal Hypoglycemia treatment: Scenario 3b. **b** Magnified chart of the monitor and adaptive states in the reification levels and vulnerable HCP mental model and corresponding real-world states in the base level of the Neonatal Hypoglycemia treatment: Scenario 3b

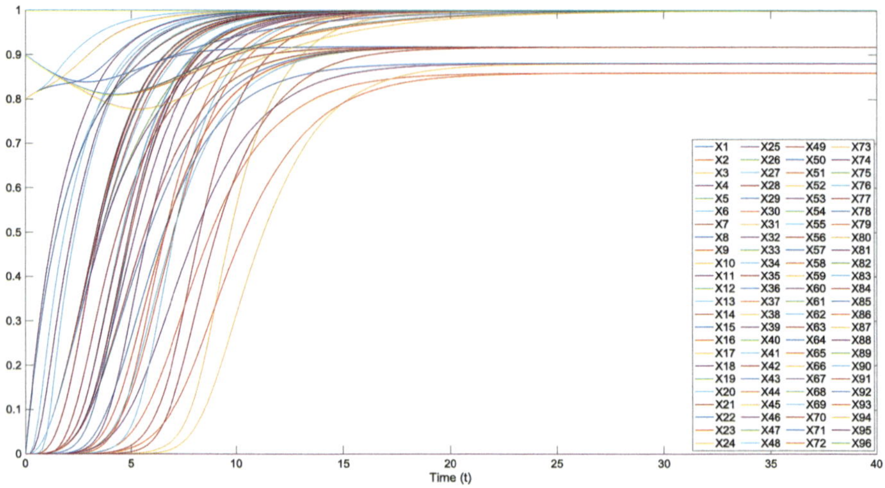

Fig. 9 Simulation results for Neonatal Hyperbilirubinemia protocol Scenario 1

W- and **M**- states starts occurring, the states executed after the vulnerable mental model states which were stagnant between 0.8 and 0.9 finally increase to their final threshold signifying their correct implementation after the previous flaw has been corrected. The entire protocol is completed by $t = 250$.

Figure 11a, b shows Scenario 3a of the Neonatal Hyperbilirubinemia treatment where the HCP performs the treatment and follows the corresponding protocol with a perfect mental model without any flaws with the additional presence of an AI Coach. It can be observed that the model follows the exact trend as scenario 1 of

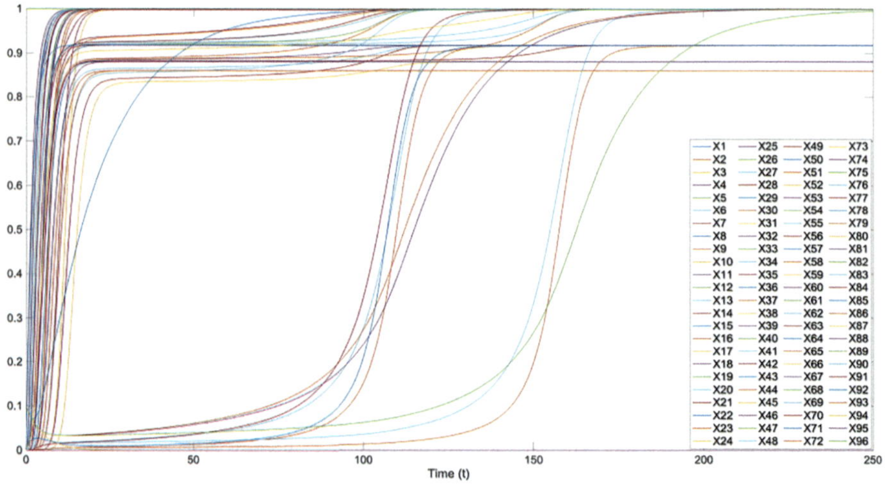

Fig. 10 Simulation results for Neonatal Hyperbilirubinemia protocol Scenario 2

the Neonatal Hyperbilirubinemia treatment. Specifically, it can be seen that the real-world state in the protocol as well as the mental model states are executed rapidly in succession without any steps taken being delayed in their execution or not being executed properly and reaches a threshold level between 0.8 and 1 signifying a high degree of completion. In addition to the protocol being followed properly the learning of HCP, can be seen to be rising to full saturation as the treatment progresses. The $\mathbf{M_W}$-states beginning from 0.8, also, do not decrease in value but only increase to full saturation as the treatment progresses, albeit in a similar fashion to the trend observed while observing the **W**-states, where $\mathbf{M_W}$-states corresponding to mental model states which are followed later on in the protocol increase afterwards when compared to $\mathbf{M_W}$-states for which the corresponding mental model states are followed earlier on in the protocol. The entire protocol is completed by $t = 25$, just as in scenario 1. It can be further noted that unlike in Scenario 2, there is not any state which increases to a certain threshold level and stays stagnant and then further increases to a final threshold level, due to other mental model states and **W**-states having a slow and gradual rise.

Upon closer inspection of the model in Fig. 11b, it can also be seen that the monitor states stay dormant due to the dynamic nature of the AIC which does not trigger support signals if the doctor does not commit a mistake and if the protocol is being followed to a high degree of satisfaction. This not only displays the dynamic nature of the AIC with respect to taking into account different scenarios of the protocol being followed but also with respect to recognizing and functioning in different environments when it comes to following different medical protocols.

Figure 12a, b below, depict Scenario 3b of the Neonatal Hyperbilirubinemia treatment where the HCP follows the corresponding protocol with an imperfect mental model having flaws in the connections between mental model states vulnerable to mistakes, however there is an added presence of the AI coach which serves to correct those deviations. It can be observed that Fig. 12a follows a very similar patterns and overall trend of the states when compared to Scenario 3b of the Neonatal Hypoglycemia protocol. This emphasizes the dynamic nature of the model irrespective of various different treatments. In Fig. 12b, unlike in Scenario 2, while there is an initial staggered execution of the vulnerable HCP mental model states and corresponding real-world and **W**- and **M**- states, (including an initial drop in the **W**-state levels which start from 0.1), this deviation is quickly fixed with the sudden and steep rise in the monitoring states around $t = 5$, which occurs few moments after the protocol has started, since the AIC recognizes mistakes committed through incomplete and faulty execution of the protocol as seen in the real-world and then further relays support signals to the mental model of the HCP. When this happens, there is a sudden and rapid rise firstly in persistence, then in the learning in **W**-states, thereafter a further increase in the correct execution of the mental model state, solidifying the correction of the mistake in the HCP mental model and lastly this translates into an increase in the threshold levels of the corresponding real-world states. The correction of the deviation in the protocol thus happens far quicker than when compared to scenario 2 where the correction happened between $t = 100$ and $t = 150$, as here the entire

(a)

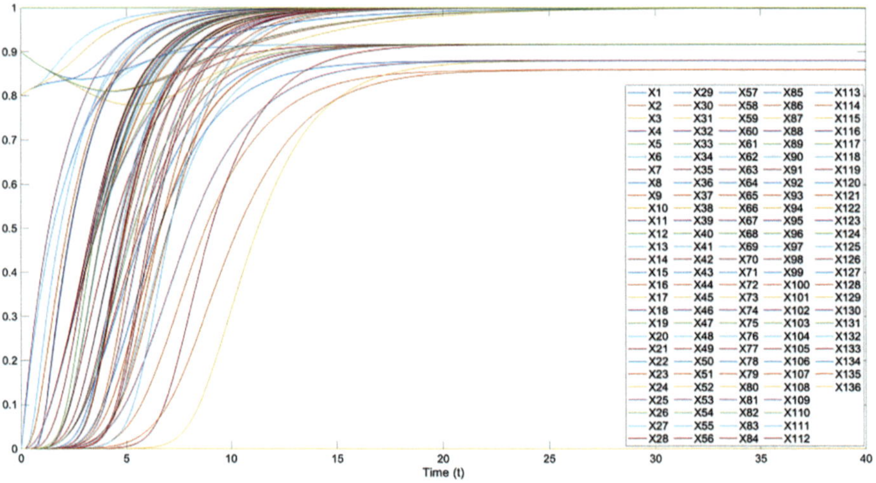

(b)

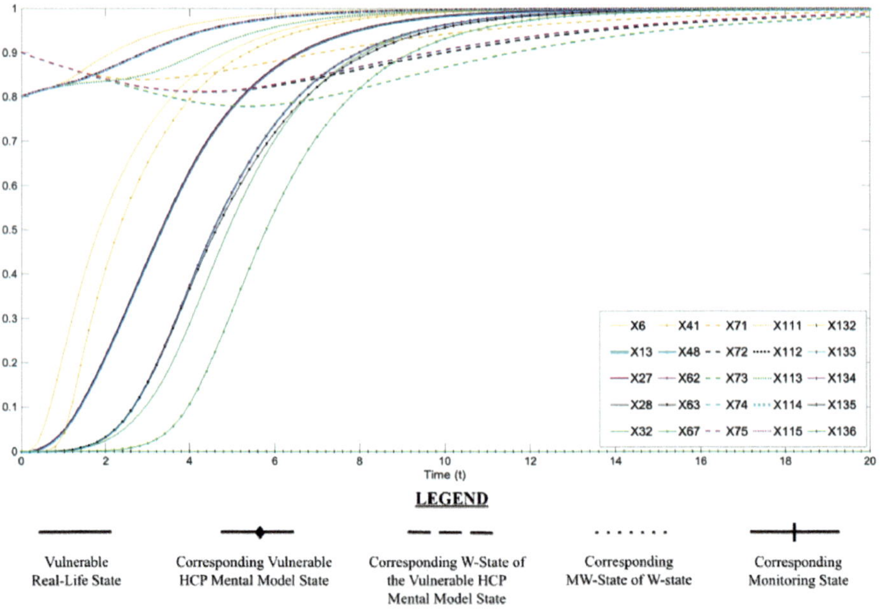

Fig. 11 Simulation results for Neonatal Hyperbilirubinemia protocol Scenario 3a. **b** Magnified chart of the monitor and adaptive states in the reification levels and vulnerable HCP mental model and corresponding real-world states in the base level of the Neonatal Hyperbilirubinemia treatment: Scenario 3a

protocol is completed by $t = 60$. This also means that unlike in scenario 2, all real-world states rise rapidly to their final thresholds, and even states executed after the vulnerable states, which have not been initially followed properly, do not experience any stagnant plateau, but only just have a slight drop in the steepness of the rise before quickly continuing to rise to their final thresholds.

5.3 Results Analysis

The above simulations show a clear common trend the AIC plays in promoting the adherence of medical health care practitioner to guidelines. Upon careful inspection of the traces in the dynamic model, it can be seen that in Scenario 1 for both simulations, the state levels which correspond to the full execution of the states and following them, increase rapidly to a certain levels and then stay stagnant, meaning completed, while the 2nd and 1st order adaptive states keep increasing, thus persistence and learning increases after a small stagnation in growth. However, with Scenario 2, a significant delay in the increase of the vulnerability states and the corresponding **W**- and **M**-states is witnessed while the other states increase to a certain level before stagnating and having to wait for the states which have incurred deviations to increase so they can further keep increasing to their required thresholds. For instance, while in Hyperbilirubenimia Scenario 1, states took 60 time units to increase fully, in Scenario 2 it took more than 250 time units. In the real world, this would play out as the HCP forgetting to perform a certain action in the protocol, or not perform it to a satisfactory level initially, and then move on to the next steps before realising something more needs to be done for the step where the mistake was made and going back to that step with significant delay and correcting it while the other states remain stagnant until the mistake is corrected. As can be seen here, persistence **M**-states traces, and **W**-state traces are very low and have a very slow rate of increase. In Scenario 3 however, with the intervention of the AIC, it can be observed that the depictions of the traces almost identically match those in Scenario 1, with the monitoring states suppressed without triggering any actions as the mental model is not flawed and the vulnerable HCP mental model states and the corresponding real-world state in the protocol reaching full saturation or their execution completion very rapidly and this knowledge of the presence of the real world states being transferred to the AIC which does not intervene. With Scenario 3b, however, the AIC detects an error in the process of the protocol being followed in the real-world states and thus intervenes by sending a signal to the monitor states to become active and help improve the HCP mental model flaws and subsequently the HCP's knowledge. The sudden and steep rise of the monitor states after being dormant till the AIC recognises a flaw and sends a corresponding signal to the monitor state, further leads to a sudden rapid and sharp rise in persistence and learning of the **W**-states of the vulnerable HCP mental model states which were flawed and caused an error while following the protocol in the real world. This further causes a fast correction of the previously seen delay in scenario 2 for the doctor to realise the error committed following the protocol and go back to

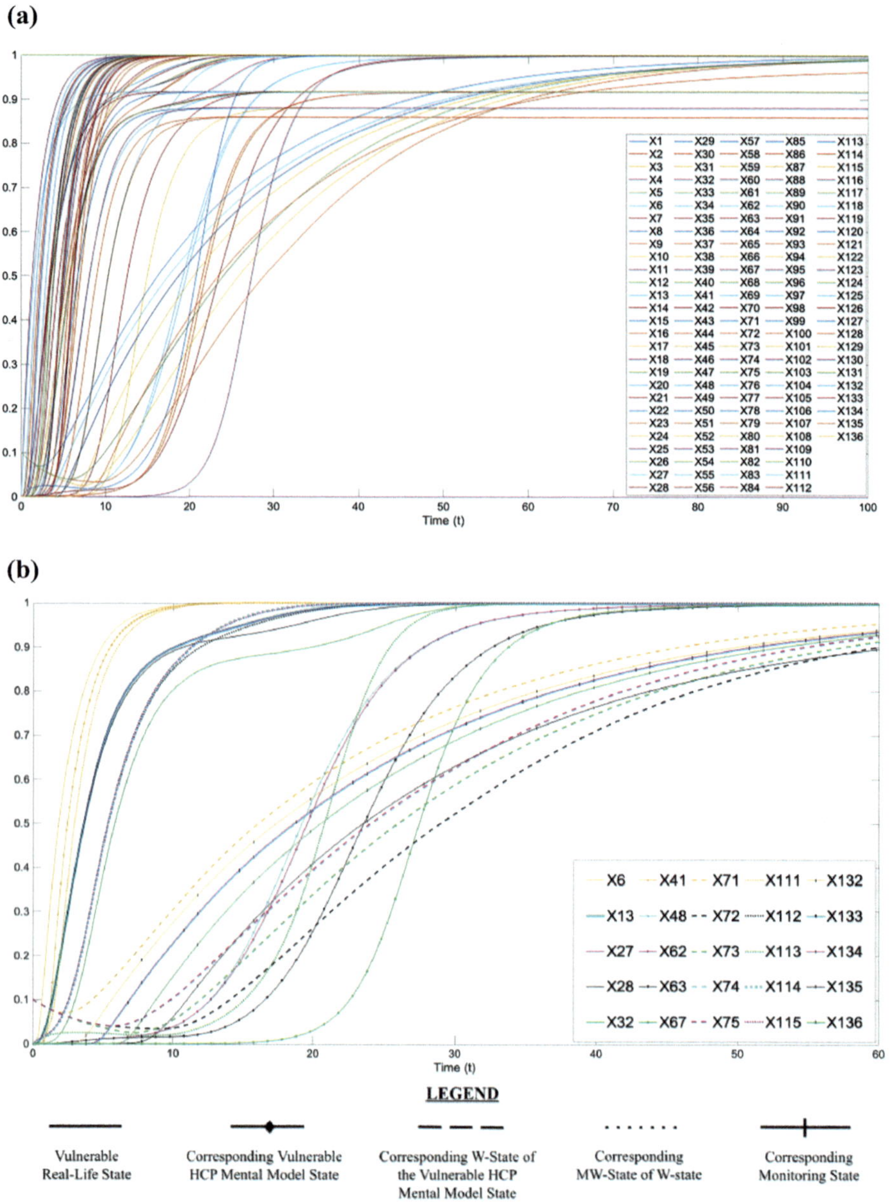

Fig. 12 Simulation results for Neonatal Hyperbilirubinemia treatment: Scenario 3b. **b** Magnified chart of the monitor and adaptive states in the reification levels and vulnerable HCP mental model and corresponding real-world states in the base level of the Neonatal Hyperbilirubinemia treatment: Scenario 3b

ensure its correct execution which then further enables other steps in the protocol to also be fully completed after being stagnated at a certain level of completion for the duration of the delay. In fact, the charts of Scenario 3b are visually the horizontally contracted version of the charts seen in Scenario 2 except the presence of monitoring states and the AIC mental model states. Thus, in Scenario 3b in both medical treatments, the vulnerable states reach full saturation by around 40 time units while in Scenario 2 of both medical treatments takes up to 150 units to reach full saturation.

6 Discussion

This chapter presented an adaptive second-order agent network model that covers three scenarios in a medical setup, where health care practitioners are following medical protocols either perfectly or imperfectly with the support of an AI Coach. These models are developed via an adaptive network-oriented modeling approach and simulated through a specialized software environment in MATLAB. The models depict the representation and processing of the actors' mental simulation, their mental model adaptability, interactions with the world, and the dynamics of these interactions. Through integrating higher-order adaptive networks, a dynamic and evolving system that caters to the changing demands and nuances of the medical profession is created and in this way this chapter addresses understanding and enforcing medical protocol adherence. Such network-oriented modeling techniques have also been applied in different domains such as addressing communication with parents and postpartum depression (Weigl et al. 2023, 2025) and addressing respiratory problems of new-born babies (Mokadem et al. 2024, 2025; Sebahi et al. 2024a, b).

In conclusion, the model successfully demonstrates how the incorporation of an AIC can effectively and dynamically intervene to prevent deviations from various different medical protocols happening in the real-world at the hands of a HCP. The model further illustrates various learning effects, including internal Hebbian learning, showcasing their influence on mental model structures. Simulation experiments highlight the role of AI-Coach in fostering adherence to guidelines as the AI-Coach monitors team actions, issuing alerts to the health care practitioners upon identifying potential mistakes.

The simulation experiments highlight how incomplete knowledge, and a flawed mental model could have severe implications in the proper execution of a protocol in the real world. They also underscore the role of an AI-coach in fostering adherence to guidelines as the AI-coach monitors the implementation of the protocol and the actions of the HCP, issuing alerts to the health care practitioners upon identifying potential mistakes. Simulation when a part of the protocol is not fully executed in the real world which in turn provides knowledge reinforcement and support. The simulation results further underscore the effects of this intervention, such as the prevention of delay in a state's full execution and increase in persistence and learning when a mistake is committed with the intervention of the AIC and the effective dynamic safeguard provided by the monitor states which act in accordance with the contexts of the different scenario's.

This research is an invaluable contribution to optimizing adherence to neonatal treatment guidelines, thereby fostering a safer, more efficient neonatal care environment. The presented model has a few limitations, namely that this is a theoretical model which does not have any practical experience. It also assumes the AI Coach monitors the medical process perfectly and also communicates with the HCP perfectly. However, in the real-world effective mechanisms to conduct the AIC's communication and monitoring will need to be developed. This could be an area of future research as well since there is a need to conduct field trials to see the practical application of this experiment, its further application to other medical protocols even outside that of the neonatal specialty, and its actual practical usefulness which further paves way for two potential directions for the evolution of cognitive AI - with applications in healthcare: monitoring and warning. It will be interesting to see if an AI coach can automatically recognise what is happening in the medical room through computer vision and recognise flaws in protocols, thus recognise if an action is performed or not, choose a pathway of treatment in its mental model which leads to the protocol being performed correctly and continue with the appropriate corresponding support signals to the health care practitioners.

References

AAP, American Academy of Pediatrics Subcommittee on Hyperbilirubinemia.: Management of hyperbilirubinemia in the newborn infant 35 or more weeks of gestation. Pediatrics **114**(1), 297–316 (2004)

Abramowski, A., Hamdan, A.H.: Neonatal Hypoglycemia. PubMed; StatPearls Publishing. https://www.ncbi.nlm.nih.gov/books/NBK537105/ (2021)

Adamkin, D.H.: Neonatal hypoglycemia. Semin. Fetal Neonatal Med. **22**(1), 36–41 (2017). https://www.sfnmjournal.com/article/S1744-165X(16)30049-X/fulltext

Canbaloğlu, G., Treur, J., Wiewiora, A. (eds.): Computational modeling of multilevel organisational learning and its control using self-modeling network models. Springer Nature (2023)

Gillott, C.: Everything you need to know about jaundice. Medicalnewstoday.com; Medical News Today (2017). https://www.medicalnewstoday.com/articles/165749#symptoms

Hawdon, J.M.: Disorders of blood glucose homeostasis in the neonate. In: Rennie and Roberton's Textbook of Neonatology (6th ed)., pp. 1065–1076. Elsevier (2019)

Hebb, D. O.: The organization of behavior: A neuropsychological theory. Psychology press (1943)

Hendrikse, S.C.F., Treur, J., Koole, S.L.: Modeling emerging interpersonal synchrony and its related adaptive short-term affiliation and long-term bonding: a second-order multi-adaptive neural agent model. Int. J. Neural Syst. **33**(7), 2350038, 1–41 (2023). https://doi.org/10.1142/S0129065723500387

https://www.aafp.org/pubs/afp/issues/2014/0601/p873.html#should-breastfeeding-be-modified-for-infants-with-jaundice-

https://www.sciencedirect.com/topics/agricultural-and-biological-sciences/bilirubin

Johnson, L., Bhutani, V.K.: The clinical syndrome of bilirubin-induced neurologic dysfunction. Semin. Perinatol. **35**(3), 101–113 (2011). https://pubmed.ncbi.nlm.nih.gov/21641482/

Kathusing, S., Samhan, N., Treur, J.: Higher-order adaptive dynamical system modeling of the role of epigenetics in anxiety disorders. Cogn. Syst. Res. **83**, 101177 (2024)

Keren, R., Tremont, K., Luan, X., Cnaan, A.: Visual assessment of jaundice in term and late preterm infants. Archiv. Dis. Childhood Fetal Neonatal Edn. **94**(5), F317–F322 (2009). https://pubmed.ncbi.nlm.nih.gov/19307221/

Kim, D.H.: A framework and methodology for linking individual and organizational learning: applications in TQM and product development (Doctoral dissertation, Massachusetts Institute of Technology) (1993)

Laxmi, J.T., Tietze, K.J., Washington, I.M., Pascual, J.M.: Bilirubin—an overview. J. Am. Vet. Med. Assoc. **252**(6) (2018). https://doi.org/10.2460/javma.252.6.667

Maisels, M.J., Bhutani, V.K., Bogen, D., Newman, T.B., Stark, A.R., Watchko, J.F.: Hyperbilirubinemia in the newborn infant $\geq$35 weeks' gestation: an update with clarifications. Pediatrics **124**(4), 1193–1198 (2009)

McKinlay, C.J., Alsweiler, J.M., Harding, J.E.: Neonatal glycemia and neurodevelopmental outcomes at 2 years. New England J. Med. **373**(16), 1507–1518 (2015). https://doi.org/10.1056/nejmoa1504909

Mokadem, N., Jabeen, F., Treur, J., Taal, H.R., Roelofsma, P.H.M.P.: An adaptive network model for AI-assisted monitoring and management of neonatal respiratory distress. Cogn. Syst. Res. **86**, 101231 (2024)

Mokadem, N., Jabeen, F., Treur, J., Taal, H.R., Roelofsma, P.H.M.P.: Increasing safety and security through cyberspace by an adaptive network model for AI-assisted risk management of neonatal respiratory distress. In: Roelofsma, P.H.M.P., Jabeen, F., Taal, H. R., Treur, J. (eds.) Using Shared Mental Models and Organisational Learning to Support Safety and Security Through Cyberspace: A Computational Analysis Approach. Springer Nature, Cham, Switzerland (this volume) (2025)

Muchowski, K.E.: Evaluation and treatment of neonatal hyperbilirubinemia. Am. Fam. Physician **89**(11), 873–878 (2014)

Narvey, M.R., Mark, S.D.: The screening and management of newborns at risk for low blood glucose. Canadian Paediatric Society.cps.ca. (2019). https://cps.ca/en/documents/position/newborns-at-risk-for-low-blood-glucose

NVK, Nederlandse Vereniging voor Kindergeneeskunde: Screening, preventie en behandeling van neonatale hypoglycemie. Richtlijnendatabase.nl (2020). https://richtlijnendatabase.nl/richtlijn/postnatale_zorg_in_de_algemene_kindergeneeskunde/screening_preventie_en_behandeling_van_neonatale_hypoglycemie/definitie_van_neonatale_hypoglycemie.html

Osmond, E., Northrop, J.: Clinical Guideline NEONATAL JAUNDICE. University Hospitals Bristol and Weston. NHS Foundation Trust (2022). https://foi.avon.nhs.uk/download.aspx?did=2986

Paul, I.M., Maisels, M.J.: Can I Stop Phototherapy for This Baby? Aap.org; American Academy of Pediatrics (2017). https://publications.aap.org/pediatrics/article-abstract/139/3/e20163832/53273/Can-I-StopPhototherapy-for-This-Baby?redirectedFrom=fulltext

Piaget, J.: The Construction of Reality in the Child. Ballantine, New York (1954)

Queensland Clinical Guidelines.: Neonatal Jaundice. Guideline No. MN22.7-V9-R27. Queensland Health (2022). Available from: https://www.health.qld.gov.au/__data/assets/pdf_file/0018/142038/g-jaundice.pdf

Rao, R.B.: Neonatal Hypoglycemia. Newbornjournal.org (2022). https://www.newbornjournal.org/abstractArticleContentBrowse/JNB/27407/JPJ/fullText

Rasmussen, A.H., Wehberg, S., Fenger-Groen, J., Christesen, H.T.: Retrospective evaluation of a national guideline to prevent neonatal hypoglycemia. Pediatr. Neonatol. **58**(5), 398–405 (2017). https://www.sciencedirect.com/science/article/pii/S1875957217300219

Sebahi, Y., Jabeen, F., Treur, J., Taal, H. R., Roelofsma, P.H.M.P.: Optimizing neonatal respiratory support through network modeling: a new approach to post-birth infant care. In: Cherifi, H., Rocha, L.M., Cherifi, C., Donduran, M. (eds.) Complex Networks and Their Applications XII. COMPLEX NETWORKS 2023. Studies in Computational Intelligence, vol. 1143, pp. 245–257. Springer Nature, Cham (2024a)

Sebahi, Y., Jabeen, F., Treur, J. Taal, H.R., Roelofsma, P.H.M.P.: Improving risk management through cyberspace: optimizing neonatal respiratory support through network oriented modeling. In: Roelofsma, P.H.M.P., Jabeen, F., Taal, H.R., Treur, J. (eds.) Using Shared Mental Models and Organisational Learning to Support Safety and Security Through Cyberspace: A Computational Analysis Approach. Springer Nature, Cham, Switzerland (this volume) (2024b)

Slusher, T.M., Zamora, T.G., Appiah, D., Stanke, J.U., Strand, M.A.: Burden of severe neonatal jaundice: a systematic review and meta-analysis. BMJ Paediatr. Open **1**(1), e000105 (2017)

Thornton, P.S., Stanley, C.A., De Leon, D.D., Harris, D., Haymond, M. W., Hussain, K. et al.: Recommendations from the pediatric endocrine society for evaluation and management of persistent hypoglycemia in neonates, infants, and children. J. Pediatr. Endocrinol. Metab. **28**(7–8), 822–830 (2015). https://www.jpeds.com/article/s0022-3476(15)00358-3/fulltext

Treur, J.: Network-Oriented Modelling: Addressing Complexity of Cognitive, Affective and Social Interactions. Springer International Publishing Switzerland (2016)

Treur, J.: Network-Oriented Modeling for Adaptive Networks: Designing Higher-Order Adaptive Biological, Mental and Social Network Models. Springer Nature, Cham (2020)

Treur, J.: On the dynamics and adaptivity of mental processes: relating adaptive dynamical systems and self-modeling network models by mathematical analysis. Cogn. Syst. Res. **70**, 93–100 (2021)

Treur, J.: On structure, dynamics, and adaptivity for biological and mental processes: a higher-order adaptive dynamical system modeling perspective. In: Proceedings of the 46th Annual Meeting of the Cognitive Science Society—Dynamics of Cognition, CogSci'24 (2024)

Treur, J., Van Ments, L. (eds.): Mental models and their dynamics, adaptation, and control. Springer Nature (2022)

Ullah, S., Rahman, K., Hedayati, M.: Hyperbilirubinemia in neonates: types, causes, clinical examinations, preventive measures and treatments: a narrative review article. Iranian J. Public Health **45**(5), 558–568 (2016). https://www.ncbi.nlm.nih.gov/pmc/articles/PMC4935699/

Wasylewicz , A.T.M., Scheepers-Hoeks, A.M.J.W.: Clinical Decision Support Systems (2018). ncbi.nlm.nih.gov/. https://www.ncbi.nlm.nih.gov/books/NBK543516/

Weigl, L.M., Jabeen, F., Treur, J., Taal, H.R., Roelofsma, P.H.M.P.: Modelling learning for a better safety culture within an organization using a virtual safety coach: reducing the risk of postpartum depression via improved communication with parents. Cogn. Syst. Res. **80**, 1–36 (2023)

Weigl, L., Jabeen, F., Treur, J., Taal, H.R., Roelofsma, P.H.M.P.: Learning for a better safety and security culture within an organization: Reducing the risk in communication with AI coaching for security communication through cyberspace. In: Roelofsma, P.H.M.P., Jabeen, F., Taal, H.R., Treur, J. (eds.) Using Shared Mental Models and Organisational Learning to Support Safety and Security Through Cyberspace: A Computational Analysis Approach. Springer Nature, Cham, Switzerland (this volume) (2025)

Yetman, R. J., Parks, D. K., Huseby, V., Mistry, K., Garcia, J.: Rebound bilirubin levels in infants receiving phototherapy. J. Pediatr. **133**(5), 705–707 (1998). https://doi.org/10.1016/s0022-3476(98)70117-9

Improving Risk Management Through Cyberspace: Optimizing Neonatal Respiratory Support Through Network-Oriented Modeling

Yassine Sebahi, Fakhra Jabeen, Jan Treur, H. Rob Taal, and Peter H. M. P. Roelofsma

Abstract This chapter presents an approach to enhancing neonatal care through the application of artificial intelligence (AI). Utilizing network-oriented modeling methodologies, the study aims to develop a network model to improve outcomes in neonatal respiratory support. The introduction sets the stage by outlining the significance of neonatal respiratory support and the challenges faced in this domain. The literature review delves into the existing body of work, highlighting the gaps and the need for a network modeling approach. The network-oriented modeling approach provides a robust framework that captures various states, such as world states, doctors' mental states, and AI coach states, facilitating a comprehensive understanding of the complex interactions in neonatal respiratory support. Through Matlab simulations, the study investigates multiple scenarios, from optimal conditions to deviations from standard protocol. The main contribution focuses on the introduction of an AI coach, which serves as a real-time intervention mechanism to fill in the doctor's knowledge gaps. The research serves as a seminal work in the intersection of

Y. Sebahi · J. Treur (✉)
Social AI Group, Vrije Universiteit Amsterdam, Amsterdam, The Netherlands
e-mail: j.treur@vu.nl; j.treur@hhs.nl

Y. Sebahi
e-mail: yassinesebahi@icloud.com

F. Jabeen
Safety and Security Section, Delft University of Technology, Delft, The Netherlands

J. Treur · P. H. M. P. Roelofsma
Risk Management and Cybersecurity Group, Centre of Expertise Cybersecurity, The Hague University of Applied Sciences, Hague, The Netherlands
e-mail: p.h.m.p.roelofsma@hhs.nl

H. Rob Taal · P. H. M. P. Roelofsma
Erasmus University Medical Center Rotterdam, Rotterdam, The Netherlands
e-mail: h.taal@erasmusmc.nl

P. H. M. P. Roelofsma et al. (eds.), *Using Shared Mental Models and Organisational Learning to Support Safety and Security Through Cyberspace: A Computational Analysis Approach*, Studies in Systems, Decision and Control 570,
https://doi.org/10.1007/978-3-031-72075-8_7

artificial intelligence and healthcare, demonstrating the potential of network-oriented modeling in improving patient outcomes and streamlining healthcare protocols.

Keywords Adaptive network model · Infant care · AI coach

1 Introduction

In the Netherlands, 166,891 babies were born in 2022. That is 457.24 per day (Cijfers over geboorte | Nederlands Jeugdinstituut 2023). Approximately 7% experience respiratory distress at birth, necessitating immediate and specialized medical intervention (Edwards and Kotecha 2013). This translates to a staggering number of infants requiring critical respiratory support each day, highlighting the urgency for effective and optimized neonatal care. Neonatal respiratory support is a critical aspect of infant care, as timely interventions can have significant impacts on survival and long-term health outcomes (Kaltsogianni et al. 2023). With advancements in technology, Artificial Intelligence (AI) has emerged as a tool with potential applications across various healthcare domains, including neonatal care (Malak et al. 2018).

This chapter explores the question, 'How can the development and analysis of a network model, representing world, AI coach, and doctors' mental states, provide insights into neonatal respiratory support through the simulation of the process of respiratory support of a newborn baby?' The research builds on principles of network-oriented modeling by adaptive self-modeling networks (Treur 2020a, b), encompassing three key states and modeling three scenarios:

1. A Successful Scenario: Reflecting optimal processes as described in neonatal care guidelines.
2. An Error Scenario: Some deviation takes place: an often occurring error or omission.
3. An AI-Coached Error Detection and Knowledge Improvement Scenario: The AI coach detects an error and improves the knowledge of the doctor if needed.

By analyzing these scenarios, this chapter aims to contribute to the understanding of the AI Coach's role in optimizing neonatal respiratory support.

2 Background Literature

Neonatal respiratory support is a vital aspect of care for newborns, particularly in the moments immediately following birth. Roehr and Bohlin (2011) state that a protective respiratory support strategy from birth is essential as it may not only reduce breathing difficulties in the immediate neonatal period, but may also influence some known triggers for the development of BPD, such as inflammation, oxidative stress and lung growth. The importance of this intervention has been highlighted in various clinical guidelines, emphasizing the need for immediate assessment and support of breathing in newborns (Anne and Murki 2021). Advances in neonatal respiratory care have led to improved survival rates and outcomes for preterm infants and those with specific respiratory conditions.

Even with recent progress, there are still some hurdles in giving the best breathing support to newborns (Kaltsogianni et al. 2023). These hurdles include identifying which babies need help, choosing the right treatments, deciding when to offer support, and avoiding mistakes. There's also inconsistency in how treatments are given, making things even more complicated. However, technology like Artificial Intelligence (AI) could help overcome some of these issues (Kaltsogianni et al. 2023). By using network models that show different situations related to breathing support, healthcare providers could get a clearer idea of how to best handle this crucial part of caring for newborns.

The integration of Artificial Intelligence (AI) into healthcare has marked a transformative era, revolutionizing various medical domains, from diagnostics to personalized treatment (Khan et al. 2022). The convergence of AI technologies with medical practices has led to improved efficiencies, enhanced patient outcomes, and the opening of new avenues for research and innovation. In the context of neonatal care, AI has demonstrated promising applications, including the analysis of complex medical data, predictive modeling for patient outcomes, and assistance in decision-making (Bajwa et al. 2021). These applications extend to neonatal respiratory support, where timely and precise interventions are crucial.

One specific area where we can investigate if AI shows potential in the process of the respiratory support of neonatal is with the use of network modeling. This approach involves the construction of network models representing various states and relationships, enabling the simulation and analysis of different scenarios related to respiratory support (Treur 2020a, b). For instance, network models can represent the world states, AI coach states, and doctors' states, each with specific roles and interactions. The development of such network models allows for a systematic exploration of neonatal respiratory support processes, including the simulation of optimal processes, common deviations, and AI-coached interventions. By leveraging the computational capabilities of AI, these models can provide insights, guide clinical decisions, and potentially optimize respiratory support for newborns. Network modeling and analysis in neonatal respiratory support offers a novel approach to understanding and enhancing care, with potential implications for both immediate neonatal outcomes and the future of technology-driven medical care.

The development of network models that represent various states and interactions is an innovative approach in healthcare, providing a computational framework to understand and analyze complex processes. In the context of neonatal respiratory support, these models can include states such as:

- World States: capturing states of the baby and the broader context and environment, including hospital settings, equipment, and external factors that may influence care.

- AI Coach states: Representing an intelligent entity that guides, monitors, and supports the healthcare process, offering insights and interventions when needed.
- Doctors' mental and action states: Reflecting the healthcare provider's actions, decisions, knowledge, and interactions with both the world and AI coach states.

These states and their interactions are covered by the network model, allowing for the simulation and analysis of different scenarios. The scenarios can include:

1. Successful Processes: Simulating the ideal process of neonatal respiratory support, serving as a baseline for understanding best practices and optimal outcomes.
2. Common Deviations: Modeling frequent errors or omissions, highlighting potential risks, and areas for improvement in care delivery.
3. AI-Coached Error Detection and Knowledge Improvement: Integrating an AI coach to detect and rectify mistakes in real-time, enhancing accuracy and safety. And utilizing AI to support healthcare workers in enhancing their knowledge, skills, and adherence to guidelines, thus improving overall care quality.

The ability to model and simulate these scenarios offers valuable insights into neonatal respiratory support, allowing for a nuanced understanding of the interactions and dependencies within the process. It opens opportunities for targeted interventions, continuous learning, and optimization of care, aligning with the broader goals of precision medicine and technology-driven healthcare.

By leveraging network modeling, this approach fosters a data-driven, evidence-based practice that transcends traditional boundaries, offering a new perspective on neonatal care and beyond.

Computational causal modeling is a powerful tool in AI that can help us understand complex healthcare situations better (Sarker 2022). In the case of helping newborns breathe, this type of modeling can map out how different factors like doctors, AI coaches, and the baby's condition interact. This approach is unique because it shows not only what directly causes what but also how changes in one area can affect the whole system (Squires and Uhler 2022). By using this method, researchers can simulate different outcomes, such as what happens when things go right, when they go wrong, how AI can spot mistakes, and how AI can help improve our knowledge (Campos and Fleury 2022).

This kind of modeling can help identify why certain treatments work or fail and point out where critical decisions should be made to improve care. The research aims to add to the growing field of network modeling in healthcare. The findings could impact not just how doctors treat newborns, but also broader healthcare policies and future studies, laying the groundwork for improving the care of newborns overall.

In conclusion, the development and application of network modeling, coupled with computational causal modeling, represent a novel and promising avenue in neonatal respiratory support. By drawing on relevant literature and innovative methodologies, this research aims to shed new light on the complexities of neonatal care and pave the way for technology-driven improvements in this vital area of healthcare.

3 Modeling Approach for Neonatal Respiratory Support

The research conducted for this chapter employs a network-oriented modeling approach to understand and analyze the complex interactions and processes in neonatal respiratory support (Weigl et al. 2023), see also Weigl et al. (2025), this volume. This methodology encompasses various states, such as the world, AI coach, and doctor states, capturing the interactions and causal impacts within the system. Key features characterize the structure of the network (Weigl et al. 2023). State are often indicated by X and Y; they have activation values (real numbers, usually in the interval [0, 1]) $X(t)$ and $Y(t)$ that vary over time t. *Connectivity Characteristics* specify connections from a state X to a state Y as defined by their weights $\boldsymbol{\omega}_{X,Y}$, symbolizing the strength of the causal impact from X to Y. *Aggregation Characteristics* are specified for any state Y by a combination function $\mathbf{c}_Y(\ldots)$ outlines the aggregation applied to the single causal impacts $\boldsymbol{\omega}_{X,Y}X(t)$ on Y from its incoming connections from states X. *Timing Characteristics* specify for each state Y a speed factor $\boldsymbol{\eta}_Y$, indicating how quickly it changes for a given causal impact.

Based on these network characteristics a standard numerical format described by the difference equation defines the dynamics of the network model:

$$Y(t + \Delta t) = Y(t) + \boldsymbol{\eta}_Y\left[\mathbf{c}_Y\left(\boldsymbol{\omega}_{X_1,Y}X_1(t), \ldots, \boldsymbol{\omega}_{X_k,Y}X_k(t)\right) - Y(t)\right]\Delta t$$

Various combination functions are available to handle the aggregation of multiple impacts, with the specific functions used here detailed in Table 1.

The modeling approach also includes the concept of network reification or self-modeling network, extending the base model by additional states, referred to as reification states or self-model states (Treur 2020a, b). Examples are self-model states $\mathbf{W}_{X,Y}$, $\mathbf{C}_Y$, $\mathbf{H}_Y$ (reification states) to represent the adaptive network structure characteristics $\boldsymbol{\omega}_{X,Y}$, $\mathbf{c}_Y$, $\boldsymbol{\eta}_Y$ for a state Y of the base network. Such self-model states are called **W**-states, **C**-states and **H**-states, respectively. This can be iterated to get higher-order self-model states. For example, the self-model state $\mathbf{W}_{\mathbf{W}_{X_1,Y_1}, \mathbf{W}_{X_2,Y_2}}$ is a second-order self-model state that indicates the weight of a communication channel from $\mathbf{W}_{X_1,Y_1}$ to $\mathbf{W}_{X_2,Y_2}$. This will be used in the introduced model for communication from AI Coach to Doctor.

In Treur (2021), Hendrikse et al. (2023), it is shown that any smooth dynamical system has a canonical representation as a temporal-causal network and any smooth (multi-order) adaptive dynamical system has a canonical representation as a

Table 1 Combination functions used

Function	Notation	Formula	Parameters
Advanced logistic sum	$\mathbf{alogistic}_{\sigma,\tau}(V_1, \ldots, V_k)$	$\left[\frac{1}{1+e^{-\sigma(V_1+\ldots+V_k-\tau)}} - \frac{1}{1+e^{\sigma\tau}}\right](1 + e^{-\sigma\tau})$	Steepness σ Threshold τ
Identity	$\mathbf{id}(V_1, \ldots, V_k)$	V_1	–

(multi-order) self-modeling temporal-causal network. Therefore, compared to adaptive dynamical systems in general, the network-oriented modeling approach used here does not introduce any fundamental limitations concerning what it can model. This has also been confirmed by applications to case studies in practice, e.g., for mental models in Treur et al. (2022) and for organizational learning in Canbaloğlu et al. (2023).

This network-oriented methodology provides a robust framework to explore and analyze scenarios related to neonatal respiratory support. The modeling and simulations are conducted using Matlab, providing a comprehensive approach to simulating and analyzing the scenarios related to neonatal care. In the development of an network model for optimizing neonatal respiratory support, two essential mathematical functions as shown in Table 1 have been employed within the MATLAB environment.

The advanced logistic sum function represents a nonlinear transformation that takes the weighted sum of the input variables $V_1,\ldots,V_k$ for incoming single causal impacts and applies a logistic function. This function is characterized by two parameters: the steepness σ and the threshold τ. The steepness parameter σ controls the slope of the logistic curve, whereas the threshold parameter τ determines the point at which the function transitions from one state to another. In the context of neonatal respiratory support, this function can be utilized to model complex relationships and transitions within the respiratory system, such as the response to different ventilatory support parameters. The identity function is a straightforward mathematical transformation that just returns the input value itself. In terms of the respiratory support model, the identity function can represent parameters or variables that are directly observed or controlled without the need for transformation or scaling. Together, these two functions serve critical roles within the network model. The advanced logistic sum provides the capability to capture nonlinear dynamics and complex relationships within the respiratory system, while the identity function ensures that certain aspects of the system can be modeled in a direct and unaltered manner. They enable the creation of sophisticated graphs that can be analyzed to better understand the underlying mechanisms of neonatal respiratory support. The ultimate goal is to leverage these insights to develop more effective and personalized interventions for post-birth infant care.

In Appendix (2023) tables with explanations for all states and for the role matrices that are used for the simulation of the scenarios can be found. Here we explain the base world states. The pathway shown in Fig. 1 is followed. The red cross represents what happens from Scenario 3, where the doctor has no knowledge that you have to follow certain instructions after gasping, or he forgets to do this.

The concept of a world state encompasses the environment in which neonatal respiratory support takes place. Understanding the world state is pivotal for the realistic simulation of scenarios that aim to optimize neonatal respiratory support. Within the network model, various states coexist to collectively influence the outcome of respiratory support for a neonate. These states can be categorized as follows see Table 2.

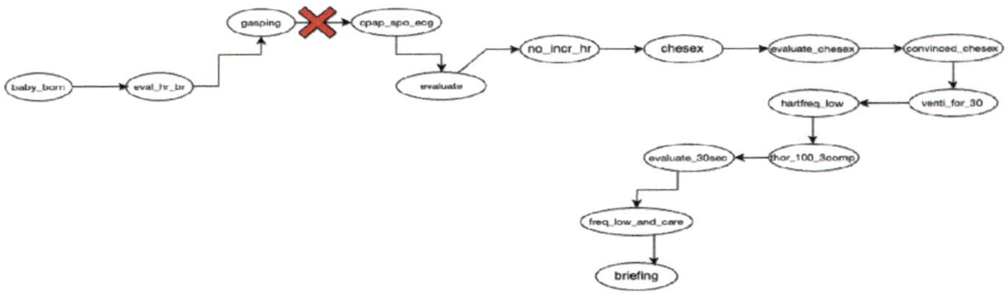

Fig. 1 The base level world states in the network model

- Context States: These states provide information on specific conditions that could influence respiratory support. For example, Context State G Indicates that the baby is gasping in this scenario.
- Evaluation States: These states, like eval_hr_br and evaluate_30sec, are pivotal for ongoing assessment of the baby's physiological parameters.
- Intervention States: These states dictate the medical interventions that should be considered, such as cpap_spo_ecg and infl_spo_ecg.
- Outcome States: These states represent the outcomes of previous actions and evaluations, like true_incr_hr and hartfreq_high.
- eval_hr_br (Evaluate heart rate, breathing, color, and muscle tone): This state is crucial for the initial evaluation post-birth. It encapsulates the assessment of multiple physiological parameters to decide the subsequent course of action.
- infl_spo_ecg (Open airway, give 5 inflation breaths (30 cm H_2O), SpO_2 and ECG monitoring): This state outlines the protocol for cases where initial assessment indicates respiratory distress, thereby requiring inflation breaths and continuous monitoring.
- evaluate_30sec (Evaluate heart rate every 30 s): This state underscores the necessity for frequent re-evaluation to adapt the treatment strategy effectively.

The states are not static but interact dynamically within the network model. For example, if the state hartfreq_low is activated, the network transition to thor_100_3comp for immediate intervention. This dynamic interplay is essential for simulating the real-world complexity of neonatal respiratory care. The granularity and complexity of these states make them ideal candidates for network modeling. By applying advanced functions like $\mathbf{alogistic}_{\sigma,\tau}$ for nonlinear relationships and **id** for direct variables, the model can simulate intricate scenarios that mimic real-life conditions. These simulations, therefore, hold the potential to significantly improve neonatal respiratory support protocols.

In this part, we'll look at the roles of doctors and AI coaches in helping newborns breathe, see Fig. 2. These roles are complex and include everything from the decisions healthcare providers make to the medical guidelines they follow. The idea is to understand how doctors think and act in these situations. For example, if a doctor knows that a baby is having trouble breathing, they would follow a specific treatment plan, known as the CPAP_SPO_ECG procedure. This decision is based on the

Table 2 World states and their explanations

State name	Description
Context State N.	Whether or not there is: no_incr_hr
Context State L.	Whether or not there is: hartfreq_low
Context State G.	Whether or not there is: gasping
Context State I.	Whether or not there is: inadequate
Context State T.	Whether or not there is: true_incr_hr
Context State H.	Whether or not there is: hartfreq_high
baby_born	Birth
eval_hr_br	Evaluate heart rate, breathing (color and muscle tone)
inadequate	Inadequate breathing
gasping	Gasps or apnea
cpap_spo_ecg	Open airway, consider CPAP SpO2 and ECG monitoring
infl_spo_ecg	Open airway, give 5 inflation breaths (30 cm H_2O) SpO2 and ECG Monitoring
evaluate	Evaluate heart rate
no_incr_hr	No increase in heart rate
true_incr_hr	Increase in heart rate
chesex	Check head and mask position. Consider alternate airway strategies. Repeat 5 inflation breaths
evaluate_chesex	Evaluate whether chest excursions had an effect on heart rate
convinced_chesex	Convinced of chest excursions
venti_for_30	Ventilation for 30 s
hartfreq_high	Heart rate is higher than 60/min
hartfreq_low	Heart rate is less than 60/min
thor_100_3comp	Start chest compressions. Increase oxygen percentage to 100%. 3 compressions on 1 breath

(continued)

Table 2 (continued)

State name	Description
evaluate_30sec	Evaluate heart rate every 30 s
freq_low_and_care	If heart rate <60/min: Provide i.v. access and give adrenaline Consider other causes (such as pneumothorax, hypovolemia, congenital abnormalities)
briefing	Inform parents, debrief with team and register

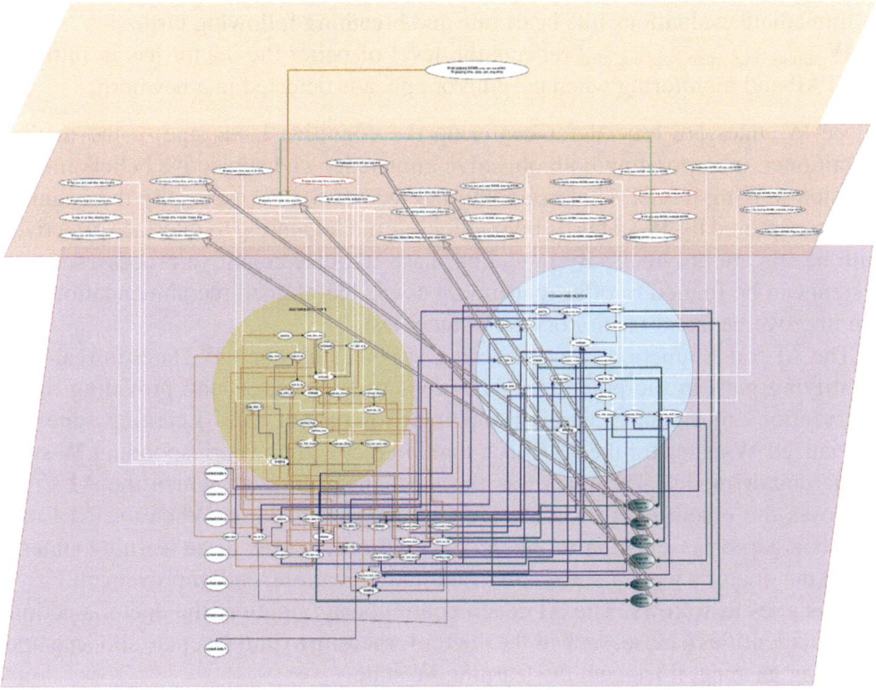

Fig. 2 The introduced overal network model

doctor's existing knowledge and experience. By breaking down the thought processes and actions of healthcare providers in this way, we can get a clearer picture of how decisions are made and treatments are administered in neonatal respiratory care.

So, if the **W**-state of this relation has no value, the doctor would not know that it needs to do CPAP_SPO_ECG after he diagnoses that the baby has gasping. The doctor states encapsulate various functionalities:

- Intervention States: Such as cpap_spo_ecg (doctor MS) and infl_spo_ecg (doctor MS), dictating specific medical actions.

- Assessment States: Focused on continuous evaluations, e.g., evaluate (doctor MS) and evaluate_30sec (doctor MS).
- Outcome States: Representing the results of interventions, like hartfreq_high (doctor MS) and hartfreq_low (doctor MS).

In our model, **W**-states serve as quantitative indicators of the doctor's level of knowledge, confidence, or belief regarding the relationship between other states. For example:

- $\mathbf{W}_{\text{baby\_born dms,eval\_hr\_br dms}}$ represents the doctor's confidence in the necessity of immediate evaluations like heart rate and breathing following birth.
- $\mathbf{W}_{\text{gasping dms, cpap\_spo\_ecg dms}}$ reflects the level of belief the doctor has in initiating CPAP and monitoring when gasping or apnea is detected in a newborn.

These **W**-states are crucial for capturing the cognitive landscape of the medical practitioner, incorporating both objective knowledge and subjective beliefs into the decision-making model. Understanding the Doctor States, especially the cognitive aspects captured by **W**-states, is pivotal for our model aiming to interface effectively with healthcare providers. By modeling these intricate cognitive processes, the AI system can be trained to offer real-time, knowledge-aligned recommendations that can improve neonatal respiratory care outcomes.

The AI coach functions by monitoring various states and **W**-states in real-time, identifying gaps in the practitioner's actions or knowledge, and providing timely interventions to enhance learning and improve patient care. Learning states are specialized **W**-states for the AI Coach that interact with the corresponding **W**-states in the doctor model. They serve as the mechanism through which the AI Coach improves the practitioner's knowledge and decision-making. When the AI Coach detects a gap or a deviation in the doctor's actions, it uses these learning states to adjust the doctor's weight states, thus facilitating learning and improvement.

This goes as follows. The AI coach continuously monitors the doctor's actions. When it identifies a lapse, such as the doctor forgetting to initiate cpap_spo_ecg upon detecting gasping, it triggers the learning **W**-state

$$\mathbf{W}_{\mathbf{W}_{\text{gasping AICMS,cpap\_spo\_ecg AICMS}},\, \mathbf{W}_{\text{gasping dms,cpap\_spo\_ecg dms}}}$$

This second-order self-model state models a communication channel from AI Coach to Doctor that adjusts the corresponding doctor's **W**-state, $\mathbf{W}_{\text{gasping dms,cpap\_spo\_ecg dms}}$, to fill in the knowledge gap. This adjustment informs the doctor of the necessary action, thereby enhancing the doctor's knowledge and improving patient outcomes. In a scenario where a newborn is detected to be gasping, and the doctor fails to initiate cpap_spo_ecg, the AI coach monitors this and intervenes. Through this learning state, the AI coach updates the doctor's corresponding **W**-state, in turn making them aware of the need to initiate CPAP, thereby facilitating immediate and appropriate medical intervention.

The incorporation of an AI coach equipped with learning states into the network model offers several advantages:

- Real-Time Intervention: The AI coach provides immediate feedback, allowing for real-time adjustments in the doctor's actions.
- Knowledge Enhancement: The learning states serve as a conduit for knowledge transfer from the AI coach to the medical practitioner, ensuring that the doctor is always updated on the best course of action.
- Adaptive Learning: The model can adapt and evolve over time, capturing the nuances of each practitioner's learning curve and adjusting its coaching strategy accordingly.

By effectively utilizing learning states, the model becomes an invaluable tool for continuous professional development, ensuring that healthcare providers are always at the forefront of medical knowledge and practice, ultimately leading to improved patient outcomes.

Monitor states serve as an integral part of the network model, capturing real-time or near-real-time observations or measurements from the system. In the context of neonatal respiratory support, these states are crucial for continuously assessing various conditions and parameters. They provide the data that informs the weight states, thus influencing the medical practitioner's decision-making process. The primary role of monitor states is to provide timely and accurate data for various attributes or conditions that are crucial in neonatal care. This data is then used to adjust the **W**-states, which represent the level of confidence or belief a medical practitioner might have in certain protocols or interventions. For example, if the doctor forgets to do the proces that comes with the state cpap_spo_ecg, the monitoring of the AI coach will make sure that the doctors knowledge about this will be updated. Examples of monitoring states are:

- MONITOR cpap_spo_ecg: This monitor state observes the effectiveness of CPAP (Continuous Positive Airway Pressure) along with SpO2 and ECG monitoring. The data collected helps in dynamically adjusting the weight state W gasping dms, cpap_spo_ecg dms, which influences the decision to initiate or continue CPAP.
- MONITOR freq_low_and_care: This state keeps track of the frequency and quality of care provided when the heart rate is below 60/min. The information is then used to adjust the weight at W evaluate_30 sec dms, freq_low_and_care dms, affecting the urgency and type of interventions considered.

Incorporating monitor states allows the AI system to make real-time adjustments based on current observations, making the model more adaptive and robust. These states serve as a bridge between the realworld conditions and the weight states, providing a dynamic feedback loop that enhances the model's predictive and decision-support capabilities. By understanding and effectively utilizing these monitor states, the model can offer more precise, timely, and context-sensitive recommendations, contributing to improved outcomes in neonatal respiratory care.

4 Findings from Network Model Simulations

This section presents the findings derived from simulations of the network model using Matlab. These simulation results are visualized as graphs, offering valuable insights into the model's effectiveness across various scenarios. The primary aim of this research is to address specific errors commonly made during the respiratory support of neonatal infants. In our Matlab simulations, we concentrated on rectifying a particular error: the omission of the cpap_spo_ecg action by doctors upon recognizing that the baby is gasping.

Importantly, all actions that a doctor can take are structured similarly within the network model. For the sake of simplicity and focus, we chose to zero in on the cpap_spo_ecg process. We posit that if our solution proves effective for this process, it should be generalizable to other processes as well. For all scenarios, see Appendix (2023). Here we focus on Scenario 3. This scenario involves an AI coach that not only detects the error or omission with the help of monitor states but also aids healthcare workers in improving their knowledge.

In this scenario, both the cpap_spo_ecg state and the corresponding **W**-state representing the doctor's knowledge remain deactivated. However, the introduced AI Coach is connected to the doctor's knowledge **W**-state. A specialized higher-order **W**-state denoted as $\mathbf{W}_{\mathbf{W}\text{gasping AICMS,cpap\_spo\_ecg AICMS}, \mathbf{W}\text{gasping dms,cpap\_spo\_ecg dms}}$ is in place to facilitate the transfer of knowledge from the AI Coach to the doctor when needed. Essentially, if the state values are already optimal (e.g., knowledge value is 1), the AI coach will not intervene, and the monitor state will remain in observational mode without triggering any actions. The inclusion of the specialized higher-order **W**-state provides an opportunity for the doctor to gain knowledge from the AI Coach. As observed in Fig. 3, the doctor's knowledge initially starts at 0 but increases to 0.9 due to the input from the AI coach. The concept underlying the connection between the AI coach's knowledge and the doctor's knowledge is that a single intervention should suffice for knowledge improvement. In other words, if the doctor receives guidance from the AI coach once, that guidance should be sufficient for future situations, thus negating the need for repeated AI interventions for the same action. The monitor state value for cpap_spo_ecg is notably low in Fig. 3, where the doctor's knowledge stands at 0. Interestingly, the monitor state value escalates significantly when the AI coach imparts knowledge to the doctor. This monitor state serves as an error-detection mechanism, intervening only when necessary. As evidenced in Fig. 4, when the doctor's knowledge level is already at 1 and the cpap_spo_ecg procedure has been performed, the monitor state peaks at 0.24. In contrast, in the absence of the doctor's knowledge, the peak is 0.52. In this instance, the monitor state only checks for errors and does not take any further action, as the necessary conditions are already met.

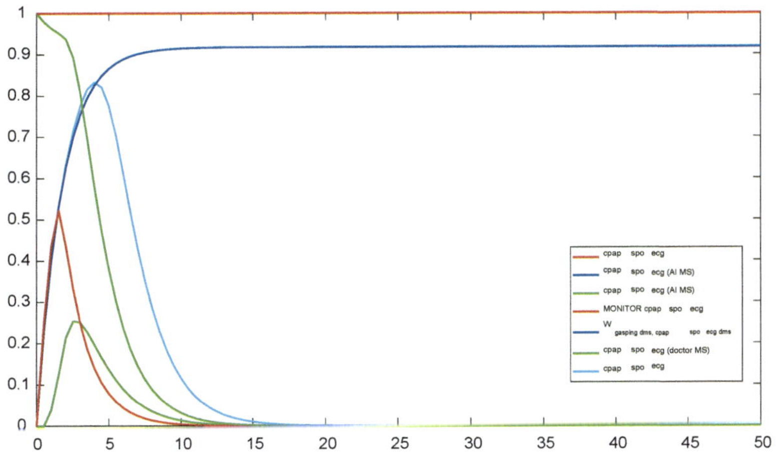

Fig. 3 Scenario 3 focused

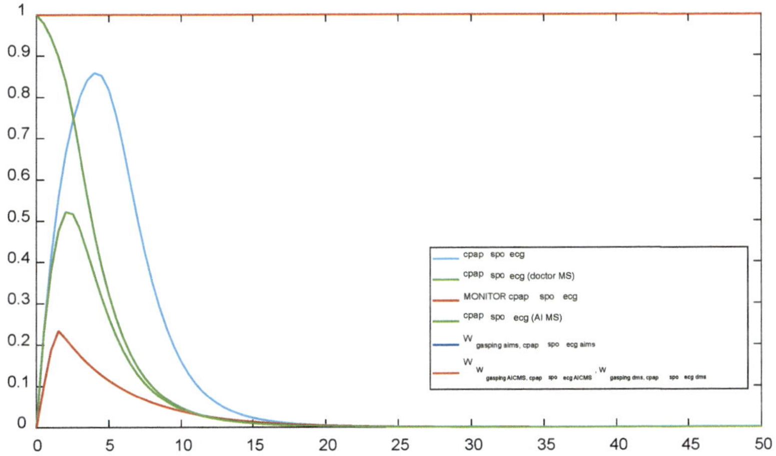

Fig. 4 cpap_spo_ecg knowledge is already adequate

5 Discussion

The primary objective of this chapter was to investigate how network modeling could optimize neonatal respiratory support protocols. The material comes from Sebahi et al. (2024). Utilizing a network-oriented modeling approach as outlined in Treur (2020a, b), various scenarios were created and analyzed within the Matlab environment. The findings from these scenarios contribute significantly to both the fields of neonatal respiratory support and network modeling. The second scenario emphasized the interconnectedness within the model. It revealed that missing links or incomplete knowledge could have far-reaching implications, affecting multiple aspects of neonatal respiratory support. This finding underscores the need for comprehensive and accurate data in models. The third scenario illustrated the potential of incorporating an AI coach into the model. The AI can not only act as a fail-safe tool but also as an educational tool. It provided real-time decision-making support and reinforced the doctor's knowledge base for future scenarios. The AI coach's intervention is a one-time requirement for each specific action or decision, equipping the doctor for future similar situations without additional AI assistance. Monitor states proved effective as safeguards, ensuring optimal performance and error minimization across different scenarios.

These findings provide evidence that an AI coach can be successfully applied to healthcare settings, particularly in the area of neonatal respiratory support. Some limitations concern that scaling up has not been addressed yet, the effectiveness has not yet been validated, and only some scenarios have been explored. Moreover, when dealing with healthcare data, you are primarily dealing with Personal Health Information (PHI), a category of data that is highly sensitive and heavily regulated to protect individuals' privacy. Health data can be a target for cyber attackers. Thus, robust security measures need to be in place to prevent unauthorized access and protect against data breaches. Future Research may address such limitations. For further details, see Appendix (2023).

References

Anne, R.P., Murki, S.: Noninvasive Respiratory support in neonates: A review of current evidence and practices. Indian J. Pediatr. **88**(7), 670–678 (2021)

Appendix (2023). Linked Data at https://www.researchgate.net/publication/373775834

Bajwa, J., Munir, U., Nori, A.V., Williams, B.: Artificial intelligence in healthcare: transforming the practice of medicine. Future Healthcare J. **8**(2), e188–e194 (2021)

Campos, A., Fleury, A.: Modeling, control strategies and design of a neonatal respiratory simulator. In: Brazilian Congress on Biomedical Engineering, pp. 563–571. Springer International Publishing, Cham (2022)

Canbaloğlu, G., Treur, J., Wiewiora, A. (eds.).: Computational Modeling of Multilevel Organisational Learning and Its Control Using Self-modeling Network Models. Springer Nature, Cham (2023)

Cijfers over geboorte | Nederlands Jeugdinstituut (2023). https://www.nji.nl/cijfers/geboorte

Edwards, M., Kotecha, S.: Respiratory distress of the term newborn infant. Paediatr. Respir. Rev. **14**(1), 29–37 (2013). https://doi.org/10.1016/j.prrv.2012.02.002

Hendrikse, S.C.F., Treur.J., Koole. S.L.: Modeling emerging interpersonal synchrony and its related adaptive short-term affiliation and long-term bonding: a second-order multi-adaptive neural agent model. Int. J. Neural Syst. **33**(7), 2350038, 1–41 (2023). https://doi.org/10.1142/S0129065723500387

Kaltsogianni, O., Dassios, T., Greenough, A.: Neonatal respiratory support strategies—short and long-term respiratory outcomes. Front. Pediatrics **11** (2023)

Khan, M., Khurshid, M., Vatsa, M., Singh, R., Duggal, M., Singh, K.: On AI approaches for promoting maternal and neonatal health in low resource settings: a review. Front. Public Health **10** (2022). https://doi.org/10.3389/fpubh.2022.880034

Malak, J.S., Zeraati, H., Nayeri, F., Safdari, R., Shahraki, A.D.: Neonatal Intensive care decision support systems using artificial intelligence techniques: a systematic review. Artif. Intell. Rev. **52**(4), 2685–2704 (2018)

Roehr, C.C., Bohlin, K.: Neonatal resuscitation and respiratory support in prevention of bronchopulmonary dysplasia. Breathe **8**(1), 14–23 (2011)

Sarker, I.H.: AI-Based modeling: techniques, applications and research issues towards automation, intelligent and smart systems. SN Comput. Sci. **3**(2), 158 (2022)

Sebahi, Y., Jabeen, F., Treur, J., Taal, H.R., Roelofsma, P.H.M.P.: Optimizing neonatal respiratory support through network modeling: a new approach to post-birth infant care. In: Cherifi, H., Rocha, L.M., Cherifi, C., Donduran, M. (eds.) Complex Networks and Their Applications XII. COMPLEX NETWORKS 2023. Studies in Computational Intelligence, vol. 1143, pp. 245–257. Springer Nature, Cham (2024)

Squires, C., Uhler, C.: Causal Structure Learning: A Combinatorial Perspective. Foundations of Computational Mathematics, pp. 1–35 (2022). https://doi.org/10.1007/s10208-022-09581-9

Treur, J.: On the dynamics and adaptivity of mental processes: relating adaptive dy-namical systems and self-modeling network models by mathematical analysis. Cogn. Syst. Res. **70**, 93–100 (2021)

Treur, J.: Network-Oriented modeling for Adaptive Networks: Designing Higher-Order Adaptive Biological, Mental and Social Network Models. Springer Nature, Cham (2020a). https://doi.org/10.1007/978-3-030-31445-3

Treur, J.: Modeling multi-order adaptive processes by self-modeling networks (Keynote Speech). In: Antonio J. Tallón-Ballesteros and Chi-Hua Chen (eds.), Proceedings of the 2nd International Conference on Machine Learning and Intelligent Systems, MLIS'20. Frontiers in Artificial Intelligence and Applications, vol. 332, pp. 206–217. IOS Press (2020b)

Treur, J., Van Ments, L. (eds.).: Mental Models and Their Dynamics, Adaptation, and Control: A Self-Modeling Network Modeling Approach. Springer Nature, Cham (2022)

Weigl, L., Jabeen, F., Treur, J., Taal, H. R., Roelofsma, P.: Modelling learning for a better safety culture within an organization using a virtual safety coach: Reducing the risk of postpartum depression via improved communication with parents. Cogn. Syst. Res. **80**, 1–36 (2023). https://doi.org/10.1016/j.cogsys.2023.01.009

Weigl, L., Jabeen, F., Treur, J., Taal, H.R., Roelofsma, P.H.M.P.: Learning for a better safety and security culture within an organization: reducing the risk in communication with AI coaching for security communication through cyberspace. In: Roelofsma, P.H.M.P., Jabeen, F., Taal, H. R., Treur, J. (eds.) Using Shared Mental Models and Organisational Learning to Support Safety and Security Through Cyberspace: A Computational Analysis Approach. Springer Nature, Cham, Switzerland (this volume) (2024)

Supporting Safety and Security for Communication with Patients and Colleagues Through Cyberspace

In the rapidly evolving landscape of healthcare, effective communication remains a cornerstone of patient safety and quality care. As technology advances, cyberspace offers new avenues for enhancing communication, particularly through the use of artificial intelligence (AI) tools. The fourth part of this book delves into the critical role of AI Coaches in supporting safety and security within the neonatal domain, focusing on communication between healthcare professionals and with patients. This part presents innovative approaches that leverage AI to foster a robust safety culture, mitigate risks, and enhance the overall quality of care. Two key studies illustrate the application of AI in promoting better communication practices: one addressing postpartum depression prevention through improved parental communication, and the other optimizing patient safety by encouraging healthcare professionals to speak up about medical errors.

What Will be Addressed in This Part

Enhancing Communication with Patients Through the Use of an AI Coach
The utilization of AI Coaches in healthcare settings aims to address persistent communication challenges that can compromise patient safety. The first study by Weigl et al. in Chapter "Learning for a Better Safety and Security Culture within an Organization: Reducing the Risk in Communication with AI Coaching for Security Communication Through Cyberspace" highlights a virtual safety coach designed to support parents and healthcare practitioners in the neonatal domain. This AI-driven model helps mitigate the risk of postpartum depression by fostering effective communication between parents and healthcare providers. By creating shared mental models, the virtual coach ensures that both individual and collective understandings are aligned, thereby improving support for parents during childbirth. The study demonstrates that timely and supportive communication can have long-lasting positive effects, even beyond the hospital environment.

Supporting Speaking up Behavior Among Colleagues Through the Use of an AI Coach
The second study by Doornkamp et al. in Chapter "Optimizing Speaking Up Behavior for Safety and Security Through Cyberspace: A Computational Network Model of

a Virtual Coach Supporting Speaking Up" explores the dynamics of speaking up among healthcare professionals in the operating theatre. Ineffective communication is a significant contributor to medical errors, and encouraging staff to voice concerns about potential errors is crucial for patient safety. This research introduces a controlled adaptive computational network model featuring a virtual coach that assists healthcare professionals in speaking up. By simulating various scenarios, the model identifies patterns that influence speaking up behavior and examines the role of an observant agent in facilitating communication. The inclusion of an AI Coach not only supports the healthcare professional in real-time but also aids in organizational learning, ultimately enhancing the safety culture within the hospital.

The Potential of AI Coaches Supporting Safety by Improving Communication and Learning

Both studies underscore the transformative potential of AI coaches in improving communication and safety in healthcare settings. The AI Coaches act as mediators and facilitators, ensuring that critical information is communicated effectively and promptly. They also contribute to a culture of continuous learning and improvement, where both individual practitioners and the organization as a whole can adapt and respond to emerging challenges. Through these models, the integration of AI in healthcare communication demonstrates significant promise in reducing risks and enhancing the overall quality of patient care.

Conclusion

The integration of AI Coaches in healthcare communication presents a groundbreaking approach to enhancing safety and security within the neonatal domain and beyond. By addressing key communication challenges between healthcare professionals and with patients, AI tools like virtual safety coaches can significantly mitigate risks and improve patient outcomes. The studies presented in this part illustrate the practical applications and benefits of such technology, from preventing postpartum depression through better parental support to optimizing patient safety by encouraging healthcare professionals to speak up about medical errors. As healthcare continues to evolve, the adoption of AI-driven communication support systems will be crucial in fostering a culture of safety, transparency, and continuous improvement. Through innovative solutions like AI Coaches, the healthcare industry can ensure that the safety and well-being of patients remain at the forefront of its mission.

Learning for a Better Safety and Security Culture Within an Organization: Reducing the Risk in Communication with AI Coaching for Security Communication Through Cyberspace

Linn-Marie Weigl, Fakhra Jabeen, Jan Treur, H. Rob Taal, and Peter H. M. P. Roelofsma

Abstract This chapter describes an extension of a safety culture within hospital organizations providing more transparency and acknowledgement of all actors, and in particular the parents. It contributes a model architecture to support a hospital to develop such an extended safety culture. It is illustrated for prevention of postpartum depression. Postpartum depression is a commonly known consequence of childbirth for both mothers and fathers. In this research, we computationally analyze the risk factors and lack of support received by fathers. Therefore, we use shared mental models to model the effects of poor and additional communication by healthcare practitioners to mitigate the development of postpartum depression in both the mother and the father. Both individual mental models and shared mental models are considered in the design of the computational model. The chapter illustrates the benefits of simple support for communication during childbirth, which has lasting

L.-M. Weigl · J. Treur (✉)
Social AI Group, Vrije Universiteit Amsterdam, Amsterdam, The Netherlands
e-mail: j.treur@vu.nl; j.treur@hhs.nl

L.-M. Weigl
e-mail: linnweigl@gmail.com

F. Jabeen
Safety and Security Science Section, Delft University of Technology, Delft, The Netherlands
e-mail: fakhra.jabeen@ru.nl

J. Treur · P. H. M. P. Roelofsma
Risk Management and Cybersecurity Group, The Hague University of Applied Sciences, Hague, The Netherlands
e-mail: p.h.m.p.roelofsma@hhs.nl

H. Rob Taal · P. H. M. P. Roelofsma
Department of Pediatrics, Division Neonatology, Erasmus University Medical Centre, Erasmus MC, University Medical Center-Sophia Children's Hospital, Rotterdam, The Netherlands
e-mail: h.taal@erasmusmc.nl

P. H. M. P. Roelofsma et al. (eds.), *Using Shared Mental Models and Organisational Learning to Support Safety and Security Through Cyberspace: A Computational Analysis Approach*, Studies in Systems, Decision and Control 570,
https://doi.org/10.1007/978-3-031-72075-8_8

effects, even outside the hospital. For the impact of additional communication, a Virtual Safety Coach is designed that intervenes when necessary to provide support, i.e., when a health care practitioner doesn't. Moreover, organizational learning is also modelled to improve the mental models of both the Safety Coach and the Health Care Practitioner.

Keywords Shared mental models · Virtual AI-Coach in healthcare · Fathers/ psychology · Depressive disorders/complications · Postpartum depression

1 Introduction

Learning is an integral part of organizations. Therefore, organizational learning (Kim 1997), has been coined as a key concept, as it facilitates further development, both within individuals, and as a result, within teams and organizations. This can lead to better governance and an adequate safety culture, also known as 'safety culture' or 'just culture'. The term refers to a culture within an organization that promotes open communication, transparency, and trust amongst the practitioners in any field (Solomon 2016). It is a tool to facilitate organizational learning.

In a healthcare organization, healthcare practitioners from multiple disciplines form healthcare teams, who strive to ensure the safety of patients. These teams coordinate and communicate throughout a certain medical procedure in a very cohesive manner to avoid all the risks that can be associated with healthcare. Moreover, these institutions maintain stability by following the protocols related to their medical procedures, and therefore, often these protocols are evaluated and modified to ensure patient safety (Crossan et al. 1999). For instance, childbirth is a sensitive and a daily occurring natural phenomenon in hospitals, that follows certain protocol or procedure. In a protocol like this, there are steps in place, for before, during and after, that healthcare practitioners must follow to ensure a safe and healthy delivery of the child. The main aim is to keep the mother and the baby safe during the whole process of delivery.

Safety culture encompasses realizing omissions and mistakes and not being scrutinized for the mistake or omission but instead learning from it. Practitioner-patient communication is a valuable step that can result in improved therapeutic outcomes (Hassan 2018). However, most of the time, delivery protocols are only in place for processes and changes or deviations in processes that may occur for the baby or the mother (Taal 2022). They fail to consider the possible adverse effects that may happen to the other-half or the family of a neonate. Bedside manners are within the curriculum of medical studies; however, the importance of adequate patient communication is often underestimated and, often considered to be a less valued step, and sometimes even a forgotten step, within the protocol. A lack of communication with both the mother and partner can result in feelings of isolation. Moreover, the different situational roles of parents and health care practitioners (HCPs) does not allow for open communication. This barrier is a factor that may develop mental health problems in parents, such as postpartum depression (PPD) (Goodman 2004). Much attention is paid to PPD of mothers. However, there is very less attention for paternal postpartum depression (Goodman 2004). The feeling of isolation often contributes to paternal postpartum depression (PPPD), since fathers are not the active actors during the childbirth process, and little attention is awarded to them during the process. PPPD is accompanied by adverse effects affecting both the mother and the child. This chapter discusses methods that could facilitate better support for a father with the aim of decreasing the frequency of fathers experiencing mental health issues (Goodman 2004, 2008).

In this chapter, we use computational models to discuss (a) how a deviation in the childbirth process may lead to poor communication towards the parent(s). Also, we present (b) how a virtual safety coach can facilitate the healthcare practitioners for effective communication with parents in the context of childbirth process. Lastly, we address (c) how organizational learning can play its role in improving the mental models underlying communication. A virtual AI Coach can also help health practitioners to keep the parents intact (i.e., psychologically and socially) during the whole process. The designed models encompass protocols with common knowledge for all actors involved, as well as their own individual knowledge. The research focuses on reducing paternal postpartum depression and illustrates the positive influence this has on maternal postpartum depression. So, we present ways to encourage open and adequate communication both from the father's side and the healthcare practitioner's side. The aim is to expand the definition of safety culture, to involve not only the medical practitioners but also the patients and their families by improved communication. In return, this implementation depicts the benefits of open communication and its subsequent effect of decreasing the risk of both the mother and father developing PPD (Goodman 2004). For more detailed references see Sect. 2.

2 Background Literature

In this section, we present the background literature of our work. We will discuss the risk factors for maternal postpartum depression (Mayers et al. 2020), together with their consequences (Rao et al. 2020). Then we explain risk factors for paternal postpartum depression along with possible outcomes (Goodman 2008). We will also discuss how postpartum depression can be prevented (Dennis 2004; Smythe et al. 2022). Specifically, how paternal postpartum depression can be prevented, which in return has significant influence on reducing maternal postpartum depression. Moreover, we will discuss how computational causal modelling can play its role preventing the related risks.

2.1 Maternal Postpartum Depression and the Related Risk Factors

The ICD-10 states that "in typical mild, moderate, or severe depressive episodes, the patient suffers from lowering of mood, reduction of energy, and decrease in activity. Capacity for enjoyment, interest, and concentration is reduced, and marked tiredness after even minimum effort is common" (World Health Organization 1993). Postpartum depression refers to "mild mental and behavioural disorders associated with the puerperium" (World Health Organization 1993). In industrialized countries 10–15% of women experience major depressive episodes after giving birth (Maternal mental health and child health and development in resource-constrained settings, 2009, January 1). This statistic encompasses women with prior and without prior mental health issues. The phenomena of developing depression after childbirth, known as maternal postpartum depression (MPPD), is among the most common mental health consequences. Determining causes, possible interventions and creating support systems for new mothers suffering from PPD is at the forefront of postnatal care research (Garthus-Niegel et al. 2022; Mayers et al. 2020; Rao et al. 2020), as it may prevail for extended periods (Rao et al. 2020; Smythe et al. 2022).

Reasons for postpartum depression include obstetric factors, psychological factors, biological factors, social factors and lifestyle (Abenova et al. 2022; Javaid et al. 2022). Women who are nulliparous generally benefit from technically oriented support such as help with breastfeeding. On the other hand, multiparous women require more mental support, as they are at a higher risk of developing PPD, due to the additional stress of having another child at home (Dunkel Schetter et al. 2016). Another factor is the risk of the birth, especially when doctors must deviate from the original birth plan. A deviation from normal protocol results in a higher risk of the mother developing PPD. Similarly, deviating (having a caesarean section) from the mothers' desires (having a natural birth), has the same effect (Ghaedrahmati et al. 2017). The mode of delivery also influences the risk of developing PPD, with a natural birth being the safest option (Silverman et al. 2017). In addition, although it is a highly debated topic, the use of epidurals has been identified to decrease the chances of mothers developing PPD (Ghaedrahmati et al. 2017). Amongst all other factors, age also plays a role. Women who are 35 years or older are at a higher risk of developing PPD. The lowest risk is for women between the ages of 25 and 29 (Silverman et al. 2017). The age factor pertains to the fact, that the riskiness of the birth is higher for the prior (Cavazos-Rehg et al. 2015). Lastly, a mother-to-be is 4–18 times more likely to develop PPD if their child is born underweight (<1500 g) (Ghaedrahmati et al. 2017).

In addition to the factors mentioned above, social factors must be acknowledged to decrease the risk of a mother developing PPD. The social network around the woman is the most important aspect regarding support. "Newly expecting mothers perceive their partner as their main support system" (Mayers et al. 2020; Smythe et al. 2022). Their partners often see signs of change and act upon them, to ensure the woman receives the necessary mental health support.

2.2 Consequences Beyond the Mother of Postpartum Depression

Mothers suffering from postpartum depression receive a great amount of attention, as the negative correlation between a mother's mental health and a child's development is widely known (Cerezo et al. 2008; Goodman 2008). However, the spillover effect onto their partners is often not addressed. At the same time, father's support has been proven to benefit a mother's mental health (Goodman 2004). "Maternal depression was identified as the strongest predictor of paternal depression during the postpartum period" (Goodman 2004). 25% of women who experience PPD have a partner who is also experiencing PPD (Goodman 2008). Paternal postpartum depression is not a commonly known consequences of childbirth, as the attention, support and communication are directed towards the mother. This inequality in mental health services provided to mothers and fathers, results in fathers reporting feelings of isolation following childbirth. Most notably, this is prevalent following a traumatic birth, or simply a deviation from the original birth plan. This isolation creates insecurity and confusion for the fathers regarding their role (Mayers et al. 2020).

2.2.1 Paternal Postpartum Depression

The responsibility of a child can be overwhelming and intimidating. Moreover, childbirth is an intense experience for both mother and father. However, while mothers are fully involved and sufficiently supported, fathers often don't get the same treatment (Goodman 2004). The lack of paternal support received, increases the difficulty, both technically and mentally, of having a child. This results in a high chance of fathers developing postpartum depression (Goodman 2004, 2008).

While most partners feel as though the mother should be receiving most of the attention, the extreme imbalance between the information available to them and information available to mothers is not justified (Mayers et al. 2020). Fathers feel as though they should be given or offered more support and information than what is currently the norm. Additionally, the quality of mental health support has been criticized. "Fathers require better recognition of their mental health needs from healthcare professionals" (Mayers et al. 2020).

Societal stigma often results in the fathers not seeking for help, as they are meant to be the providers of the family and don't allow themselves to suffer mentally. Moreover, as fathers are not briefed on PPPD, they often don't see the signs and only accept the truth when it's too late for early intervention. Similarly, women may see a change in their partner. However, they do not have the priming of it potentially being PPD and therefore ignore the signs (Abenova et al. 2022).

The chances of paternal postpartum depression increase with unexpected childbirth, as a change of procedure while in the delivery room can be traumatizing for the father. The fear that develops in their minds, when there is no proper communication between healthcare personnel and the father in the delivery room causes the fathers to think of the worst possible outcomes (Pedersen et al. 2021). A lack of support and communication from healthcare practitioners directly to the partners, results in a sensation of helplessness, isolation, and confusion regarding the partners mental health issues (Melrose 2010). Similarly to maternal PPD, paternal postpartum depression not only affects the father but also the mother and the child's development (Goodman 2004, 2008; Rohde et al. 2005).

2.2.2 Consequences of Paternal Postpartum Depression

Positive parent-infant interaction is the key to ensuring children's good psychological, cognitive and language development. Partners of women who suffer from PPD experience higher parenting stress, which results in less optimal father-infant interaction (Goodman 2008). In other words, parent-infant interaction is highly influenced by the mental well-being of the parents. Similarly, another study indicated that PPPD might have important implications on the family's mental health and wellbeing. A high correlation was found between the depression of one partner with another, producing significantly adverse effects on the other partner and child (Goodman 2004). Moreover, a study indicated that children of fathers with PPPD reported increased emotional and psychological issues (Rohde et al. 2005).

Depression is also sometimes termed as 'a preventable killer', because inadequate handling such psychiatric disorders may lead to the death of a partner, making the situation more complex (Biebel and Alikhan 2018; Kenyon 2015; Mayers et al. 2020). To ensure prosperity for a child's future life, hospitals should aim to identify signs of possible development of PPD in mothers and fathers. Moreover, they try to restore mental health as soon as possible (Melrose 2010). Tackling perinatal mental health support can reap long-term benefits for both the parents and the child (Mayers et al. 2020).

2.3 What Can Be Done More to Prevent Postpartum Depression

A study by Mayers et al. (2020) quoted fathers on their experiences and desires after childbirth.

> I read the information my wife was given.
>
> It didn't cover anything about the father and I felt lost.
>
> Leaflets on what to look out for, as you can't always remember what you are told in the immediate aftermath.
>
> Some Mental Health support, as well as social worker support and referral to a therapist.

> [I] had not given birth so had no cause for sympathy. A leaflet for my wife and a page for the fathers to read which wasn't enough (Mayers et al. 2020).

Partners of new mothers have expressed the need to a better understanding on how they can support the new mothers, as well as themselves. Specifically, clear communication and information regarding treatment and medication, to help the women, as well as to ease their own concerns. Currently, there is not enough support or information provided, and in the cases where support is given, the quality on average, is very low (Mayers et al. 2020). Due to the social stigma of a man's role in society, as well as the pressure of having to be a protector and a provider, fathers have expressed reservations about joining groups for support (Darwin et al. 2017). Darwin et al. concluded that fathers are reluctant and unable to seek help, as they "question the legitimacy of their experiences" and believe that the woman should have priority, especially when they perceive the facilities to help with mental health are under-resourced (Darwin et al. 2017). A further issue raised, was that verbally communicated information was often forgotten, especially when this information was relayed during a stressful time (Mayers et al. 2020).

Fathers have expressed their desire to receive information regarding postnatal mental illness (Pedersen et al. 2021), having someone to talk to, and direct healthcare service support that is specifically targeted towards them. The provided support currently does not seem enough, specifically in deviations that may happen during the process (Taal 2022), which in return diminishes the ability to support themselves and the mother properly. A simple solution given by a father could be "any offer of help and support" (Mayers et al. 2020). The opportunity to have someone to talk to was regarded as being very valuable, as well as simply being acknowledged. A referral to a therapist or social worker was suggested (Mayers et al. 2020).

Moreover, a standard agreed-upon method of receiving information for fathers would be in the form of readily available leaflets or various other types of written materials, as they would be easily accessible at any point in time. Emphasis was assigned to the helpfulness of having written material depicting other men's experience. This made them "realize you're not in the boat by yourself" (Darwin et al. 2017). The areas that are generally lacking within the support services are informational support and aftercare. "Early interventions to help fathers cope with stress may be needed to reduce the risk of future deterioration of emotional wellbeing" (Mayers et al. 2020; Pedersen et al. 2021).

3 Conceptual Analysis of Underlying Processes and Means to Model Them

This section presents the preliminary analysis of the underlying process that will be considered for modelling the communication and learning behavior for the safety coach, health practitioner and parents (addressed in Sect. 4). So, firstly (in Sect. 3.1), we explain the safety culture in the context of a parent seeking help and support during the procedure. Secondly (in Sect. 3.2), we explain how organizational learning can play a role in encouraging communication during the process of childbirth. Lastly, we explain the role of the mental models in Sect. 3.3, along with the cognitive architecture used to model them and the learning of them.

3.1 Safety Culture

A good safety culture facilitates learning. Safety culture encompasses realizing omissions and mistakes, realizing that one may not fully be at fault, admitting to the errors and not being scrutinized for the error, but instead learning from it. Applying a safety culture creates an open environment for further development. It decreases the frequency of errors since individuals are willing to admit to them, instead of trying to cover up which may worsen the situation (Darwin et al. 2017). Facilities, which would benefit greatly from integrating a just safety culture, are hospitals. "A blame-free, nonpunitive culture encourages clinicians to report errors and truly learn from their mistakes. It also supports organizations in efforts to better understand their errors and make improvements" (Beyea 2004). The basis of this is founded on open communication with the goal of moving away from extreme punishment or a blameless culture and instead moving towards a just culture (Boysen 2013). The scope of this chapter doesn't refer to just safety culture in the classical sense, which encompasses a medical professional speaking up to avoid immediate medical mistakes. Still, it addresses just safety culture by the father's willingness to speak up and ask for support. The awareness that all actors inside the delivery room, including medical professionals and parents, are complex beings with desires and that they possess an understanding of what is needed to ensure overall success and health, is essential (Smythe et al. 2022). In addition, the understanding and acknowledgement that the actions currently taken may not be sufficient in some cases holds similar importance. Although the action of asking for support communication itself is not an active preventative measure, the resulting support serves as a tool to avoid possible long-term mental-health consequences (Kenyon 2015; Mayers et al. 2020). This perspective aims to bridge the gap between mental and physical health and emotional health, as they are not only independent of each other, but also dependent on each other (Pinto-Foltz and Logsdon 2008). By expanding the view of the involved actors, it allows for a more well-rounded approach with various points of view and an ever-increasing just safety culture (Darwin et al. 2017; Hassan 2018; Ratnapalan and Uleryk 2014).

This approach to safety culture has the ability to extend the range of the benefits from short-term to long-term. Communication is not strictly a medical step; if a mistake occurs, it expresses itself in direct consequences. However, it should be considered a valuable medical step, as it entails a simple skill that all individuals know, and can prevent consequences that aren't immediate, but can get visible in the long term (Hassan 2018). The goal is to encourage the implementation of a well-rounded treatment where the effects don't stop when one leaves the hospital.

3.2 Organizational Learning

Crossan et al. define organizational learning as:

> a dynamic process. Not only does learning occur over time and across levels, but it also creates a tension between assimilating new learning (feed forward) and exploiting or using what has already been learned (feedback). Through feed-forward processes, new ideas and actions flow from the individual to the group to the organization levels. At the same time, what has already been learned feeds back from the organization to group and individual levels, affecting how people act and think (Crossan et al. 1999).
>
> Feed forward relates to exploration. It is the transference of learning from individuals and groups through to the learning that becomes embedded-or institutionalized-in the form of systems, structures, strategies, and procedures (Hedberg 1981; Shrivastava 1983). Feedback relates to exploitation and to the way in which institutionalized learning affects individuals and groups (Crossan et al. 1999).

The four I's of organizational learning are intuiting, interpreting, integrating, and institutionalizing. The first two relate to an individual, the second and third to a team or group and the third and last to an organization (Crossan et al. 1999). These overlaps are links between the levels. For organization learning to occur, the understanding that an "organization operates in an open system, rather than having a solely internal focus" (Crossan et al. 1999) is crucial. In other words, there is a transfer of knowledge between individuals, groups and organizations (Kim 1997). According to Peter Senge's theory successful learning has the capacity to change and manage change, where an organization can adopt system thinking, shared mental models and shared vision can help learn in teams (Ratnapalan and Uleryk 2014).

3.3 Mental Models

Mental models are at the base of an individual's actions. Kim describes them as follows:

> Mental models represent a person's view of the world, including explicit and implicit understandings. Mental models provide the context in which to view and interpret new material, and they determine how stored information is relevant to a given situation (Kim 1997).

Interesting to note is the similarity between the two I's for individuals mentioned earlier and Kim's definition of mental models. The definition encompasses both 'intuiting' ('a person's view of the world, including explicit and implicit understandings') and 'interpreting' ('interpret new material'). Mental models are useful for individuals to create internal simulations based on certain circumstances. Moreover, new mental models or additions or revisions to mental models can be learned. This is encompassed by Piaget, who assigns importance to the marriage of accommodation (adapting ones individual internal concepts based on experiences) and assimilation (integrating ones experiences into ones individual internal concepts) as the keys to learning (Kim 1997; Piaget 1970).

Shared mental models are various mental models that ideally align themselves in a manner that they become shared knowledge. Therefore, their representation may not only rely on differential knowledge of a member, but they may also represent an overlap or convergence among their mental models in relation to a task or procedure. As shared mental models are related to a mutual goal, it relies mainly on understanding requirements and steps taken during a procedure. They can be equated to the third I, 'integrating'. Integrating refers to a shared understanding and mutual adjustment, creating interactive systems (Crossan et al. 1999). To facilitate integration, interpretation must be done by the means of communication (Bouma et al. 2022).

3.4 Cognitive Architecture Used for Shared Mental Models

Network-Oriented Modelling is a technique that can be used to model interconnected and interactive causal processes that are temporal by nature (Treur 2016). Given the direct relation between dynamics and causal relations, as highlighted in Treur (2016), this classical approach is extended to contain the notion of dynamics in a network structure. These notions of causality and dynamics have been incorporated and are part of a more refined structure and semantics of the considered networks. More specifically, the nodes in a network are interpreted here as states (or state variables) that vary over time, and the connections are interpreted as causal relations that define how each state can affect other states over time. To acknowledge this perspective of dynamics and causality on networks, this type of network has been called a temporal-causal network (Treur 2016).

Temporal-causal networks are the foundation of dynamic Network-Oriented Modelling. While the popular approach of causal modelling comes with limitations, the addition of dynamic features provides a more detailed and truthful representation of the real-world (Treur 2016). This dynamic perspective is based on a continuous time dimension, represented by real numbers. The temporal dimension enables modelling by cyclic causal networks as well, and also timing of causal effects can be modelled in detail. Due to this, causal reasoning and simulation is possible for networks that inherently contain cycles, such as networks modeling mental or brain states, or networks describing social interaction (Treur 2016).

The research conducted for this chapter is performed addressing mental processes using internal mental models, which are modeled in a network-oriented manner (see Treur 2016, 2020). It encompasses a social structure, depicting how people interact amongst themselves, as well as network structures that depict an individual's mental processes, in this case also including internal mental models, (Treur and Van Ments 2022), see also Van Ments et al. (2025), this volume. These models can also be used to reflect adaptive organizational behavior (Canbaloğlu et al. 2022), see also Canbaloğlu et al. (2022), this volume. In this technique, a network can be represented by a labelled graph containing states, also known as nodes, where each state X is connected to another state Y and where X has a causal impact on Y, at a certain time, with some strength and speed. A, a temporal-causal network is characterized by Treur (2016):

Connectivity characteristics:
Connections from a state X to a state Y and their weights $\boldsymbol{\omega}_{X,Y}$.

Aggregation characteristics:
For any state Y, some combination function $\mathbf{c}_Y(\ldots)$ defines the aggregation that is applied to the single causal impacts $\boldsymbol{\omega}_{X,Y}X(t)$ on Y from its incoming connections from states X.

Timing characteristics:
Each state Y has a speed factor $\boldsymbol{\eta}_Y$ defining how fast it changes for a given causal impact.

The related difference Eq. (1) incorporates these characteristics in a standard numerical format (Treur 2016), i.e.:

$$Y(t + \Delta t) = Y(t) + \boldsymbol{\eta}_Y[\mathbf{c}_Y(\boldsymbol{\omega}_{X_1,Y}X_1(t), \ldots, \boldsymbol{\omega}_{X_k,Y}X_k(t)) - Y(t)]\Delta t \quad (1)$$

Numerous combination functions are available to address the issue of aggregating multiple impacts (Treur 2016). Those which were used for this research are shown in Table 1.

The modelling of adaptive networks is based on network reification (or self-modeling network), which entails extending the base model by reification states, also referred to as self-model states. Reification is defined as "representing something abstract as a material or concrete thing, or making something abstract more concrete or real" (Treur 2020). The self-model states are part of a higher order adaptation and are visually placed on a higher plane than the base level. Similar to how the base level can be adaptive, the first reification level, also known as first-order self-model, can also be adaptive by nature. Therefore, the higher-order adaptive levels are not limited to one, as the construction can be iterated indefinitely. Network reification for a temporal-causal network means that for the adaptive network structure characteristics $\boldsymbol{\omega}_{X,Y}$, $\mathbf{c}_Y(\ldots)$, $\boldsymbol{\eta}_Y$ for each state Y of the base network, additional network states $\mathbf{W}_{X,Y}$, $\mathbf{C}_Y$, $\mathbf{H}_Y$ (called reification states) are introduced respectively (Treur 2020). Different simulating environments are used to simulate the (shared) mental models

Table 1 Combination functions used in the self-modelling network model

	Notation	Formula	Parameters
Advanced logistic sum	$\mathbf{alogistic}_{\sigma,\tau}(V_1, \ldots, V_k)$	$\left[\frac{1}{1+e^{-\sigma(V_1+\cdots+V_k-\tau)}} - \frac{1}{1+e^{\sigma\tau}}\right](1+e^{-\sigma\tau})$	Steepness σ Excitability threshold τ
Step once	$\mathbf{steponce}_{\alpha,\beta}(V)$	1 if $\alpha \leq t \leq \beta$ for time t, else 0	Start time α End time β
Monitor	$\mathbf{monitor}_{\tau}(V_1, V_2)$	1 if $V_1 - V_2 \geq \tau$, else 0	Activation threshold τ

in Matlab and Python. In Sect. 6, we explain how these characteristics can be used to simulate a model in Python.

In Treur (2021), Hendrikse et al. (2023), it is shown that any smooth dynamical system has a canonical representation as a temporal-causal network and any smooth (multi-order) adaptive dynamical system has a canonical representation as a (multi-order) self-modeling temporal-causal network. Therefore, compared to adaptive dynamical systems in general, the network-oriented modeling approach used here does not introduce any fundamental limitations concerning what it can model. This has also been confirmed by applications to case studies in practice, e.g., for mental models in Treur and Van Ments (2022) and for organizational learning in Canbaloğlu et al. (2025).

4 The Designed Network Model

The model discussed in this chapter focuses on childbirth at a hospital, where the father is constantly present in the delivery room, even in circumstances where a deviation from a standard protocol may happen (Taal 2022). It mainly encompasses the communication process starting shortly before childbirth and ending once the parents leave the hospital. The majority of the states in the model reflect communication actions, specifically geared towards the father. The communication process may (or may not) take place during and after childbirth, even in circumstances where a deviation can occur.

The research described in this chapter extends to two levels of adaptation (two self-modelling levels) on top of the base level. Therefore, in Sect. 4.1, we present the base model, which shows the model. In Sect. 4.2, we add the adaption level through self-modelling network and the role of a healthcare practitioner and a virtual safety coach (AIC). Lastly, in Sect. 4.3, we present how organization learning can influence the learning of the safety coach. Please note that the model presented in the following sections are mainly related to the communication between the healthcare practitioner (HCP) and the father of a neonate during the process of childbirth.

4.1 Base Level of the Adaptive Network Model

The base level, depicted in Fig. 1 (yellow parallelogram), contains the world states and the mental model of the healthcare practitioner (HCP). The world states in the base level of the model can be one of five types: process states (yellow), context states (orange), emotion states (red), paternal states (light orange), or communication action states (dark yellow).

The process states define both the baby being delivered and then subsequently, for example hours or days, after birth, as well as the risk of developing maternal postpartum depression, the risk of developing paternal postpartum depression and

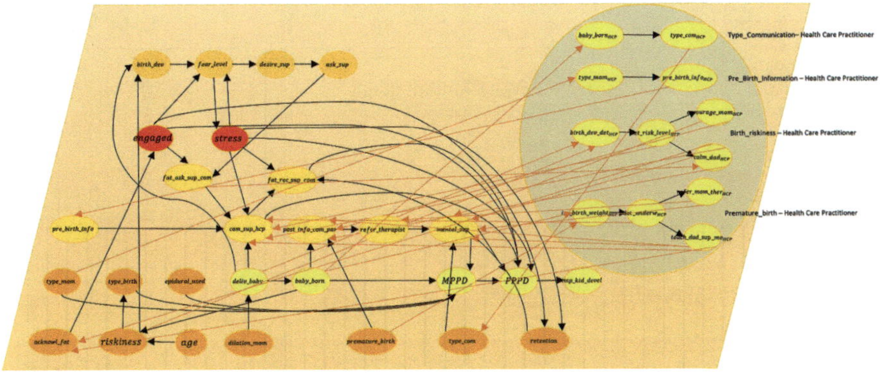

Fig. 1 Connectivity of the base model of the childbirth process

the risk of impending child development. The context states influence the process states, the emotion states and in some circumstances other context states. The emotion states represent the emotions of the actors that occur during childbirth which influence frequent communication. The paternal states indicate the process through which the father may go before reaching out for help. Lastly, the central part, the communication action states, depict the interaction between HCP and parents, specifically the father, throughout the childbirth process. The states were created based on the findings explained in Sect. 2 of this chapter. The connectivity of the base model can be seen in Fig. 1, while the detailed description of the states of the base level is enlisted in Table 2.

The mental model (light grey oval) of the HCP in the base level of the model shows which internal simulations a HCP can have during the childbirth process. There are four separate process models within the mental model, each with end states (bold), that connect to one of the communication actions and/or context states in the world states.

1. The baby is born, which affects the **type of communication** used
2. The type of mother, whether she is nulliparous or multiparous, determines whether any **special information has to be communicated before childbirth**. In the interest of space, type of mother is shortened to type of mom in the figures and tables.
3. The HCP determines that there is a deviation in the childbirth process. They determine the risk level associated with this deviation and, in return, **encourage the mother** while **calming the father**.
4. When the baby is born, especially if born prematurely, the HCP assumes the possibility of irregular birth weight of the child. Once the child is classified as being underweight, they **refer the mother to a therapist** and **teach the father how to support the mother**.

Table 2 Base level states (including HCP Mental Model)

State Number	State Name	Explanation	Level
X_1	type_mom	Pre-Birth: Determining whether the mother is nulliparous (first child) or multiparous (already has 1+ child)	Base Level
X_2	artificiality_birth	Pre-Birth: Determining the naturality of the birth: natural birth or caesarean section	
X_3	epidural_used	Pre-Birth: Determining if an epidural will be administered	
X_4	deliv_baby	Intra-Birth: Delivery of the baby; the woman is in labour	
X_5	baby_born	Post-Birth: The baby is born	
X_6	MPPD	Post-Birth: The risk of maternal postpartum depression	
X_7	PPPD	Post-Birth: The risk of paternal postpartum depression	
X_8	imp_kid_devel	Post-Birth: The risk of impeding child flourishing/development	
X_9	pre_birth_info	Pre-Birth: Information given to parents	
X_{10}	com_sup_hcp	Intra-Brith: Communication from HCP to parents	
X_{11}	fat_ask_sup_com	Intra-Birth: Father asks HCP for support communication	
X_{12}	fat_rec_sup_com	Intra-Birth: Father receives support communication from HCP	
X_{13}	post_info_com_par	Post-Birth: Information given to parents	
X_{14}	refer_therapist	Post-Birth: Referall by HCP of parent(s) to a therapist	
X_{15}	mental_sup	Post-Birth: Recieve additional mental support	
X_{16}	acknowl_fat	Acknowledgement of the father as an actor in childbirth by HCP	
X_{17}	age	Age of mother: whether she is between the ages of 25-29 or not	
X_{18}	riskiness	Riskiness of giving birth for the mother	
X_{19}	dilation_mom	Dilation of the mother	
X_{20}	premature_birth	Premature birth	
X_{21}	type_com	Type of commuication received/used: verbal or written	
X_{22}	retention	Ability to retain information received	
X_{23}	engaged	Level of active engagement	
X_{24}	stress	Level of stress	
X_{25}	birth_dev	Deviation from original birth plan	
X_{26}	fear_level	Level of fear	
X_{27}	desire_sup	Desire for support	
X_{28}	ask_sup	Willingness to ask for support	
X_{29}	birth_dev_det$_{HCP}$	Intra-Birth: Detecting a birth deviation	Base Level - HCP Mental Model
X_{30}	det_risk_level$_{HCP}$	Intra-Birth: Determining the risk of a given birth deviation	
X_{31}	encourage_mom$_{HCP}$	Intra- & Post-Birth: Encouraging the mother positively	
X_{32}	calm_dad$_{HCP}$	Intra- & Post-Birth: Calming the father	
X_{33}	irr_birth_weight$_{HCP}$	Post-Birth: Detecting irregular birth weight of the child	
X_{34}	det_underw$_{HCP}$	Post-Birth: Determining the child as underweight	
X_{35}	refer_mom_ther$_{HCP}$	Post-Birth: Referring the mother to a therapist	
X_{36}	teach_dad_sup_mom$_{HCP}$	Post-Birth: Teaching the father how to support the mother (techincally and mentally)	
X_{37}	baby_born$_{HCP}$	Post-Birth: The baby is born	
X_{38}	type_com$_{HCP}$	Type of commuication used: verbal or written	
X_{39}	type_mom$_{HCP}$	Pre-Birth: Determining whether the mother is nulliparous (first child) or multiparous (already has 1+ child)	
X_{40}	pre_birth_info$_{HCP}$	Pre-Birth: Giving information to parents	

4.2 First-Order Self-Model Level of the Adaptive Network Model

In this section, we present the first-order adaptive network level for the base model designed in Sect. 4.1. First-order adaptive networks model plasticity; changes in brain structure for learning through the 'Hebbian Learning' principle (Hebb 1949). First, we will discuss how HCP learns to communicate to the father by adding additional **W**-states on the first reification level, which determines the weight of the connection on the base level. Second, we present how a virtual safety coach (AIC) can facilitate communication using the shared mental models. Therefore, it has both **W**- and **H**-states on the first reification level. These states determine the weight of connections, and the speed factor of the connections on the base level, respectively. Third, we present how a virtual safety coach can monitor the communication during the procedure by adding both **W**-states and monitor states on the first reification level.

4.2.1 Healthcare Practitioner Learns to Communicate

Additionally, to the base level described above (see Sect. 4.1), Fig. 2, has a middle level (red parallelogram), to represent the first level or the self-model of the base model using reification level states. This level contains eight **W**-states (red), which represent the weight of the connections within the HCP mental model on the base level. The detailed description of these states can be found in Table 3.

4.2.2 The Role of the Virtual Safety Coach

This chapter aims to model and determine the usability and effectiveness in having a virtual AI Coach (AIC) present. The main goal of the coach is to reduce the risk of maternal and paternal postpartum depression. This coach is aware of deviations in the process of the communication actions, and may intervene to provide support to the health practitioner(s) to correct them. Thus the related actions can be in terms of altering the HCP of a forgotten action or communicating with the father itself (upon approval by an HCP). Thus, the Virtual AI Coach (AIC) will not only help the healthcare practitioner to monitor the process for communication, but also will facilitate them for communication, and in return avoiding adverse possible effects like PPPD and MPPD. Therefore, here, we present an extension to contain the Virtual AI Coach (Fig. 3) that intervenes when necessary in terms of communication with the father of the child. This was done by duplicating the mental model of the HCP for the Coach. The states, as well as all connections within the mental model and to the world states are identical.

The middle level was also duplicated, meaning eight **W**-states (green) were created to represent the weight of the connections within the Coach mental model. In addition,

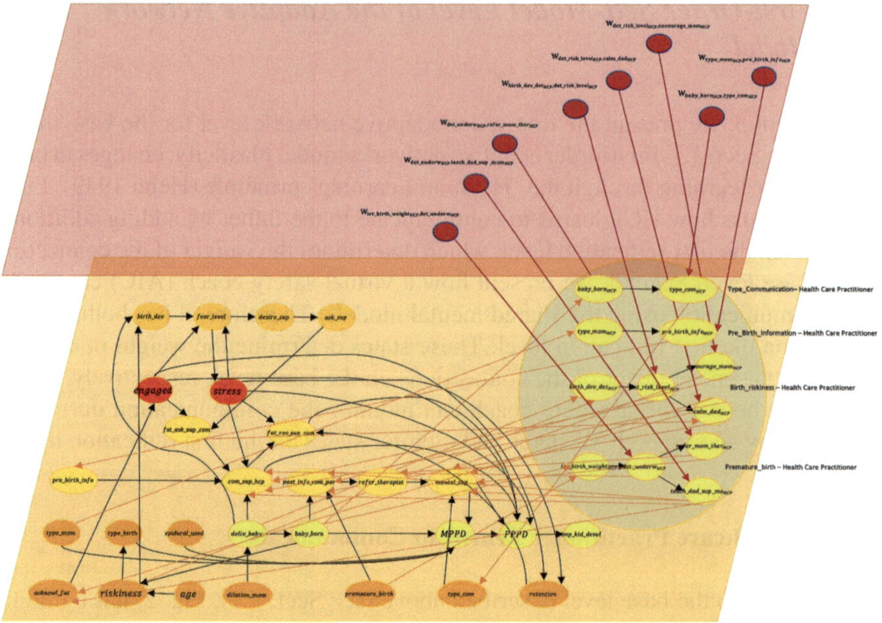

Fig. 2 Connectivity of first-order adaptive network model Case (a) first-order self-model states (or First Reification Level States)

seven new states were added to this level. Six of these additional states are **H**-states (green), which have connections to one of each of the end states within the Coach mental model. The **H**-states represent time activation of the end states and are controlled by the seventh state (blue), a context state that activates the Coach at a particular time. A detailed explanation of the additional states is given in Table 4.

4.2.3 Dynamic Monitoring by the Virtual Coach

To create a more dynamic Virtual Coach, a further model (Fig. 4) was created. The goal of this Coach is not to be activated by time; instead, we designed a 'monitor' function (see Table 1). This function is responsible for monitoring the communication and context states affected by the end states of the mental models. Therefore, if insufficient activation of those states is detected (through a parameter called threshold—τ) by the Coach, then the Coach activates the corresponding part of its mental model, which results in the Coach performing the communication action instead of the HCP.

This model of monitoring role of the virtual AI Coach differs in the number of nodes in the middle level. The 16 **W**-states representing the weight of the connections for both mental models in the base level persist. However, the seven additional states mentioned in Sect. 4.2.2 were removed. Instead, 14 new **W**-states (green) that represent the weight of connections were added. These states represent the

Table 3 Additional base level and first-order self-model states (or First Reification Level States)

State Number	State Name	Explanation	Level
X_{41}	birth_dev_det$_{AIC}$	Intra-Birth: Detecting a birth deviation	Base Level - AIC Mental Model
X_{42}	det_risk_level$_{AIC}$	Intra-Birth: Determining the risk of a given birth deviation	
X_{43}	encourage_mom$_{AIC}$	Intra- & Post-Birth: Encouraging the mother positively	
X_{44}	calm_dad$_{AIC}$	Intra- & Post-Birth: Calming the father	
X_{45}	irr_birth_weight$_{AIC}$	Post-Birth: Detecting irregular birth weight of the child	
X_{46}	det_underw$_{AIC}$	Post-Birth: Determining the child as underweight	
X_{47}	refer_mom_ther$_{AIC}$	Post-Birth: Referring the mother to a therapist	
X_{48}	teach_dad_sup_mom$_{AIC}$	Post-Birth: Teaching the father how to support the mother (techincally and mentally)	
X_{49}	baby_born$_{AIC}$	Post-Birth: The baby is born	
X_{50}	type_com$_{AIC}$	Type of commuication used: verbal or written	
X_{51}	type_mom$_{AIC}$	Pre-Birth: Determining whether the mother is nulliparous (first child) or multiparous (already has 1+ child)	
X_{52}	pre_birth_info$_{AIC}$	Pre-Birth: Giving information to parents	
X_{61}	$\mathbf{W}_{\text{birth\_dev\_det\_AIC, det\_risk\_level\_AIC}}$	Reified representation state for connection weight $\boldsymbol{\omega}_{\text{birth\_dev\_det\_AIC, det\_risk\_level\_AIC}}$	First reification level
X_{62}	$\mathbf{W}_{\text{det\_risk\_level\_AIC, encourage\_mom\_AIC}}$	Reified representation state for connection weight $\boldsymbol{\omega}_{\text{det\_risk\_level\_AIC, encourage\_mom\_AIC}}$	
X_{63}	$\mathbf{W}_{\text{det\_risk\_level\_AIC, calm\_dad\_AIC}}$	Reified representation state for connection weight $\boldsymbol{\omega}_{\text{det\_risk\_level\_AIC, calm\_dad\_AIC}}$	
X_{64}	$\mathbf{W}_{\text{irr\_birth\_weight\_AIC, det\_underw\_AIC}}$	Reified representation state for connection weight $\boldsymbol{\omega}_{\text{irr\_birth\_weight\_AIC, det\_underw\_AIC}}$	
X_{65}	$\mathbf{W}_{\text{det\_underw\_AIC, refer\_mom\_ther\_AIC}}$	Reified representation state for connection weight $\boldsymbol{\omega}_{\text{det\_underw\_AIC, refer\_mom\_ther\_AIC}}$	
X_{66}	$\mathbf{W}_{\text{det\_underw\_AIC, teach\_dad\_sup\_mom\_AIC}}$	Reified representation state for connection weight $\boldsymbol{\omega}_{\text{det\_underw\_AIC, teach\_dad\_sup\_mom\_AIC}}$	
X_{67}	$\mathbf{W}_{\text{baby\_born\_AIC, type\_com\_AIC}}$	Reified representation state for connection weight $\boldsymbol{\omega}_{\text{baby\_born\_AIC, type\_com\_AIC}}$	
X_{68}	$\mathbf{W}_{\text{type\_mom\_AIC, pre\_birth\_info\_AIC}}$	Reified representation state for connection weight $\boldsymbol{\omega}_{\text{type\_mom\_AIC, pre\_birth\_info\_AIC}}$	
X_{69}	$\mathbf{H}_{\text{encourage\_mom\_AIC}}$	Reified representation state for speed factor $\eta_{\text{encourage\_mom\_AIC}}$	
X_{70}	$\mathbf{H}_{\text{calm\_dad\_AIC}}$	Reified representation state for speed factor $\eta_{\text{calm\_dad\_AIC}}$	
X_{71}	$\mathbf{H}_{\text{refer\_mom\_ther\_AIC}}$	Reified representation state for speed factor $\eta_{\text{refer\_mom\_ther\_AIC}}$	
X_{72}	$\mathbf{H}_{\text{teach\_dad\_sup\_mom\_AIC}}$	Reified representation state for speed factor $\eta_{\text{teach\_dad\_sup\_mom\_AIC}}$	
X_{73}	$\mathbf{H}_{\text{type\_com\_AIC}}$	Reified representation state for speed factor $\eta_{\text{type\_com\_AIC}}$	
X_{74}	$\mathbf{H}_{\text{pre\_birth\_info\_AIC}}$	Reified representation state for speed factor $\eta_{\text{pre\_birth\_info\_AIC}}$	
X_{75}	con_act_AIC	Context state for determining activation of the AIC	

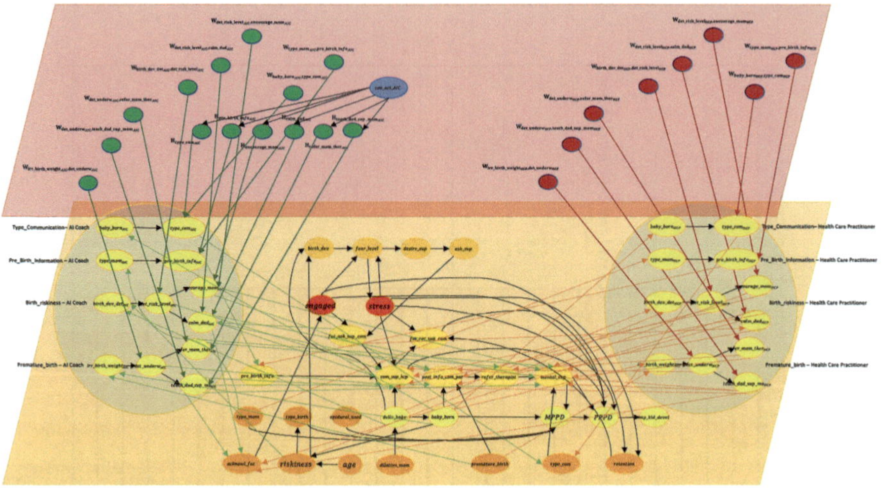

Fig. 3 Connectivity of first-order adaptive network model Case (b)

weight of connections between the end states of the mental model of the Virtual AI Coach and the world states affected by them. Moreover, 14 monitor states (blue), each with an outgoing connection to one of the new states mentioned earlier, were included that receive input from the base level and determine whether to activate their corresponding states based on the monitor combination function described in Table 1. A more in-depth explanation of the states in the middle level can be seen in Table 5.

4.3 Second-Order Self-Model Level—Organizational Learning (Feed Forward and Feedback)

The Second-Order Adaption Level adds another abstraction level of learning to the states modelled on first-order network using meta-plasticity principle. At this level, context states were added along with $\mathbf{W}_\mathbf{W}$-states on the second reification level. It is to be noted that eight more **W** states are introduced on the first-order level in relation to organizational learning. This final addition to the model was done by incorporating an expert (Fig. 5) of the organization.

The goal with adding an expert (E) was to show the possibility of the Virtual AI Coach learning from a practitioner with a complete mental model. Moreover, the Coach, having acquired this knowledge, can then pass it on to the HCP. In this case the Coach's mental model is interpreted as a shared mental model. This creates a more adaptive Virtual Coach, as it allows for the constant addition of information from an expert (as feed forward learning). This information is then relayed (feedback

Table 4 Case (c) first-order self-model states (or First Reification Level States)

State Number	State Name	Explanation	Level
X_{53}	$\mathbf{W}_{\text{birth\_dev\_det\_HCP, det\_risk\_level\_HCP}}$	Reified representation state for connection weight $\omega_{\text{birth\_dev\_det\_HCP, det\_risk\_level\_HCP}}$	First reification level
X_{54}	$\mathbf{W}_{\text{det\_risk\_level\_HCP, encourage\_mom\_HCP}}$	Reified representation state for connection weight $\omega_{\text{det\_risk\_level\_HCP, encourage\_mom\_HCP}}$	
X_{55}	$\mathbf{W}_{\text{det\_risk\_level\_HCP, calm\_dad\_HCP}}$	Reified representation state for connection weight $\omega_{\text{det\_risk\_level\_HCP, calm\_dad\_HCP}}$	
X_{56}	$\mathbf{W}_{\text{irr\_birth\_weight\_HCP, det\_underw\_HCP}}$	Reified representation state for connection weight $\omega_{\text{irr\_birth\_weight\_HCP, det\_underw\_HCP}}$	
X_{57}	$\mathbf{W}_{\text{det\_underw\_HCP, refer\_mom\_ther\_HCP}}$	Reified representation state for connection weight $\omega_{\text{det\_underw\_HCP, refer\_mom\_ther\_HCP}}$	
X_{58}	$\mathbf{W}_{\text{det\_underw\_HCP, teach\_dad\_sup\_mom\_HCP}}$	Reified representation state for connection weight $\omega_{\text{det\_underw\_HCP, teach\_dad\_sup\_mom\_HCP}}$	
X_{59}	$\mathbf{W}_{\text{baby\_born\_HCP, type\_com\_HCP}}$	Reified representation state for connection weight $\omega_{\text{baby\_born\_HCP, type\_com\_HCP}}$	
X_{60}	$\mathbf{W}_{\text{type\_mom\_HCP, pre\_birth\_info\_HCP}}$	Reified representation state for connection weight $\omega_{\text{type\_mom\_HCP, pre\_birth\_info\_HCP}}$	
X_{61}	$\mathbf{W}_{\text{birth\_dev\_det\_AIC, det\_risk\_level\_AIC}}$	Reified representation state for connection weight $\omega_{\text{birth\_dev\_det\_AIC, det\_risk\_level\_AIC}}$	
X_{62}	$\mathbf{W}_{\text{det\_risk\_level\_AIC, encourage\_mom\_AIC}}$	Reified representation state for connection weight $\omega_{\text{det\_risk\_level\_AIC, encourage\_mom\_AIC}}$	
X_{63}	$\mathbf{W}_{\text{det\_risk\_level\_AIC, calm\_dad\_AIC}}$	Reified representation state for connection weight $\omega_{\text{det\_risk\_level\_AIC, calm\_dad\_AIC}}$	
X_{64}	$\mathbf{W}_{\text{irr\_birth\_weight\_AIC, det\_underw\_AIC}}$	Reified representation state for connection weight $\omega_{\text{irr\_birth\_weight\_AIC, det\_underw\_AIC}}$	
X_{65}	$\mathbf{W}_{\text{det\_underw\_AIC, refer\_mom\_ther\_AIC}}$	Reified representation state for connection weight $\omega_{\text{det\_underw\_AIC, refer\_mom\_ther\_AIC}}$	
X_{66}	$\mathbf{W}_{\text{det\_underw\_AIC, teach\_dad\_sup\_mom\_AIC}}$	Reified representation state for connection weight $\omega_{\text{det\_underw\_AIC, teach\_dad\_sup\_mom\_AIC}}$	
X_{67}	$\mathbf{W}_{\text{baby\_born\_AIC, type\_com\_AIC}}$	Reified representation state for connection weight $\omega_{\text{baby\_born\_AIC, type\_com\_AIC}}$	
X_{68}	$\mathbf{W}_{\text{type\_mom\_AIC, pre\_birth\_info\_AIC}}$	Reified representation state for connection weight $\omega_{\text{type\_mom\_AIC, pre\_birth\_info\_AIC}}$	
X_{69}	$\mathbf{W}_{\text{pre\_birth\_info\_AIC, pre\_birth\_info}}$	Reified representation state for connection weight $\omega_{\text{pre\_birth\_info\_AIC, pre\_birth\_info}}$	
X_{70}	$\mathbf{W}_{\text{calm\_dad\_AIC, acknowl\_fat}}$	Reified representation state for connection weight $\omega_{\text{calm\_dad\_AIC, acknowl\_fat}}$	
X_{71}	$\mathbf{W}_{\text{teach\_dad\_sup\_mom\_AIC, acknowl\_fat}}$	Reified representation state for connection weight $\omega_{\text{teach\_dad\_sup\_mom\_AIC, acknowl\_fat}}$	
X_{72}	$\mathbf{W}_{\text{calm\_dad\_AIC, com\_sup\_hcp}}$	Reified representation state for connection weight $\omega_{\text{calm\_dad\_AIC, com\_sup\_hcp}}$	
X_{73}	$\mathbf{W}_{\text{teach\_dad\_sup\_mom\_AIC, com\_sup\_hcp}}$	Reified representation state for connection weight $\omega_{\text{teach\_dad\_sup\_mom\_AIC, com\_sup\_hcp}}$	
X_{74}	$\mathbf{W}_{\text{calm\_dad\_AIC, post\_info\_com\_par}}$	Reified representation state for connection weight $\omega_{\text{calm\_dad\_AIC, post\_info\_com\_par}}$	
X_{75}	$\mathbf{W}_{\text{teach\_dad\_sup\_mom\_AIC, post\_info\_com\_par}}$	Reified representation state for connection weight $\omega_{\text{teach\_dad\_sup\_mom\_AIC, post\_info\_com\_par}}$	
X_{76}	$\mathbf{W}_{\text{encourage\_mom\_AIC, post\_info\_com\_par}}$	Reified representation state for connection weight $\omega_{\text{encourage\_mom\_AIC, post\_info\_com\_par}}$	
X_{77}	$\mathbf{W}_{\text{refer\_mom\_ther\_AIC, refer\_therapist}}$	Reified representation state for connection weight $\omega_{\text{refer\_mom\_ther\_AIC, refer\_therapist}}$	
X_{78}	$\mathbf{W}_{\text{calm\_dad\_AIC, mental\_sup}}$	Reified representation state for connection weight $\omega_{\text{calm\_dad\_AIC, mental\_sup}}$	
X_{79}	$\mathbf{W}_{\text{teach\_dad\_sup\_mom\_AIC, mental\_sup}}$	Reified representation state for connection weight $\omega_{\text{teach\_dad\_sup\_mom\_AIC, mental\_sup}}$	
X_{80}	$\mathbf{W}_{\text{encourage\_mom\_AIC, mental\_sup}}$	Reified representation state for connection weight $\omega_{\text{encourage\_mom\_AIC, mental\_sup}}$	
X_{81}	$\mathbf{W}_{\text{refer\_mom\_ther\_AIC, mental\_sup}}$	Reified representation state for connection weight $\omega_{\text{refer\_mom\_ther\_AIC, mental\_sup}}$	
X_{82}	$\mathbf{W}_{\text{type\_com\_AIC, type\_com}}$	Reified representation state for connection weight $\omega_{\text{type\_com\_AIC, type\_com}}$	
X_{83}	monitor_pre_birth_info	State monitoring AIC mental model state pre_birth_info_AIC and world state pre_birth_info	
X_{84}	monitor_acknowl_fat_1	State monitoring AIC mental model state teach_dad_sup_mom_AIC and world state acknowl_fat	
X_{85}	monitor_acknowl_fat_2	State monitoring AIC mental model state calm_dad_AIC and world state acknowl_fat	
X_{86}	monitor_com_sup_hcp_1	State monitoring AIC mental model state calm_dad_AIC and world state com_sup_hcp	
X_{87}	monitor_com_sup_hcp_2	State monitoring AIC mental model state teach_dad_sup_mom_AIC and world state com_sup_hcp	
X_{88}	monitor_post_info_com_par_1	State monitoring AIC mental model state encourage_mom_AIC and world state post_info_com_par	
X_{89}	monitor_post_info_com_par_2	State monitoring AIC mental model state calm_dad_AIC and world state post_info_com_par	
X_{90}	monitor_post_info_com_par_3	State monitoring AIC mental model state teach_dad_sup_mom_AIC and world state post_info_com_par	
X_{91}	monitor_refer_therapist	State monitoring AIC mental model state refer_mom_ther_AIC and world state refer_therapist	
X_{92}	monitor_mental_sup_1	State monitoring AIC mental model state encourage_mom_AIC and world state mental_sup	
X_{93}	monitor_mental_sup_2	State monitoring AIC mental model state refer_mom_ther_AIC and world state mental_sup	
X_{94}	monitor_mental_sup_3	State monitoring AIC mental model state calm_dad_AIC and world state mental_sup	
X_{95}	monitor_mental_sup_4	State monitoring AIC mental model state teach_dad_sup_mom_AIC and world state mental_sup	
X_{96}	monitor_type_com	State monitoring AIC mental model state type_com_AIC and world state type_com	

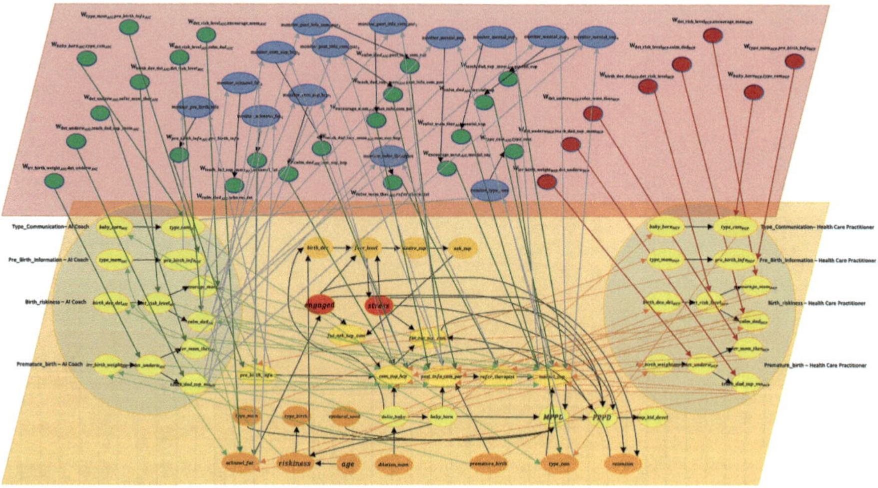

Fig. 4 Connectivity of first-order adaptive network model Case (c)

learning) to the HCPs after childbirth, making it an increasingly more valuable asset within a hospital setting and depicting organizational learning.

The set-up of the model is as follows. The base-level remains the same as in Sects. 4.2.2 and 4.2.3. The reason the expert is not being included in the actual process, is that he/she is a tool for teaching the Coach, while not being actively on the floor when the HCP and Coach work in cohesion. The middle level is the same as in Sect. 4.2.3, with the addition of eight **W**-states (purple) representing the knowledge of an expert. Each of these states has a link (light blue) from them to their counterpart of the eight original **W**-states corresponding to the Coach. Moreover, as the HCP can learn from the Coach, an additional eight links (light blue), with the same setup as mentioned before, were added from the Coach **W**-states to the HCP **W**-states.

In the second-order self-model level, two $\mathbf{W}_{\mathbf{W}}$-states -states (dark blue) were included, as well as two context states (blue). The $\mathbf{W}_{\mathbf{W}}$-states -states represent the horizontal connection weights (light blue) between the Expert (E) and the Coach, as well as between the Coach and Healthcare Practitioner (HCP). Therefore, the left most one has downward connections (dark blue) to the Coach **W**-states, and the right most one has downward connections (dark blue) to the HCP **W**-states. The context state con_feedforward creates an initial shared mental model of the knowledge necessary. The context state con_feedback updates the individual mental model, of the HCP in this case, with the information from the shared mental model. The detailed descriptions of the newly added states can be seen in Table 6.

Table 5 First- and second-order self-model states (or First and Second Reification Level States)

State Number	State Name	Explanation	Level
X_{97}	$\mathbf{W}_{\text{W\_AIC,W\_HCP}}$	Reified representation state for connection weight $\boldsymbol{\omega}_{\text{W\_AIC,W\_HCP}}$ for the connection between the reified representation states X_{61}-X_{68} and X_{53}-X_{60}, respectively	Second reification level
X_{98}	con_feedback	Context state for learning of the HCP from the AIC	
X_{99}	$\mathbf{W}_{\text{birth\_dev\_det\_E, det\_risk\_level\_E}}$	Representation state for the experts knowledge about birth_dev_det_E, det_risk_level_E	First reification level
X_{100}	$\mathbf{W}_{\text{det\_risk\_level\_E, encourage\_mom\_E}}$	Reified representation state for connection weight $\boldsymbol{\omega}_{\text{det\_risk\_level\_E, encourage\_mom\_E}}$	
X_{101}	$\mathbf{W}_{\text{det\_risk\_level\_E, calm\_dad\_E}}$	Reified representation state for connection weight $\boldsymbol{\omega}_{\text{det\_risk\_level\_E, calm\_dad\_E}}$	
X_{102}	$\mathbf{W}_{\text{irr\_birth\_weight\_E, det\_underw\_E}}$	Reified representation state for connection weight $\boldsymbol{\omega}_{\text{irr\_birth\_weight\_E, det\_underw\_E}}$	
X_{103}	$\mathbf{W}_{\text{det\_underw\_E, refer\_mom\_ther\_E}}$	Reified representation state for connection weight $\boldsymbol{\omega}_{\text{det\_underw\_E, refer\_mom\_ther\_E}}$	
X_{104}	$\mathbf{W}_{\text{det\_underw\_E, teach\_dad\_sup\_mom\_E}}$	Reified representation state for connection weight $\boldsymbol{\omega}_{\text{det\_underw\_E, teach\_dad\_sup\_mom\_E}}$	
X_{105}	$\mathbf{W}_{\text{baby\_born\_E, type\_com\_E}}$	Reified representation state for connection weight $\boldsymbol{\omega}_{\text{baby\_born\_E, type\_com\_E}}$	
X_{106}	$\mathbf{W}_{\text{type\_mom\_E, pre\_birth\_info\_E}}$	Reified representation state for connection weight $\boldsymbol{\omega}_{\text{type\_mom\_E, pre\_birth\_info\_E}}$	
X_{107}	$\mathbf{W}_{\text{W\_E,W\_AIC}}$	Reified representation state for connection weight $\boldsymbol{\omega}_{\text{W\_E,W\_AIC}}$ for the connection between the reified representation states X_{99}-X_{106} and X_{61}-X_{68}, respectively	Second reification level
X_{108}	con_feedforward	Context state for learning of the AIC from the E	

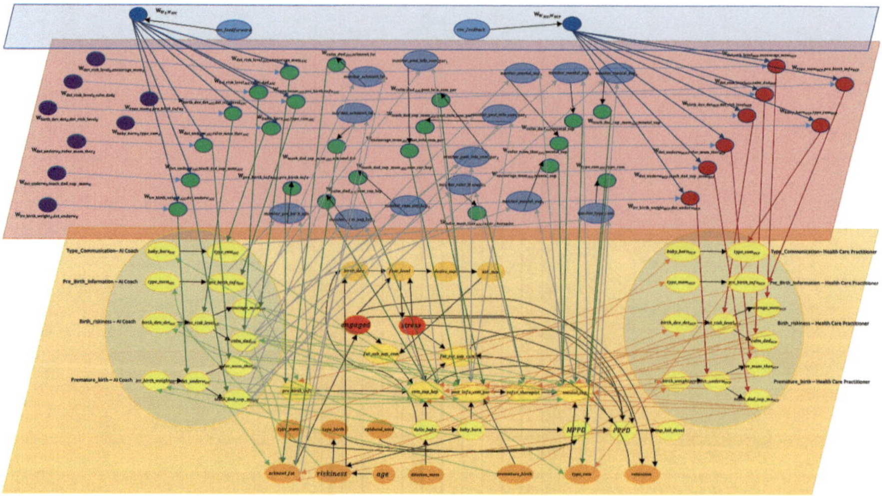

Fig. 5 Connectivity of second-order adaptive network model Case (d)

5 Simulation Results

This section presents the assumption made to simulate the model, along with the simulation results obtained from the designed model presented in Sect. 4.

5.1 *Assumptions Considered for Simulation*

While most states are dynamic, meaning they depend on the impact of other states, a handful of states within the models have manually chosen values given their binary nature.

1. The type of mother, either nulliparous or multiparous, was assigned a fixed value of 0, which corresponds to the mother being nulliparous.
2. The artificiality of birth, either being natural or caesarean section, was assigned a fixed value of 0, which corresponds to the mother having a natural birth.
3. The type of birth, either being no/little risk or high risk, was assigned a fixed value of 0, which corresponds to there being no risk.
4. The pre-birth information either being extra information is needed or not, was assigned a fixed value of 0, which corresponds to there being no need for additional information.
5. The age, either between the ages of 25–29 or not, was assigned a fixed value of 1, corresponding to the mother being between 25 and 29 years old.
6. Premature birth, either the birth being premature or not, was assigned a fixed value of 1, which corresponds to it being a premature birth.

Moreover, certain states had concrete specifications.

1. The starting value for maternal postpartum depression (MPPD) was set at 0.13, which is median of the probability of a woman developing postpartum depression (p. 526—Dennis 2004)
2. The state for dilation of the mother increases and reaches a maximum at 0.8, which represents full dilation of 10 cm (NHS 2020).
3. Once the dilation of the mother reaches 0.8, the state 'baby_born' is activated and increases quickly to 1 (Hueston 1998).

The time points (t) are referred to as unit time and have a value range between 0 and 80. This range for t is selected for an impression. However, it can also have a different range, and can be interpreted in minutes or hours. Moreover, the choice that a deviation occurs during childbirth, as well as the birth being premature, was deliberate, to show the effects of proper communication in a scenario where complications arise. In addition, the mother was set up with the best conditions to not develop MPPD. This can be seen by a slower increase of the line in the simulations compared to PPPD.

Furthermore, an active mental model is defined by the end states of the four mental processes specified earlier (See Sect. 4). Meaning all states, except the end states, may be active, as there is internal knowledge of what is occurring in front of them. However, this does not result in further communication actions. For example, the HCP is aware of there being a birth deviation but does not follow through with the further steps that then elicit the communication action.

Lastly, since the goal of this chapter is to observe and show how to reduce the risk of maternal (light blue) and paternal (red) postpartum depression and in return reducing the risk of impeding child development (dark blue), their corresponding lines were made bold.

5.2 *Communication with the Parents*

5.2.1 Adequate Communication Between Healthcare Practitioners and Parents

Here, we present a scenario of successful communication between parents and Healthcare Practitioner (HCP) during the childbirth process. The time duration is from 0 to 40 (t-axis) with a step size of $\Delta t = 0.1$. The low step size results in smoother curves in the simulation graphs.

At first, the healthcare practitioner is set to have a perfect mental model (all HCP's **W**-states are constant 1). The simulation shown in Fig. 6 depicts this scenario. The trend lines of the three bold states first increase since a deviation occurs, and risk and fear persist. However, once adequate support from the HCP to the father sets in, there is a fast decrease, eventually reaching 0. MPPD also decreases, as a result

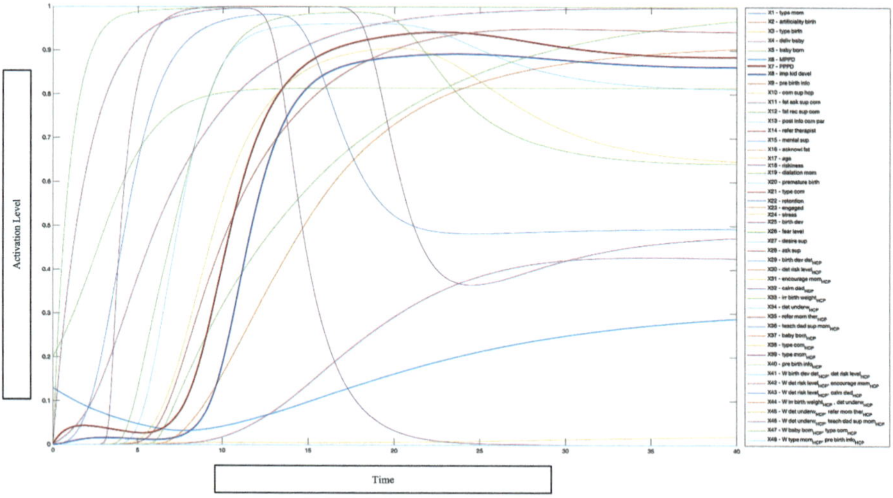

Fig. 6 Simulation results for successful communication between HCP and parents of the child

of the risk of PPPD decreasing, since the father being healthy allows for adequate support of the mother.

5.2.2 Ineffective Communication Between HealthCare Practitioners and Parents

In comparison to the simulation above, a further scenario was simulated (Fig. 7) to show the effects of imperfect communication of the HCP. When performing this simulation, all values were kept the same, except for the values of states in the mental model of the HCP. In this simulation, the HCP does not have an active mental model that facilitates proper communication (all HCP's **W**-states are constant 0).

When comparing Figs. 6 and 7, one can clearly see the effects of a lack of communication support from the HCP to the parents. While in Fig. 6, all risk-representing bold states increase, but promptly decrease again to reach 0, in Fig. 7, they all depict only an increase. At roughly $t = 25$, PPPD and imp_kid_devel correct themselves once the risk factor has subsided but remain high until the end of the time span. Moreover, MPPD has a steadily increasing trend, with no indication of decreasing whatsoever.

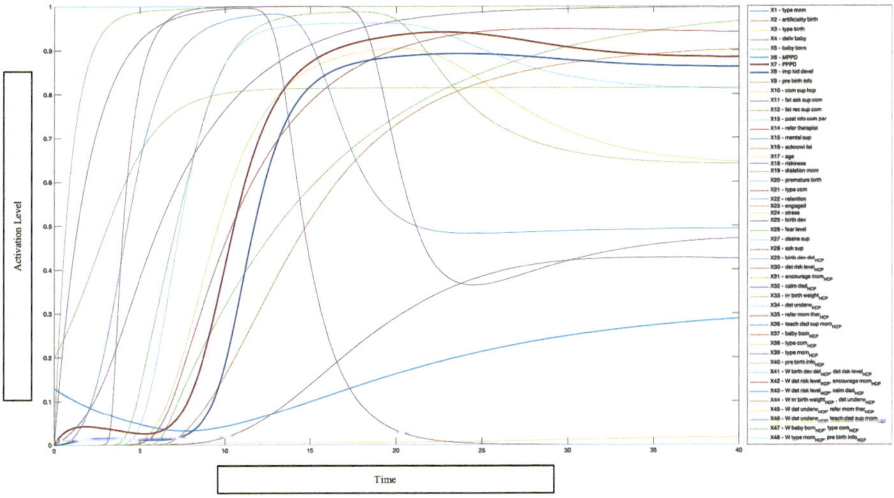

Fig. 7 Simulation results for poor communication between the HCP and the parents

5.3 *The Role of the Virtual Coach*

In this scenario, the virtual AI Coach (AIC) steps in to communicate when a healthcare practitioner doesn't attend to the father and mother appropriately during childbirth. The simulation (Fig. 8) has a time duration from 0 to 60 (t-axis) with a step size of $\Delta t = 0.1$. An increase of the time duration, in comparison to previous simulation (see Sect. 5.2), compensates for the late activation of the communication actions.

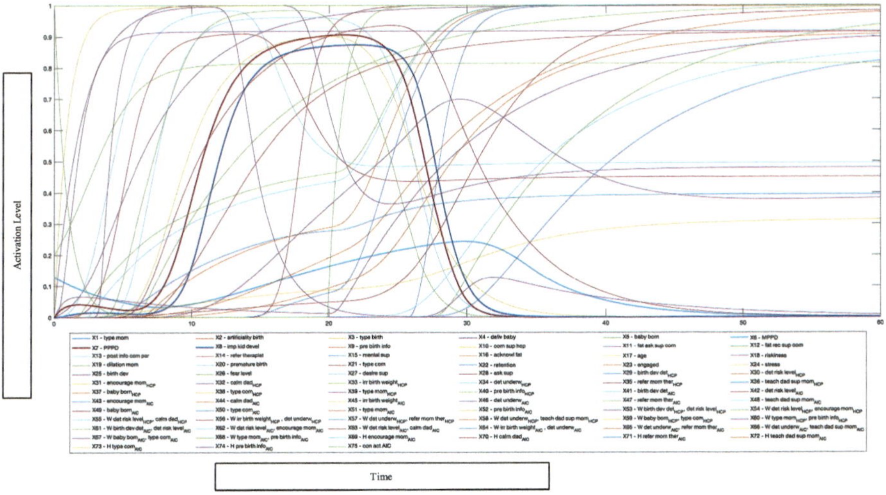

Fig. 8 Simulation results for Case (b)—AIC intervenes

The end states of the HCP never get activated and remain at 0, which results in no communication taking place from the HCP's side. This is due to the fact that the **W**-states are at 0 for the HCP. There exists an exception to this rule, as the calm_dad_HCP state gets activatedbriefly upon the fat_ask_sup_com state. Given the father approaching a HCP to ask for support, the HCP will act on being approached, even if their own initiative isn't there. This is visible by the two purple lines that reach their maximum at around $t = 30$.

On the other hand, the **W**-states for the AIC are at 1, meaning the respective mental model is active. The intervention (con_act_AIC) of the AIC starts at $t = 20$ (red line). The rapid change from 0 to 1 of this state, consequently, affects the communication action mental model states of the AIC. Before $t = 20$ these states had a slow increasing trend with little effect on the PPPD, MPPD and imp_kid_devel states. Promptly after the intervention, a rapid growth of the trend lines of the AIC mental model end states can be observed. This is a result of the **H**-states (purple line). This activation of the mental model translates into real-world support, which prompts, amongst others, the desire_sup state to decrease. The effects of this intervention can be observed by the three bold trend lines, which change course to decrease at around $t = 24$. Simultaneously to the end states of the AIC mental model reaching 1, the risk of PPPD and the risk of imp_kid_devel reach 0, followed by the risk of MPPD shortly after.

6 Dynamic Context-Sensitive Monitoring by the Virtual Coach

This is a similar scenario where a healthcare practitioner doesn't attend to the father and mother appropriately during the childbirth process. However, here, the Coach acts in a more context-sensitive manner. It is actively monitoring the relevant world states and intervenes only when it doesn't detect enough activation. The simulation (Fig. 9) has a time duration from 0 to 60 (t-axis) with a step size of $\Delta t = 0.1$.

Similarly to the role of the coach (discussed in Sects. 4.2.2 and 5.3), the HCP mental model does not get activated for the same reasons (**W**-states constant 0). Moreover, the **W**-states for AIC are also the same (constant 1). The reason for them being 1 is because the AI Coach is supposed to have perfect knowledge from the beginning. The mental model of the AIC is active starting between $t = 0$ to $t = 12$. The interesting aspect about this model are the monitoring states. These activate in the range of $t = 15$ to $t = 20$, as well as from $t = 35$ to $t = 48$.

Taking type_com (red line at $t = 20$) as an example state, the sequences of how the monitors activate will be explained. State type_com_AIC (blue line at $t = 17$), which is the mental model state of the AIC for the type of communication, activates first, since the AI Coach has a perfect mental model. This is followed by the monitor_

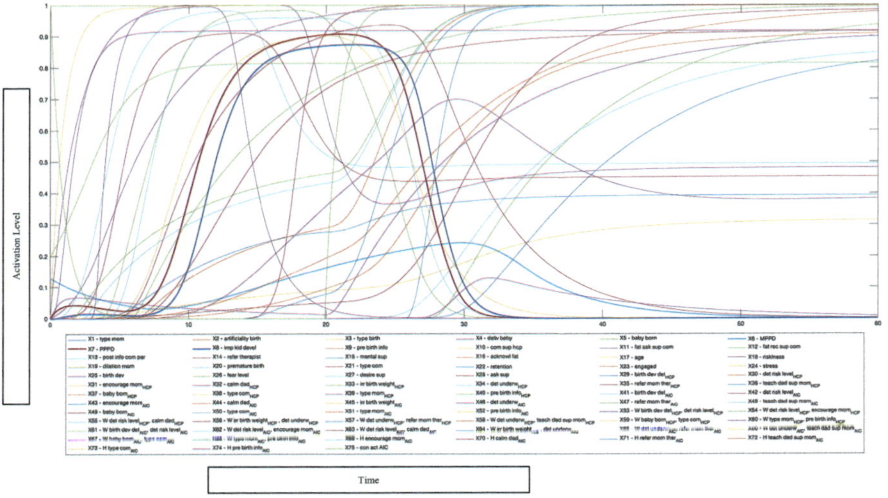

Fig. 9 Simulation results for Case (c)—AIC monitors

type_com (green high spike at $t = 20$) state, which in return activates W_type_com_AIC, type_com (green low spike at $t = 20$) to effectuate the communication intended by AIC. The process that occurs here is that the monitor state observes the type_com and type_com_AIC states and, by applying the monitor function (see Table 1), determines whether it must act to compensate for missing actions in the real world. This occurs by strengthening the W_type_com_AIC, type_com connection weight state. As a result of the strengthening, the type_com context state gets activated, which means it is being translated into a real-world action: the communication from the AIC actually takes place.

The overall time-span for the bold lines to decrease and eventually reach zero is higher than in the previous two cases, however the same outcome is eventually reached. This is a result of the increased duration due to the monitoring having to take place before the AIC can act. The speed factors for the AIC were chosen this way to make it not too fast, as it would be a disadvantage if the HCP didn't have time to start acting, before the AIC stepped in.

6.1 The Role of Organizational Learning

In this scenario, a healthcare practitioner initially has no knowledge for the appropriate mental model and doesn't attend to the father and mother appropriately during the childbirth process. However, in this case, we present the process of organizational learning by which HCP would improve its mental model, by acquiring shared knowledge. In this case, the AI Coach get on the role of maintaining the organization's shared mental knowledge. It acquires it first by feed forward learning. This

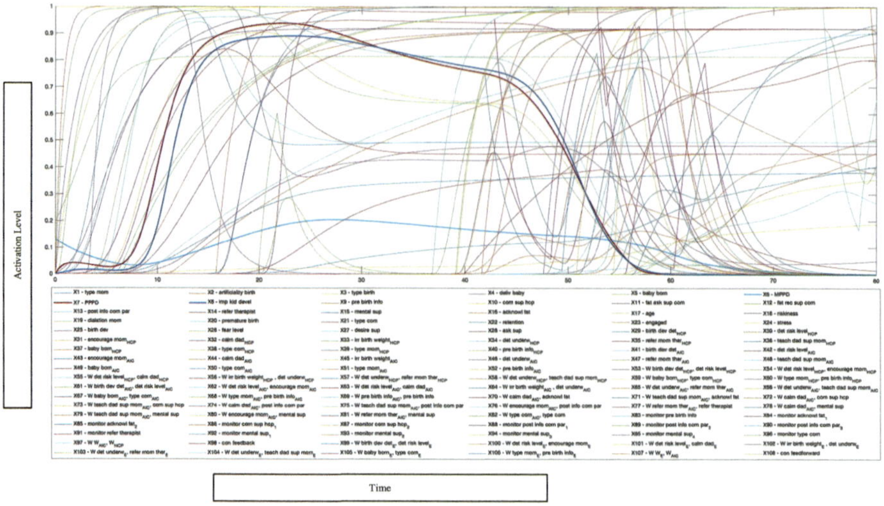

Fig. 10 Simulation results for Case (d)—expert teaches AIC, AIC monitors, AIC teaches HCP

is based on an expert teaching the AI Coach, thus creating a shared mental model. Then later in the process feedback learning takes place: this shared mental model is transferred to HCP (after the childbirth process). In this way, this simulation depicts organizational learning. The timing was a deliberate choice, as teaching the HCP during childbirth could result in increased stress levels and, in return, possibly have an adverse effect on the communication actions. Similar to our previous simulation results (Sect. 5), the AI Coach uses monitors to determine if or when to take an action. The simulation (Fig. 10) has a time duration from 0 to 80 (t-axis) with a step size of $\Delta t = 0.1$. The increased time duration is to have a full illustration of the learning phase of the HCP after the childbirth process.

This scenario follows a similar trend as dynamic monitoring (Sect. 5), given that the HCP has a non-active mental model. Meaning the actions that take place to reduce the three bold line states are identical to what is described above. The notable difference is the learning of the AIC from the expert. The left most line represents the $\mathbf{W}_{\mathbf{W}}$-states of the expert to the AIC (light red), which strengthens the connection weights, meaning the AIC acquires all the information and a shared mental model has been established. Moreover, the learning of the HCP from the AIC is activated by the con_feedback state (purple) at $t = 40$. At around $t = 48$, the $\mathbf{W}_{\mathbf{W}}$-states (light blue) of the AIC to the HCP begin to increase. This activates the **W**-states of the HCP and in return strengthens the mental model connection weights. The HCP learns between $t = 50$ and $t = 60$.

7 An Environment for the Virtual Safety Coach

The models designed and simulated in this study explain how using a Virtual Safety Coach can improve communication among healthcare practitioner and father. The Virtual AI Coach will be designed to use these models, with the aim to facilitate the medical practitioners to achieve their goals without any possible omission or other error. In other words, the Safety Coach would mainly rely on similar mental models to facilitate the healthcare practitioners.

In network models, states and their causal relations are declarative by nature and can be represented by some connection weight, speed, and a combination function values (see Sects. 3.3 and 3.4). These values are used to compute the impact of a state by the incoming states. In particular, a combination function is responsible to compute the effect of the incoming states(s), which may vary as per the nature of the combination function used for them. To reflect various behaviors different functions ranging from simple standard functions like Euclidian (i.e., eucl(V)) to more complex ones like homophily (i.e., slhomo(V_1,V_2,W)) were designed (Treur 2016, 2020). We created a library, that offers these combination functions at one place. This library can be extended further, to reflect more behaviors, which have not been addressed before. Thus the library can be considered as a collection of various functions along with information describing them, for example:

```
Library{
id,                    // unique identifier of the combination function
name,                  // the name of the function
numberOfParams,        //number of possible parameters
params                 //names of the parameters

}
```

The combination functions included in the library follows a template.

$$\mathrm{cf}_i\big(i,\,(p_1, p_2, \ldots, p_j),\,(V_1, \ldots, V_k)\big)$$

In principle, a state may have a number $k \geq 0$ of incoming causal connections. Also, a combination function in the library may have j parameters ($j = 0 \ldots m$). For instance, the identity function does not take any parameter. However, a logistic homophily function takes three parameters. Detailed information related to combination function is available online (Treur 2019). Moreover, a state is considered as a specification that can be defined as:

```
State{
initialValue,
speedFactor,
connectionWeight,
output,
cfData
}
```

For all state data (e.g., initialValue, speedFactor, weights), values range from 0 to 1, while output reflects the output of a state, that varies according to the causal impact of input states with respect to the time *t*. This output is computed using a combination function structured under the variable called 'cfData'. This consists of:

- **ImpactWeight**: How strong a specific combination function influences a state.
- **Parameters**: This contains the values of the parameters for the combination function used.

It is to be noted, that ImpactWeight facilitates the modelers to use different combination functions to study behaviors that are hybrid by nature. An example can be a state *X* which gets activated after a certain period of time, and show the behavior through **alogistic** function. In this case, a modeler can use **step** functions to reflect the period along with the alogistic function, with different impact weights for both functions (where the sum of impact weights should be 1). Parameters contain the value for the parameters used with respect to a combination function during the simulation, which is also addressed in the literature (Treur 2019). Please note, parameters may vary as per combination function used for a state. The simulation environment takes input with respect to each state. Therefore, for simulation, we provide the state specification. For instance, each state *Y* will have.

- list of incoming connections to state *Y* (e.g., states $X_1, X_2,\ldots,X_n$)
- the connection weight of each incoming state (e.g., from state $X_1, X_2,\ldots,X_n$)
- the speed of influence of the incoming states (termed as η_Y)
- initial value (iv_Y)
- cfData (cf_Y).

This input is processed to generate the simulation output. For each state, this output is computed by the formula mentioned in Eq. (1). Therefore, '*output*' is a matrix containing timestamps and values for all the states of a model with their respective timestamps. Therefore, *output* not only reflects the state value at a certain time, but it also provide reasoning for the causal interaction between the states, which can be viewed by different plotting methods. Built-in plot methods were used to generate the simulation results of the designed model, which can be viewed in Appendix B.

8 Discussion and Conclusion

This study presents adaptive computational network models covering (shared) mental models of the father and healthcare practitioner, which illustrates how they can communicate during childbirth (Weigl et al, 2023). These models are designed using multidisciplinary literature. Therefore, first, the benefits of the HCP having a mental model for communication is presented. Then it shows the utility of a Virtual Safety Coach (AIC) intervening after a given amount of time, to compensate for actions not being taken by the healthcare practitioner. This study also presents an advanced version of the Coach, illustrating a more complex system using monitoring for determining whether the Coach performs communication or not. Simulation results indicate that integrating the partners into the childbirth process and viewing them as active actors, as well as opening up communication lines between HCP and parents', aids in decreasing the risk of postpartum depression and in return ensuring positive child development. Moreover, both communication actions by AIC and HCP carry the same weight. This indicates that both are equally as effective in reaching the same end goal of having no risk of the father developing PPD.

Although the father is considered an actor, the mental model of the father was not included, as the AI Coach is a tool for HCP's and therefore, the father's internal model would have added no additional useful information to the simulations and in return, to the real-life application. However, in monitoring action of the safety coach, the willingness of the father to speak up alone, elicited a reaction in the mental model of the HCP. Therefore, promoting and applying a just safety culture into hospital organizations, where all parties involved feel welcome to speak up, has a lasting effect on the overall learning. This stems from the fact, that repeated and increasing exposure to the stimulus of the HCP being asked for support communication accelerates adaptation and can lead to long-term learning (Robinson et al. 2016). Similar to the safety culture, organizational learning allows for an ever-evolving growth of knowledge and creates stronger foundations for new and current HCPs in the medical field. This was reflected by using the notion of feedforward and feedback in the study. Applying the Virtual Safety Coach into the field and treating it as an individual reaps the benefits of using it to create a shared mental model based on the mental models existing within experts and, through this further education and improvement in medical practices. The combination of various mental models to create a shared mental model, results in optimal current knowledge.

Lastly, the strong influence of proper communication depicted in the models is not limited to the potential they depict of communication within the childbirth process. The perspective can be extended to other medical practices, which increasingly facilitates a safety culture, where open communication lies at the center.

9 Limitations and Further Work

This research has potential limitations. Given the nature of the research focusing mainly on actions that can be taken to reduce paternal postpartum depression, overcompensation may be a factor. Specifically, the bias of centering the model around necessary communication for father, may have restricted the realistic scope of the communication actions for the mother while assigning too much emphasis on those actions for the father only.

During modeling, various aspects were not discussed, for instance, how communication may influence the stress or engagement levels, while considering all the actors. Similarly, how communication can play its role in overcoming the related fears is not addressed. During simulation we only presented for a certain time duration, but, it would be interesting to investigate the impacts of communication for a longer time, i.e., when parents leave the hospital. Moreover, how equilibrium for certain states (e.g., stress or asking for support), may influence states of postpartum depression for longer time is not presented in this study.

The available literature on risk factors and intervention methods to combat paternal postpartum depression is not very extensive. Some states selected for the model are based on desires that fathers have expressed without their effectiveness being studied scientifically and determined to be successful. In future, we aim to collect more multidisciplinary information, which may help researchers to provide some analytical grounds to study the effectiveness of the modeled states. A possible improvement on this, could be conducting a study based on the Edinburgh Postpartum Depressive Scale (EPDS) and applying the values to the model created. The model is created in such a way, that does not depend on a precise definition, and it is not limited to one notion. This allows for a large application scope within the AI Coach Project.

Also, we aim to extend our model, in terms of responses of healthcare practitioners toward father and toward the virtual safety coach. In the future, we also would like to study multiple factors (e.g., culture, family structure or affiliations) regarding communication during the childbirth process (Dunkel Schetter et al. 2016; Goodman 2004). Moreover, an extension including the support during the prepartum period, would add to the effectiveness of communication in reducing the risk of postpartum depression. This stems from the fact, that developing depression is not limited to after giving birth. In addition, a relevant point that was not discussed in the scope of this chapter, is it could help parents who previously suffer from depressive episodes.

We also aim to extend the monitoring. Some world states had incoming connections from multiple mental model end states. In the models, this may have resulted in only one of these end states being activated, as one may be sufficient to pass the chosen threshold determining activation, which allows for some but not the maximum possible support.

Learning is a time-consuming task not only for individuals but also for the HCP and AIC. Therefore, in a real-world setting, this would allow for emphasis to be applied on having certain mental process over others. For example, having only post-birth communication may be as effective in decreasing the risk of PPPD as all the other states combined. This would allow for an adequate learning pace for practitioners in the field without overwhelming them regarding vast additional knowledge. Similarly, this could be applied to the AIC teaching the HCP in steps (based on importance), rather than all at once. By doing so, the level of retention of the information could be higher. Moreover, studying the aspect of trust towards AIC can also add a value while discussing the levels of communication.

In this study, the Coach (AIC) is modeled to be activated without a permission from the actors on the floor. To implement it with the current mental model into a healthcare setting, one would have to ensure high specificity and high sensitivity of the AIC. This would avoid eliciting negative emotions in the father with a false positive of, for example, a deviation occurring. A trigger initiated from the HCP could activate the Coach would be beneficial.

In the real-world, the virtual AI Coach can be considered a facility to assist healthcare practitioners. Therefore, the coach could have a different model (i.e., not similar to HCP). In this study, the virtual AI Coach is trying to monitor if communication is needed. That might be possible by analyzing the conversations or through some sensors. However, then the coach should be empathetic enough to provide support to the father. Moreover, without a trigger, it would be still questionable if the coach receives enough input to decide on whether to communicate or not in terms of accuracy and time. Lastly, how communication provided by Virtual Safety Coach will be evaluated by all the actors, is also to be considered as a future work for this research study.

Acknowledgements This research is part of the SAFECoach Project, which aims to create an AI based coach to serve hospital organization. The main aim of this coach is to assist healthcare practitioner while considering mental models and shared mental models for its implementation.

Appendix A: Role Matrices for all Model Variants

Role matrices are the numerical representations of the simulations. They provide full information about the network characteristics of the model variants. Below all role matrices for the models are visible, with each state having its own row, on which each incoming connections, as well as their impact, is specified. Moreover, they are color coordinated based on the level that the states lie on.

Communication

Health Care Practitioner Communicates with Father

See Tables 6, 7, 8, 9, 10 and 11.

Lack of Communication with Father

See Tables 12, 13, 14, 15, 16 and 17.

Role of Virtual Safety Coach

See Tables 18, 19, 20, 21, 22 and 23.

Table 6 Role matrix for best Case (a) for **mb**—base connectivity

mb - base connectivity		1	2	3	4	5	6	7	8
X_{1}	type_mom	X_{1}							
X_{2}	artificiality_birth	X_{2}	X_{18}						
X_{3}	epidural_used	X_{3}							
X_{4}	deliv_baby	X_{4}	X_{19}						
X_{5}	baby_born	X_{4}							
X_{6}	MPPD	X_{1}	X_{2}	X_{3}	X_{5}	X_{15}			
X_{7}	PPPD	X_{6}	X_{10}	X_{12}	X_{15}	X_{23}	X_{24}		
X_{8}	imp_kid_devel	X_{7}							
X_{9}	pre_birth_info	X_{9}	X_{40}						
X_{10}	com_sup_hcp	X_{4}	X_{9}	X_{11}	X_{24}	X_{30}	X_{32}	X_{34}	X_{36}
X_{11}	fat_ask_sup_com	X_{23}	X_{28}						
X_{12}	fat_rec_sup_com	X_{10}	X_{22}	X_{24}					
X_{13}	post_info_com_par	X_{5}	X_{20}	X_{31}	X_{32}	X_{36}			
X_{14}	refer_therapist	X_{13}	X_{35}						
X_{15}	mental_sup	X_{14}	X_{21}	X_{31}	X_{32}	X_{35}	X_{36}		
X_{16}	acknowl_fat	X_{9}	X_{32}	X_{36}					
X_{17}	age	X_{17}							
X_{18}	riskiness	X_{5}	X_{17}						
X_{19}	dilation_mom	X_{19}							
X_{20}	premature_birth	X_{20}							
X_{21}	type_com	X_{21}	X_{38}						
X_{22}	retention	X_{23}	X_{24}						
X_{23}	engaged	X_{16}							
X_{24}	stress	X_{26}							
X_{25}	birth_dev	X_{4}	X_{18}						
X_{26}	fear_level	X_{23}	X_{24}	X_{25}					
X_{27}	desire_sup	X_{26}							
X_{28}	ask_sup	X_{27}							
X_{29}	birth_dev_det$_{HCP}$	X_{4}	X_{18}						
X_{30}	det_risk_level$_{HCP}$	X_{29}							
X_{31}	encourage_mom$_{HCP}$	X_{30}							
X_{32}	calm_dad$_{HCP}$	X_{11}	X_{30}						
X_{33}	irr_birth_weight$_{HCP}$	X_{5}	X_{20}						
X_{34}	det_underw$_{HCP}$	X_{33}							
X_{35}	refer_mom_ther$_{HCP}$	X_{34}							
X_{36}	teach_dad_sup_mom$_{HCP}$	X_{34}							
X_{37}	baby_born$_{HCP}$	X_{5}							
X_{38}	type_com$_{HCP}$	X_{37}							
X_{39}	type_mom$_{HCP}$	X_{1}							
X_{40}	pre_birth_info$_{HCP}$	X_{39}							
X_{41}	$W_{\text{birth\_dev\_det\_HCP, det\_risk\_level\_HCP}}$	X_{41}							
X_{42}	$W_{\text{det\_risk\_level\_HCP, encourage\_mom\_HCP}}$	X_{42}							
X_{43}	$W_{\text{det\_risk\_level\_HCP, calm\_dad\_HCP}}$	X_{43}							
X_{44}	$W_{\text{irr\_birth\_weight\_HCP, det\_underw\_HCP}}$	X_{44}							
X_{45}	$W_{\text{det\_underw\_HCP, refer\_mom\_ther\_HCP}}$	X_{45}							
X_{46}	$W_{\text{det\_underw\_HCP, teach\_dad\_sup\_mom\_HCP}}$	X_{46}							
X_{47}	$W_{\text{baby\_born\_HCP, type\_com\_HCP}}$	X_{47}							
X_{48}	$W_{\text{type\_mom\_HCP, pre\_birth\_info\_HCP}}$	X_{48}							

Dynamic Monitoring of Virtual Safety Coach

See Tables 24, 25, 26, 27, 28 and 29.

Table 7 Role matrix for best Case (a) for **mcw**—connection weights

	mcw - connection weights	1	2	3	4	5	6	7	8
X_1	type_mom	1							
X_2	artificiality_birth	1	-1						
X_3	epidural_used	1							
X_4	deliv_baby	1	1						
X_5	baby_born	1							
X_6	MPPD	0.25	0.25	0.25	0.25	-1			
X_7	PPPD	1	-0.5	-0.5	-0.5	-0.5	1		
X_8	kid_devel	1							
X_9	pre_birth_info	1	1						
X_{10}	com_sup_hcp	1	1	1	-1	1	1	1	1
X_{11}	fat_ask_sup_com	1	1						
X_{12}	fat_rec_sup_com	1	1	-1					
X_{13}	post_info_com_par	1	1	1	1	1			
X_{14}	refer_therapist	1	1						
X_{15}	mental_sup	1	1	1	1	1	1		
X_{16}	acknowl_fat	1	1	1					
X_{17}	age	1							
X_{18}	riskiness	-1	1						
X_{19}	dilation_mom	1							
X_{20}	premature_birth	1							
X_{21}	type_com	1	1						
X_{22}	retention	2	-1						
X_{23}	engaged	1							
X_{24}	stress	1							
X_{25}	birth_dev	1	1						
X_{26}	fear_level	-1	1	1					
X_{27}	desire_sup	1							
X_{28}	ask_sup	1							
X_{29}	birth_dev_det$_{HCP}$	1	1						
X_{30}	det_risk_level$_{HCP}$	X_{41}							
X_{31}	encourage_mom$_{HCP}$	X_{42}							
X_{32}	calm_dad$_{HCP}$	1	X_{43}						
X_{33}	irr_birth_weight$_{HCP}$	1	2						
X_{34}	det_underw$_{HCP}$	X_{44}							
X_{35}	refer_mom_ther$_{HCP}$	X_{45}							
X_{36}	teach_dad_sup_mom$_{HCP}$	X_{46}							
X_{37}	baby_born$_{HCP}$	1							
X_{38}	type_com$_{HCP}$	X_{47}							
X_{39}	type_mom$_{HCP}$	1							
X_{40}	pre_birth_info$_{HCP}$	X_{48}							
X_{41}	$\mathbf{W}_{\text{birth\_dev\_det\_HCP, det\_risk\_level\_HCP}}$	1							
X_{42}	$\mathbf{W}_{\text{det\_risk\_level\_HCP, encourage\_mom\_HCP}}$	1							
X_{43}	$\mathbf{W}_{\text{det\_risk\_level\_HCP, calm\_dad\_HCP}}$	1							
X_{44}	$\mathbf{W}_{\text{irr\_birth\_weight\_HCP, det\_underw\_HCP}}$	1							
X_{45}	$\mathbf{W}_{\text{det\_underw\_HCP, refer\_mom\_ther\_HCP}}$	1							
X_{46}	$\mathbf{W}_{\text{det\_underw\_HCP, teach\_dad\_sup\_mom\_HCP}}$	1							
X_{47}	$\mathbf{W}_{\text{baby\_born\_HCP, type\_com\_HCP}}$	1							
X_{48}	$\mathbf{W}_{\text{type\_mom\_HCP, pre\_birth\_info\_HCP}}$	1							

Organizational Learning and Its Influence

See Tables 30, 31, 32, 33, 34 and 35.

Table 8 Role matrix for best Case (a) for **mcfw**—combination function weights

mcfw - combination function weights		1
		Alogistic
X_1	type_mom	1
X_2	artificiality_birth	1
X_3	epidural_used	1
X_4	deliv_baby	1
X_5	baby_born	1
X_6	MPPD	1
X_7	PPPD	1
X_8	kid_devel	1
X_9	pre_birth_info	1
X_{10}	com_sup_hcp	1
X_{11}	fat_ask_sup_com	1
X_{12}	fat_rec_sup_com	1
X_{13}	post_info_com_par	1
X_{14}	refer_therapist	1
X_{15}	mental_sup	1
X_{16}	acknowl_fat	1
X_{17}	age	1
X_{18}	riskiness	1
X_{19}	dilation_mom	1
X_{20}	premature_birth	1
X_{21}	type_com	1
X_{22}	retention	1
X_{23}	engaged	1
X_{24}	stress	1
X_{25}	birth_dev	1
X_{26}	fear_level	1
X_{27}	desire_sup	1
X_{28}	ask_sup	1
X_{29}	$\text{birth\_dev\_det}_{HCP}$	1
X_{30}	$\text{det\_risk\_level}_{HCP}$	1
X_{31}	$\text{encourage\_mom}_{HCP}$	1
X_{32}	calm_dad_{HCP}	1
X_{33}	$\text{irr\_birth\_weight}_{HCP}$	1
X_{34}	det_underw_{HCP}	1
X_{35}	$\text{refer\_mom\_ther}_{HCP}$	1
X_{36}	$\text{teach\_dad\_sup\_mom}_{HCP}$	1
X_{37}	baby_born_{HCP}	1
X_{38}	type_com_{HCP}	1
X_{39}	type_mom_{HCP}	1
X_{40}	$\text{pre\_birth\_info}_{HCP}$	1
X_{41}	$\mathbf{W}_{\text{birth\_dev\_det\_HCP, det\_risk\_level\_HCP}}$	1
X_{42}	$\mathbf{W}_{\text{det\_risk\_level\_HCP, encourage\_mom\_HCP}}$	1
X_{43}	$\mathbf{W}_{\text{det\_risk\_level\_HCP, calm\_dad\_HCP}}$	1
X_{44}	$\mathbf{W}_{\text{irr\_birth\_weight\_HCP, det\_underw\_HCP}}$	1
X_{45}	$\mathbf{W}_{\text{det\_underw\_HCP, refer\_mom\_ther\_HCP}}$	1
X_{46}	$\mathbf{W}_{\text{det\_underw\_HCP, teach\_dad\_sup\_mom\_HCP}}$	1
X_{47}	$\mathbf{W}_{\text{baby\_born\_HCP, type\_com\_HCP}}$	1
X_{48}	$\mathbf{W}_{\text{type\_mom\_HCP, pre\_birth\_info\_HCP}}$	1

Table 9 Role matrix for best Case (a) for **mcfp**—combination function parameters

mcfp - combination function parameters		1 Alogistic	
		Steepness	Threshold
X_1	type_mom	50	0.5
X_2	artificiality_birth	50	0.5
X_3	epidural_used	50	0.5
X_4	deliv_baby	5	0.5
X_5	baby_born	50	0.5
X_6	MPPD	5	0.3
X_7	PPPD	5	0.5
X_8	kid_devel	5	0.5
X_9	pre_birth_info	50	0.5
X_{10}	com_sup_hcp	0.5	6
X_{11}	fat_ask_sup_com	5	1
X_{12}	fat_rec_sup_com	5	0.5
X_{13}	post_info_com_par	5	4
X_{14}	refer_therapist	5	1
X_{15}	mental_sup	5	5
X_{16}	acknowl_fat	5	0.8
X_{17}	age	5	0.5
X_{18}	riskiness	50	0.5
X_{19}	dilation_mom	2.8	0
X_{20}	premature_birth	5	0.5
X_{21}	type_com	5	0.1
X_{22}	retention	5	0.3
X_{23}	engaged	5	0.5
X_{24}	stress	5	0.5
X_{25}	birth_dev	50	1
X_{26}	fear_level	5	1
X_{27}	desire_sup	5	0.3
X_{28}	ask_sup	5	0.2
X_{29}	$\text{birth\_dev\_det}_{HCP}$	5	1
X_{30}	$\text{det\_risk\_level}_{HCP}$	5	0.5
X_{31}	$\text{encourage\_mom}_{HCP}$	5	0.5
X_{32}	calm_dad_{HCP}	50	0.7
X_{33}	$\text{irr\_birth\_weight}_{HCP}$	5	1
X_{34}	det_underw_{HCP}	5	0.5
X_{35}	$\text{refer\_mom\_ther}_{HCP}$	50	0.5
X_{36}	$\text{teach\_dad\_sup\_mom}_{HCP}$	50	0.7
X_{37}	baby_born_{HCP}	5	0.5
X_{38}	type_com_{HCP}	50	0.5
X_{39}	type_mom_{HCP}	5	0.5
X_{40}	$\text{pre\_birth\_info}_{HCP}$	50	0.5
X_{41}	$\mathbf{W}_{\text{birth\_dev\_det\_HCP, det\_risk\_level\_HCP}}$	5	0.5
X_{42}	$\mathbf{W}_{\text{det\_risk\_level\_HCP, encourage\_mom\_HCP}}$	5	0.5
X_{43}	$\mathbf{W}_{\text{det\_risk\_level\_HCP, calm\_dad\_HCP}}$	5	0.5
X_{44}	$\mathbf{W}_{\text{irr\_birth\_weight\_HCP, det\_underw\_HCP}}$	5	0.5
X_{45}	$\mathbf{W}_{\text{det\_underw\_HCP, refer\_mom\_ther\_HCP}}$	5	0.5
X_{46}	$\mathbf{W}_{\text{det\_underw\_HCP, teach\_dad\_sup\_mom\_HCP}}$	5	0.5
X_{47}	$\mathbf{W}_{\text{baby\_born\_HCP, type\_com\_HCP}}$	5	0.5
X_{48}	$\mathbf{W}_{\text{type\_mom\_HCP, pre\_birth\_info\_HCP}}$	5	0.5

Table 10 Role matrix for best Case (a) for **iv**—initial values

iv - initial values		1
X_1	type_mom	0
X_2	artificiality_birth	0
X_3	epidural_used	0
X_4	deliv_baby	0
X_5	baby_born	0
X_6	MPPD	0.13
X_7	PPPD	0
X_8	kid_devel	0
X_9	pre_birth_info	0
X_{10}	com_sup_hcp	0
X_{11}	fat_ask_sup_com	0
X_{12}	fat_rec_sup_com	0
X_{13}	post_info_com_par	0
X_{14}	refer_therapist	0
X_{15}	mental_sup	0
X_{16}	acknowl_fat	0
X_{17}	age	1
X_{18}	riskiness	0
X_{19}	dilation_mom	0.2
X_{20}	premature_birth	1
X_{21}	type_com	0
X_{22}	retention	0
X_{23}	engaged	0
X_{24}	stress	0
X_{25}	birth_dev	0
X_{26}	fear_level	0
X_{27}	desire_sup	0
X_{28}	ask_sup	0
X_{29}	$\text{birth\_dev\_det}_{HCP}$	0
X_{30}	$\text{det\_risk\_level}_{HCP}$	0
X_{31}	$\text{encourage\_mom}_{HCP}$	0
X_{32}	calm_dad_{HCP}	0
X_{33}	$\text{irr\_birth\_weight}_{HCP}$	0
X_{34}	det_underw_{HCP}	0
X_{35}	$\text{refer\_mom\_ther}_{HCP}$	0
X_{36}	$\text{teach\_dad\_sup\_mom}_{HCP}$	0
X_{37}	baby_born_{HCP}	0
X_{38}	type_com_{HCP}	0
X_{39}	type_mom_{HCP}	0
X_{40}	$\text{pre\_birth\_info}_{HCP}$	0
X_{41}	$\mathbf{W}_{\text{birth\_dev\_det\_HCP, det\_risk\_level\_HCP}}$	1
X_{42}	$\mathbf{W}_{\text{det\_risk\_level\_HCP, encourage\_mom\_HCP}}$	1
X_{43}	$\mathbf{W}_{\text{det\_risk\_level\_HCP, calm\_dad\_HCP}}$	1
X_{44}	$\mathbf{W}_{\text{irr\_birth\_weight\_HCP, det\_underw\_HCP}}$	1
X_{45}	$\mathbf{W}_{\text{det\_underw\_HCP, refer\_mom\_ther\_HCP}}$	1
X_{46}	$\mathbf{W}_{\text{det\_underw\_HCP, teach\_dad\_sup\_mom\_HCP}}$	1
X_{47}	$\mathbf{W}_{\text{baby\_born\_HCP, type\_com\_HCP}}$	1
X_{48}	$\mathbf{W}_{\text{type\_mom\_HCP, pre\_birth\_info\_HCP}}$	1

Table 11 Role matrix for best Case (a) for **ms**—speed factors

	ms - speed factors	1
X_1	type_mom	1
X_2	artificiality_birth	1
X_3	epidural_used	1
X_4	deliv_baby	0.2
X_5	baby_born	0.1
X_6	MPPD	0.2
X_7	PPPD	1
X_8	kid_devel	1
X_9	pre_birth_info	0.1
X_{10}	com_sup_hcp	0.1
X_{11}	fat_ask_sup_com	0.3
X_{12}	fat_rec_sup_com	0.1
X_{13}	post_info_com_par	0.1
X_{14}	refer_therapist	0.3
X_{15}	mental_sup	0.1
X_{16}	acknowl_fat	0.1
X_{17}	age	0
X_{18}	riskiness	0.5
X_{19}	dilation_mom	1
X_{20}	premature_birth	0
X_{21}	type_com	1
X_{22}	retention	1
X_{23}	engaged	1
X_{24}	stress	0.4
X_{25}	birth_dev	1
X_{26}	fear_level	1
X_{27}	desire_sup	1
X_{28}	ask_sup	0.2
X_{29}	$birth\_dev\_det_{HCP}$	1
X_{30}	$det\_risk\_level_{HCP}$	1
X_{31}	$encourage\_mom_{HCP}$	0.1
X_{32}	$calm\_dad_{HCP}$	0.1
X_{33}	$irr\_birth\_weight_{HCP}$	1
X_{34}	det_underw_{HCP}	1
X_{35}	$refer\_mom\_ther_{HCP}$	0.3
X_{36}	$teach\_dad\_sup\_mom_{HCP}$	0.1
X_{37}	$baby\_born_{HCP}$	1
X_{38}	$type\_com_{HCP}$	1
X_{39}	$type\_mom_{HCP}$	1
X_{40}	$pre\_birth\_info_{HCP}$	1
X_{41}	$\mathbf{W}_{birth\_dev\_det\_HCP,\ det\_risk\_level\_HCP}$	0
X_{42}	$\mathbf{W}_{det\_risk\_level\_HCP,\ encourage\_mom\_HCP}$	0
X_{43}	$\mathbf{W}_{det\_risk\_level\_HCP,\ calm\_dad\_HCP}$	0
X_{44}	$\mathbf{W}_{irr\_birth\_weight\_HCP,\ det\_underw\_HCP}$	0
X_{45}	$\mathbf{W}_{det\_underw\_HCP,\ refer\_mom\_ther\_HCP}$	0
X_{46}	$\mathbf{W}_{det\_underw\_HCP,\ teach\_dad\_sup\_mom\_HCP}$	0
X_{47}	$\mathbf{W}_{baby\_born\_HCP,\ type\_com\_HCP}$	0
X_{48}	$\mathbf{W}_{type\_mom\_HCP,\ pre\_birth\_info\_HCP}$	0

Table 12 Role matrix for worst Case (a) for **mb**—base connectivity

mb - base connectivity		1	2	3	4	5	6	7	8
X_1	type_mom	X_1							
X_2	artificiality_birth	X_2	X_{18}						
X_3	epidural_used	X_3							
X_4	deliv_baby	X_4	X_{19}						
X_5	baby_born	X_4							
X_6	MPPD	X_1	X_2	X_3	X_5	X_{15}			
X_7	PPPD	X_6	X_{10}	X_{12}	X_{15}	X_{23}	X_{24}		
X_8	imp_kid_devel	X_7							
X_9	pre_birth_info	X_9	X_{40}						
X_{10}	com_sup_hcp	X_4	X_9	X_{11}	X_{24}	X_{30}	X_{32}	X_{34}	X_{36}
X_{11}	fat_ask_sup_com	X_{23}	X_{28}						
X_{12}	fat_rec_sup_com	X_{10}	X_{22}	X_{24}					
X_{13}	post_info_com_par	X_5	X_{20}	X_{31}	X_{32}	X_{36}			
X_{14}	refer_therapist	X_{13}	X_{35}						
X_{15}	mental_sup	X_{14}	X_{21}	X_{31}	X_{32}	X_{35}	X_{36}		
X_{16}	acknowl_fat	X_9	X_{32}	X_{36}					
X_{17}	age	X_{17}							
X_{18}	riskiness	X_5	X_{17}						
X_{19}	dilation_mom	X_{19}							
X_{20}	premature_birth	X_{20}							
X_{21}	type_com	X_{21}	X_{38}						
X_{22}	retention	X_{23}	X_{24}						
X_{23}	engaged	X_{16}							
X_{24}	stress	X_{26}							
X_{25}	birth_dev	X_4	X_{18}						
X_{26}	fear_level	X_{23}	X_{24}	X_{25}					
X_{27}	desire_sup	X_{26}							
X_{28}	ask_sup	X_{27}							
X_{29}	$birth\_dev\_det_{HCP}$	X_4	X_{18}						
X_{30}	$det\_risk\_level_{HCP}$	X_{29}							
X_{31}	$encourage\_mom_{HCP}$	X_{30}							
X_{32}	$calm\_dad_{HCP}$	X_{11}	X_{30}						
X_{33}	$irr\_birth\_weight_{HCP}$	X_5	X_{20}						
X_{34}	det_underw_{HCP}	X_{33}							
X_{35}	$refer\_mom\_ther_{HCP}$	X_{34}							
X_{36}	$teach\_dad\_sup\_mom_{HCP}$	X_{34}							
X_{37}	$baby\_born_{HCP}$	X_5							
X_{38}	$type\_com_{HCP}$	X_{37}							
X_{39}	$type\_mom_{HCP}$	X_1							
X_{40}	$pre\_birth\_info_{HCP}$	X_{39}							
X_{41}	$\mathbf{W}_{birth\_dev\_det\_HCP,\ det\_risk\_level\_HCP}$	X_{41}							
X_{42}	$\mathbf{W}_{det\_risk\_level\_HCP,\ encourage\_mom\_HCP}$	X_{42}							
X_{43}	$\mathbf{W}_{det\_risk\_level\_HCP,\ calm\_dad\_HCP}$	X_{43}							
X_{44}	$\mathbf{W}_{irr\_birth\_weight\_HCP,\ det\_underw\_HCP}$	X_{44}							
X_{45}	$\mathbf{W}_{det\_underw\_HCP,\ refer\_mom\_ther\_HCP}$	X_{45}							
X_{46}	$\mathbf{W}_{det\_underw\_HCP,\ teach\_dad\_sup\_mom\_HCP}$	X_{46}							
X_{47}	$\mathbf{W}_{baby\_born\_HCP,\ type\_com\_HCP}$	X_{47}							
X_{48}	$\mathbf{W}_{type\_mom\_HCP,\ pre\_birth\_info\_HCP}$	X_{48}							

Appendix B: Simulation Results in the Software Designed for the Virtual Coach

As the coach is implemented by the Python environment, the Matlab-generated simulation results have been reproduced; see examples below. They are exactly similar to the simulations by the Matlab software, which provides additional verification.

Table 13 Role matrix for worst Case (a) for **mcw**—connection weights

mcw - connection weights		1	2	3	4	5	6	7	8
X_1	type_mom	1							
X_2	artificiality_birth	1	-1						
X_3	epidural_used	1							
X_4	deliv_baby	1	1						
X_5	baby_born	1							
X_6	MPPD	0.25	0.25	0.25	0.25	-1			
X_7	PPPD	1	-0.5	-0.5	-0.5	-0.5	1		
X_8	kid_devel	1							
X_9	pre_birth_info	1	1						
X_{10}	com_sup_hcp	1	1	1	-1	1	1	1	1
X_{11}	fat_ask_sup_com	1	1						
X_{12}	fat_rec_sup_com	1	1	-1					
X_{13}	post_info_com_par	1	1	1	1	1			
X_{14}	refer_therapist	1	1						
X_{15}	mental_sup	1	1	1	1	1	1		
X_{16}	acknowl_fat	1	1	1					
X_{17}	age	1							
X_{18}	riskiness	-1	1						
X_{19}	dilation_mom	1							
X_{20}	premature_birth	1							
X_{21}	type_com	1	1						
X_{22}	retention	2	-1						
X_{23}	engaged	1							
X_{24}	stress	1							
X_{25}	birth_dev	1	1						
X_{26}	fear_level	-1	1	1					
X_{27}	desire_sup	1							
X_{28}	ask_sup	1							
X_{29}	birth_dev_det$_{HCP}$	1	1						
X_{30}	det_risk_level$_{HCP}$	X_{41}							
X_{31}	encourage_mom$_{HCP}$	X_{42}							
X_{32}	calm_dad$_{HCP}$	1	X_{43}						
X_{33}	irr_birth_weight$_{HCP}$	1	2						
X_{34}	det_underw$_{HCP}$	X_{44}							
X_{35}	refer_mom_ther$_{HCP}$	X_{45}							
X_{36}	teach_dad_sup_mom$_{HCP}$	X_{46}							
X_{37}	baby_born$_{HCP}$	1							
X_{38}	type_com$_{HCP}$	X_{47}							
X_{39}	type_mom$_{HCP}$	1							
X_{40}	pre_birth_info$_{HCP}$	X_{48}							
X_{41}	$\mathbf{W}_{\text{birth\_dev\_det\_HCP, det\_risk\_level\_HCP}}$	1							
X_{42}	$\mathbf{W}_{\text{det\_risk\_level\_HCP, encourage\_mom\_HCP}}$	1							
X_{43}	$\mathbf{W}_{\text{det\_risk\_level\_HCP, calm\_dad\_HCP}}$	1							
X_{44}	$\mathbf{W}_{\text{irr\_birth\_weight\_HCP, det\_underw\_HCP}}$	1							
X_{45}	$\mathbf{W}_{\text{det\_underw\_HCP, refer\_mom\_ther\_HCP}}$	1							
X_{46}	$\mathbf{W}_{\text{det\_underw\_HCP, teach\_dad\_sup\_mom\_HCP}}$	1							
X_{47}	$\mathbf{W}_{\text{baby\_born\_HCP, type\_com\_HCP}}$	1							
X_{48}	$\mathbf{W}_{\text{type\_mom\_HCP, pre\_birth\_info\_HCP}}$	1							

Table 14 Role matrix for worst Case (a) for **mcfw**—combination function weights

mcfw - combination function weights		1
		Alogistic
X_1	type_mom	1
X_2	artificiality_birth	1
X_3	epidural_used	1
X_4	deliv_baby	1
X_5	baby_born	1
X_6	MPPD	1
X_7	PPPD	1
X_8	kid_devel	1
X_9	pre_birth_info	1
X_{10}	com_sup_hcp	1
X_{11}	fat_ask_sup_com	1
X_{12}	fat_rec_sup_com	1
X_{13}	post_info_com_par	1
X_{14}	refer_therapist	1
X_{15}	mental_sup	1
X_{16}	acknowl_fat	1
X_{17}	age	1
X_{18}	riskiness	1
X_{19}	dilation_mom	1
X_{20}	premature_birth	1
X_{21}	type_com	1
X_{22}	retention	1
X_{23}	engaged	1
X_{24}	stress	1
X_{25}	birth_dev	1
X_{26}	fear_level	1
X_{27}	desire_sup	1
X_{28}	ask_sup	1
X_{29}	birth_dev_det$_{HCP}$	1
X_{30}	det_risk_level$_{HCP}$	1
X_{31}	encourage_mom$_{HCP}$	1
X_{32}	calm_dad$_{HCP}$	1
X_{33}	irr_birth_weight$_{HCP}$	1
X_{34}	det_underw$_{HCP}$	1
X_{35}	refer_mom_ther$_{HCP}$	1
X_{36}	teach_dad_sup_mom$_{HCP}$	1
X_{37}	baby_born$_{HCP}$	1
X_{38}	type_com$_{HCP}$	1
X_{39}	type_mom$_{HCP}$	1
X_{40}	pre_birth_info$_{HCP}$	1
X_{41}	$\mathbf{W}_{birth\_dev\_det\_HCP,\ det\_risk\_level\_HCP}$	1
X_{42}	$\mathbf{W}_{det\_risk\_level\_HCP,\ encourage\_mom\_HCP}$	1
X_{43}	$\mathbf{W}_{det\_risk\_level\_HCP,\ calm\_dad\_HCP}$	1
X_{44}	$\mathbf{W}_{irr\_birth\_weight\_HCP,\ det\_underw\_HCP}$	1
X_{45}	$\mathbf{W}_{det\_underw\_HCP,\ refer\_mom\_ther\_HCP}$	1
X_{46}	$\mathbf{W}_{det\_underw\_HCP,\ teach\_dad\_sup\_mom\_HCP}$	1
X_{47}	$\mathbf{W}_{baby\_born\_HCP,\ type\_com\_HCP}$	1
X_{48}	$\mathbf{W}_{type\_mom\_HCP,\ pre\_birth\_info\_HCP}$	1

Table 15 Role matrix for worst Case (a) for **mcfp**—combination function parameters

mcfp - combination function parameters		1 Alogistic	
		Steepness	Threshold
X_{1}	type_mom	50	0.5
X_{2}	artificiality_birth	50	0.5
X_{3}	epidural_used	50	0.5
X_{4}	deliv_baby	5	0.5
X_{5}	baby_born	50	0.5
X_{6}	MPPD	5	0.3
X_{7}	PPPD	5	0.5
X_{8}	kid_devel	5	0.5
X_{9}	pre_birth_info	50	0.5
X_{10}	com_sup_hcp	0.5	6
X_{11}	fat_ask_sup_com	5	1
X_{12}	fat_rec_sup_com	5	0.5
X_{13}	post_info_com_par	5	1
X_{14}	refer_therapist	5	1
X_{15}	mental_sup	5	5
X_{16}	acknowl_fat	5	0.8
X_{17}	age	5	0.5
X_{18}	riskiness	50	0.5
X_{19}	dilation_mom	2.8	0
X_{20}	premature_birth	5	0.5
X_{21}	type_com	5	0.1
X_{22}	retention	5	0.3
X_{23}	engaged	5	0.5
X_{24}	stress	5	0.5
X_{25}	birth_dev	50	1
X_{26}	fear_level	5	1
X_{27}	desire_sup	5	0.3
X_{28}	ask_sup	5	0.2
X_{29}	birth_dev_det$_{HCP}$	5	1
X_{30}	det_risk_level$_{HCP}$	5	0.5
X_{31}	encourage_mom$_{HCP}$	5	0.5
X_{32}	calm_dad$_{HCP}$	50	0.7
X_{33}	irr_birth_weight$_{HCP}$	5	1
X_{34}	det_underw$_{HCP}$	5	0.5
X_{35}	refer_mom_ther$_{HCP}$	50	0.5
X_{36}	teach_dad_sup_mom$_{HCP}$	50	0.7
X_{37}	baby_born$_{HCP}$	5	0.5
X_{38}	type_com$_{HCP}$	50	0.5
X_{39}	type_mom$_{HCP}$	5	0.5
X_{40}	pre_birth_info$_{HCP}$	50	0.5
X_{41}	$W_{\text{birth\_dev\_det\_HCP, det\_risk\_level\_HCP}}$	5	0.5
X_{42}	$W_{\text{det\_risk\_level\_HCP, encourage\_mom\_HCP}}$	5	0.5
X_{43}	$W_{\text{det\_risk\_level\_HCP, calm\_dad\_HCP}}$	5	0.5
X_{44}	$W_{\text{irr\_birth\_weight\_HCP, det\_underw\_HCP}}$	5	0.5
X_{45}	$W_{\text{det\_underw\_HCP, refer\_mom\_ther\_HCP}}$	5	0.5
X_{46}	$W_{\text{det\_underw\_HCP, teach\_dad\_sup\_mom\_HCP}}$	5	0.5
X_{47}	$W_{\text{baby\_born\_HCP, type\_com\_HCP}}$	5	0.5
X_{48}	$W_{\text{type\_mom\_HCP, pre\_birth\_info\_HCP}}$	5	0.5

Table 16 Role matrix for worst Case (a) for **iv**—initial values

iv - initial values		1
X_{1}	type_mom	0
X_{2}	artificiality_birth	0
X_{3}	epidural_used	0
X_{4}	deliv_baby	0
X_{5}	baby_born	0
X_{6}	MPPD	0.13
X_{7}	PPPD	0
X_{8}	kid_devel	0
X_{9}	pre_birth_info	0
X_{10}	com_sup_hcp	0
X_{11}	fat_ask_sup_com	0
X_{12}	fat_rec_sup_com	0
X_{13}	post_info_com_par	0
X_{14}	refer_therapist	0
X_{15}	mental_sup	0
X_{16}	acknowl_fat	0
X_{17}	age	1
X_{18}	riskiness	0
X_{19}	dilation_mom	0.2
X_{20}	premature_birth	1
X_{21}	type_com	0
X_{22}	retention	0
X_{23}	engaged	0
X_{24}	stress	0
X_{25}	birth_dev	0
X_{26}	fear_level	0
X_{27}	desire_sup	0
X_{28}	ask_sup	0
X_{29}	birth_dev_det$_{HCP}$	0
X_{30}	det_risk_level$_{HCP}$	0
X_{31}	encourage_mom$_{HCP}$	0
X_{32}	calm_dad$_{HCP}$	0
X_{33}	irr_birth_weight$_{HCP}$	0
X_{34}	det_underw$_{HCP}$	0
X_{35}	refer_mom_ther$_{HCP}$	0
X_{36}	teach_dad_sup_mom$_{HCP}$	0
X_{37}	baby_born$_{HCP}$	0
X_{38}	type_com$_{HCP}$	0
X_{39}	type_mom$_{HCP}$	0
X_{40}	pre_birth_info$_{HCP}$	0
X_{41}	$\mathbf{W}_{\text{birth\_dev\_det\_HCP, det\_risk\_level\_HCP}}$	0
X_{42}	$\mathbf{W}_{\text{det\_risk\_level\_HCP, encourage\_mom\_HCP}}$	0
X_{43}	$\mathbf{W}_{\text{det\_risk\_level\_HCP, calm\_dad\_HCP}}$	0
X_{44}	$\mathbf{W}_{\text{irr\_birth\_weight\_HCP, det\_underw\_HCP}}$	0
X_{45}	$\mathbf{W}_{\text{det\_underw\_HCP, refer\_mom\_ther\_HCP}}$	0
X_{46}	$\mathbf{W}_{\text{det\_underw\_HCP, teach\_dad\_sup\_mom\_HCP}}$	0
X_{47}	$\mathbf{W}_{\text{baby\_born\_HCP, type\_com\_HCP}}$	0
X_{48}	$\mathbf{W}_{\text{type\_mom\_HCP, pre\_birth\_info\_HCP}}$	0

Table 17 Role matrix for worst Case (a) for **ms**—speed factors

	ms - speed factors	1
X_1	type_mom	1
X_2	artificiality_birth	1
X_3	epidural_used	1
X_4	deliv_baby	0.2
X_5	baby_born	0.1
X_6	MPPD	0.2
X_7	PPPD	1
X_8	kid_devel	1
X_9	pre_birth_info	0.1
X_{10}	com_sup_hcp	0.1
X_{11}	fat_ask_sup_com	0.3
X_{12}	fat_rec_sup_com	0.1
X_{13}	post_info_com_par	0.1
X_{14}	refer_therapist	0.3
X_{15}	mental_sup	0.1
X_{16}	acknowl_fat	0.1
X_{17}	age	0
X_{18}	riskiness	0.5
X_{19}	dilation_mom	1
X_{20}	premature_birth	0
X_{21}	type_com	1
X_{22}	retention	1
X_{23}	engaged	1
X_{24}	stress	0.4
X_{25}	birth_dev	1
X_{26}	fear_level	1
X_{27}	desire_sup	1
X_{28}	ask_sup	0.2
X_{29}	$\text{birth\_dev\_det}_{\text{HCP}}$	1
X_{30}	$\text{det\_risk\_level}_{\text{HCP}}$	1
X_{31}	$\text{encourage\_mom}_{\text{HCP}}$	0.1
X_{32}	$\text{calm\_dad}_{\text{HCP}}$	0.1
X_{33}	$\text{irr\_birth\_weight}_{\text{HCP}}$	1
X_{34}	$\text{det\_underw}_{\text{HCP}}$	1
X_{35}	$\text{refer\_mom\_ther}_{\text{HCP}}$	0.3
X_{36}	$\text{teach\_dad\_sup\_mom}_{\text{HCP}}$	0.1
X_{37}	$\text{baby\_born}_{\text{HCP}}$	1
X_{38}	$\text{type\_com}_{\text{HCP}}$	1
X_{39}	$\text{type\_mom}_{\text{HCP}}$	1
X_{40}	$\text{pre\_birth\_info}_{\text{HCP}}$	1
X_{41}	$\mathbf{W}_{\text{birth\_dev\_det\_HCP, det\_risk\_level\_HCP}}$	0
X_{42}	$\mathbf{W}_{\text{det\_risk\_level\_HCP, encourage\_mom\_HCP}}$	0
X_{43}	$\mathbf{W}_{\text{det\_risk\_level\_HCP, calm\_dad\_HCP}}$	0
X_{44}	$\mathbf{W}_{\text{irr\_birth\_weight\_HCP, det\_underw\_HCP}}$	0
X_{45}	$\mathbf{W}_{\text{det\_underw\_HCP, refer\_mom\_ther\_HCP}}$	0
X_{46}	$\mathbf{W}_{\text{det\_underw\_HCP, teach\_dad\_sup\_mom\_HCP}}$	0
X_{47}	$\mathbf{W}_{\text{baby\_born\_HCP, type\_com\_HCP}}$	0
X_{48}	$\mathbf{W}_{\text{type\_mom\_HCP, pre\_birth\_info\_HCP}}$	0

Table 18 Role matrix for Case (b) for **mb**—base connectivity

	mb - base connectivity	1	2	3	4	5	6	7	8	9	10	11	12
X_{1}	type_mom	X_{1}											
X_{2}	artificiality_birth	X_{2}	X_{18}										
X_{3}	epidural_used	X_{3}											
X_{4}	deliv_baby	X_{4}	X_{19}										
X_{5}	baby_born	X_{4}											
X_{6}	MPPD	X_{1}	X_{2}	X_{3}	X_{5}	X_{15}							
X_{7}	PPPD	X_{6}	X_{10}	X_{12}	X_{15}	X_{23}	X_{24}						
X_{8}	imp_kid_devel	X_{7}											
X_{9}	pre_birth_info	X_{9}	X_{40}	X_{52}									
X_{10}	com_sup_hcp	X_{4}	X_{9}	X_{11}	X_{24}	X_{30}	X_{32}	X_{34}	X_{36}	X_{42}	X_{44}	X_{46}	X_{48}
X_{11}	fat_ask_sup_com	X_{23}	X_{28}										
X_{12}	fat_rec_sup_com	X_{10}	X_{22}	X_{24}									
X_{13}	post_info_com_par	X_{5}	X_{20}	X_{31}	X_{32}	X_{36}	X_{43}	X_{44}	X_{48}				
X_{14}	refer_therapist	X_{13}	X_{35}	X_{47}									
X_{15}	mental_sup	X_{14}	X_{21}	X_{31}	X_{32}	X_{35}	X_{36}	X_{43}	X_{44}	X_{47}	X_{48}		
X_{16}	acknowl_fat	X_{9}	X_{32}	X_{36}	X_{44}	X_{48}							
X_{17}	age	X_{17}											
X_{18}	riskiness	X_{5}	X_{17}										
X_{19}	dilation_mom	X_{19}											
X_{20}	premature_birth	X_{20}											
X_{21}	type_com	X_{21}	X_{38}	X_{50}									
X_{22}	retention	X_{23}	X_{24}										
X_{23}	engaged	X_{16}											
X_{24}	stress	X_{26}											
X_{25}	birth_dev	X_{4}	X_{18}										
X_{26}	fear_level	X_{23}	X_{24}	X_{25}									
X_{27}	desire_sup	X_{26}											
X_{28}	ask_sup	X_{27}											
X_{29}	birth_dev_det$_{HCP}$	X_{4}	X_{18}										
X_{30}	det_risk_level$_{HCP}$	X_{29}											
X_{31}	encourage_mom$_{HCP}$	X_{30}											
X_{32}	calm_dad$_{HCP}$	X_{11}	X_{30}										
X_{33}	irr_birth_weight$_{HCP}$	X_{5}	X_{20}										
X_{34}	det_underw$_{HCP}$	X_{33}											
X_{35}	refer_mom_ther$_{HCP}$	X_{34}											
X_{36}	teach_dad_sup_mom$_{HCP}$	X_{34}											
X_{37}	baby_born$_{HCP}$	X_{5}											
X_{38}	type_com$_{HCP}$	X_{37}											
X_{39}	type_mom$_{HCP}$	X_{1}											
X_{40}	pre_birth_info$_{HCP}$	X_{39}											
X_{41}	birth_dev_det$_{AIC}$	X_{4}	X_{18}										
X_{42}	det_risk_level$_{AIC}$	X_{41}											
X_{43}	encourage_mom$_{AIC}$	X_{42}											
X_{44}	calm_dad$_{AIC}$	X_{11}	X_{42}										
X_{45}	irr_birth_weight$_{AIC}$	X_{5}	X_{20}										
X_{46}	det_underw$_{AIC}$	X_{45}											
X_{47}	refer_mom_ther$_{AIC}$	X_{46}											
X_{48}	teach_dad_sup_mom$_{AIC}$	X_{46}											
X_{49}	baby_born$_{AIC}$	X_{5}											
X_{50}	type_com$_{AIC}$	X_{49}											
X_{51}	type_mom$_{AIC}$	X_{1}											
X_{52}	pre_birth_info$_{AIC}$	X_{51}											
X_{53}	$\mathbf{W}_{\text{birth\_dev\_det\_HCP, det\_risk\_level\_HCP}}$	X_{53}											
X_{54}	$\mathbf{W}_{\text{det\_risk\_level\_HCP, encourage\_mom\_HCP}}$	X_{54}											
X_{55}	$\mathbf{W}_{\text{det\_risk\_level\_HCP, calm\_dad\_HCP}}$	X_{55}											
X_{56}	$\mathbf{W}_{\text{irr\_birth\_weight\_HCP, det\_underw\_HCP}}$	X_{56}											
X_{57}	$\mathbf{W}_{\text{det\_underw\_HCP, refer\_mom\_ther\_HCP}}$	X_{57}											
X_{58}	$\mathbf{W}_{\text{det\_underw\_HCP, teach\_dad\_sup\_mom\_HCP}}$	X_{58}											
X_{59}	$\mathbf{W}_{\text{baby\_born\_HCP, type\_com\_HCP}}$	X_{59}											
X_{60}	$\mathbf{W}_{\text{type\_mom\_HCP, pre\_birth\_info\_HCP}}$	X_{60}											
X_{61}	$\mathbf{W}_{\text{birth\_dev\_det\_AIC, det\_risk\_level\_AIC}}$	X_{61}											
X_{62}	$\mathbf{W}_{\text{det\_risk\_level\_AIC, encourage\_mom\_AIC}}$	X_{62}											
X_{63}	$\mathbf{W}_{\text{det\_risk\_level\_AIC, calm\_dad\_AIC}}$	X_{63}											
X_{64}	$\mathbf{W}_{\text{irr\_birth\_weight\_AIC, det\_underw\_AIC}}$	X_{64}											
X_{65}	$\mathbf{W}_{\text{det\_underw\_AIC, refer\_mom\_ther\_AIC}}$	X_{65}											
X_{66}	$\mathbf{W}_{\text{det\_underw\_AIC, teach\_dad\_sup\_mom\_AIC}}$	X_{66}											
X_{67}	$\mathbf{W}_{\text{baby\_born\_AIC, type\_com\_AIC}}$	X_{67}											
X_{68}	$\mathbf{W}_{\text{type\_mom\_AIC, pre\_birth\_info\_AIC}}$	X_{68}											
X_{69}	$\mathbf{H}_{\text{encourage\_mom\_AIC}}$	X_{75}											
X_{70}	$\mathbf{H}_{\text{calm\_dad\_AIC}}$	X_{75}											
X_{71}	$\mathbf{H}_{\text{refer\_mom\_ther\_AIC}}$	X_{75}											
X_{72}	$\mathbf{H}_{\text{teach\_dad\_sup\_mom\_AIC}}$	X_{75}											
X_{73}	$\mathbf{H}_{\text{type\_com\_AIC}}$	X_{75}											
X_{74}	$\mathbf{H}_{\text{pre\_birth\_info\_AIC}}$	X_{75}											
X_{75}	con_act_AIC	X_{75}											

Table 19 Role matrix for Case (b) for **mcw**—connection weights

	mcw - connection weights	1	2	3	4	5	6	7	8	9	10	11	12
X_{1}	type_mom	1											
X_{2}	artificiality_birth	1	-1										
X_{3}	epidural_used	1											
X_{4}	deliv_baby	1	1										
X_{5}	baby_born	1											
X_{6}	MPPD	0.25	0.25	0.25	0.25	-1							
X_{7}	PPPD	1	-0.5	-0.5	-0.5	-0.5	1						
X_{8}	kid_devel	1											
X_{9}	pre_birth_info	1	1	1									
X_{10}	com_sup_hcp	1	1	1	-1	1	1	1	1	1	1	1	1
X_{11}	fat_ask_sup_com	1	1										
X_{12}	fat_rec_sup_com	1	1	-1									
X_{13}	post_info_com_par	1	1	1	1	1	1	1	1				
X_{14}	refer_therapist	1	1	1									
X_{15}	mental_sup	1	1	1	1	1	1	1	1	1	1		
X_{16}	acknowl_fat	1	1	1	1	1							
X_{17}	age	1											
X_{18}	riskiness	-1	1										
X_{19}	dilation_mom	1											
X_{20}	premature_birth	1											
X_{21}	type_com	1	1	1									
X_{22}	retention	2	-1										
X_{23}	engaged	1											
X_{24}	stress	1											
X_{25}	birth_dev	1	1										
X_{26}	fear_level	-1	1	1									
X_{27}	desire_sup	1											
X_{28}	ask_sup	1											
X_{29}	birth_dev_det$_{HCP}$	1	1										
X_{30}	det_risk_level$_{HCP}$	X_{53}											
X_{31}	encourage_mom$_{HCP}$	X_{54}											
X_{32}	calm_dad$_{HCP}$	1	X_{55}										
X_{33}	irr_birth_weight$_{HCP}$	1	2										
X_{34}	det_underw$_{HCP}$	X_{56}											
X_{35}	refer_mom_ther$_{HCP}$	X_{57}											
X_{36}	teach_dad_sup_mom$_{HCP}$	X_{58}											
X_{37}	baby_born$_{HCP}$	1											
X_{38}	type_com$_{HCP}$	X_{59}											
X_{39}	type_mom$_{HCP}$	1											
X_{40}	pre_birth_info$_{HCP}$	X_{60}											
X_{41}	birth_dev_det$_{AIC}$	1	1										
X_{42}	det_risk_level$_{AIC}$	X_{61}											
X_{43}	encourage_mom$_{AIC}$	X_{62}											
X_{44}	calm_dad$_{AIC}$	1	X_{63}										
X_{45}	irr_birth_weight$_{AIC}$	1	2										
X_{46}	det_underw$_{AIC}$	X_{64}											
X_{47}	refer_mom_ther$_{AIC}$	X_{65}											
X_{48}	teach_dad_sup_mom$_{AIC}$	X_{66}											
X_{49}	baby_born$_{AIC}$	1											
X_{50}	type_com$_{AIC}$	X_{67}											
X_{51}	type_mom$_{AIC}$	1											
X_{52}	pre_birth_info$_{AIC}$	X_{68}											
X_{53}	$\mathbf{W}_{\text{birth\_dev\_det\_HCP, det\_risk\_level\_HCP}}$	1											
X_{54}	$\mathbf{W}_{\text{det\_risk\_level\_HCP, encourage\_mom\_HCP}}$	1											
X_{55}	$\mathbf{W}_{\text{det\_risk\_level\_HCP, calm\_dad\_HCP}}$	1											
X_{56}	$\mathbf{W}_{\text{irr\_birth\_weight\_HCP, det\_underw\_HCP}}$	1											
X_{57}	$\mathbf{W}_{\text{det\_underw\_HCP, refer\_mom\_ther\_HCP}}$	1											
X_{58}	$\mathbf{W}_{\text{det\_underw\_HCP, teach\_dad\_sup\_mom\_HCP}}$	1											
X_{59}	$\mathbf{W}_{\text{baby\_born\_HCP, type\_com\_HCP}}$	1											
X_{60}	$\mathbf{W}_{\text{type\_mom\_HCP, pre\_birth\_info\_HCP}}$	1											
X_{61}	$\mathbf{W}_{\text{birth\_dev\_det\_AIC, det\_risk\_level\_AIC}}$	1											
X_{62}	$\mathbf{W}_{\text{det\_risk\_level\_AIC, encourage\_mom\_AIC}}$	1											
X_{63}	$\mathbf{W}_{\text{det\_risk\_level\_AIC, calm\_dad\_AIC}}$	1											
X_{64}	$\mathbf{W}_{\text{irr\_birth\_weight\_AIC, det\_underw\_AIC}}$	1											
X_{65}	$\mathbf{W}_{\text{det\_underw\_AIC, refer\_mom\_ther\_AIC}}$	1											
X_{66}	$\mathbf{W}_{\text{det\_underw\_AIC, teach\_dad\_sup\_mom\_AIC}}$	1											
X_{67}	$\mathbf{W}_{\text{baby\_born\_AIC, type\_com\_AIC}}$	1											
X_{68}	$\mathbf{W}_{\text{type\_mom\_AIC, pre\_birth\_info\_AIC}}$	1											
X_{69}	$\mathbf{H}_{\text{encourage\_mom\_AIC}}$	1											
X_{70}	$\mathbf{H}_{\text{calm\_dad\_AIC}}$	1											
X_{71}	$\mathbf{H}_{\text{refer\_mom\_ther\_AIC}}$	1											
X_{72}	$\mathbf{H}_{\text{teach\_dad\_sup\_mom\_AIC}}$	1											
X_{73}	$\mathbf{H}_{\text{type\_com\_AIC}}$	1											
X_{74}	$\mathbf{H}_{\text{pre\_birth\_info\_AIC}}$	1											
X_{75}	con_act_AIC	1											

Table 20 Role matrix for Case (b) for **mcfw**—combination function weights

mcfw - combination function weights		1	2
		Alogistic	Steponce
X_{1}	type_mom	1	
X_{2}	artificiality_birth	1	
X_{3}	epidural_used	1	
X_{4}	deliv_baby	1	
X_{5}	baby_born	1	
X_{6}	MPPD	1	
X_{7}	PPPD	1	
X_{8}	kid_devel	1	
X_{9}	pre_birth_info	1	
X_{10}	com_sup_hcp	1	
X_{11}	fat_ask_sup_com	1	
X_{12}	fat_rec_sup_com	1	
X_{13}	post_info_com_par	1	
X_{14}	refer_therapist	1	
X_{15}	mental_sup	1	
X_{16}	acknowl_fat	1	
X_{17}	age	1	
X_{18}	riskiness	1	
X_{19}	dilation_mom	1	
X_{20}	premature_birth	1	
X_{21}	type_com	1	
X_{22}	retention	1	
X_{23}	engaged	1	
X_{24}	stress	1	
X_{25}	birth_dev	1	
X_{26}	fear_level	1	
X_{27}	desire_sup	1	
X_{28}	ask_sup	1	
X_{29}	birth_dev_det$_{HCP}$	1	
X_{30}	det_risk_level$_{HCP}$	1	
X_{31}	encourage_mom$_{HCP}$	1	
X_{32}	calm_dad$_{HCP}$	1	
X_{33}	irr_birth_weight$_{HCP}$	1	
X_{34}	det_underw$_{HCP}$	1	
X_{35}	refer_mom_ther$_{HCP}$	1	
X_{36}	teach_dad_sup_mom$_{HCP}$	1	
X_{37}	baby_born$_{HCP}$	1	
X_{38}	type_com$_{HCP}$	1	
X_{39}	type_mom$_{HCP}$	1	
X_{40}	pre_birth_info$_{HCP}$	1	
X_{41}	birth_dev_det$_{AIC}$	1	
X_{42}	det_risk_level$_{AIC}$	1	
X_{43}	encourage_mom$_{AIC}$	1	
X_{44}	calm_dad$_{AIC}$	1	
X_{45}	irr_birth_weight$_{AIC}$	1	
X_{46}	det_underw$_{AIC}$	1	
X_{47}	refer_mom_ther$_{AIC}$	1	
X_{48}	teach_dad_sup_mom$_{AIC}$	1	
X_{49}	baby_born$_{AIC}$	1	
X_{50}	type_com$_{AIC}$	1	
X_{51}	type_mom$_{AIC}$	1	
X_{52}	pre_birth_info$_{AIC}$	1	
X_{53}	$\mathbf{W}_{\text{birth\_dev\_det\_HCP, det\_risk\_level\_HCP}}$	1	
X_{54}	$\mathbf{W}_{\text{det\_risk\_level\_HCP, encourage\_mom\_HCP}}$	1	
X_{55}	$\mathbf{W}_{\text{det\_risk\_level\_HCP, calm\_dad\_HCP}}$	1	
X_{56}	$\mathbf{W}_{\text{irr\_birth\_weight\_HCP, det\_underw\_HCP}}$	1	
X_{57}	$\mathbf{W}_{\text{det\_underw\_HCP, refer\_mom\_ther\_HCP}}$	1	
X_{58}	$\mathbf{W}_{\text{det\_underw\_HCP, teach\_dad\_sup\_mom\_HCP}}$	1	
X_{59}	$\mathbf{W}_{\text{baby\_born\_HCP, type\_com\_HCP}}$	1	
X_{60}	$\mathbf{W}_{\text{type\_mom\_HCP, pre\_birth\_info\_HCP}}$	1	
X_{61}	$\mathbf{W}_{\text{birth\_dev\_det\_AIC, det\_risk\_level\_AIC}}$	1	
X_{62}	$\mathbf{W}_{\text{det\_risk\_level\_AIC, encourage\_mom\_AIC}}$	1	
X_{63}	$\mathbf{W}_{\text{det\_risk\_level\_AIC, calm\_dad\_AIC}}$	1	
X_{64}	$\mathbf{W}_{\text{irr\_birth\_weight\_AIC, det\_underw\_AIC}}$	1	
X_{65}	$\mathbf{W}_{\text{det\_underw\_AIC, refer\_mom\_ther\_AIC}}$	1	
X_{66}	$\mathbf{W}_{\text{det\_underw\_AIC, teach\_dad\_sup\_mom\_AIC}}$	1	
X_{67}	$\mathbf{W}_{\text{baby\_born\_AIC, type\_com\_AIC}}$	1	
X_{68}	$\mathbf{W}_{\text{type\_mom\_AIC, pre\_birth\_info\_AIC}}$	1	
X_{69}	$\mathbf{H}_{\text{encourage\_mom\_AIC}}$	1	
X_{70}	$\mathbf{H}_{\text{calm\_dad\_AIC}}$	1	
X_{71}	$\mathbf{H}_{\text{refer\_mom\_ther\_AIC}}$	1	
X_{72}	$\mathbf{H}_{\text{teach\_dad\_sup\_mom\_AIC}}$	1	
X_{73}	$\mathbf{H}_{\text{type\_com\_AIC}}$	1	
X_{74}	$\mathbf{H}_{\text{pre\_birth\_info\_AIC}}$	1	
X_{75}	con_act_AIC		1

Table 21 Role matrix for Case (b) for **mcfp**—combination function parameters

mcfp - combination function parameters		1 Alogistic		2 Steponce	
		Steepness	Threshold	Start Time	End Time
X_1	type_mom	50	0.5		
X_2	artificiality_birth	50	0.5		
X_3	epidural_used	50	0.5		
X_4	deliv_baby	5	0.5		
X_5	baby_born	50	0.5		
X_6	MPPD	5	0.3		
X_7	PPPD	5	0.5		
X_8	kid_devel	5	0.5		
X_9	pre_birth_info	50	0.5		
X_{10}	com_sup_hcp	0.5	6		
X_{11}	fat_ask_sup_com	5	1		
X_{12}	fat_rec_sup_com	5	0.5		
X_{13}	post_info_com_par	5	4		
X_{14}	refer_therapist	5	1		
X_{15}	mental_sup	5	5		
X_{16}	acknowl_fat	5	0.8		
X_{17}	age	5	0.5		
X_{18}	riskiness	50	0.5		
X_{19}	dilation_mom	2.8	0		
X_{20}	premature_birth	5	0.5		
X_{21}	type_com	5	0.1		
X_{22}	retention	5	0.3		
X_{23}	engaged	5	0.5		
X_{24}	stress	5	0.5		
X_{25}	birth_dev	50	1		
X_{26}	fear_level	5	1		
X_{27}	desire_sup	5	0.3		
X_{28}	ask_sup	5	0.2		
X_{29}	birth_dev_det$_{HCP}$	5	1		
X_{30}	det_risk_level$_{HCP}$	5	0.5		
X_{31}	encourage_mom$_{HCP}$	5	0.5		
X_{32}	calm_dad$_{HCP}$	50	0.7		
X_{33}	irr_birth_weight$_{HCP}$	5	1		
X_{34}	det_underw$_{HCP}$	5	0.5		
X_{35}	refer_mom_ther$_{HCP}$	50	0.5		
X_{36}	teach_dad_sup_mom$_{HCP}$	50	0.7		
X_{37}	baby_born$_{HCP}$	5	0.5		
X_{38}	type_com$_{HCP}$	50	0.5		
X_{39}	type_mom$_{HCP}$	5	0.5		
X_{40}	pre_birth_info$_{HCP}$	50	0.5		
X_{41}	birth_dev_det$_{AIC}$	5	1		
X_{42}	det_risk_level$_{AIC}$	5	0.5		
X_{43}	encourage_mom$_{AIC}$	5	0.5		
X_{44}	calm_dad$_{AIC}$	50	0.7		
X_{45}	irr_birth_weight$_{AIC}$	5	1		
X_{46}	det_underw$_{AIC}$	5	0.5		
X_{47}	refer_mom_ther$_{AIC}$	50	0.5		
X_{48}	teach_dad_sup_mom$_{AIC}$	50	0.7		
X_{49}	baby_born$_{AIC}$	5	0.5		
X_{50}	type_com$_{AIC}$	50	0.5		
X_{51}	type_mom$_{AIC}$	5	0.5		
X_{52}	pre_birth_info$_{AIC}$	50	0.5		
X_{53}	$\mathbf{W}_{\text{birth\_dev\_det\_HCP, det\_risk\_level\_HCP}}$	5	0.5		
X_{54}	$\mathbf{W}_{\text{det\_risk\_level\_HCP, encourage\_mom\_HCP}}$	5	0.5		
X_{55}	$\mathbf{W}_{\text{det\_risk\_level\_HCP, calm\_dad\_HCP}}$	5	0.5		
X_{56}	$\mathbf{W}_{\text{irr\_birth\_weight\_HCP, det\_underw\_HCP}}$	5	0.5		
X_{57}	$\mathbf{W}_{\text{det\_underw\_HCP, refer\_mom\_ther\_HCP}}$	5	0.5		
X_{58}	$\mathbf{W}_{\text{det\_underw\_HCP, teach\_dad\_sup\_mom\_HCP}}$	5	0.5		
X_{59}	$\mathbf{W}_{\text{baby\_born\_HCP, type\_com\_HCP}}$	5	0.5		
X_{60}	$\mathbf{W}_{\text{type\_mom\_HCP, pre\_birth\_info\_HCP}}$	5	0.5		
X_{61}	$\mathbf{W}_{\text{birth\_dev\_det\_AIC, det\_risk\_level\_AIC}}$	5	0.5		
X_{62}	$\mathbf{W}_{\text{det\_risk\_level\_AIC, encourage\_mom\_AIC}}$	5	0.5		
X_{63}	$\mathbf{W}_{\text{det\_risk\_level\_AIC, calm\_dad\_AIC}}$	5	0.5		
X_{64}	$\mathbf{W}_{\text{irr\_birth\_weight\_AIC, det\_underw\_AIC}}$	5	0.5		
X_{65}	$\mathbf{W}_{\text{det\_underw\_AIC, refer\_mom\_ther\_AIC}}$	5	0.5		
X_{66}	$\mathbf{W}_{\text{det\_underw\_AIC, teach\_dad\_sup\_mom\_AIC}}$	5	0.5		
X_{67}	$\mathbf{W}_{\text{baby\_born\_AIC, type\_com\_AIC}}$	5	0.5		
X_{68}	$\mathbf{W}_{\text{type\_mom\_AIC, pre\_birth\_info\_AIC}}$	5	0.5		
X_{69}	$\mathbf{H}_{\text{encourage\_mom\_AIC}}$	5	0.5		
X_{70}	$\mathbf{H}_{\text{calm\_dad\_AIC}}$	5	0.5		
X_{71}	$\mathbf{H}_{\text{refer\_mom\_ther\_AIC}}$	5	0.5		
X_{72}	$\mathbf{H}_{\text{teach\_dad\_sup\_mom\_AIC}}$	5	0.5		
X_{73}	$\mathbf{H}_{\text{type\_com\_AIC}}$	5	0.5		
X_{74}	$\mathbf{H}_{\text{pre\_birth\_info\_AIC}}$	5	0.5		
X_{75}	con_act_AIC			20	60

Table 22 Role matrix for Case (b) for **ms**—speed factors

	ms - speed factors	1
X_{1}	type_mom	1
X_{2}	artificiality_birth	1
X_{3}	epidural_used	1
X_{4}	deliv_baby	0.2
X_{5}	baby_born	0.1
X_{6}	MPPD	0.2
X_{7}	PPPD	1
X_{8}	kid_devel	1
X_{9}	pre_birth_info	0.1
X_{10}	com_sup_hcp	0.1
X_{11}	fat_ask_sup_com	0.3
X_{12}	fat_rec_sup_com	0.1
X_{13}	post_info_com_par	0.1
X_{14}	refer_therapist	0.3
X_{15}	mental_sup	0.1
X_{16}	acknowl_fat	0.1
X_{17}	age	0
X_{18}	riskiness	0.5
X_{19}	dilation_mom	1
X_{20}	premature_birth	0
X_{21}	type_com	1
X_{22}	retention	1
X_{23}	engaged	1
X_{24}	stress	0.4
X_{25}	birth_dev	1
X_{26}	fear_level	1
X_{27}	desire_sup	1
X_{28}	ask_sup	0.2
X_{29}	birth_dev_det$_{HCP}$	1
X_{30}	det_risk_level$_{HCP}$	1
X_{31}	encourage_mom$_{HCP}$	0.1
X_{32}	calm_dad$_{HCP}$	0.1
X_{33}	irr_birth_weight$_{HCP}$	1
X_{34}	det_underw$_{HCP}$	1
X_{35}	refer_mom_ther$_{HCP}$	0.3
X_{36}	teach_dad_sup_mom$_{HCP}$	0.1
X_{37}	baby_born$_{HCP}$	1
X_{38}	type_com$_{HCP}$	1
X_{39}	type_mom$_{HCP}$	1
X_{40}	pre_birth_info$_{HCP}$	1
X_{41}	birth_dev_det$_{AIC}$	1
X_{42}	det_risk_level$_{AIC}$	1
X_{43}	encourage_mom$_{AIC}$	X_{69}
X_{44}	calm_dad$_{AIC}$	X_{70}
X_{45}	irr_birth_weight$_{AIC}$	1
X_{46}	det_underw$_{AIC}$	1
X_{47}	refer_mom_ther$_{AIC}$	X_{71}
X_{48}	teach_dad_sup_mom$_{AIC}$	X_{72}
X_{49}	baby_born$_{AIC}$	1
X_{50}	type_com$_{AIC}$	X_{73}
X_{51}	type_mom$_{AIC}$	1
X_{52}	pre_birth_info$_{AIC}$	X_{74}
X_{53}	$W_{\text{birth\_dev\_det\_HCP, det\_risk\_level\_HCP}}$	0
X_{54}	$W_{\text{det\_risk\_level\_HCP, encourage\_mom\_HCP}}$	0
X_{55}	$W_{\text{det\_risk\_level\_HCP, calm\_dad\_HCP}}$	0
X_{56}	$W_{\text{irr\_birth\_weight\_HCP, det\_underw\_HCP}}$	0
X_{57}	$W_{\text{det\_underw\_HCP, refer\_mom\_ther\_HCP}}$	0
X_{58}	$W_{\text{det\_underw\_HCP, teach\_dad\_sup\_mom\_HCP}}$	0
X_{59}	$W_{\text{baby\_born\_HCP, type\_com\_HCP}}$	0
X_{60}	$W_{\text{type\_mom\_HCP, pre\_birth\_info\_HCP}}$	0
X_{61}	$W_{\text{birth\_dev\_det\_AIC, det\_risk\_level\_AIC}}$	0
X_{62}	$W_{\text{det\_risk\_level\_AIC, encourage\_mom\_AIC}}$	0
X_{63}	$W_{\text{det\_risk\_level\_AIC, calm\_dad\_AIC}}$	0
X_{64}	$W_{\text{irr\_birth\_weight\_AIC, det\_underw\_AIC}}$	0
X_{65}	$W_{\text{det\_underw\_AIC, refer\_mom\_ther\_AIC}}$	0
X_{66}	$W_{\text{det\_underw\_AIC, teach\_dad\_sup\_mom\_AIC}}$	0
X_{67}	$W_{\text{baby\_born\_AIC, type\_com\_AIC}}$	0
X_{68}	$W_{\text{type\_mom\_AIC, pre\_birth\_info\_AIC}}$	0
X_{69}	$H_{\text{encourage\_mom\_AIC}}$	0.1
X_{70}	$H_{\text{calm\_dad\_AIC}}$	0.1
X_{71}	$H_{\text{refer\_mom\_ther\_AIC}}$	0.1
X_{72}	$H_{\text{teach\_dad\_sup\_mom\_AIC}}$	0.1
X_{73}	$H_{\text{type\_com\_AIC}}$	0.1
X_{74}	$H_{\text{pre\_birth\_info\_AIC}}$	0.1
X_{75}	con_act_AIC	1

Table 23 Role matrix for Case (b) for **iv**—initial values

iv - initial values		1
X_1	type_mom	0
X_2	artificiality_birth	0
X_3	epidural_used	0
X_4	deliv_baby	0
X_5	baby_born	0
X_6	MPPD	0.13
X_7	PPPD	0
X_8	kid_devel	0
X_9	pre_birth_info	0
X_{10}	com_sup_hcp	0
X_{11}	fat_ask_sup_com	0
X_{12}	fat_rec_sup_com	0
X_{13}	post_info_com_par	0
X_{14}	refer_therapist	0
X_{15}	mental_sup	0
X_{16}	acknowl_fat	0
X_{17}	age	1
X_{18}	riskiness	0
X_{19}	dilation_mom	0.2
X_{20}	premature_birth	1
X_{21}	type_com	0
X_{22}	retention	0
X_{23}	engaged	0
X_{24}	stress	0
X_{25}	birth_dev	0
X_{26}	fear_level	0
X_{27}	desire_sup	0
X_{28}	ask_sup	0
X_{29}	$\text{birth\_dev\_det}_{HCP}$	0
X_{30}	$\text{det\_risk\_level}_{HCP}$	0
X_{31}	$\text{encourage\_mom}_{HCP}$	0
X_{32}	calm_dad_{HCP}	0
X_{33}	$\text{irr\_birth\_weight}_{HCP}$	0
X_{34}	det_underw_{HCP}	0
X_{35}	$\text{refer\_mom\_ther}_{HCP}$	0
X_{36}	$\text{teach\_dad\_sup\_mom}_{HCP}$	0
X_{37}	baby_born_{HCP}	0
X_{38}	type_com_{HCP}	0
X_{39}	type_mom_{HCP}	0
X_{40}	$\text{pre\_birth\_info}_{HCP}$	0
X_{41}	$\text{birth\_dev\_det}_{AIC}$	0
X_{42}	$\text{det\_risk\_level}_{AIC}$	0
X_{43}	$\text{encourage\_mom}_{AIC}$	0
X_{44}	calm_dad_{AIC}	0
X_{45}	$\text{irr\_birth\_weight}_{AIC}$	0
X_{46}	det_underw_{AIC}	0
X_{47}	$\text{refer\_mom\_ther}_{AIC}$	0
X_{48}	$\text{teach\_dad\_sup\_mom}_{AIC}$	0
X_{49}	baby_born_{AIC}	0
X_{50}	type_com_{AIC}	0
X_{51}	type_mom_{AIC}	0
X_{52}	$\text{pre\_birth\_info}_{AIC}$	0
X_{53}	$\mathbf{W}_{\text{birth\_dev\_det\_HCP, det\_risk\_level\_HCP}}$	0
X_{54}	$\mathbf{W}_{\text{det\_risk\_level\_HCP, encourage\_mom\_HCP}}$	0
X_{55}	$\mathbf{W}_{\text{det\_risk\_level\_HCP, calm\_dad\_HCP}}$	0
X_{56}	$\mathbf{W}_{\text{irr\_birth\_weight\_HCP, det\_underw\_HCP}}$	0
X_{57}	$\mathbf{W}_{\text{det\_underw\_HCP, refer\_mom\_ther\_HCP}}$	0
X_{58}	$\mathbf{W}_{\text{det\_underw\_HCP, teach\_dad\_sup\_mom\_HCP}}$	0
X_{59}	$\mathbf{W}_{\text{baby\_born\_HCP, type\_com\_HCP}}$	0
X_{60}	$\mathbf{W}_{\text{type\_mom\_HCP, pre\_birth\_info\_HCP}}$	0
X_{61}	$\mathbf{W}_{\text{birth\_dev\_det\_AIC, det\_risk\_level\_AIC}}$	1
X_{62}	$\mathbf{W}_{\text{det\_risk\_level\_AIC, encourage\_mom\_AIC}}$	1
X_{63}	$\mathbf{W}_{\text{det\_risk\_level\_AIC, calm\_dad\_AIC}}$	1
X_{64}	$\mathbf{W}_{\text{irr\_birth\_weight\_AIC, det\_underw\_AIC}}$	1
X_{65}	$\mathbf{W}_{\text{det\_underw\_AIC, refer\_mom\_ther\_AIC}}$	1
X_{66}	$\mathbf{W}_{\text{det\_underw\_AIC, teach\_dad\_sup\_mom\_AIC}}$	1
X_{67}	$\mathbf{W}_{\text{baby\_born\_AIC, type\_com\_AIC}}$	1
X_{68}	$\mathbf{W}_{\text{type\_mom\_AIC, pre\_birth\_info\_AIC}}$	1
X_{69}	$\mathbf{H}_{\text{encourage\_mom\_AIC}}$	0
X_{70}	$\mathbf{H}_{\text{calm\_dad\_AIC}}$	0
X_{71}	$\mathbf{H}_{\text{refer\_mom\_ther\_AIC}}$	0
X_{72}	$\mathbf{H}_{\text{teach\_dad\_sup\_mom\_AIC}}$	0
X_{73}	$\mathbf{H}_{\text{type\_com\_AIC}}$	0
X_{74}	$\mathbf{H}_{\text{pre\_birth\_info\_AIC}}$	0
X_{75}	con_act_AIC	1

Table 24 Role matrix for Case (c) for **mb**—base connectivity

	mb - base connectivity	1	2	3	4	5	6	7	8	9	10	11	12
X_{1}	type_mom	X_{1}											
X_{2}	artificiality_birth	X_{2}	X_{18}										
X_{3}	epidural_used	X_{3}											
X_{4}	deliv_baby	X_{4}	X_{19}										
X_{5}	baby_born	X_{4}											
X_{6}	MPPD	X_{1}	X_{2}	X_{3}	X_{5}	X_{15}							
X_{7}	PPPD	X_{6}	X_{10}	X_{12}	X_{15}	X_{23}	X_{24}						
X_{8}	imp_kid_devel	X_{7}											
X_{9}	pre_birth_info	X_{9}	X_{40}	X_{52}									
X_{10}	com_sup_hcp	X_{4}	X_{9}	X_{11}	X_{24}	X_{30}	X_{32}	X_{34}	X_{36}	X_{42}	X_{44}	X_{46}	X_{48}
X_{11}	fat_ask_sup_com	X_{23}	X_{28}										
X_{12}	fat_rec_sup_com	X_{10}	X_{22}	X_{24}									
X_{13}	post_info_com_par	X_{5}	X_{20}	X_{31}	X_{32}	X_{36}	X_{43}	X_{44}	X_{48}				
X_{14}	refer_therapist	X_{13}	X_{35}	X_{47}									
X_{15}	mental_sup	X_{14}	X_{21}	X_{31}	X_{32}	X_{35}	X_{36}	X_{43}	X_{44}	X_{47}	X_{48}		
X_{16}	acknowl_fat	X_{9}	X_{32}	X_{36}	X_{44}	X_{48}							
X_{17}	age	X_{17}											
X_{18}	riskiness	X_{5}	X_{17}										
X_{19}	dilation_mom	X_{19}											
X_{20}	premature_birth	X_{20}											
X_{21}	type_com	X_{21}	X_{38}	X_{50}									
X_{22}	retention	X_{23}	X_{24}										
X_{23}	engaged	X_{16}											
X_{24}	stress	X_{26}											
X_{25}	birth_dev	X_{4}	X_{18}										
X_{26}	fear_level	X_{23}	X_{24}	X_{25}									
X_{27}	desire_sup	X_{26}											
X_{28}	ask_sup	X_{27}											
X_{29}	birth_dev_det$_{HCP}$	X_{4}	X_{18}										
X_{30}	det_risk_level$_{HCP}$	X_{29}											
X_{31}	encourage_mom$_{HCP}$	X_{30}											
X_{32}	calm_dad$_{HCP}$	X_{11}	X_{30}										
X_{33}	irr_birth_weight$_{HCP}$	X_{5}	X_{20}										
X_{34}	det_underw$_{HCP}$	X_{33}											
X_{35}	refer_mom_ther$_{HCP}$	X_{34}											
X_{36}	teach_dad_sup_mom$_{HCP}$	X_{34}											
X_{37}	baby_born$_{HCP}$	X_{5}											
X_{38}	type_com$_{HCP}$	X_{37}											
X_{39}	type_mom$_{HCP}$	X_{1}											
X_{40}	pre_birth_info$_{HCP}$	X_{39}											
X_{41}	birth_dev_det$_{AIC}$	X_{4}	X_{18}										
X_{42}	det_risk_level$_{AIC}$	X_{41}											
X_{43}	encourage_mom$_{AIC}$	X_{42}											
X_{44}	calm_dad$_{AIC}$	X_{11}	X_{42}										
X_{45}	irr_birth_weight$_{AIC}$	X_{5}	X_{20}										
X_{46}	det_underw$_{AIC}$	X_{45}											
X_{47}	refer_mom_ther$_{AIC}$	X_{46}											
X_{48}	teach_dad_sup_mom$_{AIC}$	X_{46}											
X_{49}	baby_born$_{AIC}$	X_{5}											
X_{50}	type_com$_{AIC}$	X_{49}											
X_{51}	type_mom$_{AIC}$	X_{1}											
X_{52}	pre_birth_info$_{AIC}$	X_{51}											
X_{53}	$W_{\text{birth\_dev\_det\_HCP, det\_risk\_level\_HCP}}$	X_{53}											
X_{54}	$W_{\text{det\_risk\_level\_HCP, encourage\_mom\_HCP}}$	X_{54}											
X_{55}	$W_{\text{det\_risk\_level\_HCP, calm\_dad\_HCP}}$	X_{55}											
X_{56}	$W_{\text{irr\_birth\_weight\_HCP, det\_underw\_HCP}}$	X_{56}											
X_{57}	$W_{\text{det\_underw\_HCP, refer\_mom\_ther\_HCP}}$	X_{57}											
X_{58}	$W_{\text{det\_underw\_HCP, teach\_dad\_sup\_mom\_HCP}}$	X_{58}											
X_{59}	$W_{\text{baby\_born\_HCP, type\_com\_HCP}}$	X_{59}											
X_{60}	$W_{\text{type\_mom\_HCP, pre\_birth\_info\_HCP}}$	X_{60}											
X_{61}	$W_{\text{birth\_dev\_det\_AIC, det\_risk\_level\_AIC}}$	X_{61}											
X_{62}	$W_{\text{det\_risk\_level\_AIC, encourage\_mom\_AIC}}$	X_{62}											
X_{63}	$W_{\text{det\_risk\_level\_AIC, calm\_dad\_AIC}}$	X_{63}											
X_{64}	$W_{\text{irr\_birth\_weight\_AIC, det\_underw\_AIC}}$	X_{64}											
X_{65}	$W_{\text{det\_underw\_AIC, refer\_mom\_ther\_AIC}}$	X_{65}											
X_{66}	$W_{\text{det\_underw\_AIC, teach\_dad\_sup\_mom\_AIC}}$	X_{66}											
X_{67}	$W_{\text{baby\_born\_AIC, type\_com\_AIC}}$	X_{67}											
X_{68}	$W_{\text{type\_mom\_AIC, pre\_birth\_info\_AIC}}$	X_{68}											
X_{69}	$W_{\text{pre\_birth\_info\_AIC, pre\_birth\_info}}$	X_{83}											
X_{70}	$W_{\text{calm\_dad\_AIC, acknowl\_fat}}$	X_{85}											
X_{71}	$W_{\text{teach\_dad\_sup\_mom\_AIC, acknowl\_fat}}$	X_{84}											
X_{72}	$W_{\text{calm\_dad\_AIC, com\_sup\_hcp}}$	X_{86}											
X_{73}	$W_{\text{teach\_dad\_sup\_mom\_AIC, com\_sup\_hcp}}$	X_{87}											
X_{74}	$W_{\text{calm\_dad\_AIC, post\_info\_com\_par}}$	X_{89}											
X_{75}	$W_{\text{teach\_dad\_sup\_mom\_AIC, post\_info\_com\_par}}$	X_{90}											
X_{76}	$W_{\text{encourage\_mom\_AIC, post\_info\_com\_par}}$	X_{88}											
X_{77}	$W_{\text{refer\_mom\_ther\_AIC, refer\_therapist}}$	X_{91}											
X_{78}	$W_{\text{calm\_dad\_AIC, mental\_sup}}$	X_{94}											
X_{79}	$W_{\text{teach\_dad\_sup\_mom\_AIC, mental\_sup}}$	X_{95}											
X_{80}	$W_{\text{encourage\_mom\_AIC, mental\_sup}}$	X_{92}											
X_{81}	$W_{\text{refer\_mom\_ther\_AIC, mental\_sup}}$	X_{93}											
X_{82}	$W_{\text{type\_com\_AIC, type\_com}}$	X_{96}											
X_{83}	monitor_pre_birth_info	X_{52}	X_{9}										
X_{84}	monitor_acknowl_fat$_{1}$	X_{48}	X_{16}										
X_{85}	monitor_acknowl_fat$_{2}$	X_{44}	X_{16}										
X_{86}	monitor_com_sup_hcp$_{1}$	X_{44}	X_{10}										
X_{87}	monitor_com_sup_hcp$_{2}$	X_{48}	X_{10}										
X_{88}	monitor_post_info_com_par$_{1}$	X_{43}	X_{13}										
X_{89}	monitor_post_info_com_par$_{2}$	X_{44}	X_{13}										
X_{90}	monitor_post_info_com_par$_{3}$	X_{48}	X_{13}										
X_{91}	monitor_refer_therapist	X_{47}	X_{14}										
X_{92}	monitor_mental_sup$_{1}$	X_{43}	X_{15}										
X_{93}	monitor_mental_sup$_{2}$	X_{47}	X_{15}										
X_{94}	monitor_mental_sup$_{3}$	X_{44}	X_{15}										
X_{95}	monitor_mental_sup$_{4}$	X_{48}	X_{15}										
X_{96}	monitor_type_com	X_{50}	X_{21}										

Table 25 Role matrix for Case (c) for **mcw**—connection weights

mcw - connection weights		1	2	3	4	5	6	7	8	9	10	11	12
X_{1}	type_mom	1											
X_{2}	artificiality_birth	1	-1										
X_{3}	epidural_used	1											
X_{4}	deliv_baby	1	1										
X_{5}	baby_born	1											
X_{6}	MPPD	0.25	0.25	0.25	0.25	-1							
X_{7}	PPPD	1	-0.5	-0.5	-0.5	-0.5	1						
X_{8}	kid_devel	1											
X_{9}	pre_birth_info	1	1	X_{69}									
X_{10}	com_sup_hcp	1	1	1	-1	1	1	1	1	1	X_{72}	1	X_{73}
X_{11}	fat_ask_sup_com	1	1										
X_{12}	fat_rec_sup_com	1	1	-1									
X_{13}	post_info_com_par	1	1	1	1	1	X_{76}	X_{74}	X_{75}				
X_{14}	refer_therapist	1	1	X_{77}									
X_{15}	mental_sup	1	1	1	1	1	1	X_{80}	X_{78}	X_{81}	X_{79}		
X_{16}	acknowl_fat	1	1	1	X_{70}	X_{71}							
X_{17}	age	1											
X_{18}	riskiness	-1	1										
X_{19}	dilation_mom	1											
X_{20}	premature_birth	1											
X_{21}	type_com	1	1	X_{82}									
X_{22}	retention	2	-1										
X_{23}	engaged	1											
X_{24}	stress	1											
X_{25}	birth_dev	1	1										
X_{26}	fear_level	-1	1	1									
X_{27}	desire_sup	1											
X_{28}	ask_sup	1											
X_{29}	birth_dev_det$_{HCP}$	1	1										
X_{30}	det_risk_level$_{HCP}$	X_{53}											
X_{31}	encourage_mom$_{HCP}$	X_{54}											
X_{32}	calm_dad$_{HCP}$	1	X_{55}										
X_{33}	irr_birth_weight$_{HCP}$	1	2										
X_{34}	det_underw$_{HCP}$	X_{56}											
X_{35}	refer_mom_ther$_{HCP}$	X_{57}											
X_{36}	teach_dad_sup_mom$_{HCP}$	X_{58}											
X_{37}	baby_born$_{HCP}$	1											
X_{38}	type_com$_{HCP}$	X_{59}											
X_{39}	type_mom$_{HCP}$	1											
X_{40}	pre_birth_info$_{HCP}$	X_{60}											
X_{41}	birth_dev_det$_{AIC}$	1	1										
X_{42}	det_risk_level$_{AIC}$	X_{61}											
X_{43}	encourage_mom$_{AIC}$	X_{62}											
X_{44}	calm_dad$_{AIC}$	1	X_{63}										
X_{45}	irr_birth_weight$_{AIC}$	1	2										
X_{46}	det_underw$_{AIC}$	X_{64}											
X_{47}	refer_mom_ther$_{AIC}$	X_{65}											
X_{48}	teach_dad_sup_mom$_{AIC}$	X_{66}											
X_{49}	baby_born$_{AIC}$	1											
X_{50}	type_com$_{AIC}$	X_{67}											
X_{51}	type_mom$_{AIC}$	1											
X_{52}	pre_birth_info$_{AIC}$	X_{68}											
X_{53}	$\mathbf{W}_{\text{birth\_dev\_det\_HCP, det\_risk\_level\_HCP}}$	1											
X_{54}	$\mathbf{W}_{\text{det\_risk\_level\_HCP, encourage\_mom\_HCP}}$	1											
X_{55}	$\mathbf{W}_{\text{det\_risk\_level\_HCP, calm\_dad\_HCP}}$	1											
X_{56}	$\mathbf{W}_{\text{irr\_birth\_weight\_HCP, det\_underw\_HCP}}$	1											
X_{57}	$\mathbf{W}_{\text{det\_underw\_HCP, refer\_mom\_ther\_HCP}}$	1											
X_{58}	$\mathbf{W}_{\text{det\_underw\_HCP, teach\_dad\_sup\_mom\_HCP}}$	1											
X_{59}	$\mathbf{W}_{\text{baby\_born\_HCP, type\_com\_HCP}}$	1											
X_{60}	$\mathbf{W}_{\text{type\_mom\_HCP, pre\_birth\_info\_HCP}}$	1											
X_{61}	$\mathbf{W}_{\text{birth\_dev\_det\_AIC, det\_risk\_level\_AIC}}$	1											
X_{62}	$\mathbf{W}_{\text{det\_risk\_level\_AIC, encourage\_mom\_AIC}}$	1											
X_{63}	$\mathbf{W}_{\text{det\_risk\_level\_AIC, calm\_dad\_AIC}}$	1											
X_{64}	$\mathbf{W}_{\text{irr\_birth\_weight\_AIC, det\_underw\_AIC}}$	1											
X_{65}	$\mathbf{W}_{\text{det\_underw\_AIC, refer\_mom\_ther\_AIC}}$	1											
X_{66}	$\mathbf{W}_{\text{det\_underw\_AIC, teach\_dad\_sup\_mom\_AIC}}$	1											
X_{67}	$\mathbf{W}_{\text{baby\_born\_AIC, type\_com\_AIC}}$	1											
X_{68}	$\mathbf{W}_{\text{type\_mom\_AIC, pre\_birth\_info\_AIC}}$	1											
X_{69}	$\mathbf{W}_{\text{pre\_birth\_info\_AIC, pre\_birth\_info}}$	1											
X_{70}	$\mathbf{W}_{\text{calm\_dad\_AIC, acknowl\_fat}}$	1											
X_{71}	$\mathbf{W}_{\text{teach\_dad\_sup\_mom\_AIC, acknowl\_fat}}$	1											
X_{72}	$\mathbf{W}_{\text{calm\_dad\_AIC, com\_sup\_hcp}}$	1											
X_{73}	$\mathbf{W}_{\text{teach\_dad\_sup\_mom\_AIC, com\_sup\_hcp}}$	1											
X_{74}	$\mathbf{W}_{\text{calm\_dad\_AIC, post\_info\_com\_par}}$	1											
X_{75}	$\mathbf{W}_{\text{teach\_dad\_sup\_mom\_AIC, post\_info\_com\_par}}$	1											
X_{76}	$\mathbf{W}_{\text{encourage\_mom\_AIC, post\_info\_com\_par}}$	1											
X_{77}	$\mathbf{W}_{\text{refer\_mom\_ther\_AIC, refer\_therapist}}$	1											
X_{78}	$\mathbf{W}_{\text{calm\_dad\_AIC, mental\_sup}}$	1											
X_{79}	$\mathbf{W}_{\text{teach\_dad\_sup\_mom\_AIC, mental\_sup}}$	1											
X_{80}	$\mathbf{W}_{\text{encourage\_mom\_AIC, mental\_sup}}$	1											
X_{81}	$\mathbf{W}_{\text{refer\_mom\_ther\_AIC, mental\_sup}}$	1											
X_{82}	$\mathbf{W}_{\text{type\_com\_AIC, type\_com}}$	1											
X_{83}	monitor_pre_birth_info	1	1										
X_{84}	monitor_acknowl_fat$_1$	1	1										
X_{85}	monitor_acknowl_fat$_2$	1	1										
X_{86}	monitor_com_sup_hcp$_1$	1	1										
X_{87}	monitor_com_sup_hcp$_2$	1	1										
X_{88}	monitor_post_info_com_par$_1$	1	1										
X_{89}	monitor_post_info_com_par$_2$	1	1										
X_{90}	monitor_post_info_com_par$_3$	1	1										
X_{91}	monitor_refer_therapist	1	1										
X_{92}	monitor_mental_sup$_1$	1	1										
X_{93}	monitor_mental_sup$_2$	1	1										
X_{94}	monitor_mental_sup$_3$	1	1										
X_{95}	monitor_mental_sup$_4$	1	1										
X_{96}	monitor_type_com	1	1										

Table 26 Role matrix for Case (c) for **mcfw**—combination function weights

mcfw - combination function weights		1	2
		Alogistic	Monitor
X_{1}	type_mom	1	
X_{2}	artificiality_birth	1	
X_{3}	epidural_used	1	
X_{4}	deliv_baby	1	
X_{5}	baby_born	1	
X_{6}	MPPD	1	
X_{7}	PPPD	1	
X_{8}	kid_devel	1	
X_{9}	pre_birth_info	1	
X_{10}	com_sup_hcp	1	
X_{11}	fat_ask_sup_com	1	
X_{12}	fat_rec_sup_com	1	
X_{13}	post_info_com_par	1	
X_{14}	refer_therapist	1	
X_{15}	mental_sup	1	
X_{16}	acknowl_fat	1	
X_{17}	age	1	
X_{18}	riskiness	1	
X_{19}	dilation_mom	1	
X_{20}	premature_birth	1	
X_{21}	type_com	1	
X_{22}	retention	1	
X_{23}	engaged	1	
X_{24}	stress	1	
X_{25}	birth_dev	1	
X_{26}	fear_level	1	
X_{27}	desire_sup	1	
X_{28}	ask_sup	1	
X_{29}	birth_dev_det$_{HCP}$	1	
X_{30}	det_risk_level$_{HCP}$	1	
X_{31}	encourage_mom$_{HCP}$	1	
X_{32}	calm_dad$_{HCP}$	1	
X_{33}	irr_birth_weight$_{HCP}$	1	
X_{34}	det_underw$_{HCP}$	1	
X_{35}	refer_mom_ther$_{HCP}$	1	
X_{36}	teach_dad_sup_mom$_{HCP}$	1	
X_{37}	baby_born$_{HCP}$	1	
X_{38}	type_com$_{HCP}$	1	
X_{39}	type_mom$_{HCP}$	1	
X_{40}	pre_birth_info$_{HCP}$	1	
X_{41}	birth_dev_det$_{AIC}$	1	
X_{42}	det_risk_level$_{AIC}$	1	
X_{43}	encourage_mom$_{AIC}$	1	
X_{44}	calm_dad$_{AIC}$	1	
X_{45}	irr_birth_weight$_{AIC}$	1	
X_{46}	det_underw$_{AIC}$	1	
X_{47}	refer_mom_ther$_{AIC}$	1	
X_{48}	teach_dad_sup_mom$_{AIC}$	1	
X_{49}	baby_born$_{AIC}$	1	
X_{50}	type_com$_{AIC}$	1	
X_{51}	type_mom$_{AIC}$	1	
X_{52}	pre_birth_info$_{AIC}$	1	
X_{53}	$\mathbf{W}_{\text{birth\_dev\_det\_HCP, det\_risk\_level\_HCP}}$	1	
X_{54}	$\mathbf{W}_{\text{det\_risk\_level\_HCP, encourage\_mom\_HCP}}$	1	
X_{55}	$\mathbf{W}_{\text{det\_risk\_level\_HCP, calm\_dad\_HCP}}$	1	
X_{56}	$\mathbf{W}_{\text{irr\_birth\_weight\_HCP, det\_underw\_HCP}}$	1	
X_{57}	$\mathbf{W}_{\text{det\_underw\_HCP, refer\_mom\_ther\_HCP}}$	1	
X_{58}	$\mathbf{W}_{\text{det\_underw\_HCP, teach\_dad\_sup\_mom\_HCP}}$	1	
X_{59}	$\mathbf{W}_{\text{baby\_born\_HCP, type\_com\_HCP}}$	1	
X_{60}	$\mathbf{W}_{\text{type\_mom\_HCP, pre\_birth\_info\_HCP}}$	1	
X_{61}	$\mathbf{W}_{\text{birth\_dev\_det\_AIC, det\_risk\_level\_AIC}}$	1	
X_{62}	$\mathbf{W}_{\text{det\_risk\_level\_AIC, encourage\_mom\_AIC}}$	1	
X_{63}	$\mathbf{W}_{\text{det\_risk\_level\_AIC, calm\_dad\_AIC}}$	1	
X_{64}	$\mathbf{W}_{\text{irr\_birth\_weight\_AIC, det\_underw\_AIC}}$	1	
X_{65}	$\mathbf{W}_{\text{det\_underw\_AIC, refer\_mom\_ther\_AIC}}$	1	
X_{66}	$\mathbf{W}_{\text{det\_underw\_AIC, teach\_dad\_sup\_mom\_AIC}}$	1	
X_{67}	$\mathbf{W}_{\text{baby\_born\_AIC, type\_com\_AIC}}$	1	
X_{68}	$\mathbf{W}_{\text{type\_mom\_AIC, pre\_birth\_info\_AIC}}$	1	
X_{69}	$\mathbf{W}_{\text{pre\_birth\_info\_AIC, pre\_birth\_info}}$	1	
X_{70}	$\mathbf{W}_{\text{calm\_dad\_AIC, acknowl\_fat}}$	1	
X_{71}	$\mathbf{W}_{\text{teach\_dad\_sup\_mom\_AIC, acknowl\_fat}}$	1	
X_{72}	$\mathbf{W}_{\text{calm\_dad\_AIC, com\_sup\_hcp}}$	1	
X_{73}	$\mathbf{W}_{\text{teach\_dad\_sup\_mom\_AIC, com\_sup\_hcp}}$	1	
X_{74}	$\mathbf{W}_{\text{calm\_dad\_AIC, post\_info\_com\_par}}$	1	
X_{75}	$\mathbf{W}_{\text{teach\_dad\_sup\_mom\_AIC, post\_info\_com\_par}}$	1	
X_{76}	$\mathbf{W}_{\text{encourage\_mom\_AIC, post\_info\_com\_par}}$	1	
X_{77}	$\mathbf{W}_{\text{refer\_mom\_ther\_AIC, refer\_therapist}}$	1	
X_{78}	$\mathbf{W}_{\text{calm\_dad\_AIC, mental\_sup}}$	1	
X_{79}	$\mathbf{W}_{\text{teach\_dad\_sup\_mom\_AIC, mental\_sup}}$	1	
X_{80}	$\mathbf{W}_{\text{encourage\_mom\_AIC, mental\_sup}}$	1	
X_{81}	$\mathbf{W}_{\text{refer\_mom\_ther\_AIC, mental\_sup}}$	1	
X_{82}	$\mathbf{W}_{\text{type\_com\_AIC, type\_com}}$	1	
X_{83}	monitor_pre_birth_info		1
X_{84}	monitor_acknowl_fat$_{1}$		1
X_{85}	monitor_acknowl_fat$_{2}$		1
X_{86}	monitor_com_sup_hcp$_{1}$		1
X_{87}	monitor_com_sup_hcp$_{2}$		1
X_{88}	monitor_post_info_com_par$_{1}$		1
X_{89}	monitor_post_info_com_par$_{2}$		1
X_{90}	monitor_post_info_com_par$_{3}$		1
X_{91}	monitor_refer_therapist		1
X_{92}	monitor_mental_sup$_{1}$		1
X_{93}	monitor_mental_sup$_{2}$		1
X_{94}	monitor_mental_sup$_{3}$		1
X_{95}	monitor_mental_sup$_{4}$		1
X_{96}	monitor_type_com		1

Table 27 Role matrix for Case (c) for **mcfp**—combination function parameters

mcfp - combination function parameters		1 Alogistic		2 Monitor
		Steepness	Threshold	Threshold
X_1	type_mom	50	0.5	
X_2	artificiality_birth	50	0.5	
X_3	epidural_used	50	0.5	
X_4	deliv_baby	5	0.5	
X_5	baby_born	50	0.5	
X_6	MPPD	5	0.3	
X_7	PPPD	5	0.5	
X_8	kid_devel	5	0.5	
X_9	pre_birth_info	50	0.5	
X_{10}	com_sup_hcp	0.5	6	
X_{11}	fat_ask_sup_com	5	1	
X_{12}	fat_rec_sup_com	5	0.5	
X_{13}	post_info_com_par	5	4	
X_{14}	refer_therapist	5	1	
X_{15}	mental_sup	5	5	
X_{16}	acknowl_fat	5	0.8	
X_{17}	age	5	0.5	
X_{18}	riskiness	50	0.5	
X_{19}	dilation_mom	2.8	0	
X_{20}	premature_birth	5	0.5	
X_{21}	type_com	5	0.1	
X_{22}	retention	5	0.3	
X_{23}	engaged	5	0.5	
X_{24}	stress	5	0.5	
X_{25}	birth_dev	50	1	
X_{26}	fear_level	5	1	
X_{27}	desire_sup	5	0.3	
X_{28}	ask_sup	5	0.2	
X_{29}	birth_dev_det$_{HCP}$	5	1	
X_{30}	det_risk_level$_{HCP}$	5	0.5	
X_{31}	encourage_mom$_{HCP}$	5	0.5	
X_{32}	calm_dad$_{HCP}$	50	0.7	
X_{33}	irr_birth_weight$_{HCP}$	5	1	
X_{34}	det_underw$_{HCP}$	5	0.5	
X_{35}	refer_mom_ther$_{HCP}$	50	0.5	
X_{36}	teach_dad_sup_mom$_{HCP}$	50	0.7	
X_{37}	baby_born$_{HCP}$	5	0.5	
X_{38}	type_com$_{HCP}$	50	0.5	
X_{39}	type_mom$_{HCP}$	5	0.5	
X_{40}	pre_birth_info$_{HCP}$	50	0.5	
X_{41}	birth_dev_det$_{AIC}$	5	1	
X_{42}	det_risk_level$_{AIC}$	5	0.5	
X_{43}	encourage_mom$_{AIC}$	5	0.5	
X_{44}	calm_dad$_{AIC}$	50	0.7	
X_{45}	irr_birth_weight$_{AIC}$	5	1	
X_{46}	det_underw$_{AIC}$	5	0.5	
X_{47}	refer_mom_ther$_{AIC}$	50	0.5	
X_{48}	teach_dad_sup_mom$_{AIC}$	50	0.7	
X_{49}	baby_born$_{AIC}$	5	0.5	
X_{50}	type_com$_{AIC}$	50	0.5	
X_{51}	type_mom$_{AIC}$	5	0.5	
X_{52}	pre_birth_info$_{AIC}$	50	0.5	
X_{53}	$W_{birth\_dev\_det\_HCP,\ det\_risk\_level\_HCP}$	5	0.5	
X_{54}	$W_{det\_risk\_level\_HCP,\ encourage\_mom\_HCP}$	5	0.5	
X_{55}	$W_{det\_risk\_level\_HCP,\ calm\_dad\_HCP}$	5	0.5	
X_{56}	$W_{irr\_birth\_weight\_HCP,\ det\_underw\_HCP}$	5	0.5	
X_{57}	$W_{det\_underw\_HCP,\ refer\_mom\_ther\_HCP}$	5	0.5	
X_{58}	$W_{det\_underw\_HCP,\ teach\_dad\_sup\_mom\_HCP}$	5	0.5	
X_{59}	$W_{baby\_born\_HCP,\ type\_com\_HCP}$	5	0.5	
X_{60}	$W_{type\_mom\_HCP,\ pre\_birth\_info\_HCP}$	5	0.5	
X_{61}	$W_{birth\_dev\_det\_AIC,\ det\_risk\_level\_AIC}$	5	0.5	
X_{62}	$W_{det\_risk\_level\_AIC,\ encourage\_mom\_AIC}$	5	0.5	
X_{63}	$W_{det\_risk\_level\_AIC,\ calm\_dad\_AIC}$	5	0.5	
X_{64}	$W_{irr\_birth\_weight\_AIC,\ det\_underw\_AIC}$	5	0.5	
X_{65}	$W_{det\_underw\_AIC,\ refer\_mom\_ther\_AIC}$	5	0.5	
X_{66}	$W_{det\_underw\_AIC,\ teach\_dad\_sup\_mom\_AIC}$	5	0.5	
X_{67}	$W_{baby\_born\_AIC,\ type\_com\_AIC}$	5	0.5	
X_{68}	$W_{type\_mom\_AIC,\ pre\_birth\_info\_AIC}$	5	0.5	
X_{69}	$W_{pre\_birth\_info\_AIC,\ pre\_birth\_info}$	5	0.5	
X_{70}	$W_{calm\_dad\_AIC,\ acknowl\_fat}$	5	0.5	
X_{71}	$W_{teach\_dad\_sup\_mom\_AIC,\ acknowl\_fat}$	5	0.5	
X_{72}	$W_{calm\_dad\_AIC,\ com\_sup\_hcp}$	5	0.5	
X_{73}	$W_{teach\_dad\_sup\_mom\_AIC,\ com\_sup\_hcp}$	5	0.5	
X_{74}	$W_{calm\_dad\_AIC,\ post\_info\_com\_par}$	5	0.5	
X_{75}	$W_{teach\_dad\_sup\_mom\_AIC,\ post\_info\_com\_par}$	5	0.5	
X_{76}	$W_{encourage\_mom\_AIC,\ post\_info\_com\_par}$	5	0.5	
X_{77}	$W_{refer\_mom\_ther\_AIC,\ refer\_therapist}$	5	0.5	
X_{78}	$W_{calm\_dad\_AIC,\ mental\_sup}$	5	0.5	
X_{79}	$W_{teach\_dad\_sup\_mom\_AIC,\ mental\_sup}$	5	0.5	
X_{80}	$W_{encourage\_mom\_AIC,\ mental\_sup}$	5	0.5	
X_{81}	$W_{refer\_mom\_ther\_AIC,\ mental\_sup}$	5	0.5	
X_{82}	$W_{type\_com\_AIC,\ type\_com}$	5	0.5	
X_{83}	monitor_pre_birth_info			0.6
X_{84}	monitor_acknowl_fat$_1$			0.6
X_{85}	monitor_acknowl_fat$_2$			0.6
X_{86}	monitor_com_sup_hcp$_1$			0.6
X_{87}	monitor_com_sup_hcp$_2$			0.6
X_{88}	monitor_post_info_com_par$_1$			0.6
X_{89}	monitor_post_info_com_par$_2$			0.6
X_{90}	monitor_post_info_com_par$_3$			0.6
X_{91}	monitor_refer_therapist			0.6
X_{92}	monitor_mental_sup$_1$			0.6
X_{93}	monitor_mental_sup$_2$			0.6
X_{94}	monitor_mental_sup$_3$			0.6
X_{95}	monitor_mental_sup$_4$			0.6
X_{96}	monitor_type_com			0.6

Table 28 Role matrix for Case (c) for **iv**—initial values

iv - initial values		1
X_{1}	type_mom	0
X_{2}	artificiality_birth	0
X_{3}	epidural_used	0
X_{4}	deliv_baby	0
X_{5}	baby_born	0
X_{6}	MPPD	0.13
X_{7}	PPPD	0
X_{8}	kid_devel	0
X_{9}	pre_birth_info	0
X_{10}	com_sup_hcp	0
X_{11}	fat_ask_sup_com	0
X_{12}	fat_rec_sup_com	0
X_{13}	post_info_com_par	0
X_{14}	refer_therapist	0
X_{15}	mental_sup	0
X_{16}	acknowl_fat	0
X_{17}	age	1
X_{18}	riskiness	0
X_{19}	dilation_mom	0.2
X_{20}	premature_birth	1
X_{21}	type_com	0
X_{22}	retention	0
X_{23}	engaged	0
X_{24}	stress	0
X_{25}	birth_dev	0
X_{26}	fear_level	0
X_{27}	desire_sup	0
X_{28}	ask_sup	0
X_{29}	birth_dev_det$_{HCP}$	0
X_{30}	det_risk_level$_{HCP}$	0
X_{31}	encourage_mom$_{HCP}$	0
X_{32}	calm_dad$_{HCP}$	0
X_{33}	irr_birth_weight$_{HCP}$	0
X_{34}	det_underw$_{HCP}$	0
X_{35}	refer_mom_ther$_{HCP}$	0
X_{36}	teach_dad_sup_mom$_{HCP}$	0
X_{37}	baby_born$_{HCP}$	0
X_{38}	type_com$_{HCP}$	0
X_{39}	type_mom$_{HCP}$	0
X_{40}	pre_birth_info$_{HCP}$	0
X_{41}	birth_dev_det$_{AIC}$	0
X_{42}	det_risk_level$_{AIC}$	0
X_{43}	encourage_mom$_{AIC}$	0
X_{44}	calm_dad$_{AIC}$	0
X_{45}	irr_birth_weight$_{AIC}$	0
X_{46}	det_underw$_{AIC}$	0
X_{47}	refer_mom_ther$_{AIC}$	0
X_{48}	teach_dad_sup_mom$_{AIC}$	0
X_{49}	baby_born$_{AIC}$	0
X_{50}	type_com$_{AIC}$	0
X_{51}	type_mom$_{AIC}$	0
X_{52}	pre_birth_info$_{AIC}$	0
X_{53}	$\mathbf{W}_{birth\_dev\_det\_HCP,\ det\_risk\_level\_HCP}$	0
X_{54}	$\mathbf{W}_{det\_risk\_level\_HCP,\ encourage\_mom\_HCP}$	0
X_{55}	$\mathbf{W}_{det\_risk\_level\_HCP,\ calm\_dad\_HCP}$	0
X_{56}	$\mathbf{W}_{irr\_birth\_weight\_HCP,\ det\_underw\_HCP}$	0
X_{57}	$\mathbf{W}_{det\_underw\_HCP,\ refer\_mom\_ther\_HCP}$	0
X_{58}	$\mathbf{W}_{det\_underw\_HCP,\ teach\_dad\_sup\_mom\_HCP}$	0
X_{59}	$\mathbf{W}_{baby\_born\_HCP,\ type\_com\_HCP}$	0
X_{60}	$\mathbf{W}_{type\_mom\_HCP,\ pre\_birth\_info\_HCP}$	0
X_{61}	$\mathbf{W}_{birth\_dev\_det\_AIC,\ det\_risk\_level\_AIC}$	1
X_{62}	$\mathbf{W}_{det\_risk\_level\_AIC,\ encourage\_mom\_AIC}$	1
X_{63}	$\mathbf{W}_{det\_risk\_level\_AIC,\ calm\_dad\_AIC}$	1
X_{64}	$\mathbf{W}_{irr\_birth\_weight\_AIC,\ det\_underw\_AIC}$	1
X_{65}	$\mathbf{W}_{det\_underw\_AIC,\ refer\_mom\_ther\_AIC}$	1
X_{66}	$\mathbf{W}_{det\_underw\_AIC,\ teach\_dad\_sup\_mom\_AIC}$	1
X_{67}	$\mathbf{W}_{baby\_born\_AIC,\ type\_com\_AIC}$	1
X_{68}	$\mathbf{W}_{type\_mom\_AIC,\ pre\_birth\_info\_AIC}$	1
X_{69}	$\mathbf{W}_{pre\_birth\_info\_AIC,\ pre\_birth\_info}$	0
X_{70}	$\mathbf{W}_{calm\_dad\_AIC,\ acknowl\_fat}$	0
X_{71}	$\mathbf{W}_{teach\_dad\_sup\_mom\_AIC,\ acknowl\_fat}$	0
X_{72}	$\mathbf{W}_{calm\_dad\_AIC,\ com\_sup\_hcp}$	0
X_{73}	$\mathbf{W}_{teach\_dad\_sup\_mom\_AIC,\ com\_sup\_hcp}$	0
X_{74}	$\mathbf{W}_{calm\_dad\_AIC,\ post\_info\_com\_par}$	0
X_{75}	$\mathbf{W}_{teach\_dad\_sup\_mom\_AIC,\ post\_info\_com\_par}$	0
X_{76}	$\mathbf{W}_{encourage\_mom\_AIC,\ post\_info\_com\_par}$	0
X_{77}	$\mathbf{W}_{refer\_mom\_ther\_AIC,\ refer\_therapist}$	0
X_{78}	$\mathbf{W}_{calm\_dad\_AIC,\ mental\_sup}$	0
X_{79}	$\mathbf{W}_{teach\_dad\_sup\_mom\_AIC,\ mental\_sup}$	0
X_{80}	$\mathbf{W}_{encourage\_mom\_AIC,\ mental\_sup}$	0
X_{81}	$\mathbf{W}_{refer\_mom\_ther\_AIC,\ mental\_sup}$	0
X_{82}	$\mathbf{W}_{type\_com\_AIC,\ type\_com}$	0
X_{83}	monitor_pre_birth_info	0
X_{84}	monitor_acknowl_fat$_{1}$	0
X_{85}	monitor_acknowl_fat$_{2}$	0
X_{86}	monitor_com_sup_hcp$_{1}$	0
X_{87}	monitor_com_sup_hcp$_{2}$	0
X_{88}	monitor_post_info_com_par$_{1}$	0
X_{89}	monitor_post_info_com_par$_{2}$	0
X_{90}	monitor_post_info_com_par$_{3}$	0
X_{91}	monitor_refer_therapist	0
X_{92}	monitor_mental_sup$_{1}$	0
X_{93}	monitor_mental_sup$_{2}$	0
X_{94}	monitor_mental_sup$_{3}$	0
X_{95}	monitor_mental_sup$_{4}$	0
X_{96}	monitor_type_com	0

Table 29 Role matrix for Case (c) for **ms**—speed factors

	ms - speed factors	1
X_{1}	type_mom	1
X_{2}	artificiality_birth	1
X_{3}	epidural_used	1
X_{4}	deliv_baby	0.2
X_{5}	baby_born	0.1
X_{6}	MPPD	0.2
X_{7}	PPPD	1
X_{8}	kid_devel	1
X_{9}	pre_birth_info	0.1
X_{10}	com_sup_hcp	0.1
X_{11}	fat_ask_sup_com	0.3
X_{12}	fat_rec_sup_com	0.1
X_{13}	post_info_com_par	0.1
X_{14}	refer_therapist	0.3
X_{15}	mental_sup	0.1
X_{16}	acknowl_fat	0.1
X_{17}	age	0
X_{18}	riskiness	0.5
X_{19}	dilation_mom	1
X_{20}	premature_birth	0
X_{21}	type_com	1
X_{22}	retention	1
X_{23}	engaged	1
X_{24}	stress	0.4
X_{25}	birth_dev	1
X_{26}	fear_level	1
X_{27}	desire_sup	1
X_{28}	ask_sup	0.2
X_{29}	birth_dev_det$_{HCP}$	1
X_{30}	det_risk_level$_{HCP}$	1
X_{31}	encourage_mom$_{HCP}$	0.1
X_{32}	calm_dad$_{HCP}$	0.1
X_{33}	irr_birth_weight$_{HCP}$	1
X_{34}	det_underw$_{HCP}$	1
X_{35}	refer_mom_ther$_{HCP}$	0.3
X_{36}	teach_dad_sup_mom$_{HCP}$	0.1
X_{37}	baby_born$_{HCP}$	1
X_{38}	type_com$_{HCP}$	1
X_{39}	type_mom$_{HCP}$	1
X_{40}	pre_birth_info$_{HCP}$	1
X_{41}	birth_dev_det$_{AIC}$	0.33
X_{42}	det_risk_level$_{AIC}$	0.33
X_{43}	encourage_mom$_{AIC}$	0.033
X_{44}	calm_dad$_{AIC}$	0.033
X_{45}	irr_birth_weight$_{AIC}$	0.33
X_{46}	det_underw$_{AIC}$	0.33
X_{47}	refer_mom_ther$_{AIC}$	0.1
X_{48}	teach_dad_sup_mom$_{AIC}$	0.033
X_{49}	baby_born$_{AIC}$	0.33
X_{50}	type_com$_{AIC}$	0.33
X_{51}	type_mom$_{AIC}$	0.33
X_{52}	pre_birth_info$_{AIC}$	0.33
X_{53}	$W_{\text{birth\_dev\_det\_HCP, det\_risk\_level\_HCP}}$	0
X_{54}	$W_{\text{det\_risk\_level\_HCP, encourage\_mom\_HCP}}$	0
X_{55}	$W_{\text{det\_risk\_level\_HCP, calm\_dad\_HCP}}$	0
X_{56}	$W_{\text{irr\_birth\_weight\_HCP, det\_underw\_HCP}}$	0
X_{57}	$W_{\text{det\_underw\_HCP, refer\_mom\_ther\_HCP}}$	0
X_{58}	$W_{\text{det\_underw\_HCP, teach\_dad\_sup\_mom\_HCP}}$	0
X_{59}	$W_{\text{baby\_born\_HCP, type\_com\_HCP}}$	0
X_{60}	$W_{\text{type\_mom\_HCP, pre\_birth\_info\_HCP}}$	0
X_{61}	$W_{\text{birth\_dev\_det\_AIC, det\_risk\_level\_AIC}}$	0
X_{62}	$W_{\text{det\_risk\_level\_AIC, encourage\_mom\_AIC}}$	0
X_{63}	$W_{\text{det\_risk\_level\_AIC, calm\_dad\_AIC}}$	0
X_{64}	$W_{\text{irr\_birth\_weight\_AIC, det\_underw\_AIC}}$	0
X_{65}	$W_{\text{det\_underw\_AIC, refer\_mom\_ther\_AIC}}$	0
X_{66}	$W_{\text{det\_underw\_AIC, teach\_dad\_sup\_mom\_AIC}}$	0
X_{67}	$W_{\text{baby\_born\_AIC, type\_com\_AIC}}$	0
X_{68}	$W_{\text{type\_mom\_AIC, pre\_birth\_info\_AIC}}$	0
X_{69}	$W_{\text{pre\_birth\_info\_AIC, pre\_birth\_info}}$	0
X_{70}	$W_{\text{calm\_dad\_AIC, acknowl\_fat}}$	0.5
X_{71}	$W_{\text{teach\_dad\_sup\_mom\_AIC, acknowl\_fat}}$	0.5
X_{72}	$W_{\text{calm\_dad\_AIC, com\_sup\_hcp}}$	0.5
X_{73}	$W_{\text{teach\_dad\_sup\_mom\_AIC, com\_sup\_hcp}}$	0.5
X_{74}	$W_{\text{calm\_dad\_AIC, post\_info\_com\_par}}$	0.5
X_{75}	$W_{\text{teach\_dad\_sup\_mom\_AIC, post\_info\_com\_par}}$	0.5
X_{76}	$W_{\text{encourage\_mom\_AIC, post\_info\_com\_par}}$	0.5
X_{77}	$W_{\text{refer\_mom\_ther\_AIC, refer\_therapist}}$	0.5
X_{78}	$W_{\text{calm\_dad\_AIC, mental\_sup}}$	0.5
X_{79}	$W_{\text{teach\_dad\_sup\_mom\_AIC, mental\_sup}}$	0.5
X_{80}	$W_{\text{encourage\_mom\_AIC, mental\_sup}}$	0.5
X_{81}	$W_{\text{refer\_mom\_ther\_AIC, mental\_sup}}$	0.5
X_{82}	$W_{\text{type\_com\_AIC, type\_com}}$	0.5
X_{83}	monitor_pre_birth_info	0.5
X_{84}	monitor_acknowl_fat$_{1}$	0.5
X_{85}	monitor_acknowl_fat$_{2}$	0.5
X_{86}	monitor_com_sup_hcp$_{1}$	0.5
X_{87}	monitor_com_sup_hcp$_{2}$	0.5
X_{88}	monitor_post_info_com_par$_{1}$	0.5
X_{89}	monitor_post_info_com_par$_{2}$	0.5
X_{90}	monitor_post_info_com_par$_{3}$	0.5
X_{91}	monitor_refer_therapist	0.5
X_{92}	monitor_mental_sup$_{1}$	0.5
X_{93}	monitor_mental_sup$_{2}$	0.5
X_{94}	monitor_mental_sup$_{3}$	0.5
X_{95}	monitor_mental_sup$_{4}$	0.5
X_{96}	monitor_type_com	0.5

Table 30 Role matrix for Case (d) for **mb**—base connectivity

	mb - base connectivity	1	2	3	4	5	6	7	8	9	10	11	12
X_{1}	type_mom	X_{1}											
X_{2}	artificiality_birth	X_{2}	X_{18}										
X_{3}	epidural_used	X_{3}											
X_{4}	deliv_baby	X_{4}	X_{19}										
X_{5}	baby_born	X_{4}											
X_{6}	MPPD	X_{1}	X_{2}	X_{9}	X_{5}	X_{15}							
X_{7}	PPPD	X_{6}	X_{10}	X_{12}	X_{15}	X_{23}	X_{24}						
X_{8}	imp_kid_devel	X_{7}											
X_{9}	pre_birth_info	X_{9}	X_{40}	X_{52}									
X_{10}	com_sup_hcp	X_{4}	X_{9}	X_{11}	X_{24}	X_{30}	X_{32}	X_{34}	X_{36}	X_{42}	X_{44}	X_{46}	X_{48}
X_{11}	fat_ask_sup_com	X_{23}	X_{28}										
X_{12}	fat_rec_sup_com	X_{10}	X_{22}	X_{24}									
X_{13}	post_info_com_par	X_{5}	X_{20}	X_{31}	X_{32}	X_{36}	X_{43}	X_{44}	X_{48}				
X_{14}	refer_therapist	X_{13}	X_{35}	X_{47}									
X_{15}	mental_sup	X_{14}	X_{21}	X_{31}	X_{32}	X_{35}	X_{36}	X_{43}	X_{44}	X_{47}	X_{48}		
X_{16}	acknowl_fat	X_{9}	X_{32}	X_{36}	X_{44}	X_{48}							
X_{17}	age	X_{17}											
X_{18}	riskiness	X_{5}	X_{17}										
X_{19}	dilation_mom	X_{19}											
X_{20}	premature_birth	X_{20}											
X_{21}	type_com	X_{21}	X_{38}	X_{50}									
X_{22}	retention	X_{23}	X_{24}										
X_{23}	engaged	X_{16}											
X_{24}	stress	X_{26}											
X_{25}	birth_dev	X_{4}	X_{18}										
X_{26}	fear_level	X_{23}	X_{24}	X_{25}									
X_{27}	desire_sup	X_{26}											
X_{28}	ask_sup	X_{27}											
X_{29}	birth_dev_det$_{HCP}$	X_{4}	X_{18}										
X_{30}	det_risk_level$_{HCP}$	X_{29}											
X_{31}	encourage_mom$_{HCP}$	X_{30}											
X_{32}	calm_dad$_{HCP}$	X_{11}	X_{30}										
X_{33}	irr_birth_weight$_{HCP}$	X_{5}	X_{20}										
X_{34}	det_underw$_{HCP}$	X_{33}											
X_{35}	refer_mom_ther$_{HCP}$	X_{34}											
X_{36}	teach_dad_sup_mom$_{HCP}$	X_{34}											
X_{37}	baby_born$_{HCP}$	X_{5}											
X_{38}	type_com$_{HCP}$	X_{37}											
X_{39}	type_mom$_{HCP}$	X_{1}											
X_{40}	pre_birth_info$_{HCP}$	X_{39}											
X_{41}	birth_dev_det$_{AIC}$	X_{4}	X_{18}										
X_{42}	det_risk_level$_{AIC}$	X_{41}											
X_{43}	encourage_mom$_{AIC}$	X_{42}											
X_{44}	calm_dad$_{AIC}$	X_{11}	X_{42}										
X_{45}	irr_birth_weight$_{AIC}$	X_{5}	X_{20}										
X_{46}	det_underw$_{AIC}$	X_{45}											
X_{47}	refer_mom_ther$_{AIC}$	X_{46}											
X_{48}	teach_dad_sup_mom$_{AIC}$	X_{46}											
X_{49}	baby_born$_{AIC}$	X_{5}											
X_{50}	type_com$_{AIC}$	X_{49}											
X_{51}	type_mom$_{AIC}$	X_{1}											
X_{52}	pre_birth_info$_{AIC}$	X_{51}											
X_{53}	$\mathbf{W}_{birth\_dev\_det\_HCP,\ det\_risk\_level\_HCP}$	X_{53}	X_{61}										
X_{54}	$\mathbf{W}_{det\_risk\_level\_HCP,\ encourage\_mom\_HCP}$	X_{54}	X_{62}										
X_{55}	$\mathbf{W}_{det\_risk\_level\_HCP,\ calm\_dad\_HCP}$	X_{55}	X_{63}										
X_{56}	$\mathbf{W}_{irr\_birth\_weight\_HCP,\ det\_underw\_HCP}$	X_{56}	X_{64}										
X_{57}	$\mathbf{W}_{det\_underw\_HCP,\ refer\_mom\_ther\_HCP}$	X_{57}	X_{65}										
X_{58}	$\mathbf{W}_{det\_underw\_HCP,\ teach\_dad\_sup\_mom\_HCP}$	X_{58}	X_{66}										
X_{59}	$\mathbf{W}_{baby\_born\_HCP,\ type\_com\_HCP}$	X_{59}	X_{67}										
X_{60}	$\mathbf{W}_{type\_mom\_HCP,\ pre\_birth\_info\_HCP}$	X_{60}	X_{68}										
X_{61}	$\mathbf{W}_{birth\_dev\_det\_AIC,\ det\_risk\_level\_AIC}$	X_{61}	X_{99}										
X_{62}	$\mathbf{W}_{det\_risk\_level\_AIC,\ encourage\_mom\_AIC}$	X_{62}	X_{100}										
X_{63}	$\mathbf{W}_{det\_risk\_level\_AIC,\ calm\_dad\_AIC}$	X_{63}	X_{101}										
X_{64}	$\mathbf{W}_{irr\_birth\_weight\_AIC,\ det\_underw\_AIC}$	X_{64}	X_{102}										
X_{65}	$\mathbf{W}_{det\_underw\_AIC,\ refer\_mom\_ther\_AIC}$	X_{65}	X_{103}										
X_{66}	$\mathbf{W}_{det\_underw\_AIC,\ teach\_dad\_sup\_mom\_AIC}$	X_{66}	X_{104}										
X_{67}	$\mathbf{W}_{baby\_born\_AIC,\ type\_com\_AIC}$	X_{67}	X_{105}										
X_{68}	$\mathbf{W}_{type\_mom\_AIC,\ pre\_birth\_info\_AIC}$	X_{68}	X_{106}										
X_{69}	$\mathbf{W}_{pre\_birth\_info\_AIC,\ pre\_birth\_info}$	X_{83}											
X_{70}	$\mathbf{W}_{calm\_dad\_AIC,\ acknowl\_fat}$	X_{85}											
X_{71}	$\mathbf{W}_{teach\_dad\_sup\_mom\_AIC,\ acknowl\_fat}$	X_{84}											
X_{72}	$\mathbf{W}_{calm\_dad\_AIC,\ com\_sup\_hcp}$	X_{86}											
X_{73}	$\mathbf{W}_{teach\_dad\_sup\_mom\_AIC,\ com\_sup\_hcp}$	X_{87}											
X_{74}	$\mathbf{W}_{calm\_dad\_AIC,\ post\_info\_com\_par}$	X_{89}											
X_{75}	$\mathbf{W}_{teach\_dad\_sup\_mom\_AIC,\ post\_info\_com\_par}$	X_{90}											
X_{76}	$\mathbf{W}_{encourage\_mom\_AIC,\ post\_info\_com\_par}$	X_{88}											
X_{77}	$\mathbf{W}_{refer\_mom\_ther\_AIC,\ refer\_therapist}$	X_{91}											
X_{78}	$\mathbf{W}_{calm\_dad\_AIC,\ mental\_sup}$	X_{94}											
X_{79}	$\mathbf{W}_{teach\_dad\_sup\_mom\_AIC,\ mental\_sup}$	X_{95}											
X_{80}	$\mathbf{W}_{encourage\_mom\_AIC,\ mental\_sup}$	X_{92}											
X_{81}	$\mathbf{W}_{refer\_mom\_ther\_AIC,\ mental\_sup}$	X_{93}											
X_{82}	$\mathbf{W}_{type\_com\_AIC,\ type\_com}$	X_{96}											
X_{83}	monitor_pre_birth_info	X_{52}	X_{9}										
X_{84}	monitor_acknowl_fat$_{1}$	X_{48}	X_{16}										
X_{85}	monitor_acknowl_fat$_{2}$	X_{44}	X_{16}										
X_{86}	monitor_com_sup_hcp$_{1}$	X_{44}	X_{10}										
X_{87}	monitor_com_sup_hcp$_{2}$	X_{48}	X_{10}										
X_{88}	monitor_post_info_com_par$_{1}$	X_{43}	X_{13}										
X_{89}	monitor_post_info_com_par$_{2}$	X_{44}	X_{13}										
X_{90}	monitor_post_info_com_par$_{3}$	X_{48}	X_{13}										
X_{91}	monitor_refer_therapist	X_{47}	X_{14}										
X_{92}	monitor_mental_sup$_{1}$	X_{43}	X_{15}										
X_{93}	monitor_mental_sup$_{2}$	X_{47}	X_{15}										
X_{94}	monitor_mental_sup$_{3}$	X_{44}	X_{15}										
X_{95}	monitor_mental_sup$_{4}$	X_{48}	X_{15}										
X_{96}	monitor_type_com	X_{50}	X_{21}										
X_{97}	$\mathbf{W}_{W\_AIC,W\_HCP}$	X_{98}											
X_{98}	con_feedback	X_{98}											
X_{99}	$\mathbf{W}_{birth\_dev\_det\_E,\ det\_risk\_level\_E}$												
X_{100}	$\mathbf{W}_{det\_risk\_level\_E,\ encourage\_mom\_E}$												
X_{101}	$\mathbf{W}_{det\_risk\_level\_E,\ calm\_dad\_E}$												
X_{102}	$\mathbf{W}_{irr\_birth\_weight\_E,\ det\_underw\_E}$												
X_{103}	$\mathbf{W}_{det\_underw\_E,\ refer\_mom\_ther\_E}$												
X_{104}	$\mathbf{W}_{det\_underw\_E,\ teach\_dad\_sup\_mom\_E}$												
X_{105}	$\mathbf{W}_{baby\_born\_E,\ type\_com\_E}$												
X_{106}	$\mathbf{W}_{type\_mom\_E,\ pre\_birth\_info\_E}$												
X_{107}	$\mathbf{W}_{W\_E,W\_AIC}$	X_{108}											
X_{108}	con_feedforward	X_{108}											

Table 31 Role matrix for Case (d) for **mcw**—connection weights

	mcw - connection weights	1	2	3	4	5	6	7	8	9	10	11	12
X_1	type_mom	1											
X_2	artificiality_birth	1	-1										
X_3	epidural_used	1											
X_4	deliv_baby	1	1										
X_5	baby_born	1											
X_6	MPPD	0.25	0.25	0.25	0.25	-1							
X_7	PPPD	1	-0.5	-0.5	-0.5	-0.5	1						
X_8	kid_devel	1											
X_9	pre_birth_info	1	1	X_{69}									
X_{10}	com_sup_hcp	1	1	1	-1	1	1	1	1	1	X_{72}	1	X_{73}
X_{11}	fat_ask_sup_com	1	1										
X_{12}	fat_rec_sup_com	1	1	-1									
X_{13}	post_info_com_par	1	1	1	1	1	X_{76}	X_{74}	X_{75}				
X_{14}	refer_therapist	1	1	X_{77}									
X_{15}	mental_sup	1	1	1	1	1	1	X_{80}	X_{78}	X_{81}	X_{79}		
X_{16}	acknowl_fat	1	1	1	X_{70}	X_{71}							
X_{17}	age	1											
X_{18}	riskiness	-1	1										
X_{19}	dilation_mom	1											
X_{20}	premature_birth	1											
X_{21}	type_com	1	1	X_{82}									
X_{22}	retention	2	-1										
X_{23}	engaged	1											
X_{24}	stress	1											
X_{25}	birth_dev	1	1										
X_{26}	fear_level	-1	1	1									
X_{27}	desire_sup	1											
X_{28}	ask_sup	1											
X_{29}	birth_dev_det$_{HCP}$	1	1										
X_{30}	det_risk_level$_{HCP}$	X_{53}											
X_{31}	encourage_mom$_{HCP}$	X_{54}											
X_{32}	calm_dad$_{HCP}$	1	X_{55}										
X_{33}	irr_birth_weight$_{HCP}$	1	2										
X_{34}	det_underw$_{HCP}$	X_{56}											
X_{35}	refer_mom_ther$_{HCP}$	X_{57}											
X_{36}	teach_dad_sup_mom$_{HCP}$	X_{58}											
X_{37}	baby_born$_{HCP}$	1											
X_{38}	type_com$_{HCP}$	X_{59}											
X_{39}	type_mom$_{HCP}$	1											
X_{40}	pre_birth_info$_{HCP}$	X_{60}											
X_{41}	birth_dev_det$_{AIC}$	1	1										
X_{42}	det_risk_level$_{AIC}$	X_{61}											
X_{43}	encourage_mom$_{AIC}$	X_{62}											
X_{44}	calm_dad$_{AIC}$	1	X_{63}										
X_{45}	irr_birth_weight$_{AIC}$	1	2										
X_{46}	det_underw$_{AIC}$	X_{64}											
X_{47}	refer_mom_ther$_{AIC}$	X_{65}											
X_{48}	teach_dad_sup_mom$_{AIC}$	X_{66}											
X_{49}	baby_born$_{AIC}$	1											
X_{50}	type_com$_{AIC}$	X_{67}											
X_{51}	type_mom$_{AIC}$	1											
X_{52}	pre_birth_info$_{AIC}$	X_{68}											
X_{53}	$W_{birth\_dev\_det\_HCP,\ det\_risk\_level\_HCP}$	1	X_{97}										
X_{54}	$W_{det\_risk\_level\_HCP,\ encourage\_mom\_HCP}$	1	X_{97}										
X_{55}	$W_{det\_risk\_level\_HCP,\ calm\_dad\_HCP}$	1	X_{97}										
X_{56}	$W_{irr\_birth\_weight\_HCP,\ det\_underw\_HCP}$	1	X_{97}										
X_{57}	$W_{det\_underw\_HCP,\ refer\_mom\_ther\_HCP}$	1	X_{97}										
X_{58}	$W_{det\_underw\_HCP,\ teach\_dad\_sup\_mom\_HCP}$	1	X_{97}										
X_{59}	$W_{baby\_born\_HCP,\ type\_com\_HCP}$	1	X_{97}										
X_{60}	$W_{type\_mom\_HCP,\ pre\_birth\_info\_HCP}$	1	X_{97}										
X_{61}	$W_{birth\_dev\_det\_AIC,\ det\_risk\_level\_AIC}$	1	X_{107}										
X_{62}	$W_{det\_risk\_level\_AIC,\ encourage\_mom\_AIC}$	1	X_{107}										
X_{63}	$W_{det\_risk\_level\_AIC,\ calm\_dad\_AIC}$	1	X_{107}										
X_{64}	$W_{irr\_birth\_weight\_AIC,\ det\_underw\_AIC}$	1	X_{107}										
X_{65}	$W_{det\_underw\_AIC,\ refer\_mom\_ther\_AIC}$	1	X_{107}										
X_{66}	$W_{det\_underw\_AIC,\ teach\_dad\_sup\_mom\_AIC}$	1	X_{107}										
X_{67}	$W_{baby\_born\_AIC,\ type\_com\_AIC}$	1	X_{107}										
X_{68}	$W_{type\_mom\_AIC,\ pre\_birth\_info\_AIC}$	1	X_{107}										
X_{69}	$W_{pre\_birth\_info\_AIC,\ pre\_birth\_info}$	1											
X_{70}	$W_{calm\_dad\_AIC,\ acknowl\_fat}$	1											
X_{71}	$W_{teach\_dad\_sup\_mom\_AIC,\ acknowl\_fat}$	1											
X_{72}	$W_{calm\_dad\_AIC,\ com\_sup\_hcp}$	1											
X_{73}	$W_{teach\_dad\_sup\_mom\_AIC,\ com\_sup\_hcp}$	1											
X_{74}	$W_{calm\_dad\_AIC,\ post\_info\_com\_par}$	1											
X_{75}	$W_{teach\_dad\_sup\_mom\_AIC,\ post\_info\_com\_par}$	1											
X_{76}	$W_{encourage\_mom\_AIC,\ post\_info\_com\_par}$	1											
X_{77}	$W_{refer\_mom\_ther\_AIC,\ refer\_therapist}$	1											
X_{78}	$W_{calm\_dad\_AIC,\ mental\_sup}$	1											
X_{79}	$W_{teach\_dad\_sup\_mom\_AIC,\ mental\_sup}$	1											
X_{80}	$W_{encourage\_mom\_AIC,\ mental\_sup}$	1											
X_{81}	$W_{refer\_mom\_ther\_AIC,\ mental\_sup}$	1											
X_{82}	$W_{type\_com\_AIC,\ type\_com}$	1											
X_{83}	monitor_pre_birth_info	1	1										
X_{84}	monitor_acknowl_fat$_1$	1	1										
X_{85}	monitor_acknowl_fat$_2$	1	1										
X_{86}	monitor_com_sup_hcp$_1$	1	1										
X_{87}	monitor_com_sup_hcp$_2$	1	1										
X_{88}	monitor_post_info_com_par$_1$	1	1										
X_{89}	monitor_post_info_com_par$_2$	1	1										
X_{90}	monitor_post_info_com_par$_3$	1	1										
X_{91}	monitor_refer_therapist	1	1										
X_{92}	monitor_mental_sup$_1$	1	1										
X_{93}	monitor_mental_sup$_2$	1	1										
X_{94}	monitor_mental_sup$_3$	1	1										
X_{95}	monitor_mental_sup$_4$	1	1										
X_{96}	monitor_type_com	1	1										
X_{97}	W_{W_AIC,W_HCP}	1											
X_{98}	con_feedback	1											
X_{99}	$W_{birth\_dev\_det\_E,\ det\_risk\_level\_E}$												
X_{100}	$W_{det\_risk\_level\_E,\ encourage\_mom\_E}$												
X_{101}	$W_{det\_risk\_level\_E,\ calm\_dad\_E}$												
X_{102}	$W_{irr\_birth\_weight\_E,\ det\_underw\_E}$												
X_{103}	$W_{det\_underw\_E,\ refer\_mom\_ther\_E}$												
X_{104}	$W_{det\_underw\_E,\ teach\_dad\_sup\_mom\_E}$												
X_{105}	$W_{baby\_born\_E,\ type\_com\_E}$												
X_{106}	$W_{type\_mom\_E,\ pre\_birth\_info\_E}$												
X_{107}	W_{W_E,W_AIC}	1											
X_{108}	con_feedforward	1											

Table 32 Role matrix for Case (d) for **mcfw**—combination function weights

mcfw - combination function weights		1	2	3
		Alogistic	Monitor	Steponce
X_{1}	type_mom	1		
X_{2}	artificiality_birth	1		
X_{3}	epidural_used	1		
X_{4}	deliv_baby	1		
X_{5}	baby_born	1		
X_{6}	MPPD	1		
X_{7}	PPPD	1		
X_{8}	kid_devel	1		
X_{9}	pre_birth_info	1		
X_{10}	com_sup_hcp	1		
X_{11}	fat_ask_sup_com	1		
X_{12}	fat_rec_sup_com	1		
X_{13}	post_info_com_par	1		
X_{14}	refer_therapist	1		
X_{15}	mental_sup	1		
X_{16}	acknowl_fat	1		
X_{17}	age	1		
X_{18}	riskiness	1		
X_{19}	dilation_mom	1		
X_{20}	premature_birth	1		
X_{21}	type_com	1		
X_{22}	retention	1		
X_{23}	engaged	1		
X_{24}	stress	1		
X_{25}	birth_dev	1		
X_{26}	fear_level	1		
X_{27}	desire_sup	1		
X_{28}	ask_sup	1		
X_{29}	birth_dev_det$_{HCP}$	1		
X_{30}	det_risk_level$_{HCP}$	1		
X_{31}	encourage_mom$_{HCP}$	1		
X_{32}	calm_dad$_{HCP}$	1		
X_{33}	irr_birth_weight$_{HCP}$	1		
X_{34}	det_underw$_{HCP}$	1		
X_{35}	refer_mom_ther$_{HCP}$	1		
X_{36}	teach_dad_sup_mom$_{HCP}$	1		
X_{37}	baby_born$_{HCP}$	1		
X_{38}	type_com$_{HCP}$	1		
X_{39}	type_mom$_{HCP}$	1		
X_{40}	pre_birth_info$_{HCP}$	1		
X_{41}	birth_dev_det$_{AIC}$	1		
X_{42}	det_risk_level$_{AIC}$	1		
X_{43}	encourage_mom$_{AIC}$	1		
X_{44}	calm_dad$_{AIC}$	1		
X_{45}	irr_birth_weight$_{AIC}$	1		
X_{46}	det_underw$_{AIC}$	1		
X_{47}	refer_mom_ther$_{AIC}$	1		
X_{48}	teach_dad_sup_mom$_{AIC}$	1		
X_{49}	baby_born$_{AIC}$	1		
X_{50}	type_com$_{AIC}$	1		
X_{51}	type_mom$_{AIC}$	1		
X_{52}	pre_birth_info$_{AIC}$	1		
X_{53}	$\mathbf{W}_{\text{birth\_dev\_det\_HCP, det\_risk\_level\_HCP}}$	1		
X_{54}	$\mathbf{W}_{\text{det\_risk\_level\_HCP, encourage\_mom\_HCP}}$	1		
X_{55}	$\mathbf{W}_{\text{det\_risk\_level\_HCP, calm\_dad\_HCP}}$	1		
X_{56}	$\mathbf{W}_{\text{irr\_birth\_weight\_HCP, det\_underw\_HCP}}$	1		
X_{57}	$\mathbf{W}_{\text{det\_underw\_HCP, refer\_mom\_ther\_HCP}}$	1		
X_{58}	$\mathbf{W}_{\text{det\_underw\_HCP, teach\_dad\_sup\_mom\_HCP}}$	1		
X_{59}	$\mathbf{W}_{\text{baby\_born\_HCP, type\_com\_HCP}}$	1		
X_{60}	$\mathbf{W}_{\text{type\_mom\_HCP, pre\_birth\_info\_HCP}}$	1		
X_{61}	$\mathbf{W}_{\text{birth\_dev\_det\_AIC, det\_risk\_level\_AIC}}$	1		
X_{62}	$\mathbf{W}_{\text{det\_risk\_level\_AIC, encourage\_mom\_AIC}}$	1		
X_{63}	$\mathbf{W}_{\text{det\_risk\_level\_AIC, calm\_dad\_AIC}}$	1		
X_{64}	$\mathbf{W}_{\text{irr\_birth\_weight\_AIC, det\_underw\_AIC}}$	1		
X_{65}	$\mathbf{W}_{\text{det\_underw\_AIC, refer\_mom\_ther\_AIC}}$	1		
X_{66}	$\mathbf{W}_{\text{det\_underw\_AIC, teach\_dad\_sup\_mom\_AIC}}$	1		
X_{67}	$\mathbf{W}_{\text{baby\_born\_AIC, type\_com\_AIC}}$	1		
X_{68}	$\mathbf{W}_{\text{type\_mom\_AIC, pre\_birth\_info\_AIC}}$	1		
X_{69}	$\mathbf{W}_{\text{pre\_birth\_info\_AIC, pre\_birth\_info}}$	1		
X_{70}	$\mathbf{W}_{\text{calm\_dad\_AIC, acknowl\_fat}}$	1		
X_{71}	$\mathbf{W}_{\text{teach\_dad\_sup\_mom\_AIC, acknowl\_fat}}$	1		
X_{72}	$\mathbf{W}_{\text{calm\_dad\_AIC, com\_sup\_hcp}}$	1		
X_{73}	$\mathbf{W}_{\text{teach\_dad\_sup\_mom\_AIC, com\_sup\_hcp}}$	1		
X_{74}	$\mathbf{W}_{\text{calm\_dad\_AIC, post\_info\_com\_par}}$	1		
X_{75}	$\mathbf{W}_{\text{teach\_dad\_sup\_mom\_AIC, post\_info\_com\_par}}$	1		
X_{76}	$\mathbf{W}_{\text{encourage\_mom\_AIC, post\_info\_com\_par}}$	1		
X_{77}	$\mathbf{W}_{\text{refer\_mom\_ther\_AIC, refer\_therapist}}$	1		
X_{78}	$\mathbf{W}_{\text{calm\_dad\_AIC, mental\_sup}}$	1		
X_{79}	$\mathbf{W}_{\text{teach\_dad\_sup\_mom\_AIC, mental\_sup}}$	1		
X_{80}	$\mathbf{W}_{\text{encourage\_mom\_AIC, mental\_sup}}$	1		
X_{81}	$\mathbf{W}_{\text{refer\_mom\_ther\_AIC, mental\_sup}}$	1		
X_{82}	$\mathbf{W}_{\text{type\_com\_AIC, type\_com}}$	1		
X_{83}	monitor_pre_birth_info		1	
X_{84}	monitor_acknowl_fat$_{1}$		1	
X_{85}	monitor_acknowl_fat$_{2}$		1	
X_{86}	monitor_com_sup_hcp$_{1}$		1	
X_{87}	monitor_com_sup_hcp$_{2}$		1	
X_{88}	monitor_post_info_com_par$_{1}$		1	
X_{89}	monitor_post_info_com_par$_{2}$		1	
X_{90}	monitor_post_info_com_par$_{3}$		1	
X_{91}	monitor_refer_therapist		1	
X_{92}	monitor_mental_sup$_{1}$		1	
X_{93}	monitor_mental_sup$_{2}$		1	
X_{94}	monitor_mental_sup$_{3}$		1	
X_{95}	monitor_mental_sup$_{4}$		1	
X_{96}	monitor_type_com		1	
X_{97}	$\mathbf{W}_{\text{W\_AIC,W\_HCP}}$	1		
X_{98}	con_feedback			1
X_{99}	$\mathbf{W}_{\text{birth\_dev\_det\_E, det\_risk\_level\_E}}$	1		
X_{100}	$\mathbf{W}_{\text{det\_risk\_level\_E, encourage\_mom\_E}}$	1		
X_{101}	$\mathbf{W}_{\text{det\_risk\_level\_E, calm\_dad\_E}}$	1		
X_{102}	$\mathbf{W}_{\text{irr\_birth\_weight\_E, det\_underw\_E}}$	1		
X_{103}	$\mathbf{W}_{\text{det\_underw\_E, refer\_mom\_ther\_E}}$	1		
X_{104}	$\mathbf{W}_{\text{det\_underw\_E, teach\_dad\_sup\_mom\_E}}$	1		
X_{105}	$\mathbf{W}_{\text{baby\_born\_E, type\_com\_E}}$	1		
X_{106}	$\mathbf{W}_{\text{type\_mom\_E, pre\_birth\_info\_E}}$	1		
X_{107}	$\mathbf{W}_{\text{W\_E,W\_AIC}}$	1		
X_{108}	con_feedforward	1		

Table 33 Role matrix for Case (d) for **mcfp**—combination function parameters

mcfp - combination function parameters		1 Alogistic		2 Monitor		3 Steponce	
		Steepness	Threshold	Threshold		Start Time	End Time
X_{1}	type_mom	50	0.5				
X_{2}	artificiality_birth	50	0.5				
X_{3}	epidural_used	50	0.5				
X_{4}	deliv_baby	5	0.5				
X_{5}	baby_born	50	0.5				
X_{6}	MPPD	5	0.3				
X_{7}	PPPD	5	0.5				
X_{8}	kid_devel	5	0.5				
X_{9}	pre_birth_info	50	0.5				
X_{10}	com_sup_hcp	0.5	8				
X_{11}	fat_ask_sup_com	5	1				
X_{12}	fat_rec_sup_com	5	0.5				
X_{13}	post_info_com_par	5	4				
X_{14}	refer_therapist	5	1				
X_{15}	mental_sup	0.5	5				
X_{16}	acknowl_fat	5	0.8				
X_{17}	age	5	0.5				
X_{18}	riskiness	50	0.5				
X_{19}	dilation_mom	2.8	0				
X_{20}	premature_birth	5	0.5				
X_{21}	type_com	5	0.1				
X_{22}	retention	5	0.3				
X_{23}	engaged	5	0.5				
X_{24}	stress	5	0.5				
X_{25}	birth_dev	50	1				
X_{26}	fear_level	5	1				
X_{27}	desire_sup	5	0.3				
X_{28}	ask_sup	5	0.2				
X_{29}	birth_dev_det$_{\text{HCP}}$	5	1				
X_{30}	det_risk_level$_{\text{HCP}}$	5	0.5				
X_{31}	encourage_mom$_{\text{HCP}}$	5	0.6				
X_{32}	calm_dad$_{\text{HCP}}$	50	0.7				
X_{33}	irr_birth_weight$_{\text{HCP}}$	5	1				
X_{34}	det_underw$_{\text{HCP}}$	5	0.5				
X_{35}	refer_mom_ther$_{\text{HCP}}$	50	0.5				
X_{36}	teach_dad_sup_mom$_{\text{HCP}}$	50	0.7				
X_{37}	baby_born$_{\text{HCP}}$	5	0.5				
X_{38}	type_com$_{\text{HCP}}$	50	0.5				
X_{39}	type_mom$_{\text{HCP}}$	5	0.5				
X_{40}	pre_birth_info$_{\text{HCP}}$	50	0.5				
X_{41}	birth_dev_det$_{\text{AIC}}$	5	1				
X_{42}	det_risk_level$_{\text{AIC}}$	5	0.5				
X_{43}	encourage_mom$_{\text{AIC}}$	5	0.5				
X_{44}	calm_dad$_{\text{AIC}}$	50	0.7				
X_{45}	irr_birth_weight$_{\text{AIC}}$	5	1				
X_{46}	det_underw$_{\text{AIC}}$	5	0.5				
X_{47}	refer_mom_ther$_{\text{AIC}}$	50	0.5				
X_{48}	teach_dad_sup_mom$_{\text{AIC}}$	50	0.7				
X_{49}	baby_born$_{\text{AIC}}$	5	0.5				
X_{50}	type_com$_{\text{AIC}}$	50	0.5				
X_{51}	type_mom$_{\text{AIC}}$	5	0.5				
X_{52}	pre_birth_info$_{\text{AIC}}$	50	0.5				
X_{53}	$\mathbf{W}_{\text{birth\_dev\_det\_HCP, det\_risk\_level\_HCP}}$	5	0.5				
X_{54}	$\mathbf{W}_{\text{det\_risk\_level\_HCP, encourage\_mom\_HCP}}$	5	0.5				
X_{55}	$\mathbf{W}_{\text{det\_risk\_level\_HCP, calm\_dad\_HCP}}$	5	0.5				
X_{56}	$\mathbf{W}_{\text{irr\_birth\_weight\_HCP, det\_underw\_HCP}}$	5	0.5				
X_{57}	$\mathbf{W}_{\text{det\_underw\_HCP, refer\_mom\_ther\_HCP}}$	5	0.5				
X_{58}	$\mathbf{W}_{\text{det\_underw\_HCP, teach\_dad\_sup\_mom\_HCP}}$	5	0.5				
X_{59}	$\mathbf{W}_{\text{baby\_born\_HCP, type\_com\_HCP}}$	5	0.5				
X_{60}	$\mathbf{W}_{\text{type\_mom\_HCP, pre\_birth\_info\_HCP}}$	5	0.5				
X_{61}	$\mathbf{W}_{\text{birth\_dev\_det\_AIC, det\_risk\_level\_AIC}}$	5	0.5				
X_{62}	$\mathbf{W}_{\text{det\_risk\_level\_AIC, encourage\_mom\_AIC}}$	5	0.5				
X_{63}	$\mathbf{W}_{\text{det\_risk\_level\_AIC, calm\_dad\_AIC}}$	5	0.5				
X_{64}	$\mathbf{W}_{\text{irr\_birth\_weight\_AIC, det\_underw\_AIC}}$	5	0.5				
X_{65}	$\mathbf{W}_{\text{det\_underw\_AIC, refer\_mom\_ther\_AIC}}$	5	0.5				
X_{66}	$\mathbf{W}_{\text{det\_underw\_AIC, teach\_dad\_sup\_mom\_AIC}}$	5	0.5				
X_{67}	$\mathbf{W}_{\text{baby\_born\_AIC, type\_com\_AIC}}$	5	0.5				
X_{68}	$\mathbf{W}_{\text{type\_mom\_AIC, pre\_birth\_info\_AIC}}$	5	0.5				
X_{69}	$\mathbf{W}_{\text{pre\_birth\_info\_AIC, pre\_birth\_info}}$	5	0.5				
X_{70}	$\mathbf{W}_{\text{calm\_dad\_AIC, acknowl\_fat}}$	5	0.5				
X_{71}	$\mathbf{W}_{\text{teach\_dad\_sup\_mom\_AIC, acknowl\_fat}}$	5	0.5				
X_{72}	$\mathbf{W}_{\text{calm\_dad\_AIC, com\_sup\_hcp}}$	5	0.5				
X_{73}	$\mathbf{W}_{\text{teach\_dad\_sup\_mom\_AIC, com\_sup\_hcp}}$	5	0.5				
X_{74}	$\mathbf{W}_{\text{calm\_dad\_AIC, post\_info\_com\_par}}$	5	0.5				
X_{75}	$\mathbf{W}_{\text{teach\_dad\_sup\_mom\_AIC, post\_info\_com\_par}}$	5	0.5				
X_{76}	$\mathbf{W}_{\text{encourage\_mom\_AIC, post\_info\_com\_par}}$	5	0.5				
X_{77}	$\mathbf{W}_{\text{refer\_mom\_ther\_AIC, refer\_therapist}}$	5	0.5				
X_{78}	$\mathbf{W}_{\text{calm\_dad\_AIC, mental\_sup}}$	5	0.5				
X_{79}	$\mathbf{W}_{\text{teach\_dad\_sup\_mom\_AIC, mental\_sup}}$	5	0.5				
X_{80}	$\mathbf{W}_{\text{encourage\_mom\_AIC, mental\_sup}}$	5	0.5				
X_{81}	$\mathbf{W}_{\text{refer\_mom\_ther\_AIC, mental\_sup}}$	5	0.5				
X_{82}	$\mathbf{W}_{\text{type\_com\_AIC, type\_com}}$	5	0.5				
X_{83}	monitor_pre_birth_info			0.6			
X_{84}	monitor_acknowl_fat$_{1}$			0.6			
X_{85}	monitor_acknowl_fat$_{2}$			0.6			
X_{86}	monitor_com_sup_hcp$_{1}$			0.6			
X_{87}	monitor_com_sup_hcp$_{2}$			0.6			
X_{88}	monitor_post_info_com_par$_{1}$			0.6			
X_{89}	monitor_post_info_com_par$_{2}$			0.6			
X_{90}	monitor_post_info_com_par$_{3}$			0.6			
X_{91}	monitor_refer_therapist			0.6			
X_{92}	monitor_mental_sup$_{1}$			0.6			
X_{93}	monitor_mental_sup$_{2}$			0.6			
X_{94}	monitor_mental_sup$_{3}$			0.6			
X_{95}	monitor_mental_sup$_{4}$			0.6			
X_{96}	monitor_type_com			0.6			
X_{97}	$\mathbf{W}_{\text{W\_AIC,W\_HCP}}$	30	0.5				
X_{98}	con_feedback					40	80
X_{99}	$\mathbf{W}_{\text{birth\_dev\_det\_E, det\_risk\_level\_E}}$	50	0.5				
X_{100}	$\mathbf{W}_{\text{det\_risk\_level\_E, encourage\_mom\_E}}$	50	0.5				
X_{101}	$\mathbf{W}_{\text{det\_risk\_level\_E, calm\_dad\_E}}$	50	0.5				
X_{102}	$\mathbf{W}_{\text{irr\_birth\_weight\_E, det\_underw\_E}}$	50	0.5				
X_{103}	$\mathbf{W}_{\text{det\_underw\_E, refer\_mom\_ther\_E}}$	50	0.5				
X_{104}	$\mathbf{W}_{\text{det\_underw\_E, teach\_dad\_sup\_mom\_E}}$	50	0.5				
X_{105}	$\mathbf{W}_{\text{baby\_born\_E, type\_com\_E}}$	50	0.5				
X_{106}	$\mathbf{W}_{\text{type\_mom\_E, pre\_birth\_info\_E}}$	50	0.5				
X_{107}	$\mathbf{W}_{\text{W\_E,W\_AIC}}$	50	0.5				
X_{108}	con_feedforward	5	0.5				

Table 34 Role matrix for Case (d) for **iv**——initial values

	iv - initial values	1
X_1	type_mom	0
X_2	artificiality_birth	0
X_3	epidural_used	0
X_4	deliv_baby	0
X_5	baby_born	0
X_6	MPPD	0.13
X_7	PPPD	0
X_8	kid_devel	0
X_9	pre_birth_info	0
X_{10}	com_sup_hcp	0
X_{11}	fat_ask_sup_com	0
X_{12}	fat_rec_sup_com	0
X_{13}	post_info_com_par	0
X_{14}	refer_therapist	0
X_{15}	mental_sup	0
X_{16}	acknowl_fat	0
X_{17}	age	1
X_{18}	riskiness	0
X_{19}	dilation_mom	0.2
X_{20}	premature_birth	1
X_{21}	type_com	0
X_{22}	retention	0
X_{23}	engaged	0
X_{24}	stress	0
X_{25}	birth_dev	0
X_{26}	fear_level	0
X_{27}	desire_sup	0
X_{28}	ask_sup	0
X_{29}	birth_dev_det$_{HCP}$	0
X_{30}	det_risk_level$_{HCP}$	0
X_{31}	encourage_mom$_{HCP}$	0
X_{32}	calm_dad$_{HCP}$	0
X_{33}	irr_birth_weight$_{HCP}$	0
X_{34}	det_underw$_{HCP}$	0
X_{35}	refer_mom_ther$_{HCP}$	0
X_{36}	teach_dad_sup_mom$_{HCP}$	0
X_{37}	baby_born$_{HCP}$	0
X_{38}	type_com$_{HCP}$	0
X_{39}	type_mom$_{HCP}$	0
X_{40}	pre_birth_info$_{HCP}$	0
X_{41}	birth_dev_det$_{AIC}$	0
X_{42}	det_risk_level$_{AIC}$	0
X_{43}	encourage_mom$_{AIC}$	0
X_{44}	calm_dad$_{AIC}$	0
X_{45}	irr_birth_weight$_{AIC}$	0
X_{46}	det_underw$_{AIC}$	0
X_{47}	refer_mom_ther$_{AIC}$	0
X_{48}	teach_dad_sup_mom$_{AIC}$	0
X_{49}	baby_born$_{AIC}$	0
X_{50}	type_com$_{AIC}$	0
X_{51}	type_mom$_{AIC}$	0
X_{52}	pre_birth_info$_{AIC}$	0
X_{53}	$\mathbf{W}_{\text{birth\_dev\_det\_HCP, det\_risk\_level\_HCP}}$	0
X_{54}	$\mathbf{W}_{\text{det\_risk\_level\_HCP, encourage\_mom\_HCP}}$	0
X_{55}	$\mathbf{W}_{\text{det\_risk\_level\_HCP, calm\_dad\_HCP}}$	0
X_{56}	$\mathbf{W}_{\text{irr\_birth\_weight\_HCP, det\_underw\_HCP}}$	0
X_{57}	$\mathbf{W}_{\text{det\_underw\_HCP, refer\_mom\_ther\_HCP}}$	0
X_{58}	$\mathbf{W}_{\text{det\_underw\_HCP, teach\_dad\_sup\_mom\_HCP}}$	0
X_{59}	$\mathbf{W}_{\text{baby\_born\_HCP, type\_com\_HCP}}$	0
X_{60}	$\mathbf{W}_{\text{type\_mom\_HCP, pre\_birth\_info\_HCP}}$	0
X_{61}	$\mathbf{W}_{\text{birth\_dev\_det\_AIC, det\_risk\_level\_AIC}}$	0
X_{62}	$\mathbf{W}_{\text{det\_risk\_level\_AIC, encourage\_mom\_AIC}}$	0
X_{63}	$\mathbf{W}_{\text{det\_risk\_level\_AIC, calm\_dad\_AIC}}$	0
X_{64}	$\mathbf{W}_{\text{irr\_birth\_weight\_AIC, det\_underw\_AIC}}$	0
X_{65}	$\mathbf{W}_{\text{det\_underw\_AIC, refer\_mom\_ther\_AIC}}$	0
X_{66}	$\mathbf{W}_{\text{det\_underw\_AIC, teach\_dad\_sup\_mom\_AIC}}$	0
X_{67}	$\mathbf{W}_{\text{baby\_born\_AIC, type\_com\_AIC}}$	0
X_{68}	$\mathbf{W}_{\text{type\_mom\_AIC, pre\_birth\_info\_AIC}}$	0
X_{69}	$\mathbf{W}_{\text{pre\_birth\_info\_AIC, pre\_birth\_info}}$	0
X_{70}	$\mathbf{W}_{\text{calm\_dad\_AIC, acknowl\_fat}}$	0
X_{71}	$\mathbf{W}_{\text{teach\_dad\_sup\_mom\_AIC, acknowl\_fat}}$	0
X_{72}	$\mathbf{W}_{\text{calm\_dad\_AIC, com\_sup\_hcp}}$	0
X_{73}	$\mathbf{W}_{\text{teach\_dad\_sup\_mom\_AIC, com\_sup\_hcp}}$	0
X_{74}	$\mathbf{W}_{\text{calm\_dad\_AIC, post\_info\_com\_par}}$	0
X_{75}	$\mathbf{W}_{\text{teach\_dad\_sup\_mom\_AIC, post\_info\_com\_par}}$	0
X_{76}	$\mathbf{W}_{\text{encourage\_mom\_AIC, post\_info\_com\_par}}$	0
X_{77}	$\mathbf{W}_{\text{refer\_mom\_ther\_AIC, refer\_therapist}}$	0
X_{78}	$\mathbf{W}_{\text{calm\_dad\_AIC, mental\_sup}}$	0
X_{79}	$\mathbf{W}_{\text{teach\_dad\_sup\_mom\_AIC, mental\_sup}}$	0
X_{80}	$\mathbf{W}_{\text{encourage\_mom\_AIC, mental\_sup}}$	0
X_{81}	$\mathbf{W}_{\text{refer\_mom\_ther\_AIC, mental\_sup}}$	0
X_{82}	$\mathbf{W}_{\text{type\_com\_AIC, type\_com}}$	0
X_{83}	monitor_pre_birth_info	0
X_{84}	monitor_acknowl_fat$_1$	0
X_{85}	monitor_acknowl_fat$_2$	0
X_{86}	monitor_com_sup_hcp$_1$	0
X_{87}	monitor_com_sup_hcp$_2$	0
X_{88}	monitor_post_info_com_par$_1$	0
X_{89}	monitor_post_info_com_par$_2$	0
X_{90}	monitor_post_info_com_par$_3$	0
X_{91}	monitor_refer_therapist	0
X_{92}	monitor_mental_sup$_1$	0
X_{93}	monitor_mental_sup$_2$	0
X_{94}	monitor_mental_sup$_3$	0
X_{95}	monitor_mental_sup$_4$	0
X_{96}	monitor_type_com	0
X_{97}	$\mathbf{W}_{\text{W\_AIC,W\_HCP}}$	0
X_{98}	con_feedback	0
X_{99}	$\mathbf{W}_{\text{birth\_dev\_det\_E, det\_risk\_level\_E}}$	1
X_{100}	$\mathbf{W}_{\text{det\_risk\_level\_E, encourage\_mom\_E}}$	1
X_{101}	$\mathbf{W}_{\text{det\_risk\_level\_E, calm\_dad\_E}}$	1
X_{102}	$\mathbf{W}_{\text{irr\_birth\_weight\_E, det\_underw\_E}}$	1
X_{103}	$\mathbf{W}_{\text{det\_underw\_E, refer\_mom\_ther\_E}}$	1
X_{104}	$\mathbf{W}_{\text{det\_underw\_E, teach\_dad\_sup\_mom\_E}}$	1
X_{105}	$\mathbf{W}_{\text{baby\_born\_E, type\_com\_E}}$	1
X_{106}	$\mathbf{W}_{\text{type\_mom\_E, pre\_birth\_info\_E}}$	1
X_{107}	$\mathbf{W}_{\text{W\_E,W\_AIC}}$	0
X_{108}	con_feedforward	1

Table 35 Role matrix for Case (d) for **ms**—speed factors

ms - speed factors		1
X_{1}	type_mom	1
X_{2}	artificiality_birth	1
X_{3}	epidural_used	1
X_{4}	deliv_baby	0.2
X_{5}	baby_born	0.1
X_{6}	MPPD	0.2
X_{7}	PPPD	1
X_{8}	kid_devel	1
X_{9}	pre_birth_info	0.1
X_{10}	com_sup_hcp	0.1
X_{11}	fat_ask_sup_com	0.3
X_{12}	fat_rec_sup_com	0.1
X_{13}	post_info_com_par	0.1
X_{14}	refer_therapist	0.3
X_{15}	mental_sup	0.1
X_{16}	acknowl_fat	0.1
X_{17}	age	0
X_{18}	riskiness	0.5
X_{19}	dilation_mom	1
X_{20}	premature_birth	0
X_{21}	type_com	1
X_{22}	retention	1
X_{23}	engaged	1
X_{24}	stress	0.4
X_{25}	birth_dev	1
X_{26}	fear_level	1
X_{27}	desire_sup	1
X_{28}	ask_sup	0.2
X_{29}	birth_dev_det$_{HCP}$	1
X_{30}	det_risk_level$_{HCP}$	1
X_{31}	encourage_mom$_{HCP}$	0.1
X_{32}	calm_dad$_{HCP}$	0.1
X_{33}	irr_birth_weight$_{HCP}$	1
X_{34}	det_underw$_{HCP}$	1
X_{35}	refer_mom_ther$_{HCP}$	0.3
X_{36}	teach_dad_sup_mom$_{HCP}$	0.1
X_{37}	baby_born$_{HCP}$	1
X_{38}	type_com$_{HCP}$	1
X_{39}	type_mom$_{HCP}$	1
X_{40}	pre_birth_info$_{HCP}$	1
X_{41}	birth_dev_det$_{AIC}$	0.33
X_{42}	det_risk_level$_{AIC}$	0.33
X_{43}	encourage_mom$_{AIC}$	0.033
X_{44}	calm_dad$_{AIC}$	0.033
X_{45}	irr_birth_weight$_{AIC}$	0.33
X_{46}	det_underw$_{AIC}$	0.33
X_{47}	refer_mom_ther$_{AIC}$	0.1
X_{48}	teach_dad_sup_mom$_{AIC}$	0.033
X_{49}	baby_born$_{AIC}$	0.33
X_{50}	type_com$_{AIC}$	0.33
X_{51}	type_mom$_{AIC}$	0.33
X_{52}	pre_birth_info$_{AIC}$	0.33
X_{53}	$\mathbf{W}_{birth\_dev\_det\_HCP,\ det\_risk\_level\_HCP}$	1
X_{54}	$\mathbf{W}_{det\_risk\_level\_HCP,\ encourage\_mom\_HCP}$	1
X_{55}	$\mathbf{W}_{det\_risk\_level\_HCP,\ calm\_dad\_HCP}$	1
X_{56}	$\mathbf{W}_{irr\_birth\_weight\_HCP,\ det\_underw\_HCP}$	1
X_{57}	$\mathbf{W}_{det\_underw\_HCP,\ refer\_mom\_ther\_HCP}$	1
X_{58}	$\mathbf{W}_{det\_underw\_HCP,\ teach\_dad\_sup\_mom\_HCP}$	1
X_{59}	$\mathbf{W}_{baby\_born\_HCP,\ type\_com\_HCP}$	1
X_{60}	$\mathbf{W}_{type\_mom\_HCP,\ pre\_birth\_info\_HCP}$	1
X_{61}	$\mathbf{W}_{birth\_dev\_det\_AIC,\ det\_risk\_level\_AIC}$	0.5
X_{62}	$\mathbf{W}_{det\_risk\_level\_AIC,\ encourage\_mom\_AIC}$	0.5
X_{63}	$\mathbf{W}_{det\_risk\_level\_AIC,\ calm\_dad\_AIC}$	0.5
X_{64}	$\mathbf{W}_{irr\_birth\_weight\_AIC,\ det\_underw\_AIC}$	0.5
X_{65}	$\mathbf{W}_{det\_underw\_AIC,\ refer\_mom\_ther\_AIC}$	0.5
X_{66}	$\mathbf{W}_{det\_underw\_AIC,\ teach\_dad\_sup\_mom\_AIC}$	0.5
X_{67}	$\mathbf{W}_{baby\_born\_AIC,\ type\_com\_AIC}$	0.5
X_{68}	$\mathbf{W}_{type\_mom\_AIC,\ pre\_birth\_info\_AIC}$	0.5
X_{69}	$\mathbf{W}_{pre\_birth\_info\_AIC,\ pre\_birth\_info}$	0.5
X_{70}	$\mathbf{W}_{calm\_dad\_AIC,\ acknowl\_fat}$	0.5
X_{71}	$\mathbf{W}_{teach\_dad\_sup\_mom\_AIC,\ acknowl\_fat}$	0.5
X_{72}	$\mathbf{W}_{calm\_dad\_AIC,\ com\_sup\_hcp}$	0.5
X_{73}	$\mathbf{W}_{teach\_dad\_sup\_mom\_AIC,\ com\_sup\_hcp}$	0.5
X_{74}	$\mathbf{W}_{calm\_dad\_AIC,\ post\_info\_com\_par}$	0.5
X_{75}	$\mathbf{W}_{teach\_dad\_sup\_mom\_AIC,\ post\_info\_com\_par}$	0.5
X_{76}	$\mathbf{W}_{encourage\_mom\_AIC,\ post\_info\_com\_par}$	0.5
X_{77}	$\mathbf{W}_{refer\_mom\_ther\_AIC,\ refer\_therapist}$	0.5
X_{78}	$\mathbf{W}_{calm\_dad\_AIC,\ mental\_sup}$	0.5
X_{79}	$\mathbf{W}_{teach\_dad\_sup\_mom\_AIC,\ mental\_sup}$	0.5
X_{80}	$\mathbf{W}_{encourage\_mom\_AIC,\ mental\_sup}$	0.5
X_{81}	$\mathbf{W}_{refer\_mom\_ther\_AIC,\ mental\_sup}$	0.5
X_{82}	$\mathbf{W}_{type\_com\_AIC,\ type\_com}$	0.5
X_{83}	monitor_pre_birth_info	0.5
X_{84}	monitor_acknowl_fat$_{1}$	0.5
X_{85}	monitor_acknowl_fat$_{2}$	0.5
X_{86}	monitor_com_sup_hcp$_{1}$	0.5
X_{87}	monitor_com_sup_hcp$_{2}$	0.5
X_{88}	monitor_post_info_com_par$_{1}$	0.5
X_{89}	monitor_post_info_com_par$_{2}$	0.5
X_{90}	monitor_post_info_com_par$_{3}$	0.5
X_{91}	monitor_refer_therapist	0.5
X_{92}	monitor_mental_sup$_{1}$	0.5
X_{93}	monitor_mental_sup$_{2}$	0.5
X_{94}	monitor_mental_sup$_{3}$	0.5
X_{95}	monitor_mental_sup$_{4}$	0.5
X_{96}	monitor_type_com	0.5
X_{97}	$\mathbf{W}_{W\_AIC,W\_HCP}$	1
X_{98}	con_feedback	0.05
X_{99}	$\mathbf{W}_{birth\_dev\_det\_E,\ det\_risk\_level\_E}$	0
X_{100}	$\mathbf{W}_{det\_risk\_level\_E,\ encourage\_mom\_E}$	0
X_{101}	$\mathbf{W}_{det\_risk\_level\_E,\ calm\_dad\_E}$	0
X_{102}	$\mathbf{W}_{irr\_birth\_weight\_E,\ det\_underw\_E}$	0
X_{103}	$\mathbf{W}_{det\_underw\_E,\ refer\_mom\_ther\_E}$	0
X_{104}	$\mathbf{W}_{det\_underw\_E,\ teach\_dad\_sup\_mom\_E}$	0
X_{105}	$\mathbf{W}_{baby\_born\_E,\ type\_com\_E}$	0
X_{106}	$\mathbf{W}_{type\_mom\_E,\ pre\_birth\_info\_E}$	0
X_{107}	$\mathbf{W}_{W\_E,W\_AIC}$	1.5
X_{108}	con_feedforward	0

Communication

Adequate Communication

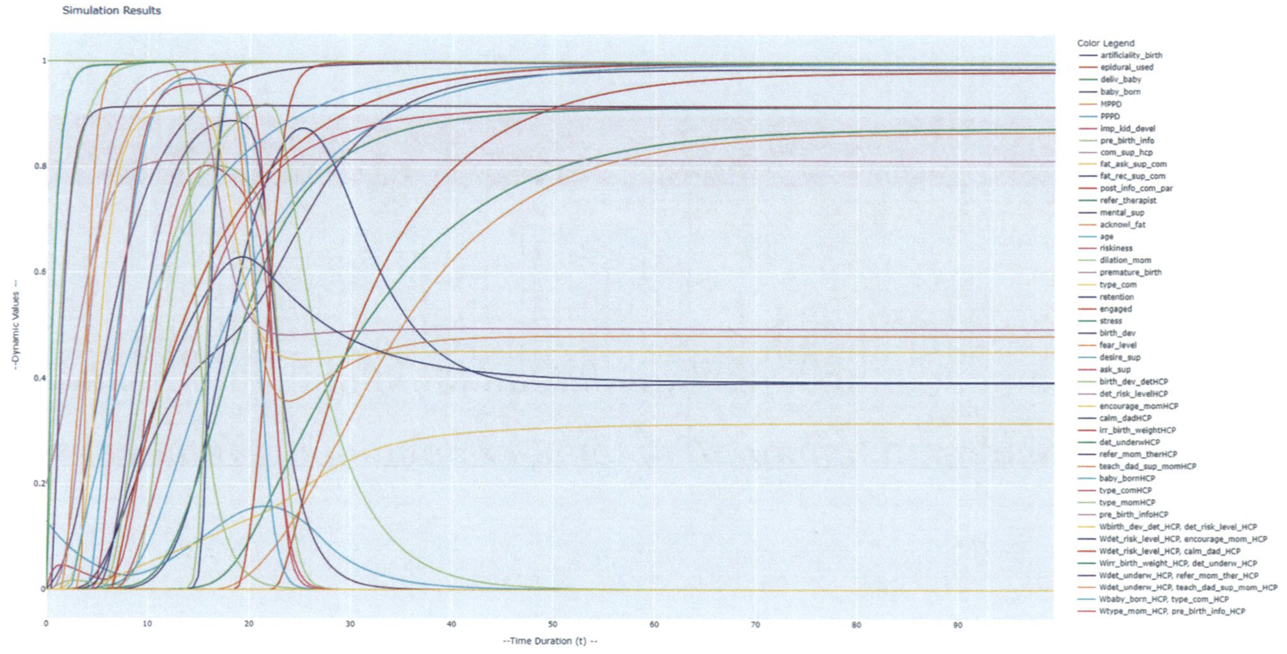

Ineffective Communication

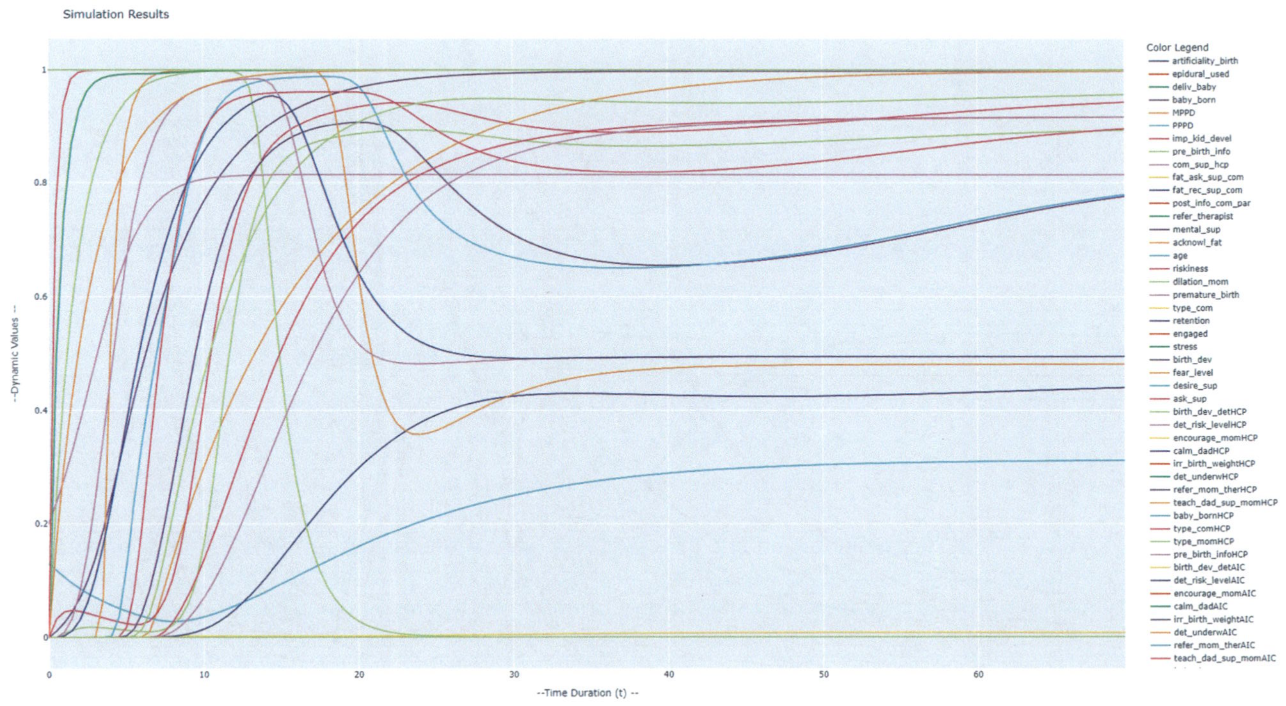

Role of Virtual Coach

References

Abenova, M., Myssayev, A., Kanya, L., Turliuc, M.N., Jamedinova, U.: Prevalence of postpartum depression and its associated factors within a year after birth in Semey, Kazakhstan: a cross sectional study. Clin. Epidemiol. Global Health **16**, 101103 (2022). https://doi.org/10.1016/j.cegh.2022.101103

Beyea, S.C.: Creating a just safety culture. AORN J. **79**(2), 412–414 (2004). https://doi.org/10.1016/S0001-2092(06)60618-2

Biebel, K., Alikhan, S.: University of Massachusetts Medical School. Paternal Postpartum Depression. J. Parent Family Mental Health **1**(1) (2018). https://doi.org/10.7191/parentandfamily.1000

Bouma, D., Canbaloğlu, G., Treur, J., Wiewiora, A.: Adaptive network modeling of the influence of leadership and communication on learning within an organization. Cogn. Syst. Res. **79**, 55–70 (2022)

Boysen, P.G.: Just culture: a foundation for balanced accountability and patient safety. Ochsner J. **13**(3), 400–406. PubMed (2013)

Canbaloğlu, G., Treur, J., Roelofsma, P.H.M.P.: Computational modeling of organisational learning by self-modeling networks. Cogn. Syst. Res. **73**, 51–64 (2022). https://doi.org/10.1016/j.cogsys.2021.12.003

Canbaloğlu, G., Treur, J., Roelofsma, P.H.M.P.: Learning of safety and security management through cyberspace: an adaptive self-modeling network model for multilevel organizational learning. In: Roelofsma, P.H.M.P., Jabeen, F., Taal, H. R., Treur, J. (eds.) Using Shared Mental Models and Organisational Learning to Support Safety and Security through Cyberspace: A Computational Analysis Approach. Springer Nature, Cham, Switzerland (this volume) (2025)

Cavazos-Rehg, P.A., Krauss, M.J., Spitznagel, E.L., Bommarito, K., Madden, T., Olsen, M.A., Subramaniam, H., Peipert, J.F., Bierut, L.J.: Maternal age and risk of labor and delivery complications. Matern. Child Health J. **19**(6), 1202–1211. PubMed (2015). https://doi.org/10.1007/s10995-014-1624-7

Cerezo, M.A., Pons-Salvador, G., Trenado, R.M.: Mother–infant interaction and children's socio-emotional development with high- and low-risk mothers. Infant Behav. Dev. **31**(4), 578–589 (2008). https://doi.org/10.1016/j.infbeh.2008.07.010

Crossan, M.M., Lane, H.W., White, R.E.: An organizational learning framework: from intuition to institution. Acad. Manag. Rev. **24**, 522–537 (1999)

Darwin, Z., Galdas, P., Hinchliff, S., Littlewood, E., McMillan, D., McGowan, L., Gilbody, S., Born, Bred in Yorkshire, team.: Fathers' views and experiences of their own mental health during pregnancy and the first postnatal year: A qualitative interview study of men participating in the UK Born and Bred in Yorkshire (BaBY) cohort. BMC Pregnancy Childbirth **17**(1), 45. Medline (2017). https://doi.org/10.1186/s12884-017-1229-4

Dennis, C.-L.E.: Preventing postpartum depression Part II: a critical review of nonbiological interventions. Can. J. Psychiatry **49**(8), 526–538 (2004). https://doi.org/10.1177/070674370404900804

Dunkel Schetter, C., Saxbe, D.E., Cheadle, A.C.D., Guardino, C.M.: Postpartum depressive symptoms following consecutive pregnancies: stability, change, and mechanisms. Clin. Psychol. Sci. **4**(5), 909–918 (2016). https://doi.org/10.1177/2167702616644894

Garthus-Niegel, S., Radoš, S. N., Horsch, A.: Perinatal depression and beyond—implications for research design and clinical management. JAMA Netw. Open **5**(6), e2218978 (2022). https://doi.org/10.1001/jamanetworkopen.2022.18978

Ghaedrahmati, M., Kazemi, A., Kheirabadi, G., Ebrahimi, A., Bahrami, M.: Postpartum depression risk factors: a narrative review. J. Educ. Health Promot. **6**, 60. PubMed-not-MEDLINE (2017). https://doi.org/10.4103/jehp.jehp_9_16

Goodman, J.H.: Paternal postpartum depression, its relationship to maternal postpartum depression, and implications for family health. J. Adv. Nursing **45**(1), 26–35 (2004). https://doi.org/10.1046/j.1365-2648.2003.02857.x

Goodman, J.H.: Influences of maternal postpartum depression on fathers and on father-infant interaction. Infant. Ment. Health J. **29**(6), 624–643. PubMed-not-MEDLINE (2008). https://doi.org/10.1002/imhj.20199

Hassan, I.: Avoiding medication errors through effective communication in healthcare environment. SSRN Electron. J. (2018). https://doi.org/10.2139/ssrn.3573437

Hebb, D.: The Organization of Behavior. Wiley, New York (1949)

Hedberg, B.: How organizations learn and unlearn. In: Nystrom, P.C. and Starbuck, W.H. (eds.) The Handbook of Organizational Design, vol. 1. Oxford University Press, New York, NY (1981)

Hendrikse, S.C.F., Treur.J., Koole. S.L.: Modeling emerging interpersonal synchrony and its related adaptive short-term affiliation and long-term bonding: a second-order multi-adaptive neural agent model. Int. J. Neural Syst. **33**(7), 2350038, 1–41 (2023). https://doi.org/10.1142/S0129065723500387

Hueston, W.: Preterm contractions in community settings: II. predicting preterm birth in women with preterm contractions. Obstet. Gynecol. **92**(1), 43–46 (1998). https://doi.org/10.1016/S0029-7844(98)00108-2

Javaid, I., Majeed, Z., Ahmad, A., Komal, M., Khan, H.: Postpartum depression: a comprehensive review. Int. Clin. Img. Med. Rew. **2**(3), 1071 (2022)

Kenyon, S.: Saving lives, improving mothers' care: surveillance of maternal deaths in the UK 2011–13 and lessons learned to inform maternity care from the UK and Ireland confidential enquiries into maternal deaths and morbidity 2009–2013. In: Knight, M., Tuffnell, D., Shakespeare, J., Grey, R., Kurinczuk, J. (eds.) National Perinatal Epidemiology Unit. University of Oxford (2015). https://commed.vcu.edu/IntroPH/MCH/2016/UKRptMatHlth.pdf

Kim, D.H.: The link between individual and organizational learning. In: Klein, D.A. (ed.) The Strategic Management of Intellectual Capital: Vol. 1st Edition. Routledge (1997). https://doi.org/10.4324/9780080517926

Mayers, A., Hambidge, S., Bryant, O., Arden-Close, E.: Supporting women who develop poor postnatal mental health: What support do fathers receive to support their partner and their own mental health? BMC Pregnancy Childbirth **20**(1), 359 (2020). https://doi.org/10.1186/s12884-020-03043-2

Melrose, S.: Paternal postpartum depression: how can nurses begin to help? Contemp. Nurse **34**(2), 199–210 (2010). https://doi.org/10.5172/conu.2010.34.2.199

Maternal mental health and child health and development in resource-constrained settings. Who.int; World Health Organization (2009). https://www.who.int/publications/i/item/WHO-RHR-09.24

NHS, N.H.S.: The stages of labour and birth (2020). https://www.nhs.uk/pregnancy/labour-and-birth/what-happens/the-stages-of-labour-and-birth/

Pedersen, S.C., Maindal, H.T., Ryom, K.: I wanted to be there as a father, but i couldn't: a qualitative study of fathers' experiences of postpartum depression and their help-seeking behavior. Am. J. Mens. Health **15**(3), 15579883211024376. Medline (2021). https://doi.org/10.1177/15579883211024375

Piaget, J.: Structuralism. Basic Books (1970). https://books.google.nl/books?id=GSnxHZZOZmQC

Pinto-Foltz, M.D., Logsdon, M.C.: Stigma towards mental illness: a concept analysis using postpartum depression as an exemplar. Issues Mental Health Nursing **29**(1), 21–36 (2008). https://doi.org/10.1080/01612840701748698

Rao, W.-W., Zhu, X.-M., Zong, Q.-Q., Zhang, Q., Hall, B.J., Ungvari, G.S., Xiang, Y.-T.: Prevalence of prenatal and postpartum depression in fathers: A comprehensive meta-analysis of observational surveys. J. Affect. Disord. **263**, 491–499 (2020). https://doi.org/10.1016/j.jad.2019.10.030

Ratnapalan, S., Uleryk, E.: Organizational learning in health care organizations. System **2**(1), 24–33 (2014)

Robinson, B.L., Harper, N.S., McAlpine, D.: Meta-adaptation in the auditory midbrain under cortical influence. Nat. Commun. **7**(1), 13442 (2016). https://doi.org/10.1038/ncomms13442

Rohde, P., Lewinsohn, P.M., Klein, D.N., Seeley, J.R.: Association of parental depression with psychiatric course from adolescence to young adulthood among formerly depressed individuals. J. Abnorm. Psychol. **114**(3), 409–420 (2005). https://doi.org/10.1037/0021-843X.114.3.409

Shrivastava, P.: A typology of organizational learning systems. J. Manag. Stud. **20**(1), 7–28 (1983)

Silverman, M.E., Reichenberg, A., Savitz, D.A., Cnattingius, S., Lichtenstein, P., Hultman, C.M., Larsson, H., Sandin, S.: The risk factors for postpartum depression: a population-based study. Depression Anxiety **34**(2), 178–187 (2017). https://doi.org/10.1002/da.22597

Smythe, K.L., Petersen, I., Schartau, P.: Prevalence of perinatal depression and anxiety in both parents: a systematic review and meta-analysis. JAMA Netw. Open **5**(6), e2218969–e2218969 (2022). https://doi.org/10.1001/jamanetworkopen.2022.18969

Solomon, A.: A Road Map to Just Culture. TiPS Top. Patient Saf. **16**(2) (2016). https://www.patientsafety.va.gov/docs/TIPS/TIPS_AprilMayJune16.pdf

Taal, H.R.: Inside the Delivery Room [Interview] (2022). https://www.youtube.com/channel/UCsNe6b1bgEHNKNXibXaLaYw

Treur, J.: Network-Oriented Modeling: Addressing Complexity of Cognitive, Affective and Social Interactions. Springer International Publishing (2016)

Treur, J.: Overview of the Combination Function Library (2019)

Treur, J.: Network-Oriented Modeling for Adaptive Networks: Designing Higher-Order Adaptive Biological, Mental and Social Network Models. Springer Nature (2020)

Treur, J.: On the dynamics and adaptivity of mental processes: relating adaptive dynamical systems and self-modeling network models by mathematical analysis. Cogn. Syst. Res. **70**, 93–100 (2021)

Treur, J., Van Ments, L.: Mental Models and Their Dynamics, Adaptation, and Control: A Self-Modeling Network Modeling Approach. Springer Nature (2022)

Van Ments, L., Treur, J., Klein, J. Roelofsma, P.H.M.P.: Developing a safety and security AI coach: A second-order adaptive network model of shared mental models in hospital teamwork. In: Roelofsma, P.H.M.P., Jabeen, F., Taal, H.R., Treur, J. (eds.) Using Shared Mental Models and Organisational Learning to Support Safety and Security through Cyberspace: A Computational Analysis Approach. Springer Nature, Cham, Switzerland (this volume) (2025)

Weigl, L. M., Jabeen, F., Treur, J., Taal, H. R., Roelofsma, P. H.: Modelling learning for a better safety culture within an organization using a virtual safety coach: Reducing the risk of postpartum depression via improved communication with parents. Cogn. Syst. Res. **80**, 1–36 (2023)

World Health Organization (WHO).: The ICD-10 classification of mental and behavioural disorders: Diagnostic criteria for research. World Health Organization (1993)

Optimizing Speaking up Behavior for Safety and Security Through Cyberspace: A Computational Network Model of a Virtual Coach Supporting Speaking up

Shaney Doornkamp, Fakhra Jabeen, Jan Treur, H. Rob Taal, and Peter H. M. P. Roelofsma

Abstract Previous reports show that a substantial proportion of (near) medical errors in the operating theatre is attributable to ineffective communication between health-care professionals. Speaking up about observed medical errors is a safety behaviour which promotes effective communication between health care professionals, consequently optimising patient care by reducing medical error risk. Speaking up by health care professionals (e.g., nurses, residents) remains difficult to execute in practice despite increasing awareness of its importance. Therefore, this chapter introduces a computational model concerning the mechanisms known from psychological, observational, and medical literature which underlie the speaking up behaviour of a health care professional. It also addresses how a doctor may respond to the communicated

S. Doornkamp · F. Jabeen
Safety and Security Science Section, Delft University of Technology, Delft, The Netherlands
e-mail: shaneymdoornkamp@outlook.com

F. Jabeen
e-mail: fakhra.jabeen@ru.nl

J. Treur (✉)
Social AI Group, Department of Computer Science, Vrije Universiteit Amsterdam, Amsterdam, The Netherlands
e-mail: j.treur@vu.nl; j.treur@hhs.nl

J. Treur · P. H. M. P. Roelofsma
Risk Management and Cybersecurity Group, Centre of Expertise on Cybersecurity, The Hague University of Applied Sciences, Hague, The Netherlands
e-mail: p.h.m.p.roelofsma@hhs.nl

H. Rob Taal · P. H. M. P. Roelofsma
Department of Pediatrics, Division Neonatology, Erasmus MC, Rotterdam, The Netherlands
e-mail: h.taal@erasmusmc.nl

H. Rob Taal
University Medical Center, Sophia Children's Hospital, Rotterdam, The Netherlands

P. H. M. P. Roelofsma et al. (eds.), *Using Shared Mental Models and Organisational Learning to Support Safety and Security Through Cyberspace: A Computational Analysis Approach*, Studies in Systems, Decision and Control 570,
https://doi.org/10.1007/978-3-031-72075-8_9

message. Through several scenarios, we illustrate what pattern of factors causes a healthcare professional to speak up when witnessing a (near) medical error. We moreover demonstrate how introducing an observant agent can facilitate effective communication and helps to ensure patient safety through speaking up when a nurse can not. In conclusion, the current chapter introduces a computational model which predicts speaking up behaviour from the perspective of the speaker and receiver, with the addition of a virtual coach to further optimise patient safety when a patient could be in harm's way.

Keywords Network modelling · Patient safety · Speaking up · Psychological safety

1 Introduction

Healthcare professionals in hospitals operate in a high-risk environment as, even slight, medical errors or unsafe events are hazardous to patient safety (Bienefeld and Grote 2014; Schwappach and Gehring 2014; Vogus and Iacobucci 2016). Because of the threat to patients' safety, prevention of medical errors is considered an important challenge in healthcare (Alyahya et al. 2021). The likelihood of medical errors is even greater in intensive care units and the operating theatre (OT) (Nacioglu 2016). Prior studies have identified ineffective communication about preventable medical errors in OTs as one of the most frequent causes of patient morbidity (James 2013; Pattni et al. 2019). Ineffective communication in this context comprises the hesitancy of health care professionals to address an aberration which could lead to harming the patient (Okuyama et al. 2014; Palatnik 2016), thus jeopardising patient safety. Effective communication, on the other hand, in the form of speaking up is considered a safety behaviour associated with increased technical performance in the OT (Kolbe et al. 2012; Nacioglu 2016). In line with the existing literature, we define speaking up in the current chapter as addressing concerns with patient safety in mind through clear communication about observations, the need for clarification, or suggestions for corrective measurements during the OT (Morrison 2011; Okuyama et al. 2014).

Despite growing acknowledgement in healthcare concerning the importance of speaking up relative to error prevention and optimisation of care, consistent implementation is reportedly difficult in practice (Maxfield et al. 2011). There seems to be a hesitancy to speaking up, as healthcare professionals noticing a (near) deviation of the medical protocol are often ignored when speaking up or choose to stay silent (Okuyama et al. 2014). Considering that healthcare professionals are increasingly aware of the importance and benefits of speaking up yet hesitate to speak up, we can be assume that motivations to speak up about patient safety are complex. Indeed, several lines of research have showed that speaking up may depend on specific combinations of individual, contextual, and interpersonal factors (Alyahya et al. 2021; Bienefeld and Grote 2014). These studies mostly scrutinise the determining factors which cause the inability to speak up, while studies which include the state of the receiver of speaking up are scarce. Since effective communication is dyadic, after all it is important to consider the response of the receiver as it can affect the perception of the sender regarding the acceptance of speaking up (Lemke et al. 2021; Long et al. 2020).

With the increasing interest in scientific literature concerning the cornerstones of speaking up behaviour among healthcare professionals (Alingh et al. 2019; Lemke et al. 2021; Pattni et al. 2019; Peadon et al. 2020), several efforts were made to delineate speaking up through theoretical frameworks using (simulated) experiments and literature reviews (Long et al. 2020; Umoren et al. 2022). A framework which analyses these findings that specifies how constituents of speaking up interact as well as how individuals interact with speaking up in a high-risk environment in a formalised causal computational manner is lacking within this research domain. The current chapter therefore aims to outline existing sources to design a network model using a computational modelling approach capable of representing such complex processes in a generative and causal manner (Treur 2016, 2020b). Because of the reported hesitancy to speaking up, the chapter introduces a mechanism named the Safe Coach (SFC) to control to prevent medical errors in the circumstance where a health care professional stays silent.

Literature from a variety of research domains (organisation, psychology, communication sciences) were explored to illustrate which causal pathways and processes underlie speaking up as well as the constituents to the receptivity to speaking up of the receiver, to design a computational model substantiating the effects hypothesised (Sect. 2). Next, based on this exploration, we provide a brief description of the adaptive causal network modelling approach (Sect. 3) followed by a description of the proposed network model (Sect. 4). Through simulation experiments we show how the combination of individual and contextual factors facilitates speaking up (Sect. 5). Connections are furthermore exploited to depict the internal processes for the response of the receiver, as well as portraying a circumstance where the nurse chooses silence for the SFC to intervene. In the silence circumstance, internal processes of the SFC intervention will also be presented. At last, in this chapter we briefly discuss how the model was verified by mathematical analysis.

2 Background Information

This section contains a brief description of cornerstones to speaking up behaviour to delineate what causes the suggested hesitancy to speak up. Despite the scarcity of existing research about the impact and receptiveness to speaking up by the receiver, we will attempt to illustrate what factors cause a negative or positive response provided the argued significance of the leaders' response during the OT.

2.1 Speaking up Behaviour

Psychological safety is the degree to which people in their work environment feel comfortable to take interpersonal risks without negative implications based on perceived team climate (O'Donovan and McAuliffe 2020b; Roussin et al. 2018). In a work environment where the perceived psychological safety is high, it is likely that these individuals would be more willing to share their concerns and opinions on task-related and organisational matters. Indeed, higher psychological safety has been associated with increased speaking up behaviour and a subsequent decline of silence behaviours (Edmondson 1999; Newman et al. 2017; O'Donovan and McAuliffe 2020a). Psychological safety is a concept that can be perceived by an individual, the team (team psychological safety), and at organisational level (organisational psychological safety) (Bienefeld and Grote 2014). At the individual level, it can be assumed that a variety of factors influence the degree to which psychological safety is felt with top-to-bottom influence from perceived team and organisational psychological safety. Study findings suggest that fear is one facilitator for silence and is negatively associated with psychological safety (Hémon et al. 2020; Kish-Gephart et al. 2009; Moore and McAuliffe 2012; Okuyama et al. 2014). It is rooted in the fear of punishment or negative responses from others, which can be interpreted as the perceived fear of repercussion (Bienefeld and Grote 2014). Other individual factors that affect the likelihood of speaking up are seniority and experience, with junior members being prone to low psychological safety (Nacioglu 2016; O'Donovan and McAuliffe 2020b), certainty about their own knowledge and skills (i.e., task-related efficacy) (Beament and Mercer 2016; Okuyama et al. 2014), and prior training regarding communication within teams (Sayre et al. 2012).

As a medical event occurs, the risk for patient safety increases: the healthcare professional perceives an increase in (possible) patient harm which can, at the very worst, lead to patient death. For instance, a newborn in need of resuscitation immediately after birth is typically hypoxic (i.e., low levels of oxygen in bodily tissues) and needs ventila tion (Boldingh et al. 2018). Inefficient ventilation may lead to infant death due to neonatal hypoxia–ischaemia or life-long (persistent) motor, sensory, and cognitive impairments (Boldingh et al. 2018; Millar et al. 2017; Raju et al. 2011). This perception of increased patient harm is suggested to be a motivation to speaking up, regardless of psychological safety (Okuyama et al. 2014). Increased stress, while assumed to be a constant mental state (Nacioglu 2016), related to decision-making (i.e., speaking up or silence) can have a negative impact on the choice to speak up (Pabst et al. 2013) and may cause the healthcare professionals to stay silent instead.

2.2 *Contextual and Interpersonal Influences on Speaking up Behaviour*

Despite the scarcity of literature scrutinising the effects of the response of the receiver, it is still an aspect worthwhile to discuss. As described before, positive relationships and responses can facilitate higher degrees of psychological safety while negative consequences to behaviour such as speaking up can be detrimental to psychological safety. Therefore, positive responses are beneficial to psychological safety of the team members.

A study by Long et al. (2020) interviewed fully qualified surgeons, anaesthetists, nurses, and anaesthetic technicians in New Zealand. Participants expressed that the valence of their response was determined by stress and fatigue, with some insights about the respect and familiarity they held towards the sender. Specifically, participants have showed that when they respected the sender, they were more likely to respond in a positive manner. Similarly, when they were familiar with the sender and, notably, when they had worked together previously, they were more often positive in demeanour. In other words, the elements that influence receptivity to speaking up are (perceived) trust, familiarity, stress, and fatigue.

Positive interpersonal relationships, such as the doctor-nurse relationship, can furthermore affect the perception of psychological safety, in particular team psychological safety. Responses of health care professionals should therefore be highlighted. When the health care professional has a positive attitude, therefore approach, to communication with the nurse, chances for effective communication may advance (Tan et al. 2017). These positive attitudes promote feelings of trust, (team) support, and familiarity within teams which are accordingly pertinent to effective communication (O'Donovan and McAuliffe 2020b). Literature suggests that these interpersonal factors to (effective) communication between health care professionals cultivate a positive relationship, which is beneficial to increasing team psychological safety (O'Donovan and McAuliffe 2020b), subsequently perceived psychological safety by an individual.

3 The Modelling Approach Used

An adaptive causal network-oriented modelling approach (Treur 2016, 2020b) was used to illustrate the causal relationships and pathways between contextual and psychological factors, speaking up as an outcome behaviour, and related situational consequences. This method allows for a generic, dynamic, and declarative depiction of various processes; see (Treur 2020b) for more details.

A temporal-causal network comprises *states*, the *connectivity* between respective states, indicated by respectively X and Y, for which activation values $X(t)$ and $Y(t)$ are assumed (usually with an interval of [0, 1]) that can change over time t. A network model is defined by its network characteristics. One type of these characteristics are *connection weights* $\boldsymbol{\omega}_{X,Y}$ from state X to state Y, usually with an interval between $[-1, 1]$ denote the strength of the connections. When there are multiple incoming connections to a state Y, *combination functions* $\boldsymbol{c}_Y$ address the aggregated effect on Y. There is a wide variety of combination functions available in a library depending on the type of effect to represent in the model. Table 1 presents which combination functions were applied for states X_i in the introduced model (see Appendix, Supplementary Materials for the full state description). The last network characteristic to consider is the *speed factor* $\boldsymbol{\eta}_Y$ which characterises the timing of the causal effects. The numerical representation of a temporal-causal network model is described as:

- **Single causal impacts** are defined by $\mathbf{impact}_{X,Y}(t) = \boldsymbol{\omega}_{X,Y}X(t)$ for the effect from X on Y at time t.
- Aggregation of multiple causal effects is expressed by

$$\mathbf{aggimpact}_Y(t) = \mathbf{c}_Y\left(\boldsymbol{\omega}_{X_1,Y}X_1(t), \ldots, \boldsymbol{\omega}_{X_k,Y}X_k(t)\right) \tag{1}$$

  where Y gets its incoming connections from the X_i.
- Timing of the causal effect is expressed by

$$Y(t + \Delta t) = Y(t) + \boldsymbol{\eta}_Y\left[\mathbf{c}_Y\left(\boldsymbol{\omega}_{X_1 Y}X_1(t), \ldots, \boldsymbol{\omega}_{X_k,Y}X_k(t)\right) - Y(t)\right]\Delta t \tag{2}$$

To model adaptive dynamical systems, adaptive network models can be used: in that case not only a network's state activation levels but also some of the network structure characteristics can change over time. By using the self-modelling principle to a network, a network-oriented conceptualization can also be applied to adaptive networks; see Treur (2020a, b). This can be done through the addition of new states to the network (called *self-model* or *reification* states) which represent (adaptive) network characteristics. By changing the activation values of such self-model states over time, the corresponding network characteristics become adaptive. In a graphical 3D-format such additional states are depicted at a next level (called *self-model* level or *reification* level), where the original network is at the *base level.* As an example, the weight $\boldsymbol{\omega}_{X,Y}$ of a connection from state X to state Y can be represented (at a next self-model level) by a self-model state named $\mathbf{W}_{X,Y}$. In this way learning can be modelled. In such a case, in Eq. (1), for the connectivity characteristics $\boldsymbol{\omega}_{X_i,Y}$ the values $\mathbf{W}_{X_i,Y}(t)$ are used. The self-model states are handled in the same way as other states: they have their own network characteristics and for their dynamics the canonical equations are applied to them as well.

As the outcome of a process of network self-modeling results also in a temporal-causal network model itself (Treur 2020a, 2020b, Chap. 10), this self-modelling network construction can easily be applied iteratively to obtain multiple orders of self-models at multiple (first-order, second-order, …) self-model levels. For example, a second-order self-model may include a second-order self-model state $\mathbf{W}_{Z,\mathbf{W}_{X,Y}}$ representing the adaptive connection weight $\boldsymbol{\omega}_{Z,\mathbf{W}_{X,Y}}$ of a connection from context state Z to $\mathbf{W}_{X,Y}$ which in turn represents the adaptive connection weight $\boldsymbol{\omega}_{X,Y}$.

Table 1 Combination functions used in the network model

Description	Formula	Parameters	Used for
Advanced logistic $\mathbf{alogistic}_{\sigma,\tau}(V_1, \ldots, V_k)$	$\left[\frac{1}{1+e^{-\sigma(V_1+\ldots+V_k-\tau)}} - \frac{1}{1+e^{\mathbf{st}}}\right](1+e^{-\mathbf{st}})$	Steepness $\boldsymbol{\sigma}$ Excitability threshold $\boldsymbol{\tau}$	X_2–X_{24}, X_{28}–X_{50}, X_{54}–X_{79}, X_{83}–X_{88}
Identity function	$\mathbf{id}(V) = V$	None	X_{25}–X_{27}, X_{51}–X_{53}, X_{80}–X_{82}, X_{89}
Step once $\mathbf{steponce}_{\alpha,\beta}(\ldots)$	1 if $\boldsymbol{\alpha} \leq t \leq \boldsymbol{\beta}$ else 0 (time t)	Start time $\boldsymbol{\alpha}$ End time $\boldsymbol{\beta}$	X_1

Any smooth dynamical system has a canonical representation as a network and any smooth (multi-order) adaptive dynamical system has a canonical representation as a (multi-order) self-modeling network, as is shown in Treur (2021), Hendrikse et al. (2023). Therefore, compared to adaptive dynamical systems in general, the network-oriented modeling approach used does not introduce any fundamental limitations concerning what it can model. This has also been confirmed by applications in other disciplines, e.g., Treur and Van Ments (2022), Canbaloğlu et al. (2023), Hendrikse et al. (2025).

The literature on speaking up behaviour described in Sect. 2 was used to render the relevant processes in the network characteristics of the network-oriented modelling approach.

4 The Proposed Computational Network Model

Figure 1 presents a graphical representation of the complete temporal-causal network model containing causal connectivity between the states described in Table 2 in the Appendix. This model illustrates the assumed conjunction of contextual and psychological elements a cognitive process of evaluative decision-making regarding whether to speak up or stay silent when witnessing a (near) medical error. A similar evaluative decision-making system introduced the receptiveness towards speaking up from the recipient.

4.1 The Overall Structure of the Network Model

The network model consists of the world state model (**WS**), individual mental models (Treur and Van Ments 2022) for the nurse (**N**), doctor (**D**), and the safe coach (**SFC**), and the mental and social processes interacting with them and with each other, see also Van Ments et al. (2025), this volume. The world state model signifies the sequence of events that occurs over time in the real world. It is based on descriptions of actions taken by the healthcare professionals and their effects in the world, particularly when they need to perform interventions if a neonate is not transitioning adequately (Madar et al. 2021). Interventions are initiated when the vital checkpoints of the neonate (tone, heart rate, breathing, and colour) are considered insufficient. The context state $\mathbf{WS1}_{\text{call\_intervention}}$ triggers activation of the model for which the assumption holds that the responsible healthcare professional calls the start of intervention. To design episodes in the model, the function $\mathbf{steponce}_{\alpha,\beta}(\ldots)$ was used. The subsequent states are to provide a general sequence of an intervention wherein both doctor and nurse execute their designated tasks. Provided that medical errors often occur through failure of planned action (Raju et al. 2011), such as using an insufficient technique or intervention, the patient physical wellbeing is stagnant despite the intervention and is reflected in $\mathbf{WS6}_{\text{intervention\_insufficient}}$. While the same intervention can be performed

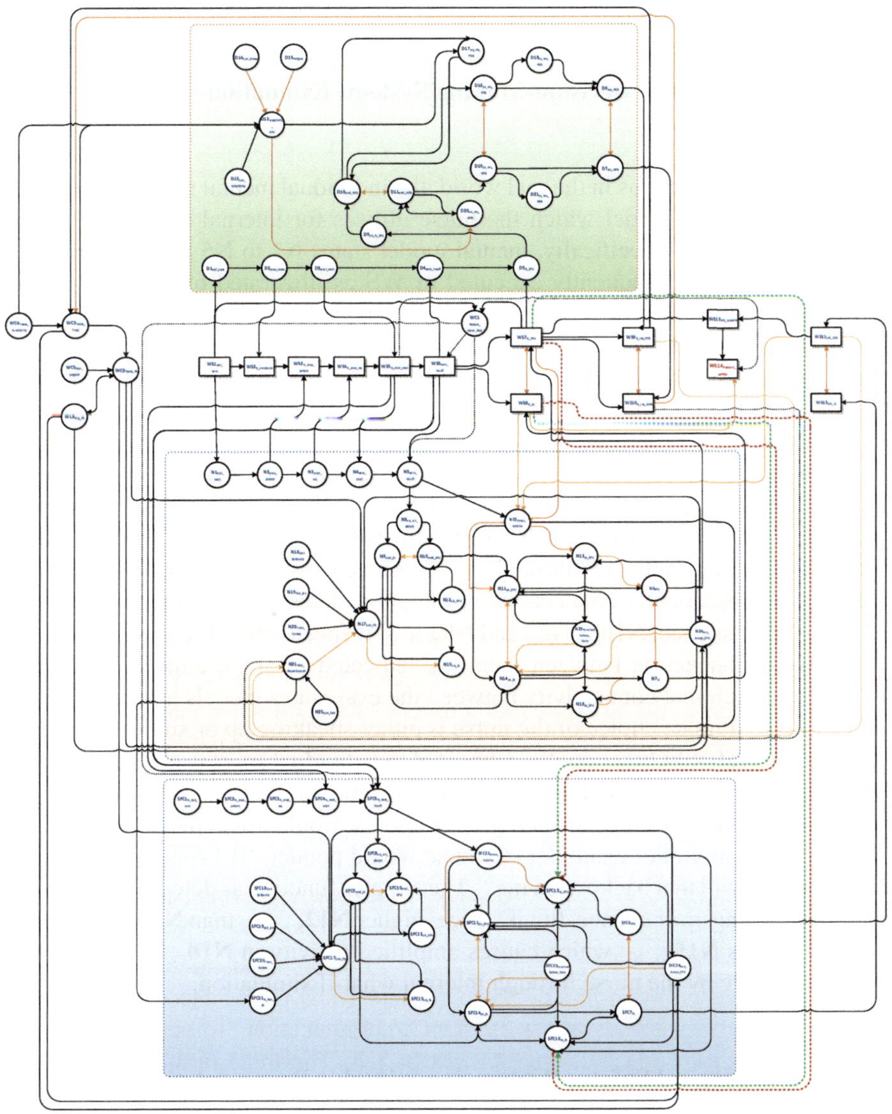

Fig. 1 Complete proposed temporal-causal network model for speaking up behaviour and related response

multiple times (with the same result), this specific part is not modelled, considering that it is not the focus of the current chapter. In other words, $\mathbf{WS6}_{\text{intervention\_insufficient}}$ is a broad proxy for a (near) medical event. Next, patient general risk increases as the intervention continues to be insufficient, explaining the bidirectional connection between $\mathbf{WS6}_{\text{intervention\_insufficient}}$ to the world's context state $\mathbf{WC1}_{\text{Patient\_Centr\_risk}}$.

4.2 The Nurse Model

4.2.1 The Evaluative Decision-Making System: Evaluation—Stimulus Loop

While **WS** models actions in the real world, the individual mental model of the nurse comprises a mental model which the nurse utilises for internal simulation within cognitive processes. Specifically, mental model states **N1** to **N5** represent medical actions which are subsequently executed in **WS** as illustrated by the connections between the **WS** and **N** models. The individual **N** mental model also receives information from the **WS**, which models observation. For speaking up to occur, the information about the failed intervention prompts the detection of a medical event, as denoted by the outgoing connection from $\mathbf{WS6}_{\text{intervention\_insufficient}}$ to $\mathbf{N5}_{\text{intervention\_insufficient}}$. This incoming connection activates the cognitive process of evaluative decision-making for speaking up by the nurse. Its design is based on a what-if analysis approach making use of internal simulation of simulated action and perception pathways (Hesslow 2002). First, the stimulus representation state (srs) for the error detection $\mathbf{N8}_{\text{srs\_error\_detect}}$ activates through the perceived insufficiency of the intervention, which subsequently connects two evaluation states representing the choice of speaking up or silence, $\mathbf{N10}_{\text{eval\_SPU}}$ and $\mathbf{N9}_{\text{eval\_SL}}$ respectively, which have a negative bidirectional connection between them (as indicated by the orange arrows). This negative bidirectional connectivity between the eval-states models an exclusiveness effect, so the ultimate choice of the nurse is either speaking up or silence. The eval-states are thereafter bidirectionally connected to individual srs-states; $\mathbf{N12}_{\text{srs\_SPU}}$ and $\mathbf{N15}_{\text{srs\_SL}}$. They encompass the concept of a person (in this case, the nurse) internally envisioning how either speaking up or silence would affect themselves and the real-world situation. For example, the nurse would ponder, 'If I speak up about this situation, how will that backfire on me?' These states function as determining components of speaking up or silence. For instance, higher $\mathbf{N12}_{\text{srs\_SPU}}$ than $\mathbf{N15}_{\text{srs\_SL}}$ activity would suppress $\mathbf{N15}_{\text{srs\_SL}}$ which causes amplified activity in $\mathbf{N10}_{\text{eval\_SPU}}$ rendering the choice made by the nurse through internal what-if simulation.

4.2.2 Psychological Safety—Enhancing Stimulus Representation

The srs-states address aspects of psychological safety, as illustrated in Fig. 2, considering that psychological safety is hypothesised to be a mediator to speaking up. For the sake of modelling, psychological safety is defined as follows. While the nurse is deciding, there is a sense of safety to speaking up within the context, hence contextual psychological safety; $\mathbf{N17}_{\text{con\_PS}}$. Team psychological safety and individual factors directly affect contextual psychological safety, such as confidence in skills and knowledge, years of experience in position, prior training in communication, and (mainly) fear of repercussion to speaking up. Team psychological safety is a context state which depends on organisational psychological safety, team trust, peer support and

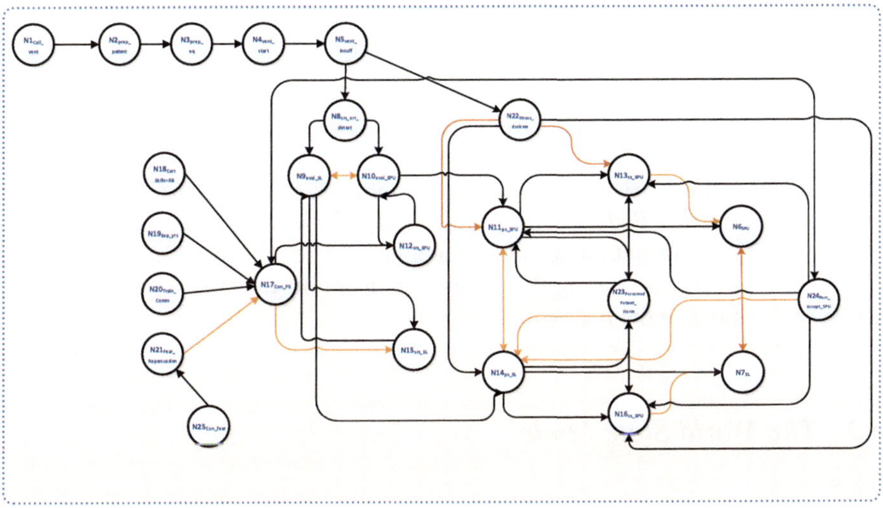

Fig. 2 Nurse individual model

team familiarity. Fear of repercussion, while affected by the outcomes of speaking up, has a memory state representing prior experiences with speaking up which induced the fear of repercussion. The assumption is that high contextual psychological safety means that the nurse feels safe to speak up. This state henceforth amplifies $\mathbf{N12}_{\text{srs\_SPU}}$, the choice to speaking up, whilst the negative outgoing connection to $\mathbf{N15}_{\text{srs\_SL}}$ suppresses activity for the choice to stay silent. Low contextual psychological safety amplifies $\mathbf{N15}_{\text{srs\_SL}}$ through the negative feed-forward connection.

4.2.3 From Evaluation to Preparation for Action

As activation from the srs-states are fed into the eval-states, the eval-states consequently activate the preparation states (ps) which entail the nurse internally preparing themselves to speaking up or be silent after they made the choice. Based on the all-or-nothing exclusiveness principle of this choice system, there is a bidirectional negative connection between $\mathbf{N11}_{\text{ps\_SPU}}$ and $\mathbf{N14}_{\text{ps\_SL}}$, indicating that the nurse cannot prepare for both situations as they are to choose only one of the two actions. While preparing for the chosen action, there are still factors that influence the ultimate choice. For instance, stress for deciding to speak up negatively affects the nurse while preparing for action, which could eventually lead to choosing silence regardless of the evaluation, as illustrated by the positive connection to $\mathbf{N14}_{\text{ps\_SL}}$ and negative connection to $\mathbf{N11}_{\text{ps\_SPU}}$. A similar principle was used to model the effect of perceived patient harm on the choice made, with higher perceived patient harm amplifying the choice to speaking up. Through modelling a loop between the state of perceived patient harm and the ps-states, the choice is further evaluated given that patient harm can be dynamic provided the link to the patient state's vitals. The last state to influence the ps-states is the perceived acceptability to speaking up. Along with contextual psychological safety, this state encompasses a historical context to prior experiences that shape how acceptable the individual perceives speaking up. The ps-states are consequently connected to control states (cs-states) which are connected to the execution states (es-states) $\mathbf{N6}_{SPU}$ and $\mathbf{N7}_{SL}$. These control states model general assumed hesitancy for making a risky choice, therefore these cs-states have a slight repression effect on the es-states. Patient harm, acceptability of speaking up, and decision stress provide additional weight for hesitance for either choice. The es-states are ultimately connected to the corresponding **WS**-states demonstrating the execution of the chosen action of the nurse in the real world.

4.3 The World State Model

As described before, the **WS** model includes the actual actions of both the nurse and the doctor in the world. As two of these actions, Fig. 1 also depicts the **WS** for the nurse either having spoken up or having kept silent. When the nurse chooses speaking up, this is consequently being perceived by the doctor via mental model state $\mathbf{D5}_{N\_SPU}$ which activates the evaluation decision-making system for responding to speaking up. This model assumes that speaking up is a safety behaviour with consequences for patient safety. For the sake of simplicity, the assumption holds that the doctor always incorporates the input of the nurse if the nurse chooses speaking up, regardless of the valence to the doctor's response. Therefore, speaking up by the nurse in the **WS** causes actions by the doctor to prevent the medical event from happening; $\mathbf{WS13}_{D\_ME\_Prevention}$, consequently increasing patient safety. In contrast, when the nurse chooses silence, it causes patient safety to decrease.

4.4 The Doctor Model

4.4.1 The Evaluative Decision-Making System for Responding to Speaking up

The **D** individual model was designed similarly to the model of the nurse, see Fig. 3. The doctor also uses the mental model to process both internal simulation of medical actions through mental model states and evaluation decision-making system for responding to speaking up. Indeed, this pattern will only activate if the nurse chose to speak up, illustrated by the connection from $\mathbf{WS6}_{N\_speaking\ up}$ to $\mathbf{D5}_{N\_speaking\ up}$. The abovementioned connection activates the doctor's internal representation of the nurse speaking up. Following $\mathbf{D5}_{N\_speaking\ up}$, the evaluation decision-making system for responding to speaking up denotes a similar eval-srs loop and eval-ps connectivity explained for the **N** individual model. These systems, however, differ in the

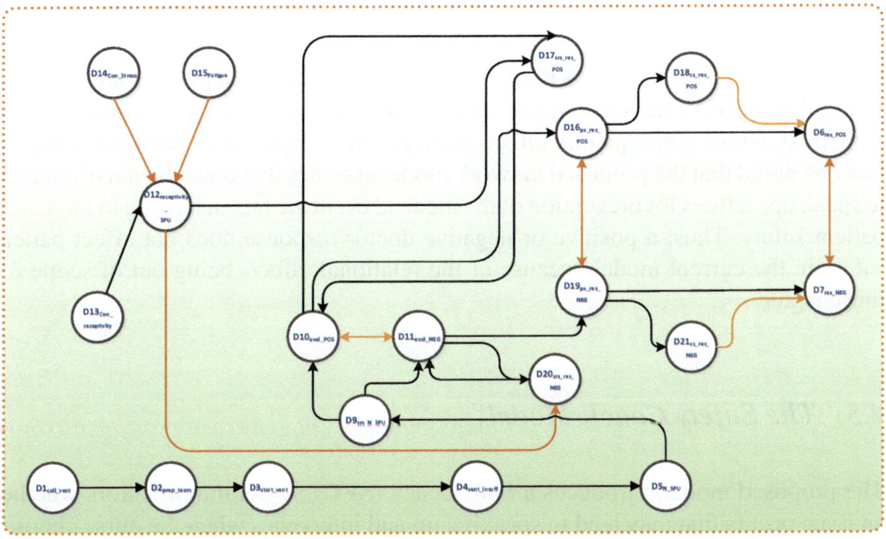

Fig. 3 Doctor individual model

choices for the agent. The doctor chooses whether to respond in a positive or negative manner. In like manner to the influence of contextual psychological safety on the srs-states in the N individual model, the state receptivity to speaking up entails how receptive the doctor is to a colleague speaking up in the moment. Contextual stress, fatigue, and a context state of receptivity that represents the individual value for receptiveness to speaking up by the doctor negatively affect this state. Receptivity to speaking up is further influenced by two context factors, however: team trust and team familiarity, as presented by literature (O'Donovan and McAuliffe 2020b). The receptivity to speaking up and srs-state connectivity assumes that a high value of receptivity to speaking up amplifies $\mathbf{D17}_{\text{srs\_res\_POS}}$ while attenuating $\mathbf{D20}_{\text{srs\_res\_NEG}}$. As described before, activity feeding back from srs-states to eval-states with the bidirectional negative connectivity between eval-states result in the doctor choosing either a negative or positive response. After activation of the ps-states, assumed hesitancy for response was modelled through subsequent cs-states which slightly repress the es-states to ultimately activate the corresponding **WS**-states.

4.4.2 Consequences of Response

The doctor choosing to respond positively adversely affects the state fear of repercussion within the **N** individual model. When the doctor replies positively to the nurse speaking up, the fear felt for repercussion will decrease. A negative response from the doctor causes the reverse; the nurse will feel more fear for repercussion as they can consider the negative response as a detrimental consequence. The fear of repercussion and doctor response connectivity illustrates the influence the doctor has concerning psychological safety for both the team and the individual. To further emphasise the significance of the doctor's response on team psychological safety, the doctor response in the **WS** model both influence team trust (Weller and Webster 2021). The positive response increases team trust and the negative response reduces team trust, which consequently affects team psychological safety for future cases. It must be noted that the proposed network model assumes that once the nurse chooses to speak up, actions for prevention of the medical event are taken, leading to increased patient safety. Thus, a positive or negative doctor response does not affect patient safety in the current model because of the relational effects being out of scope for this chapter.

4.5 The Safety Coach Model

The proposed model introduces a Safe Coach (**SFC**) model that monitors whether an issue occurs that may lead to speaking up and intervenes when the nurse chooses to not speak up. The assumption holds that this SFC individual model represents the 'perfect' nurse model, which entails the nurse's mental model where speaking up is always chosen whenever relevant. The Safe Coach consistently chooses speaking up in such cases. Hence the individual model of the Safe Coach is a near identical copy of the nurse's; see Fig. 1 for an overview of the **SFC** mental model relative to the **N** mental model. The Safe Coach only contains a copy of the individual mental model of the nurse due to the coach being proposed as a mechanism to control speaking up behaviour. Speaking up is less common than would be expected for health care and explicates a greater likelihood that nurses and residents do not always possess the 'perfect' mental model for speaking up in the real world. Nurses and residents exercise a higher threshold to speaking up than the Safe Coach implied by the modelled hesitancy through cs-states. The SFC model therefore functions to predict the speaking up behaviour of the nurse by comparing the N mental model outcomes (i.e., $\mathbf{WS7}_{\text{speaking up}}$ and $\mathbf{WS8}_{\text{SL}}$) to its own outcomes. To facilitate comparison, the SFC initially needs the same input as the nurse. This initial input is assumed to be $\mathbf{WS6}_{\text{intervention\_insufficient}}$ as the SFC is presumed to perceive a (near) medical error in like manner to the nurse. Diverging model outcomes prompt the Safe Coach to intervene by executing a form of speaking up related to the medical error at hand. This comparison is modelled through the connection between $\mathbf{WS8}_{\text{SL}}$ and the cs-state for speaking up by the SFC; $\mathbf{SFC13}_{\text{cs\_SPU}}$. This cs-state functions slightly differently compared to the **N** mental model. Since the Safe Coach does not 'hesitate', it controls for activation of the subsequent es-states by functioning as a gateway. If and only if the activation for this cs-state is sufficient, the es-state for speaking up by the SFC can activate. The excitability threshold $\boldsymbol{\tau}$ of this cs-state is set to a high value to prevent the SFC intervention if the nurse does choose to speak up but is slower than the SFC. For the sake of completeness, $\mathbf{WS7}_{\text{SPU}}$ is negatively connected to $\mathbf{SFC13}_{\text{os\_SPU}}$ and positively connected to $\mathbf{WS16}_{\text{os\_SL}}$ so the SFC does not intervene when the nurse chooses to speak up. To ensure the SFC intervenes properly, $\mathbf{WS8}_{\text{SL}}$ has an additional negative connection to $\mathbf{WS16}_{\text{os\_SL}}$. Furthermore, the intervention of the SFC leads to medical event prevention, subsequently improving patient safety, as well as reducing decision stress and fear of repercussion by the nurse. The assumption holds that this intervention models speaking up behaviour, which could lead to fear reduction for the nurse. However, due to the artificial nature of the SFC, this connection is rather weak.

5 Simulation Experiments Without the Safety Coach

In this section, some of the multiple simulation experiments performed are discussed. They are based on the following case:

> After delivery, the neonate is not transitioning well. After birth, there is a very low heart rate, the tone is floppy, its colour is blue, and the neonate is not breathing, despite drying and stimulating the neonate. Following the assessment by the doctor, they call for mechanical ventilation, as is recommended by the neonatal support during the transition and resuscitation protocol. The subsequent two ventilation attempts did not help the neonate stabilise. The nurse notices by this point what the error is. As this suggestion pertains to the doctor, the nurse will have to choose to speak up or remain silent.

5.1 Baseline: No Speaking up Required

The following situation that is modelled serves as a baseline for the narrative at hand. This simulation demonstrates the behaviours of the actors when there is no (near) medical error, which therefore renders speaking up from both the nurse or Safe Coach unnecessary. To model this lack of necessity, it is assumed that ventilation is sufficient, causing the incoming weight from $\mathbf{WS5}_{\text{intervention\_start}}$ to $\mathbf{WS6}_{\text{intervention\_insufficient}}$ to be zero. For the sake of clarity, only the relevant graphs relating to the description are displayed. Figure 4 demonstrates the states and actions in the world. As the call for intervention is made ($\mathbf{WS1}_{\text{call\_intervention}}$), both D and N act according to their position. However, under the presumption that no error occurs, the neonate stabilises because of the intervention. The state patient safety illustrates the stabilisation. Figure 4 shows the steady increase of patient safety after ventilation starts. This simulation above all demonstrates that the states $\mathbf{WS7}_{\text{speaking up}}$ and $\mathbf{WS8}_{\text{SL}}$ do not activate as the evaluation decision-making system is not activated by $\mathbf{WS6}_{\text{intervention\_insufficient}}$. In a subsequent manner, neither evaluation-decision making systems for either **D** or **SFC** are activated, which results in neither corresponding **WS**-state to be active.

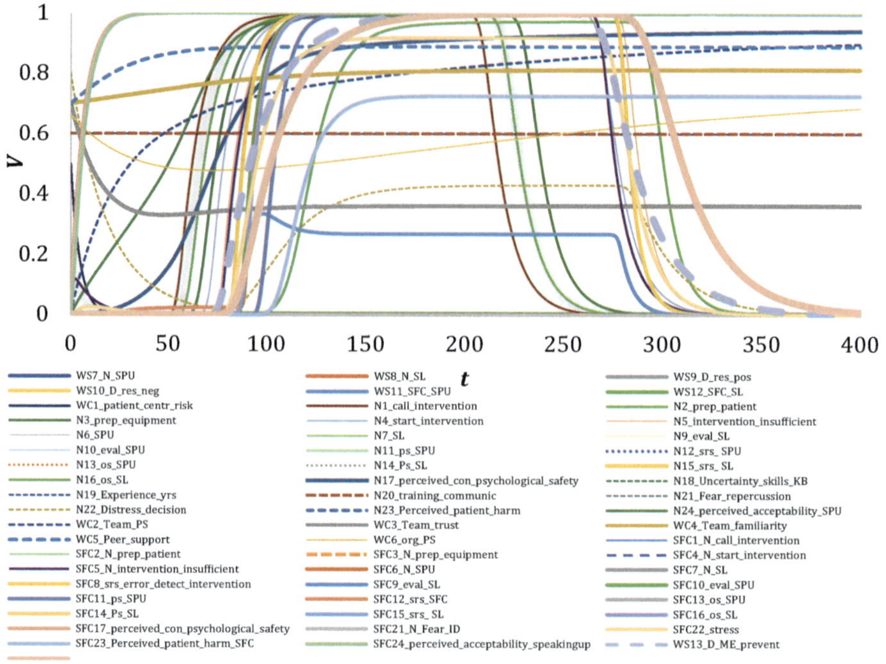

Fig. 4 Baseline simulation: no speaking up necessary

5.2 *Speaking up*

To model speaking up behaviour by the nurse, initial values of states 'certainty of skills and knowledge', 'prior training in communication', and 'experience in years' were all set to 0.6 given the direct connection to perceived contextual psychological safety. Contextual fear was set at 0.45 causing fear for repercussion to suppress perceived contextual psychological safety for a more realistic simulation of speaking up. Figure 5 depicts the situation where the nurse chooses to speak up due to the initial values mentioned. The figure includes the same sequence of WS-states as simulated in the baseline. In this case, however, ventilation is not sufficient. To model this activation, the connection weight $\boldsymbol{\omega}_{\text{intervention\_start, intervention\_insufficient}}$ was changed to 1. Because of this $\mathbf{WS6}_{\text{intervention\_insufficient}}$ becomes active at $t = 73$ and the evaluation-decision making system for speaking up in the N individual model was activated through the connectivity from $\mathbf{N5}_{\text{ventilation\_insufficient}}$ to $\mathbf{N8}_{\text{srs\_error\_detect}}$. Subsequently, 'patient risk' also increases at a similar time point. Figure 5 illustrates the subsequent activation of the eval-states after $\mathbf{N8}_{\text{srs\_error\_detect}}$ starting around $t = 87.5$. These eval-states trigger the activation for the corresponding srs-states at $t = 104.5$. Through the relatively high activity from states relating to 'team psychological safety', 'perceived acceptability to speaking up', 'experience in years', 'certainty of skills and knowledge', 'prior training in communication', and low 'fear for repercussion' ($t = 113.5$;

$V = 0.28$) the value for 'perceived contextual psychological safety' amounts to 0.67 at $t = 106.5$. Due to the higher activation value for 'perceived contextual psychological safety', the $\mathbf{N12}_{srs\_SPU}$ activation increases while $\mathbf{N15}_{srs\_SL}$ decreases because of the eval-srs loop. As depicted in the simulation, $\mathbf{N9}_{eval\_SL}$ peaks at $t = 109$ due to continued suppression by 'perceived contextual psychological safety' through $\mathbf{N15}_{srs\_SL}$ and consequently decreases. In contrast, $\mathbf{N10}_{eval\_SPU}$ increases at the same time point. Activation for this eval-state causes $\mathbf{N11}_{ps\_SPU}$ to activate at $t = 107.5$ with consecutive activation of $\mathbf{N13}_{cs\_SPU}$. With high values of 'perceived patient harm' and 'perceived acceptability to speaking up' at $t = 106$, $\mathbf{N11}_{ps\_SPU}$ is further augmented with slight suppression from $\mathbf{N13}_{cs\_SPU}$, whilst the opposite states continue to be repressed. $\mathbf{N6}_{SPU}$ then activates at $t = 121$, followed by activation of $\mathbf{WS7}_{N\_SPU}$ (i.e., the nurse speaking out in the real world). The nurse choosing speaking up causes an expected decrease in 'decision stress' which had been increasing at $t = 79.5$ after the nurse assessed insufficient ventilation ($\mathbf{N5}_{intervention\_insufficient}$) and decreases after activation of $\mathbf{N6}_{SPU}$ (i.e., the eventual choice). At last, the nurse speaking up in the real world causes actions for medical event prevention ($t = 139.5$) which leads to increased patient safety ($t = 148.5$), see Fig. 6.

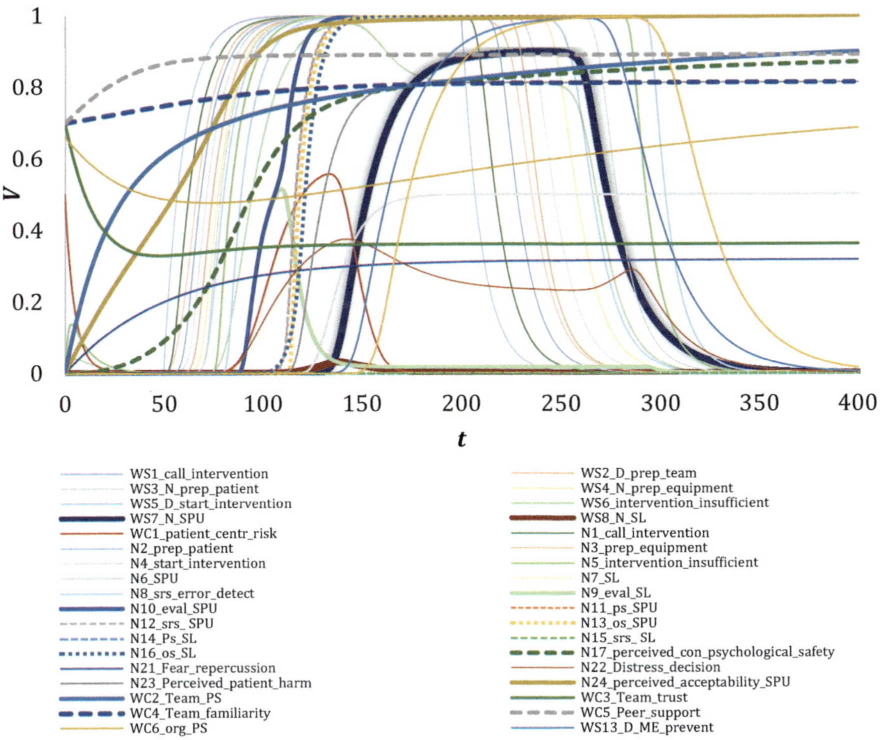

Fig. 5 Simulation for speaking up

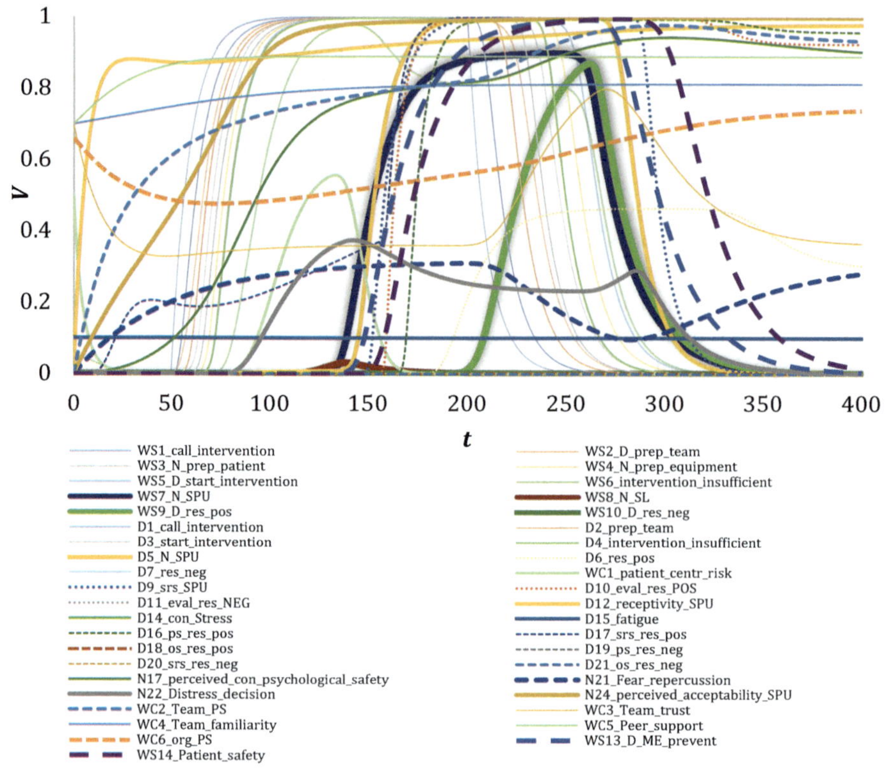

Fig. 6 Speaking up and a positive response from the doctor

5.2.1 Speaking up with a Positive Response

The next simulation highlights the response of the doctor when the nurse chooses to speak up. To achieve this simulation, initial values of the states pertaining to receptivity to speaking up were set to the following: contextual receptivity = 0.7, contextual stress = 0.1, fatigue = 0.1. The evaluation-decision system for valence of response by the doctor functions very similarly to the evaluation-decision system for speaking up of the nurse. State $\mathbf{D5}_{\mathrm{N\_SPU}}$ activates at $t = 138.5$, right after the nurse speaks up in the world, which depicts the sensory information to the doctor that the nurse spoke up. This activation prompts activation the eval-states. Because of the high receptivity for speaking up, $\mathbf{D10}_{\mathrm{srs\_POS}}$ activity amplifies, whilst $\mathbf{D11}_{\mathrm{srs\_NEG}}$ decreases. In like manner to the srs-eval loop previously described for the evaluation-decision system for speaking up, $\mathbf{D10}_{\mathrm{srs\_POS}}$ activity receives more activation than $\mathbf{D11}_{\mathrm{srs\_NEG}}$. In turn, the eval-state activates the related ps-state subsequently. The ps-state obtains additional activation due to the relatively high perceived patient harm ($t = 164$; $V = 0.79$) with slight repression through hesitancy by the cs-state. Due to the related ps-state, the es-state for the doctor responding positive $\mathbf{D6}_{res\_POS}$ activates. The states relating to choosing a negative response were suppressed in accordance with the literature. The doctor promptly responds positively in the world $\mathbf{WS9}_{res\_POS}$ causing the predicted decrease of fear of repercussion in the **N** model at $t = 206.5$. Because of $\mathbf{WS9}_{res\_POS}$ activity, team trust and the states related to psychological safety see an increase in activity given the expectation that a positive response by the doctor applies to an environment where individuals feel safe to take interpersonal risks. Team trust, however, decreases as $\mathbf{WS9}_{res\_POS}$ halts activity. An explanation for this lack of retention pertains to model design as the activity of this state is not persistent but can in the future be addressed by reification of the state persistence.

5.2.2 Speaking up with a Negative Response

Provided that receptivity to speaking up by the doctor largely depends on their physiological and mental state (i.e., fatigue and stress), the next simulation sets the initial values of contextual stress and fatigue at 0.65. In addition, contextual receptivity to speaking up is set to 0.2 to model initial low receptiveness. Figure 7 shows low receptiveness to speaking up with a highly agitated mental state of the doctor (i.e., high stress and fatigue). Therefore, the srs-eval loop in the evaluation-decision system amplifies $\mathbf{D11}_{eval\_NEG}$ instead due to the negative connectivity between the contextual receptivity to speaking up and $\mathbf{D20}_{srs\_res\_NEG}$. This leads to choosing a negative response as the doctor chooses the ps-states for the negative response, causing activation of $\mathbf{WS10}_{res\_NEG}$. The doctor responding negatively leads to an inverse effect compared to the doctor responding positively. Specifically, fear for repercussion increases and related psychological safety states and team trust decreases. The comparison between the positive and negative response illustrates some interpersonal consequences, considering the importance of the attitude of the doctor.

5.3 Silence

The next simulation experiment shows silence behaviour. To model this behaviour, changes in initial values of the following states were made: certainty skills and knowledge is set at 0.01, experience in years at 0.2, prior training in communication at 0.01, contextual fear for repercussion at 0.7. Figure 8 indicates the consequences of low contextual psychological safety on speaking up behaviour. Given that the mentioned states were low, with higher suppression of fear for repercussion, contextual psychological safety activates slightly. Due to this low activity, the $\mathbf{N15}_{srs\_SL}$ state increases in activity given the negative connection between contextual psychological safety and the srs-state. Contrarily, low contextual psychological safety does not provide adequate activation for $\mathbf{N12}_{srs\_SPU}$, despite the relatively high perceived patient harm. The evaluative-decision making system accordingly causes activation for the choice of silence. As expected, however, decision stress decreases once the nurse has made

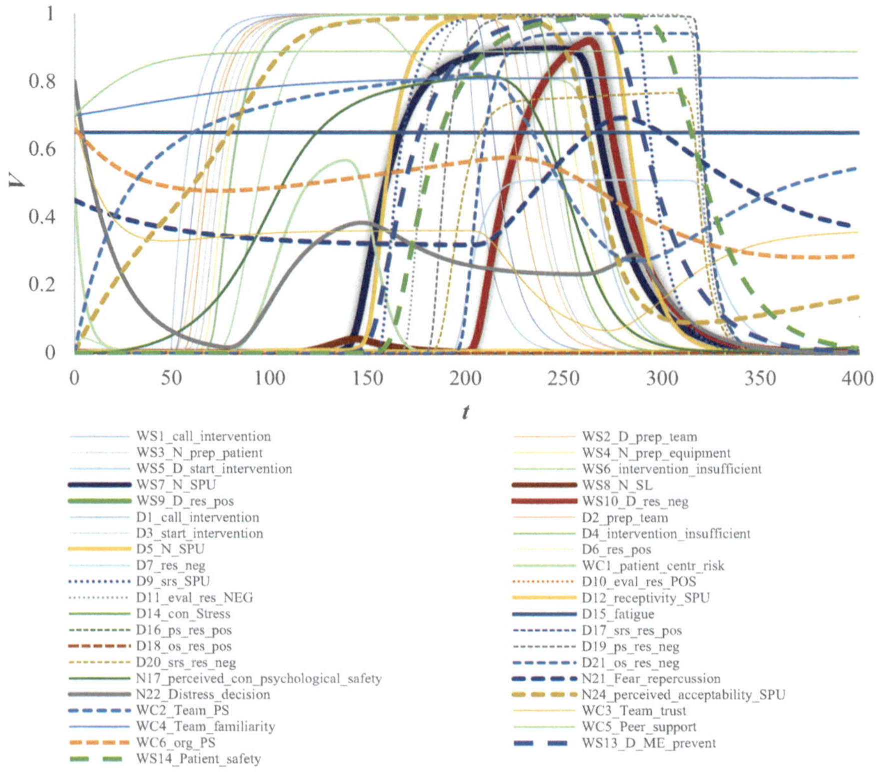

Fig. 7 Speaking up and a negative response from the doctor

the choice to keep silent which solidifies that the stress felt is mainly due to the choice in the provided model. Because the nurse chose silence, there is no prompt from the doctor, as $\mathbf{WS8}_{SL}$ is not connected to the doctor's individual model. Consequently, no preventative measures are taken to apprehend possible medical error, which results in patient safety to be stagnant and low as depicted in Fig. 8.

6 Simulation Experiments Using the Safety Coach

The last simulation experiment discussed addresses the situation wherein the Safety Coach intervenes to prevent the medical event from occurring or deteriorating, leading to an increase in patient safety. The Safety Coach contains a copy of the 'perfect' nurse model for speaking up and will therefore always choose to speak up whenever relevant. Hence, contextual psychological safety and perceived acceptability to speaking up within the SFC individual model are high in activity, as illustrated in Fig. 9. In this situation, the nurse situation is not 'perfect', which

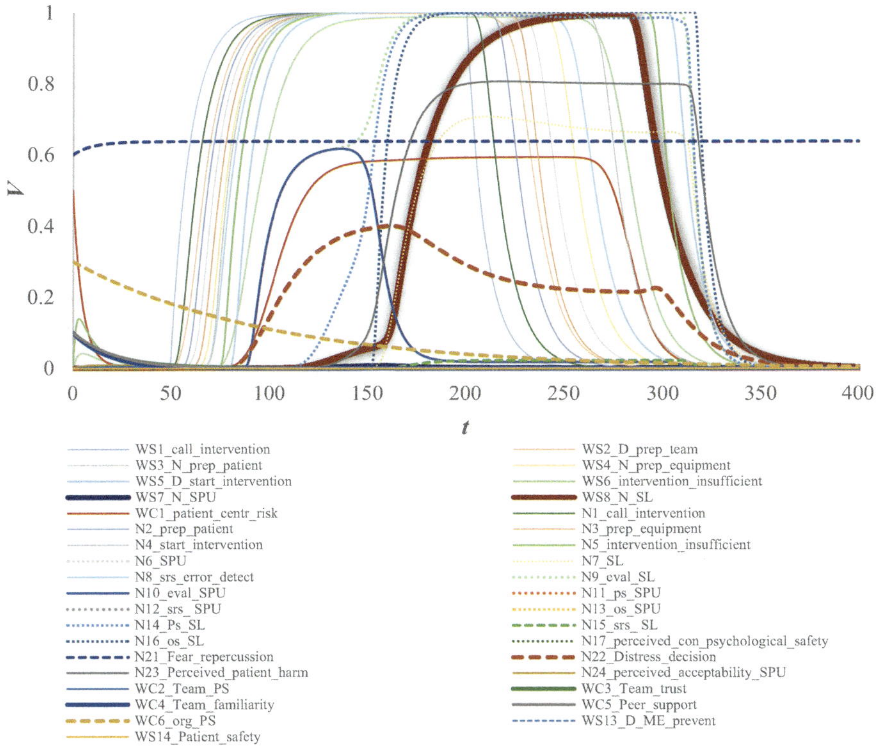

Fig. 8 Simulation experiment for silence by the nurse

is likely the case, and therefore chooses silence. For the **SFC** mental model to activate, it receives activation from $\mathbf{WS6}_{\text{intervention\_insufficient}}$ to $\mathbf{SFC5}_{\text{intervention\_insufficient}}$. To monitor behaviour of the nurse, the SFC model receives the same activity to the copied states. In Fig. 9 the activation of $\mathbf{SFC5}_{\text{intervention\_insufficient}}$ after $\mathbf{WS6}_{\text{intervention\_insufficient}}$ for this is illustrated, with an amplification of this WS-state because of the inherent increased patient risk. The Safe Coach makes the choice to speak up in a similar sequence of state activation as the nurse if they choose to speak up. However, the simulation shows that only if the nurse chooses silence, then $\mathbf{SFC13}_{\text{cs\_SPU}}$ receives adequate activation to intervene by speaking up itself, depicted by $\mathbf{WS11}_{\text{SFC\_SPU}}$ in the simulation through activation of the cs-states. In Fig. 9 the consequences of the SFC intervention are displayed. Because of intervention, the Safety Coach allows for an increase of patient safety on account of all healthcare professionals in the delivery room now being aware of the problem, hence leading the team to take measures against medical error deterioration. With the Safety Coach controlling the situation and essentially modelling speaking up behaviour, the Safety Coach decreases fear of repercussion in the nurse as portrayed by the simulation.

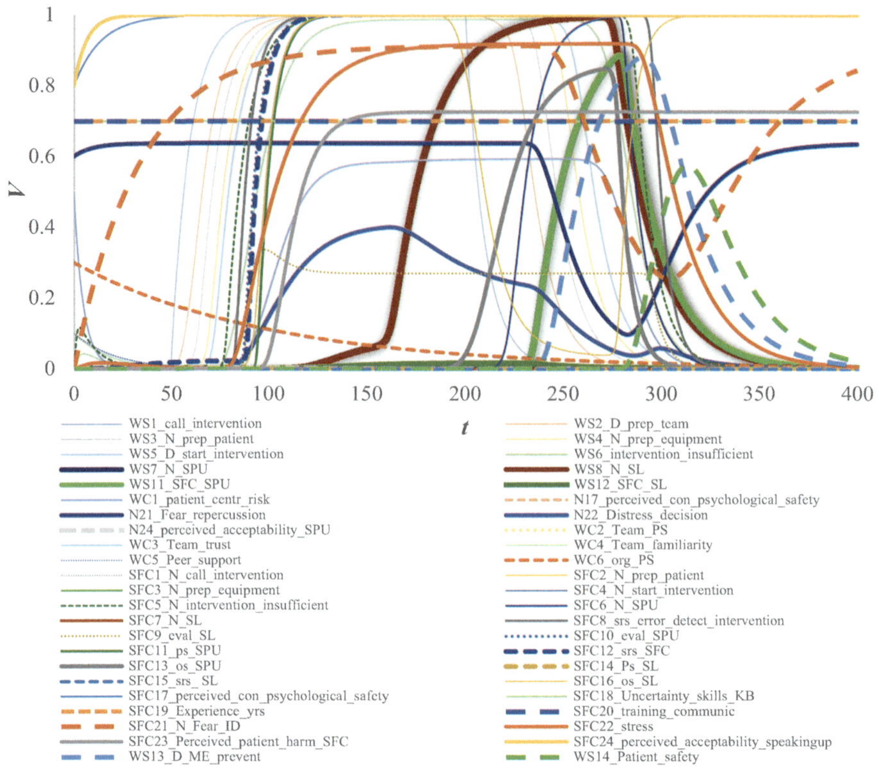

Fig. 9 Simulation experiment for silence by the nurse with intervention of the SFC

7 Discussion

In this chapter, we have proposed an initial network model for speaking up behaviour during delivery room management. The material was taken from (Doornkamp et al. 2023). The aim was to identify and demonstrate the complexities of speaking up behaviour of healthcare professionals, specifically nurses, in healthcare systems using available literature and network modelling. The proposed model shows how speaking up behaviour in nurses or residents can be influenced and what the presence or absence of this behaviour may cause for patient safety in accordance with available literature. In addition, efforts were made to highlight the importance of the attitude of the doctor or person in charge on perceived psychological safety at individual and contextual level. Last, the simulation experiments demonstrated how a monitoring and controlling system may be beneficial as intervention when healthcare professionals hesitate to speak up to prevent the patient from deteriorating due to (preventable) medical errors.

The model was verified by using mathematical stationary point analysis. These points occur for state Y when $\mathrm{d}Y(t)/\mathrm{d}t = 0$; by (1) and (2) of Sect. 3, in terms of the network characteristics, the criterion for $\mathrm{d}Y(t)/\mathrm{d}t = 0$ is as follows:

$$\mathbf{aggimpact}_Y(t) = Y(t) \text{ or } \eta_Y = 0 \tag{3}$$

where

$$\mathbf{aggimpact}_Y(t) = \mathbf{c}_Y(\boldsymbol{\omega}_{X_1,Y}X_1(t), \ldots, \boldsymbol{\omega}_{X_k,Y}, X_k(t)) \tag{4}$$

For four states stationary points were verified; see Table 2. The deviations |$\mathbf{aggimpact}_Y$(t)—Y(t)| obtained were small, which provides an indication that the proposed model is correct in terms of design.

The proposed model includes psychological safety as a mediator for speaking up by the nurse. Whilst several factors were taken into consideration based on literature, the model largely depends on fear of repercussion as determining input for psychological safety. This factor was chosen specifically due to direct associations made in the existing literature on psychological safety (Kish-Gephart et al. 2009). Future studies, however, may consider scrutinization of social contextual and group biases (Jones and Roelofsma 2000), leadership (Bienefeld and Grote 2014; Weiss et al. 2018), hierarchy and interpersonal relations (Beament and Mercer 2016; Bould et al. 2015; O'Donovan et al. 2020a, 2020b; Peadon et al. 2020), influence of speaking up content and tone (Lemke et al. 2021). Leadership style relative to speaking up is one of the more well-studied topics in known literature. There are lines of evidence suggesting inclusive leadership to be a predictor for increased psychological safety of both the team and individual (Pattni et al. 2017; Weiss et al. 2018) and, subsequently, speaking up behaviour. A related predictor of speaking up to leadership in teams and the organisation on a grander scale is the hierarchical structure between team members (Hémon et al. 2020; Long et al. 2020). In other words, the impact of the doctor or team leader on the perceived acceptability to speaking up is a valuable domain in understanding how a work culture can be cultivated that prompts speaking up instead of silence. It is worthwhile for future studies to extend the proposed initial model for speaking up, as addressing these topics was out of the scope of the current chapter.

In addition, the chapter briefly introduced the dyadic communication between doctor and nurse by including the doctor individual model for receptivity to speaking up. However, effective communication is reliant on a multitude of factors not included in the current model for it is out of the scope of the study. In the future, studies

Table 2 Mathematical verification using stationary point analysis results

State Y	$\mathrm{WS6}_{\mathrm{ventilation\_insufficient}}$	$\mathrm{WS7}_{\mathrm{N\_speaking\ up}}$	$\mathrm{WS9}_{\mathrm{D\_res\_POS}}$	$\mathrm{N}_{\mathrm{fear\_repercussion}}$
Time point	142.5	243.5	262	199.5
$Y(t)$	0.977839	0.899415	0.871426	0.310906
$\mathrm{aggimpact}_Y(t)$	0.976556	0.899060	0.850113	0.309844903
Deviation	-0.001283	−0.000356	−0.021313	−0.001061

related to speaking up can emphasise the complexity of this dyadic communication by addressing (e.g.) communication styles (Tan et al. 2017), the tone and content of speaking up and its consequences on doctor-nurse interactions (Lemke et al. 2021), and cultural and professional differences in norms (Long et al. 2020). These themes can be addressed specifically in terms of how the nurse-doctor relationship is affected as well as how this may consequently influence future speaking up behaviour.

Another proposition for future studies with the purpose of achieving completeness is the added dimension of organisational learning (Canbaloğlu et al. 2023) relative to psychological safety and its effects on speaking up among healthcare professionals, see also (Canbaloğlu et al. 2024), this volume. Organisational learning pertains to the collective learning by groups or teams within an organisation, which has a positive influence on organisational performance through shared knowledge between workers (Ratnapalan and Uleryk 2014). Existing literature links healthcare professionals operating in a psychologically safe environment to, on average, being more inclined to engage in team and individual learning regarding taking initiatives for care optimisation (Newman et al. 2017). The prevention of medical errors, to be specific. Therefore, encouraging a psychologically safe climate of transparency about safety concerns allows for organisational learning (O'Donovan and McAuliffe 2020a). Future models can address this dimension of organisational learning through the addition of adaptivity, which entails modelling complex and dynamic processes; processes that adaptively change over time (Treur 2020a), see also (Canbaloğlu et al. 2024), this volume. Predictors introduced in this chapter are familiarity and trust among team members, which also change over time and thus affect psychological safety dynamically. For future studies, it is therefore worthwhile to introduce this adaptivity level to the emergence of speaking up among healthcare professionals.

This chapter addresses the complexity of speaking up within the healthcare setting by introducing an adaptive computational model. Despite considering fewer factors compared to the literature, the model illustrates in a causal and generic manner the following findings with the use of experimental simulations. The model demonstrates that high perceived psychological safety, affected by individual, contextual, and interpersonal factors, prompts the likelihood of speaking up behaviour of healthcare professionals. In like manner, the model predicts how the valence of response from the receiver affects the perceived psychological safety. Furthermore, the virtual coach within the model intervenes to speak up in order to optimize patient safety when it predicts silent behaviour from the healthcare professionals. In conclusion, we introduce the first computational model to predict speaking up behaviour regarding speaker and receiver, as well as an intervention to further optimize care in healthcare organisations.

References

Alingh, C.W., van Wijngaarden, J.D.H., van de Voorde, K., Paauwe, J., Huijsman, R.: Speaking up about patient safety concerns: the influence of safety management approaches and climate on

nurses' willingness to speak up. BMJ Qual. Saf. **28**(1), 39–48 (2019). https://doi.org/10.1136/BMJQS-2017-007163
Alyahya, M.S., Hijazi, H.H., Alolayyan, M.N., Ajayneh, F.J., Khader, Y.S., Al-Sheyab, N.A.: The association between cognitive medical errors and their contributing organizational and individual factors. Risk Manag. Healthcare Policy **14**, 415–430 (2021). https://doi.org/10.2147/RMHP.S293110
Beament, T., Mercer, S.J.: Speak up! Barriers to challenging erroneous decisions of seniors in anaesthesia. Anaesthesia **71**(11), 1332–1340 (2016). https://doi.org/10.1111/ANAE.13546
Bienefeld, N., Grote, G.: Speaking up in ad hoc multiteam systems: Individual-level effects of psychological safety, status, and leadership within and across teams. Euro. J. Work Organ. Psychol. **23**(6), 930–945 (2014). https://doi.org/10.1080/1359432X.2013.808398
Boldingh, A.M., Solevåg, A.L., Nakstad, B.: Outcomes following neonatal cardiopulmonary resuscitation. Tidsskrift for Den Norske Legeforening **138**(9) (2018). 10.4045AIDSSKR.17.0358
Bould, M.D., Sutherland, S., Sydor, D., Naik, V., Friedman, Z.: Residents' reluctance to challenge negative hierarchy in the operating room: a qualitative study. Can. Aneth. Soc. **62**, 576–586 (2015). https://doi.org/10.1007/s12630-015-0364-5
Canbaloğlu, G., Treur, J., Wiewiora, A. (eds.).: Computational Modeling of Multilevel Organisational Learning and Its Control Using Self-Modeling Network Models. Springer Nature (2023). https://doi.org/10.1007/978-3-031-28735-0
Canbaloğlu, G., Treur, J., Roelofsma, P.H.M.P.: Learning of safety and security management through cyberspace: an adaptive self-modeling network model for multi-level organizational learning. In: Roelofsma, P.H.M.P., Jabeen, F., Taal, H. R., Treur, J. (eds.) Using Shared Mental Models and Organisational Learning to Support Safety and Security through Cyberspace: A Computational Analysis Approach. Springer Nature, Cham, Switzerland (this volume) (2024)
Doornkamp, S., Jabeen, F., Treur, J., Taal, H.R., Roelofsma, P.H.M.P.: A controlled adaptive computational network model of a virtual coach supporting speaking up by healthcare professionals to optimise patient safety. Cogn. Syst. Res. **81**, 37–49 (2023)
Edmondson, A.: Psychological safety and learning behavior in work teams. Adm. Sci. q. **44**, 350–383 (1999)
Hémon, B., Michinov, E., Guy, D., Mancheron, P., Scipion, A.: Speaking up about errors in routine clinical practice: a simulation-based intervention with nursing students. Clin. Simul. Nursing **45**, 32–41 (2020). https://doi.org/10.1016/J.ECNS.2020.03.003
Hendrikse, S.C.F., Treur, J., Koole, S.: Modeling emerging interpersonal synchrony and its related adaptive short-term affiliation and long-term bonding: a second-order multi-adaptive neural agent model. Int. J. Neural Syst. **33**(7), 2350038, 41 (2023)
Hendrikse, S.C.F., Treur, J., Koole, S.L. (eds.).: New Analysis and Modeling Directions in Social Interaction Science: Emergent Multimodal Interpersonal Synchrony, Affiliation, and Bonding. Springer Nature, to appear (2025)
Hesslow, G.: Conscious thought as simulation of behaviour and perception. Trends Cogn. Sci. **6**, 242–247 (2002)
James, J.T.: A new, evidence-based estimate of patient harms associated with hospital care. J. Patient Saf. **9**(3), 122–128 (2013). https://doi.org/10.1097/PTS.0B013E3182948A69
Jones, P.E., Roelofsma, P.H.M.P.: The potential for social contextual and group biases in team decision-making: biases, conditions and psychological mechanisms. Ergonomics **43**(8), 1129–1152 (2000). https://doi.org/10.1080/00140130050084914
Kish-Gephart, J.J., Detert, J.R., Klebe Treviño, L., Edmondson, A.C.: Silenced by Fear: The Nature, Sources, and Consequences of Fear at Work (2009). https://doi.org/10.1016/j.riob.2009.07.002
Kolbe, M., Burtscher, M.J., Wacker, J., Grande, B., Nohynkova, R., Manser, T., Spahn, D.R., Grote, G.: Speaking up is related to better team performance in simulated anesthesia inductions: an observational study. Anesth. Analg. **115**(5), 1099–1108 (2012). https://doi.org/10.1213/ANE.0B013E318269CD32

Lemke, R., Burtscher, M.J., Seelandt, J.C., Grande, B., Kolbe, M.: Associations of Form and Function of Speaking Up in Anaesthesia: A Prospective Observational Study (2021). https://doi.org/10.1016/j.bja.2021.08.014

Long, J., Jowsey, T., Garden, A., Henderson, K., Weller, J.: The flip side of speaking up: a new model to facilitate positive responses to speaking up in the operating theatre. Brit. J. Anaesth. **125**(6), 1099–1106 (2020). https://doi.org/10.1016/J.BJA.2020.08.025

Madar, J., Roehr, C.C., Ainsworth, S., Ersdal, H., Morley, C., Rüdiger, M., Skåre, C., Szczapa, T., Te Pas, A., Trevisanuto, D., Urlesberger, B., Wilkinson, D., Wyllie, J.P.: European resuscitation council guidelines 2021: newborn resuscitation and support of transition of infants at birth resuscitation introduction and scope. Resuscitation **161**, 291–326 (2021). https://doi.org/10.1016/j.resuscitation.2021.02.014

Maxfield, D., Grenny, J., Lavandero, R., Groah, L.: Safety tools and organizational silence-story collector findings. Occup. Health **57**(12) (2011)

Millar, L.J., Shi, L, Hoerder-Suabedissen, A., Molnår, Z.: Neonatal hypoxia ischaemia: mechanisms, models, and therapeutic challenges. Front. Cellular Neurosci. **11** (2017)

Moore, L., McAuliffe, E.: To report or not to report? why some nurses are reluctant to whistleblow. Clin. Govern. **17**(4), 332–342 (2012). https://doi.org/10.1108/14777271211273215/FULL/PDF

Morrison, E.W.: Employee voice behavior: integration and directions for future research. Acad. Manag. Ann. **5**(1), 373–412 (2011). https://doi.org/10.1080/19416520.2011.574506

Nacioglu, A.: As a critical behavior to improve quality and patient safety in health care: speaking up! Saf. Health **2**(1), 1–25 (2016). https://doi.org/10.1186/S40886-016-0021-X

Newman, A., Donohue, R., Eva, N.: Psychological safety: a systematic review of the literature. Human Resour. Manag. Rev. **27**(3), 521–535 (2017). https://doi.org/10.1016/J.HRMR.2017.01.001

O'Donovan, R., McAuliffe, E.: A systematic review exploring the content and outcomes of interventions to improve psychological safety, speaking up and voice behaviour. BMC Health Serv. Res. **20**(1), 1–11 (2020a). https://doi.org/10.1186/S12913-020-4931-2/FIGURES/2

O'Donovan, R., McAuliffe, E.: Exploring psychological safety in healthcare teams to inform the development of interventions: combining observational, survey and interview data. BMC Health Serv. Res. **20**(1), 1–16 (2020b). https://doi.org/10.1186/S12913-020-05646-Z/FIGURES/1

Okuyama, A., Wagner, C., Bijnen, B.: Speaking up for patient safety by hospital-based health care professionals: a literature review. BMC Health Serv. Res. **14**(1), 1–8 (2014). https://doi.org/10.1186/1472-6963-14-61/FIGURES/2

Pabst, S., Brand, M., Wolf, O.T.: Stress and decision making: A few minutes make all the difference. Behav. Brain Res. **250**, 39–45 (2013). https://doi.org/10.1016/J.BBR.2013.04.046

Palatnik, A.M.: Speak up for patient safety. Nursing Crit. Care **11**(6), 4 (2016). https://doi.org/10.1097/01.CCN.0000503425.05594.02

Pattni, N., Bould, M.D., Hayter, M.A., McLuckie, D., Noble, L.M.K., Malavade, A., Friedman, Z.: Gender, power and leadership: the effect of a superior's gender on respiratory therapists' ability to challenge leadership during a life-threatening emergency. Brit. J. Anaesth. **119**(4), 697–702 (2017). https://doi.org/10.1093/BJA/AEX246

Pattni, N., Arzola, C., Malavade, A., Varmani, S., Krimus, L., Friedman, Z.: Challenging authority and speaking up in the operating room environment: a narrative synthesis. Brit. J. Anaesth. **122**(2), 233–244 (2019). https://doi.org/10.1016/J.BJA.2018.10.056

Peadon, R., Hurley, J., Hutchinson, M.: Hierarchy and medical error: Speaking up when witnessing an error. Saf. Sci. **125**, 104648 (2020). https://doi.org/10.1016/J.SSCI.2020.104648

Raju, T.N.K., Suresh, G., Higgins, R.D.: Patient safety in the context of neonatal intensive care: research and educational opportunities. Pediatric Res. **70**(1), 109–115 (2011). https://doi.org/10.1203/pdr.0b013e3182182853

Ratnapalan, S., Uleryk, E.: Organizational learning in health care organizations. Systems **2**(1), 24–33 (2014). https://doi.org/10.3390/SYSTEMS2010024

Roussin, C.J., Larraz, E., Jamieson, K., Maestre, J.M.: Psychological safety, self-efficacy, and speaking up in interprofessional health care simulation. Clin. Simul. Nursing **17**, 38–46 (2018). https://doi.org/10.1016/J.ECNS.2017.12.002

Sayre, M.M., McNeese-Smith, D., Leach, L.S., Phillips, L.R.: An educational intervention to increase "speaking-up" behaviors in nurses and improve patient safety. J. Nursing Care Qual. **27**(2), 154–160 (2012). https://doi.org/10.1097/NCQ.0B013E318241D9FF

Schwappach, D.L.B., Gehring, K.: Saying it without words: a qualitative study of oncology staff's experiences with speaking up about safety concerns. BMJ Open **4**, 4740 (2014). https://doi.org/10.1136/bmjopen-2013

Tan, T.C., Zhou, H., Kelly, M.: Nurse–physician communication—an integrated review. J. Clin. Nursing **26**(23–24), 3974–3989 (2017). https://doi.org/10.1111/JOCN.13832

Treur, J.: Network-Oriented Modeling: Addressing Complexity of Cognitive, Affective and Social Interactions. Springer, Berlin (2016)

Treur, J.: Modeling higher order adaptivity of a network by multilevel network reification. Netw. Sci. **8**(S1), S110-S144 (2020a). https://doi.org/10.1017/NWS.2019.56

Treur, J.: Network-Oriented Modeling for Adaptive Networks: Designing Higher-Order Adaptive Biological, Mental and Social Network Models. Springer Nature, Cham (2020b)

Treur, J.: On the dynamics and adaptivity of mental processes: relating adaptive dynamical systems and self-modeling network models by mathematical analysis. Cogn. Syst. Res. **70**, 93–100 (2021)

Treur, J., Van Ments, L.: Mental Models and Their Dynamics, Adaptation, and Control. Springer Nature, Cham (2022). https://doi.org/10.1007/978-3-030-85821-6

Umoren, R., Kim, S., Gray, M.M., Best, J.A., Robins, L.: Interprofessional model on speaking up behaviour in healthcare professionals: a qualitative study. BMJ Leader **6**(1), 15–19 (2022). https://doi.org/10.1136/LEADER-2020-000407

Van Ments, L., Treur, J., Klein, J. Roelofsma, P.H.M.P.: Developing a safety and security AI coach: A second-order adaptive network model of shared mental models in hospital teamwork. In: Roelofsma, P.H.M.P., Jabeen, F., Taal, H. R., Treur, J. (eds.) Using Shared Mental Models and Organisational Learning to Support Safety and Security through Cyberspace: A Computational Analysis Approach. Springer Nature, Cham, Switzerland (this volume) (2025)

Vogus, T.J., Iacobucci, D.: Creating highly reliable health care: how reliability-enhancing work practices affect patient safety in hospitals. **69**(4), 911–938 (2016). https://doi.org/10.1177/0019793916642759

Weiss, M., Kolbe, M., Grote, G., Spahn, D.R., Grande, B.: We can do it! Inclusive leader language promotes voice behavior in multi-professional teams. Leadersh. Q. **29**(3), 389–402 (2018). https://doi.org/10.1016/J.LEAQUA.2017.09.002

Weller, J.M., Webster, C.S.: Normalising good communication in hospital teams. Brit. J. Anaesth. **126**(4), 758–760 (2021). https://doi.org/10.1016/J.BJA.2020.12.036

Organisational Learning for Safety and Security Through Cyberspace

The fifth part of this book explores the pivotal role of organizational learning in enhancing safety and security through cyberspace. This part delves into the intricacies of learning from mistakes within organizations and the development of strategies that foster a culture of continuous improvement for safety and security. In the modern digital era, where the stakes of operational failures are higher, understanding and implementing effective organizational learning mechanisms are crucial for maintaining robust safety and security standards. This part highlights how adaptive network models and self-modeling approaches can facilitate organizational learning, with a particular emphasis on addressing and leveraging mistakes to drive better outcomes.

The chapters in this part provide comprehensive insights into the dynamics of organizational learning. Chapter "Learning from Mistakes within Organizations: An Adaptive Network-Oriented Model for a Double Bias Perspective for Safety and Security through Cyberspace" by Hossein et al., introduces an adaptive network model that examines the double bias perspective on learning from mistakes within organizations. It underscores the challenges posed by the fear of mistakes and the influence of leadership mindsets on organizational behavior. Chapter "Organisational Learning for Safety and Security Through Cyberspace: Adaptive Modelling of the Implementation of Environment Health and Safety Standards" by van den Hurk et al., shifts focus to the implementation of Environment Health and Safety (EHS) standards, using a self-modeling network approach to analyze and predict organizational behavior in real-world scenarios. Together, these chapters offer a nuanced understanding of how organizations can harness mistakes and learning processes to enhance safety and security.

What Will be Addressed in This Part

Organizational learning is a multifaceted process that significantly impacts safety and security within any institution. The integration of cyberspace tools and adaptive models provides innovative ways to foster a learning culture that is both proactive and resilient. The chapters in Part V illustrate the transformative potential of these approaches in addressing the complexities of organizational behavior and learning.

Learning from Mistakes

Chapter "Learning from Mistakes within Organizations: An Adaptive Network-Oriented Model for a Double Bias Perspective for Safety and Security through Cyberspace" addresses a critical aspect of organizational learning: the role of mistakes and the attitudes towards them. The adaptive network model presented by Hosseini, Treur, and Kucharska highlights the detrimental impact of a fear-based approach to mistakes. Leaders who avoid complex tasks to maintain a facade of perfection inadvertently stifle learning and growth within their teams. This chapter contrasts two distinct leadership mindsets—one that shuns mistakes and one that embraces them as learning opportunities. The former creates an environment where mistakes are hidden and unaddressed, leading to stagnation and missed opportunities for improvement. In contrast, the latter promotes a culture of transparency and continuous learning, where mistakes are viewed as valuable feedback for enhancing overall performance.

The chapter also emphasizes the importance of leadership in shaping organizational attitudes towards mistakes. Leaders who model a learning-oriented mindset encourage their employees to take on challenges and learn from their errors, fostering an environment of trust and growth. This approach not only improves individual performance but also enhances the organization's capacity to adapt and innovate.

Implementing EHS Standards

Chapter "Organisational Learning for Safety and Security Through Cyberspace: Adaptive Modelling of the Implementation of Environment Health and Safety Standards" extends the discussion by focusing on the implementation of Environmental Health and Safety (EHS) standards in a global organization. The study by van den Hurk, Treur, and Roelofsma utilizes a self-modeling network approach to analyze how organizations can effectively implement and learn from EHS standards. The real-world scenario presented in the chapter demonstrates the complexity of applying uniform safety standards across diverse geographical and legislative contexts. The dynamic computational modeling technique used in the study allows for the simulation of various implementation scenarios, providing valuable insights into the challenges and opportunities associated with organizational learning in this context.

The chapter highlights the importance of cooperation and open communication among different stakeholders within the organization. Successful implementation of EHS standards requires a collaborative effort between the system owner, EHS experts, and departmental teams. The self-modeling network approach facilitates this collaboration by enabling continuous feedback and feed-forward learning, ensuring that lessons learned in pilot phases are effectively applied across the organization. This iterative learning process is crucial for maintaining high safety standards and adapting to new challenges as they arise.

Organizational Learning and Safety Culture
Both chapters underscore the critical role of organizational learning in building a robust safety culture. A key takeaway is the necessity of creating an environment where mistakes are not only tolerated but are actively used as learning tools. This requires a shift in mindset from viewing mistakes as failures to seeing them as opportunities for improvement. By leveraging adaptive network models and self-modeling approaches, organizations can better understand the dynamics of learning from mistakes and implement strategies that promote continuous improvement.

Furthermore, the integration of cyberspace tools enhances the ability of organizations to collect, analyze, and disseminate information related to safety and security. These tools enable real-time feedback and adaptive learning, allowing organizations to respond swiftly to new information and evolving challenges. This proactive approach to organizational learning is essential for maintaining high standards of safety and security in an increasingly complex and interconnected world.

Conclusion

This part of this book offers valuable insights into the mechanisms of organizational learning and its impact on safety and security. The chapters highlight the importance of embracing mistakes as learning opportunities and the role of leadership in fostering a culture of continuous improvement. The adaptive network models and self-modeling approaches presented in this part provide innovative frameworks for analyzing and enhancing organizational learning processes.

By focusing on real-world scenarios and practical applications, this part underscores the potential of adaptive learning models to transform organizational behavior and improve safety outcomes. The discussions emphasize the need for open communication, collaboration, and a proactive approach to learning from mistakes. As organizations continue to navigate the challenges of the digital age, the insights and models presented in this part offer a roadmap for building resilient and adaptive safety cultures.

In conclusion, the integration of organizational learning strategies supported by advanced computational models is crucial for enhancing safety and security in cyberspace. By leveraging the power of mistakes and fostering a learning-oriented mindset, organizations can achieve better performance, innovation, and resilience. The continued exploration and implementation of these strategies hold significant promise for the future of organizational safety and security, paving the way for more effective and sustainable practices.

Learning from Mistakes Within Organizations: An Adaptive Network-Oriented Model for a Double Bias Perspective for Safety and Security Through Cyberspace

Mojgan Hosseini, Jan Treur, and Wioleta Kucharska

Abstract Although making mistakes is a crucial part of learning, it is still often being avoided in companies as it is considered as a shameful incident. This goes hand in hand with a mindset of a boss who dominantly believes that mistakes usually have negative consequences and therefore avoids them by only accepting simple tasks. Thus, there is no mechanism to learn from mistakes. Employees working for and being influenced by such a boss also strongly believe that mistakes usually have negative consequences but in addition they believe that the boss never makes mistakes, it is often believed that only those who never make mistakes can be bosses and hold power. That's the problem, such kinds of bosses do not learn. So, on the one hand, we have bosses who select simple tasks to be always seen as perfect. Therefore, also they believe they should avoid mistakes. On the other hand, there exists a mindset of a boss who is not limited to simple tasks, he/she accepts more complex tasks and therefore in the end has better general performance by learning from mistakes. This then also affects the mindset and actions of employees in the same direction. This chapter investigates the consequences of both attitudes for the organizations. It does so by computational analysis based on an adaptive dynamical

M. Hosseini · J. Treur (✉)
Department of Computer Science, Vrije Universiteit Amsterdam, Social AI Group, Amsterdam, The Netherlands
e-mail: j.treur@vu.nl; j.treur@hhs.nl

M. Hosseini
e-mail: m.hosseini@student.vu.nl

J. Treur
Risk Management and Cybersecurity Group, The Hague University of Applied Sciences, Hague, The Netherlands

W. Kucharska
Faculty of Management and Economics, Gdansk University of Technology, Gdansk, Poland
e-mail: wioleta.kucharska@pg.edu.pl

P. H. M. P. Roelofsma et al. (eds.), *Using Shared Mental Models and Organisational Learning to Support Safety and Security Through Cyberspace: A Computational Analysis Approach*, Studies in Systems, Decision and Control 570,
https://doi.org/10.1007/978-3-031-72075-8_10

systems modeling approach represented in a network format using the self-modeling network modeling principle.

Keywords Learning from mistakes · Double bias · Adaptive network model

1 Introduction

Mistakes, a natural part of being human, evoke both positive learning experiences and negative consequences. This duality gives rise to the framing effect, a cognitive bias that influences how mistakes are perceived and impacts decision-making. Often beliefs that mistakes have negative consequences dominate within both bosses and employees. However, if due to this biased type of belief a boss avoids situations where mistakes can be made, for employees another belief occurs, namely that a boss never makes mistakes and therefore they should behave in a similar way. This is called the double bias effect (Cannon and Edmondson 2004).

In organizational settings, understanding this double bias is crucial to unlock untapped potential and foster a culture of growth and resilience. The transformative learning theory emphasizes the importance of critical reflection on personal experiences for profound adult learning. However, the negative framing effect can hinder individuals from gaining valuable insights (Mishra 2012). Avoiding or overlooking mistakes leads to missed opportunities for growth and triggering negative resource spirals. Addressing this double bias is essential as it significantly affects collective intelligence within organizations. The presence of a double bias regarding mistakes leads to a cultivation of fixed mindsets in society. Mindsets are psychological constructs that shape individuals' perceptions and understanding of the world. Growth mindsets are learning-oriented, valuing progress and resilience, while fixed mindsets are image-oriented, seeking constant validation of self-perfection.

Leaders with fixed mindsets avoid the risk of making mistakes to maintain their positive self-image, leading to a crisis of transformational leadership. Organizations dominated by fixed mindsets and a refusal to accept mistakes as learning opportunities face challenges in developing shared collective intelligence. This organizational shared mindset significantly influences the ability to learn and adapt to changes, which is considered organizational intelligence (Treur 2020b). Understanding and addressing the double bias and promoting growth mindsets can foster a culture of learning, adaptability, and transformational leadership within organizations. In this chapter, we highlight the value of embracing mistakes as powerful learning resources and make a computational analysis of it using an adaptive network modeling format.

2 Modeling Adaptive Networks as Self-modeling Networks

Network-oriented modelling is a useful approach to analyse adaptive and dynamic aspects of networks and pathways of causal connections. In a network nodes X and Y (also called states) are used and connections between them. These nodes serve as state variables with real number values $X(t)$ and $Y(t)$ varying over time t. A temporal-causal network model is defined by the following network characteristics:

- **Connectivity characteristics**: connections from a state X to a state Y have a connection weight $\boldsymbol{\omega}_{X,Y}$ which indicates the strength of the connection.
- **Timing characteristics**: for all states Y, a speed factor $\boldsymbol{\eta}_Y$ indicates how fast the state is changing upon causal impact.
- **Aggregation characteristics**: for all Y states, a combination function $\mathbf{c}_Y(..)$ is referred to which aggregates how the causal impacts from other states affect Y.

The canonical difference equation used for any state Y to compute from time t the impact by all incoming connections for Y at later time $t + \Delta t$ is:

$$Y(t + \Delta t) = Y(t) + \boldsymbol{\eta}_Y\left[\mathbf{c}_Y\left(\boldsymbol{\omega}_{X_1,Y}X_1(t), \ldots, \boldsymbol{\omega}_{X_k,Y}X_k(t)\right) - Y(t)\right]\Delta t \tag{1}$$

where the X_i are the states from which state Y gets incoming connections. So, from a dynamical system modeling perspective, for the dynamics of each of the state variables Y of the model a difference equation is used which is an instantiation (by substitution of the relevant network characteristics $\boldsymbol{\omega}$, $\boldsymbol{\eta}$, and $\mathbf{c}$) of the canonical difference Eq. (1). For example, the model introduced in Sect. 3 will have 64 states Y, therefore it models a dynamical system by using 64 (interacting) difference equations for its 64 state variables which all are instantiations of (1) for the specific network characteristics $\boldsymbol{\omega}$, $\boldsymbol{\eta}$, and $\mathbf{c}$ for the states.

For adaptive networks, some of these network characteristics can change over time. For example, a connection weight $\boldsymbol{\omega}_{X,Y}$ can change due to learning. Using the self-modelling principle (Treur 2020a, b, c), a self-model state $\mathbf{W}_{X,Y}$ can be added to the network that represents the value of $\boldsymbol{\omega}_{X,Y}$. Similarly, the excitability threshold $\boldsymbol{\tau}_Y$ of a logistic sum combination function can change over time, making a state Y more sensitive to less sensitive in responding to causal impact. Again by the self-modelling principle a self-model state $\mathbf{T}_Y$ can be added to the network that represents the value of $\boldsymbol{\tau}_Y$. Or a speed factor $\boldsymbol{\eta}_Y$ can change, for example due to changing contextual circumstances. By the self-modelling principle, a self-model state $\mathbf{H}_Y$ can be added to the network that represents the value of $\boldsymbol{\eta}_Y$. In this way the network gets a self-model of part of its structure; this is also called network reification (Treur 2020a, b, c). In such a case, in Eq. (1) for adaptive characteristics the values at time t of the self-model states are used, for example, $\mathbf{W}_{X,Y}(t)$ instead of $\boldsymbol{\omega}_{X,Y}$. A dedicated software environment based on these network characteristics and the self-modelling principle was used to conduct simulation experiments; see Chap 9 of Treur (2020b).

Two combination functions are used in the model, the **alogistic** function and the function **hebb** for hebbian learning. See Table 1 for these functions.

Table 1 The two combination functions used and their parameters

Name	Notation	Formula	Parameters
Advanced logistic sum	$\textbf{alogistic}_{\sigma,\tau}(V_1, \ldots, V_k)$	$\left[\frac{1}{1+e^{-\sigma(V_1+\cdots+V_k-\tau)}} - \frac{1}{1+e^{\sigma\tau}}\right](1+e^{-\sigma\tau})$	p(1) = steepness σ p(2) = threshold τ
Hebbian learning	$\textbf{hebb}_{\mu}(V_1, \ldots, V_k)$	$V_1 V_2(1 - V_3) + \mu\, V_3$	p(1) = persistence factor μ

In Treur (2021), Hendrikse et al. (2023), it is shown that any smooth dynamical system has a canonical representation as a temporal-causal network and any smooth (multi-order) adaptive dynamical system has a canonical representation as a (multi-order) self-modeling temporal-causal network. Therefore, compared to adaptive dynamical systems in general, the network-oriented modeling approach used here does not introduce any fundamental limitations concerning what it can model.

3 Setup of the Computational Analysis

The presence of the 'double bias of leaders' mistakes' can have profound implications for an organization's learning, innovativeness, and sustainability. The positive attitude towards learning and the negative attitude towards mistakes cause trouble with learning from mistakes. Such contradictory perspectives create a cognitive bias that is doubled in organizations by the belief that bosses never make mistakes—that altogether jeopardizes organizational learning. This analysis refers to the existence of two opposing belief when leaders make mistakes: the belief that bosses should never make mistakes and the opposite belief that they are prone to errors like anyone else. These beliefs can trigger the construction of a fixed mindset among leaders, leading to a culture of domination and hindered growth. As a result, the root of sustainability problems within the organization can be traced back to how these biases impact leaders' learning behaviors. Sustainability achievement in organizations requires constant learning and adaptivity (Kucharska et al. 2023).

The computational analysis is based on modeling the mindset of the boss by a set of context factors for their beliefs and attitudes and using two mental models for performing a task: mental model 1 for simple (mistake avoiding) performance and mental model 2 for more complex performance (with more risk for mistakes). The knowledge for these two mental models is in their connections which are represented by self-model states, following Treur (2020c). In order to show the connectivity of the model between them better, they are shown in an architecture to describe and distinguish the dynamics within the network: base level (lower plane), first-order self-model level (middle plane) and second-order self-model level (upper plane);

see Fig. 1. This is an architecture with two variants, one for the boss B and one for employees E, with similar states on the planes.

Mindset of the boss and of employees

A set of context factors describes either a limited or expanded mindset of the boss, addressing attitudes and beliefs for mistakes, acceptance of mistakes, openness for learning from them, and so on; see Table 2. Both boss B and employees E have these states for their personal characteristics. They are represented in their base planes, one for the boss and the other for the employee. To indicate the effect of boss's mindset

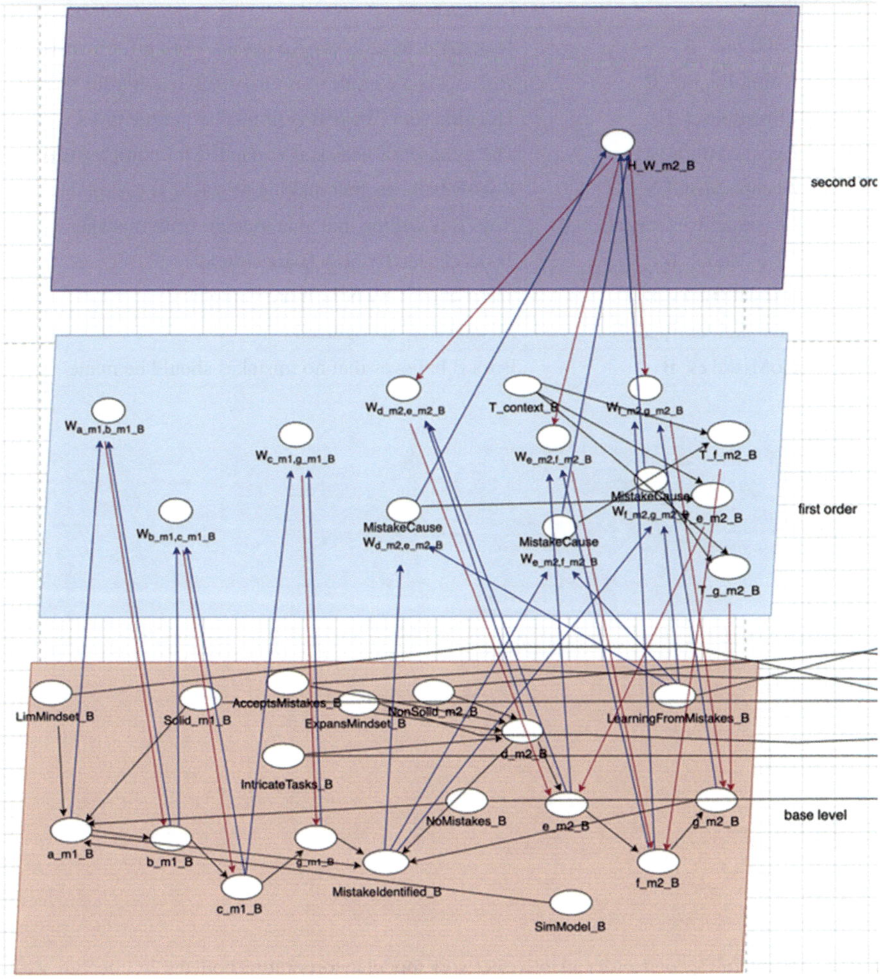

Fig. 1 The second-order adaptive network architecture used for boss and for employees

on the employees' mindset by contagion from boss to employees, the ones for the boss have connections to the ones for the employees as can be seen in Fig. 2.

Mental model 1: simple tasks

Base level states a_m1, b_m1, c_m1 and g_m1 are mental model states in mental model m1 that represent tasks to achieve result g which are simple for the boss (B) and employees (E), no mistakes are made and therefore nothing is learnt. For the sake of simplicity, these mental model states are assumed connected as follows:

Table 2 States for context factors defining a mindset

Nr.	State	Explanation
X_1	Solid_m1_B	Task m1 is an easy task to obtain result g for boss B
X_2	NonSolid_m2_B	Task m2 is a complex task to obtain result g for boss B
X_3	LimMindset_B	The mindset of boss B is limited to simple tasks
X_4	ExpandMindset_B	The mindset of boss B is expanded for complex tasks
X_5	AcceptsMistakes_B	Boss B believes that making mistakes is acceptable
X_6	LearningFromMistakes_B	Boss B is willing and able to learn from mistakes
X_7	SimpModel_B	Tasks chosen by boss B are simple
X_8	IntricateTasks_B	Tasks chosen by boss B are complex
X_9	MistakeIdentified_B	Mistakes are recognized
X_{10}	NoMistakes_B	Boss B believes that no mistakes should be made

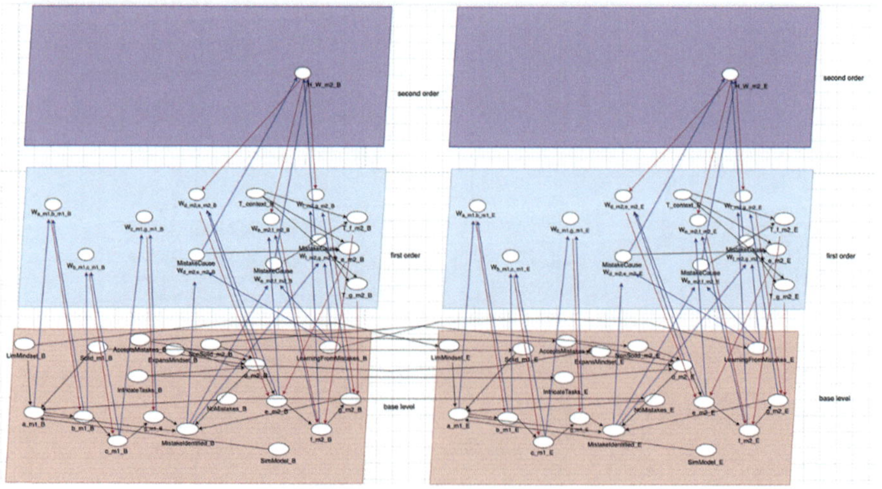

Fig. 2 The overall second-order adaptive network model incorporating both the architecture variant for the boss and for an employee and connections between them for the contagion of the mindset from boss to employee

$$a\_m1 \rightarrow b\_m1 \rightarrow c\_m1 \rightarrow g\_m1$$

Mental model 2: complex tasks

Base level states d_m2, e_m2, f_m2 and g_m2 are mental model states in mental model m2 representing tasks to achieve result g which are more complex, with more risk for making mistakes therefore it can lead to learning. These mental states are assumed connected as follows (Table 3).

$$d\_m2 \rightarrow e\_m2 \rightarrow f\_m2 \rightarrow g\_m2$$

The quality of the knowledge of these mental models is represented by the weights of their connections, where weight value 1 is perfect knowledge. To personalize these mental model states, a name indicating the person is added, for example for mental model m2:

$$d\_m2\_B \rightarrow e\_m2\_B \rightarrow f\_m2\_B \rightarrow g\_m2\_B \text{for boss B}$$

$$d\_m2\_E \rightarrow e\_m2\_E \rightarrow f\_m2\_E \rightarrow g\_m2\_E \text{for employee E}$$

Adaptive knowledge for the mental models

As the knowledge for the mental models is assumed to be adaptive, their connection weights are represented by self-model **W**-states as discussed in Sect. 2, for example,

Table 3 Base level mental model states

Nr.	State	Explanation
X_{11}	a_m1_B	Simple task in mental model m1 for boss B, no mistakes and therefore no learning from mistakes
X_{12}	b_m1_B	Simple task b in mental model m1 for boss B, no mistakes and therefore no learning from mistakes
X_{13}	c_m1_B	Simple task c in mental model m1 for boss B, no mistakes and therefore no learning from mistakes
X_{14}	g_m1_B	Task g in mental model m1 for boss B
X_{15}	d_m2_B	Complex task d in mental model m2 for boss B, risk of making mistakes and learning
X_{15}	e_m2_B	Complex task e in mental model m2 for boss B, risk of making mistakes and learning
X_{17}	f_m2_B	Complex task f in mental model m2 for boss B, risk of making mistakes and learning
X_{18}	g_m2_B	Task g in mental model m2 for boss B

for mental model m2 of the boss B and of an employee E:

$$\mathbf{W}_{d\_m2\_B,e\_m2\_B}, \mathbf{W}_{e\_m2\_B,f\_m2\_B}, \mathbf{W}_{f\_m2\_B,g\_m2\_B}\text{knowledge of boss B}$$

$$\mathbf{W}_{d\_m2\_E,e\_m2\_E}, \mathbf{W}_{e\_m2\_E,f\_m2\_E}, \mathbf{W}_{f\_m2\_E,g\_m2\_E}\text{knowledge of employee E}$$

See Table 4 for further explanations of the first- and second-order self-model states. In Fig. 1 the network model is shown in 3D according to three planes. On the base level (pink plane), base states and base connections are depicted. On first-order self-model level (blue plane), simple and complex tasks from base level are connected to their **W**-states. On the second-order self-model level (purple plane) **H**-states represent speed factors for the adaptation of the knowledge, connected upward and downward to **W**-states; see the last row of Table 4. The **T**-states are used to boost the Hebbian learning (by temporarely lowered excitability thresholds for the mental model states of m2) after a mistake is identified, see also Table 4.

The contagion of the boss b's mindset to an employee E's mindset

As can be seen in Fig. 2 the states for B's mindset have connections to the corresponding states for E's mindset. Given that B's mindset is considered static with values from the beginning, by the indicated connections these values transfer to E's mindset as well.

4 Simulation Experiments

In order to simulate the processes and the learning according to the two alternative mindsets, different variants were made and simulated in Matlab. These variants were specified by role matrices (Treur 2020b): **mb** (base connectivity), **mcw** (connection weights), **mcfw** (combination function weights), **mcfp** (combination function parameters), **msv** (speed factors), and **iv** (initial values). See the Appendix (2023) for them. Two scenarios have been addressed, the first one is a limited mindset for boss B (Fig. 3) and the second is extended mindset (Fig. 4). They have constant values as specified in Table 5. They show what happens when simple and complex tasks are done by the bosses B and their effects on employees E. They also present that when mistakes are identified, and learning happens on the extended mindset while nothing happens on the limited mindset.

Figure 5 indicates **T**-states for the mental model states in mental model m2 which shows that learning is happening as boosted by lower excitability threshold values (the curves starting at 0.8 and going down to around 0.5) when doing complex tasks and mistakes are identified.

Figure 6 shows **W**- and $\mathbf{H}_\mathbf{W}$-states which represent learning effects of boss B and employees E. The learning speed is shown via $\mathbf{H}_\mathbf{W}$-states, learning happens for boss B first (purple line starting from 0) and then for employee E (blue line starting from

Table 4 First- and second-order self-model states

Nr.	State	Explanation
X_{19}	$\mathbf{W}_{a\_m1\_B,b\_m1\_B}$	Self-model state representing the weight of the connection from task a to task b within mental model m1 of boss B
X_{20}	$\mathbf{W}_{b\_m1\_B,c\_m21\_B}$	Self-model state representing the weight of the connection from task b to task c within mental model m1 of boss B
X_{21}	$\mathbf{W}_{c\_m1\_B,g\_m1\_B}$	Self-model state representing the weight of the connection from task c to task g within mental model m1 of boss B
X_{22}	$\mathbf{W}_{d\_m2\_B,e\_m2\_B}$	Self-model state representing the weight of the connection from task d to task e within mental model m2 of boss B
X_{23}	$\mathbf{W}_{e\_m2\_B,f\_m2\_B}$	Self-model state representing the weight of the connection from task e to task f within mental model m2 of boss
X_{24}	$\mathbf{W}_{f\_m2\_B,g\_m2\_B}$	Self-model state representing the weight of the connection from task f to task g within mental model m2 of boss
X_{25}	Mistake cause $\mathbf{W}_{d\_m2\_B,e\_m2\_B}$	The cause of the mistake is the connection from task d to task e in mental model m2 of boss B
X_{26}	Mistake cause $\mathbf{W}_{e\_m2\_B,f\_m2\_B}$	The cause of the mistake is the connection from task e to task f in mental model m2 of boss B
X_{27}	Mistake cause $\mathbf{W}_{f\_m2\_B,g\_m2\_B}$	The cause of the mistake is the connection from task f to task g in mental model m2 of boss B
X_{28}	T_context_B	Threshold context state for boss B
X_{29}	$\mathbf{T}_{e\_m2\_B}$	Self-model state representing the threshold of complex task e in mental model m2 of boss B
X_{30}	$\mathbf{T}_{f\_m2\_B}$	Self-model state representing the threshold of complex task f in mental model m2 of boss B
X_{31}	$\mathbf{T}_{g\_m2\_B}$	Self-model state representing the threshold of complex task g in mental model m2 of boss B
X_{32}	$\mathbf{H}_{\mathbf{W}_{m2\_B}}$	Second-order self-model state for the speed factor of change in the **W**-states for mental model m2 of boss B

0) as the employee takes the effect from the boss; it is shown by the dotted lines. The two lines which start from value 1 on the vertical axis indicate the simple tasks which are not used and therefore slowly forgotten.

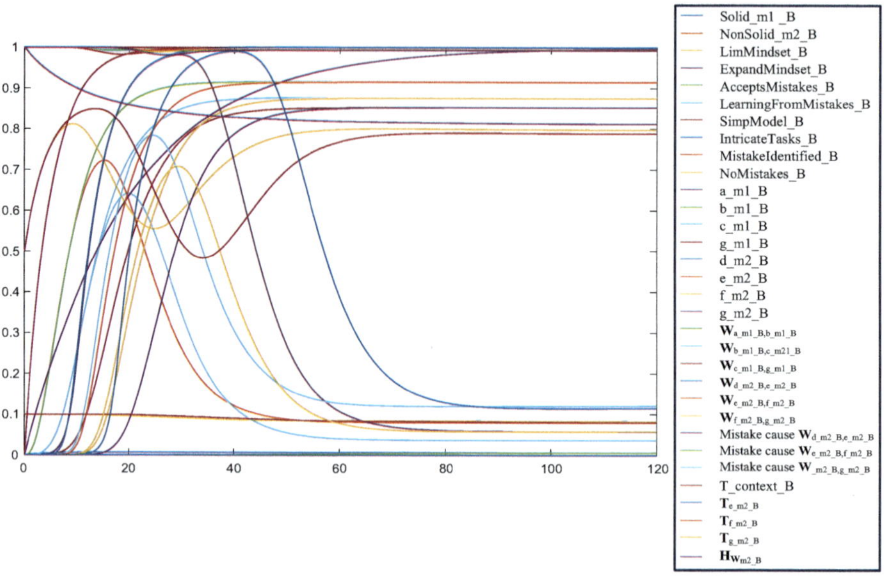

Fig. 3 Simulation for the limited mindset scenario. For the legend of employee E's states, see Fig. 4

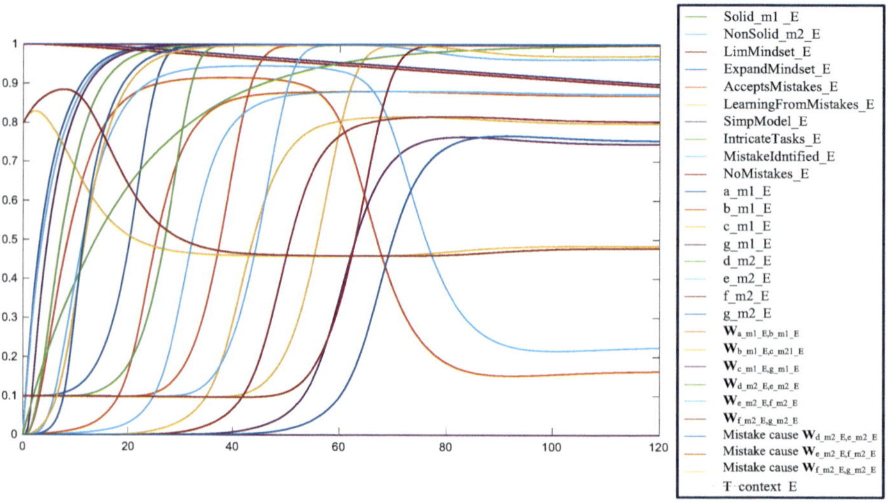

Fig. 4 Simulation for the extended mindset scenario. For the legend of boss B's states, see Fig. 3

Table 5 Values for the mindset context factors for the two scenarios

Nr.	State	Limited mindset	Extended mindset
X_1	Solid_m1 _B	1	1
X_2	NonSolid_m2 _B	1	1
X_3	LimMindset_B	1	0
X_4	ExpandMindset _B	0	1
X_5	AcceptsMistakes _B	0	1
X_6	LearningFromMistakes _B	0	1
X_7	SimpModel _B	1	0
X_8	IntricateTasks _B	0	1
X_9	MistakeIdentified _B	–	–
X_{10}	NoMistakes_B	1	0

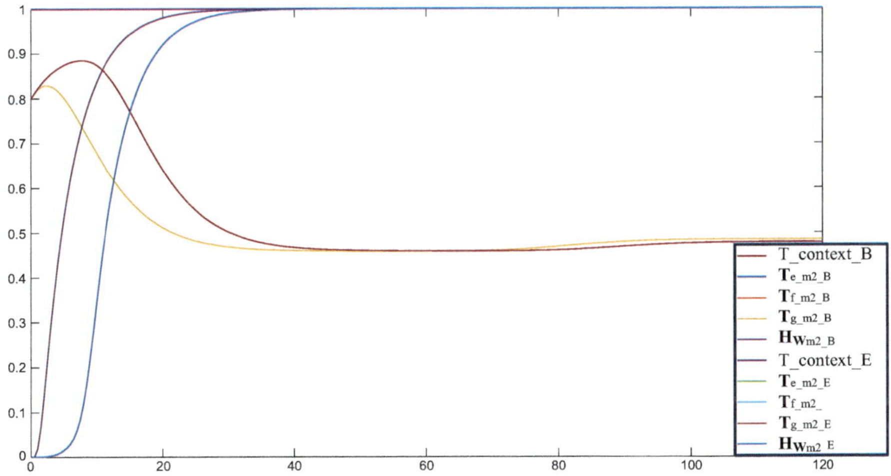

Fig. 5 The **T**-states for the extended mindset scenario: decreasing of excitability thresholds when identifying mistakes for complex tasks

5 Discussion

When leaders make mistakes and those mistakes are considered as a negative incident, it leads to a fixed way of thinking of avoiding mistakes. This is the main cause of not achieving long-term success. However, when leaders have the confidence of identifying their mistakes and learning from them, it effects the learning process exponentially and as a result progress happens (Kucharska et al. 2023). The objective of this chapter was to analyze the simulation outcomes for the different mindsets in a comparative manner and illustrate the differences via graphs of these simulation experiments.

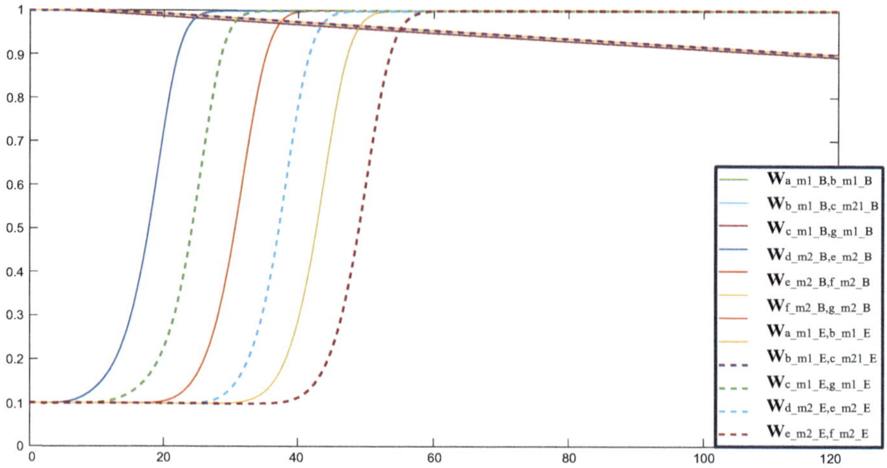

Fig. 6 The **W**-states for the extended mindset scenario: learning

Boss B's willingness to tackle challenging tasks influences the employees, encouraging them to also take on and learn from such tasks. Conversely, if the boss avoids complex assignments, the entire team might refrain from engaging in such a learning process, leading to a lack of progress. This contagion behavior is shown in the model design depicted in Figs. 1 and 2.

Via Matlab simulations, the study explored two contrasting mindsets using role matrices (**mb**, **mcw**, **mcfw**, **mcfp**, **msv**, **iv** that can be found in Appendix (2023). Findings illuminated how bosses' (B) engagement in simple and complex tasks affected employees (E). Crucially, extended mindsets facilitated learning from mistakes, contrasting with limited mindsets. Figure 5 revealed heightened learning (boosted by lower excitability thresholds triggered by identified mistakes) during complex tasks and mistake recognition (m2). Figure 6 showcased boss-employee learning dynamics (**W**-states), indicating boss-initiated learning followed by employee influence. Notably, neglected tasks exhibited a slow knowledge decay. This simulation-driven approach unraveled intricate leadership-learning interplays, underscoring the pivotal role of mindsets and task complexity in shaping organizational learning dynamics.

No other computational models of this double bias phenomenon can be found in the literature. Somewhat related are Samhan et al. (2022) and its extended version (Samhan et al. 2023). But in that case the learning from mistakes has been covered in a more simplified and more abstracted form, whereas no double bias and no contagion from managers to employees is addressed. For the network model introduced in the current chapter, still extensions can be incorporated in order to include further effects of the boss who accepts making mistakes and learning from them on employees' personal productivity on tasks, plus the effect on reaching goals.

6 Conclusion

In conclusion, mistakes hold a dual role as both learning opportunities and sources of negativity. This influences how we perceive them and make decisions, known as the framing effect. This outlook affects bosses and by contagion also their employees, creating a loop called the double bias effect. This plays a big role in organizations, the theory of transformative learning emphasizes learning from personal experiences, but negativity can block this. Ignoring mistakes means missing chances to learn and grow. The double bias limits growth mindsets and transformational leadership. It's a roadblock to shared learning in organizations and affects how well a company can adapt to change. Fixing this requires understanding and changing this bias, promoting mindsets that prioritize learning and adaptability. This study used computational analysis by an adaptive network model based on material from Hosseini et al. (2024) to show how important it is to allow mistakes and learn from them. By addressing the double bias, organizations can build a culture of continuous learning and flexible leadership (Shallenberger 2015).

The self-modeling network modeling approach used is a generic adaptive dynamical system modeling approach as it has been shown that any smooth adaptive dynamical system can be represented in a canonical manner as a self-modeling network model (Treur 2021; Hendrikse et al. 2023). Therefore, choosing it does not introduce fundamental limitations.

References

Appendix of An Adaptive Network Model for a Double Bias Perspective on Learning from Mistakes within Organizations (2023). See Linked Data at https://www.researchgate.net/publication/373554900

Cannon, D.M., Edmondson, C.A.: Failing to learn and learning to fail (intelligently): how great organizations put failure to work to innovate and improve. Long Range Plan. **38**(3), 299–319 (2004)

Hendrikse, S.C.F., Treur, J., Koole, S.L.: Modeling emerging interpersonal synchrony and its related adaptive short-term affiliation and long-term bonding: a second-order multi-adaptive neural agent model. Int. J. Neural Syst. **33**(7), 350038 (41 pages) (2023). https://doi.org/10.1142/S0129065723500387

Hosseini, M., Treur, J., Kucharska, W.: An adaptive network model for a double bias perspective on learning from mistakes within organizations. In: Cherifi, H., Rocha, L.M., Cherifi, C., Donduran, M. (eds.). Complex Networks and Their Applications XII. COMPLEX NETWORKS 2023. Studies in Computational Intelligence, vol. 1143, pp. 91–103. Springer Nature Switzerland, Cham (2024)

Kucharska, W., Bedford, D.A.D., Kopytko, A.: The double cognitive bias of mistakes: a measurement method. In: Proceedings of the 22nd European Conference on Research Methodology for Business and Management Studies, vol. 22(1), pp. 93–102 (2023). https://doi.org/10.34190/ecrm.22.1.1372

Mishra, A.: Learning from mistakes. J. Motilal Rastogi School Manag. **5**(2), 22–31 (2012). https://www.researchgate.net/publication/323695816

Samhan, N., Treur, J., Kucharska, W., Wiewiora, A.: An adaptive network model simulating the effects of different culture types and leader qualities on mistake handling and organisational learning. In: Complex Networks and Their Applications XI. Proc. of the 11th International Conference on Complex Networks and Their Applications. Publisher: Studies in Computational Intelligence, vol. 1077, pp. 224–238. Springer Nature, Cham (2022)

Samhan, N., Treur, J., Kucharska, W., Wiewiora, A.: Computational simulation of the effects of different culture types and leader qualities on mistake handling and organisational learning. In: Canbaloğlu, G., Treur, J., Wiewiora, A. (eds.), Computational Modeling of Multilevel Organisational Learning and Its Control Using Self-modeling Network Models, Chap. 14, pp. 363–408. Springer Nature, Cham (2023)

Shallenberger, D.: Learning from our mistakes: international educators reflect. Front. Interdisc. J. Study Abroad **26**(1), 248–263 (2015)

Treur, J.: Modeling higher-order adaptivity of a network by multilevel network reification. Netw. Sci. **8**, 110–144 (2020a)

Treur, J.: On the dynamics and adaptivity of mental processes: relating adaptive dynamical systems and self-modeling network models by mathematical analysis. Cogn. Syst. Res. Syst. Res. **70**, 93–100 (2021)

Treur, J.: Network-Oriented Modeling for Adaptive Networks: Designing Higher-Order Adaptive Biological, Mental and Social Network Models. Springer Nature, Cham (2020b)

Treur, J.: Modeling multi-order adaptive processes by self-modeling networks. In: Tallón-Ballesteros, A.J., Chen, C.-H. (eds.), Proceedings of the 2nd International Conference on Machine Learning and Intelligent Systems, MLIS'20. Frontiers in Artificial Intelligence and Applications, vol. 332, pp. 206–217. IOS Press (2020c)

Treur, J., Van Ments, L, (eds.).: Mental Models and Their Dynamics, Adaptation, and Control: A Self-Modeling Network Modeling Approach. Springer Nature, Cham (2022)

Organisational Learning for Safety and Security Through Cyberspace: Adaptive Modelling of the Implementation of Environment Health and Safety Standards

Nick Van den Hurk, Jan Treur, and Peter H. M. P. Roelofsma

Abstract This chapter focusses on using a self-modelling network modeling approach to analyse and predict the behavior of people within organizations, specifically on the implementation of EHS standards. The study focusses on a real-world scenario in which EHS standards are implemented in a multi-level organizational learning scenario. To ensure the health and safety of employees worldwide, the company provides standards for every EHS aspect of activities. The standards are fed by state-of-the-art technologies and science. They require local implementation at every site, also incorporating legislation for the country where it is based. The technique used in this study is dynamic computational modelling wielding the self-modelling network modeling approach. In the base scenario, the organization started with the implementation in a pilot phase. After a while the organization acknowledges it needs to take the lessons learned in the pilot and apply them to all other standards, enabling feed-forward and feedback learning within the organisation. Subsequently five alternative scenarios are described in which the organization is challenged with issues complicating the implementation process. The study shows the possibility of

N. Van den Hurk
Merck & Co., Inc., Rahway, NJ, USA
e-mail: nick.hurk.van.den@merck.com

J. Treur (✉)
Social AI Group, Vrije Universiteit Amsterdam, Amsterdam, The Netherlands
e-mail: j.treur@vu.nl; j.treur@hhs.nl

J. Treur · P. H. M. P. Roelofsma
Risk Management and Cybersecurity Group, Centre of Expertise Cybersecurity, The Hague University of Applied Sciences, Hague, The Netherlands
e-mail: p.h.m.p.roelofsma@tudelft.nl

P. H. M. P. Roelofsma
Erasmus University Medical Center Rotterdam, Rotterdam, The Netherlands

P. H. M. P. Roelofsma et al. (eds.), *Using Shared Mental Models and Organisational Learning to Support Safety and Security Through Cyberspace: A Computational Analysis Approach*, Studies in Systems, Decision and Control 570,
https://doi.org/10.1007/978-3-031-72075-8_11

using a self-modelling social network in the execution of EHS activities. It demonstrates the importance of cooperation and open communication between the system owner, EHS expert and department. This is needed to be able to learn about processes and efficiently safeguard the organization and its members.

1 Introduction

To be able to keep up with the Environmental, Health and Safety (EHS) standards of industry, a company's ability to implement EHS knowledge on the shopfloor is important for its survival and its employees wellbeing. Several industries have developed standards describing how to perform activities without risking an EHS event (i.e. spill, injury, exposure). An EHS department or officer is often appointed to implement these standards. Together with other EHS professionals and the departmental experts they assess how to implement standards and create a safe working environment. When executed in the right manner, the formed team will learn. By documenting the process, the organization will also learn. Organizational learning is a shared knowledge development process involving individuals, groups and the organization. Organizational learning occurs through formation of shared mental models and common believes developed by organizational members institutionalized for future use (Abraham and Bear 1996). The multilevel and context-dependent characteristic of this organizational learning makes it a difficult phenomenon to observe and analyze. Practical examples of organizational learning are the drafting of instructions, creating structures/hierarchies and keeping records.

The issue when studying an EHS change is doing a proper intervention study. It is difficult because interventions rarely start at a set timeframe or are followed completely through to assess its effectiveness (i.e. project aborted early if there is no direct effect). This is a problem encountered in many EHS related studies, especially those on structure and culture changes within companies. Therefor this thesis focusses on Behaviour Dynamics in Social Networks (Bogenrieder 2002; Bouma et al. 2023) using a self-modelling network approach to predict the behavior of people within organizations, specifically on the implementation of EHS standards. The study focusses on a real-world scenario in which EHS standards are implemented in a multi-level organizational learning scenario. To show this is a team and, on a higher level, organizational effort, different historical and hypothetical scenario's are described. The goal of the study is to show the importance of proper collaboration and the conditions necessary to properly implement EHS standards. Previous studies in this area have shown the applicability of the a self-modelling network for organizational learning in project-based organizations (Canbaloğlu et al. 2021). This chapter addresses a base scenario for EHS standard implementation with different variations assessing the predictability of success in alternate scenarios.

Problem Statement

To implement EHS standards, input of several parts of the organization is needed. Leadership needs to support the implementing team, which in its turn needs to work together and create resources. While EHS standards are often clear cut and very prescriptive; the way of collaborating is not. The rules of communication and teamwork are mostly unwritten. The resulting social networking however, is a phenomenon which we are able to model. This thesis describes the social network involved in EHS standard implementation with several scenarios. In this way we want to predict how standards will be implemented and how the organization will learn.

2 Literature Review

To be able to survive, organizations need to learn. If acquired knowledge is not shared, documented and retained properly, a lot of waste of resource occurs because members of the organization keep re-inventing the wheel. Organizational members work in a complex social network, which gives them the opportunity to exchange diverse mental models and create new learnings (Canbaloğlu et al. 2022a). Mental models are organized knowledge structures that enable individuals to interact with their environment (Canbaloğlu et al. 2022b). If the parties involved do not have like-minded or (partially) overlapping mental models, this can have significant consequences for effective and adequate cooperation (Canbaloğlu et al. 2022c). Providing insight into the mental models and the possible differences or inequalities in them, will contribute to improving collaboration and thus effectively and efficiently deploy EHS capacity and knowledge. In order to be able to adapt well to the changing circumstances within an organization, employees need some predictability in how colleagues and organizations react to certain situations. This predictability is necessary in team dynamics as well as for the tasks that have to be performed. Intermediary agents like implementation teams are a key factor in the process of organizational learning. Organizational learning is a shared development process involving individuals, groups and the organization. It takes place on a team level by exchange of information between teams or project members. On an organizational level it takes place by institutionalizing the outcome of the team level for future reference. This is called feed-forward learning. Feedback learning also occurs when the institutionalized information is called upon by a different team or project to use to their advantage. The process of organizational learning can be influenced by contextual factors like leadership, culture and structure or hierarchy (Canbaloğlu et al. 2023).

Research on teamwork has shown that in environments where one cannot afford to make mistakes, minimizing the number of mistakes depends more on improving non-technical factors than on technical skills (Crossan et al. 1999). This also reflects to the field of EHS studies. When addressing EHS risk management, it is important to collaborate with each other and have a clear way of communication.

3 Method

This chapter describes a what-if-analysis studying a baseline scenario and variants to that scenario. The base scenario is described in the next chapter. The technique used to study these variants is dynamic computational modelling through self-modelling networks as described in Treur (2020a, b, c). Self-modeling temporal-causal networks are specified by nodes (also called states) X and Y having activation levels over time t varying between 0 and 1. Nodes being either base level nodes for parties/organizations performing actions or higher-order self-model states describing learning processes and their control. A dedicated software environment with templates and a combination function library can be found on ResearchGate (Bogenrieder 2002). We used this software environment in MATLAB R2021b.

Network characteristics which specify a network model are:

- **Connectivity characteristics**:
  Network characteristics describing which nodes X connect to which nodes Y, and with what connection weight $\boldsymbol{\omega}_{X,Y}$ indicating how strongly nodes are connected
- **Aggregation characteristics**:
  Network characteristics which describe the way in which causal impacts have combined effects on node Y by assigning a combination function $\mathbf{c}_Y$ to Y
- **Timing characteristics**:
  Network characteristics specifying speed factors $\boldsymbol{\eta}_Y$ for all nodes Y BY which certain changes of node activations take place, for example used for timing of enabling or disabling certain MEchanisms in the model; moreover, initial values are needed for all of the nodes.
- **Standard difference equation**:
  Based on these network characteristics, the following difference equation is used for simulation and analysis.

  Here the X_i are all states from which state Y has incoming connections:

$$Y(t+\Delta t) = Y(t) + \boldsymbol{\eta}_Y\left[\mathbf{c}_Y\left(\boldsymbol{\omega}_{X_1,Y}X_1(t), \ldots, \boldsymbol{\omega}_{X_k,Y}X_k(t)\right) - Y(t)\right]\Delta t$$

  The network characteristics are specified in a standard table format called role matrices. These are input for the dedicated software environment, after which the above difference equation is automatically generated and simulated within this software environment.
- **Self-model nodes for adaptation**:
  To model adaptivity, self-model nodes are added. For example, if a connection weight $\boldsymbol{\omega}_{X,Y}$ changes over time by learning or forgetting, then a self-model node $\mathbf{W}_{X,Y}$ is added that represents the dynamic value of $\boldsymbol{\omega}_{X,Y}$. In the processing, the activation value of $\mathbf{W}_{X,Y}$ is used for $\boldsymbol{\omega}_{X,Y}$. This is applied to obtain an enriched form of causal modelling where not only node activations but also the underlying causal relations themselves can change over time. Self-model nodes are depicted on a higher level according to a 3D structure as shown in Fig. 1. They are integrated

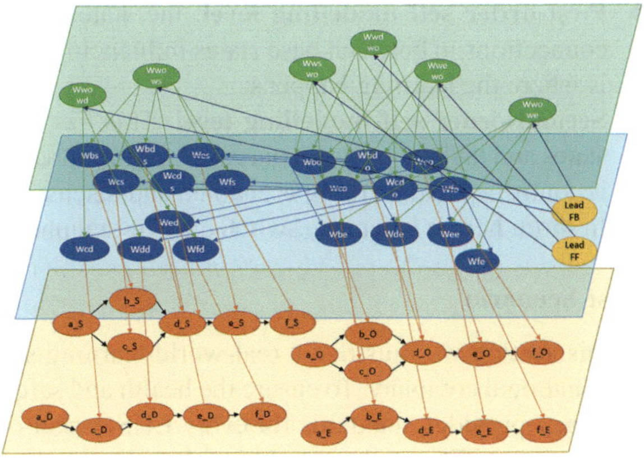

Fig. 1 3D graphical representation of the adaptive self-modeling network model

in the network structure by specifying connections to them and connections from them and by also specifying the other types of network characteristics such as combination functions (in this case declarative specifications of the adaptation or learning principles modeled) and speed factors (learning rates in this case). For example, a connection from $\mathbf{W}_{a\_A,b\_A}$ to $\mathbf{W}_{a\_B,b\_B}$ can be used to specify a communication channel for person A communicating to person B knowledge $\mathbf{W}_{a\_A,b\_A}$ that task a is to be followed by task b, thus after communication resulting in knowledge $\mathbf{W}_{a\_B,b\_B}$ of B.

- **Higher-order self-model nodes for control**:
  The self-modeling principle can also be used in an iterative manner. For example, the control decision for person A to communicate knowledge $\mathbf{W}_{a\_A,b\_A}$ to person B, can be modelled by a (second-order) self-model node $\mathbf{W}_{\mathbf{W}_{a\_A,b\_A},\mathbf{W}_{a\_B,b\_B}}$ representing the weight $\boldsymbol{\omega}_{\mathbf{W}_{a\_A,b\_A},\mathbf{W}_{a\_B,b\_B}}$ of the connection from (first-order) self-model node $\mathbf{W}_{a\_A,b\_A}$ to (first-order) self-model node $\mathbf{W}_{a\_B,b\_B}$. As long as this second-order self-model node has activation level 0, no communication takes place (communication channel not open). At the moment that this activation level is changed into 1, communication takes place (communication channel opened). In this way, it can be controlled whether and when such communication takes place.

4 Modeling Multilevel Organizational Learning

The implementation of EHS standards is a learning process which has several levels. We distinguish the following:

- **The base level**: this is the lowest level describing the activities or states which are needed in the work floor process modelled. In short: this is where the work is done;

- **First-order self-modelling level**: the states in this level represent weights of connections in between base states influencing the learning process. In short: this is where the learning happens;
- **Second-order self-modelling level**: This has states which control the learning states and their in-between interactions. In short: this is where the learning process is controlled and learning is enabled and discussed. Part of this is general control. In short: here is where we switch on different phases within the modeled scenario.

Base Scenario

In this section we illustrate a real-world learning scenario that occurred in a large multinational company. To ensure the health and safety of employees worldwide, the company provides standards for every EHS aspect of activities. Examples of these standards are: Elevated work, Fire safety, Ergonomics and Process safety management. The standards are fed by state-of-the-art technologies and science. They require local implementation at every site, also incorporating legislation for the country where it is based. To do this, System Owners (SO's) are appointed. These SO's are responsible for implementing the standard locally at all relevant departments within a site. SO's are supported by leadership members who carry the accountability of having a standard implemented. A member of the local EHS team is assigned to provide expert input, connect to the global organization and support the SO. Implementation of standards needs to be done within every department of the organization, therefor the SO has contact with the shopfloor, usually with a safety-coordinator. This coordinator manages all EHS tasks locally, next to his/her normal tasks. These tasks could include: standard implementation, incident investigation, awareness training and general safety culture improvement. Typically implementation starts with getting the team together and setting the goals and deliverables. Subsequently there are two streams of work: the gathering of EHS specific information, guidelines and legislation, and, getting specific shop-floor information to which this all applies. The combination of these data leads to a local operating procedure which needs to be documented. After this, employees need to be trained on the subject and internal auditing will take place to ensure that proper execution of the new standard way of working is done. In the base scenario, the organization started with the implementation in a pilot phase. After a while the organization acknowledges it needs to take the lessons learned in the pilot and apply them to all other standards, enabling feed-forward and feedback learning. Subsequently five alternative scenario's are described in which the organization is challenged with issues complicating the implementation process.

As stated: standard implementation needs to incorporate legislation, local authority guidance, and site specific information into tools, instructions and recordkeeping. For the purpose of this chapter, the process of standard implementation is broken down into six steps:

1. Start of the implementation,
2. Having subject matter information,
3. Having department specific information,
4. Writing local instructions incorporating gathered information,
5. Training employees on the instructions,
6. Doing internal audit.

The network model was designed showing:

- Base states for the System Owner (a_S, b_S, etc.)
- Base states for the Department (a_D, b_D, etc.)
- Base states for the EHS specialist (a_E, b_E, etc.)
- Base states for the Organisation (a_O, b_O, etc.)
- First order states for connections of the System Owner ($\mathbf{W}_{bS}$, $\mathbf{W}_{bdS}$, etc.)
- First-order self-model states for connections of the Department ($\mathbf{W}_{cD}$, $\mathbf{W}_{dD}$, etc.)
- First-order self-model states for connections of the EHS specialist ($\mathbf{W}_{bE}$, $\mathbf{W}_{dE}$, etc.)
- First-order self-model states for connections of the Organisation ($\mathbf{W}_{bO}$, $\mathbf{W}_{bdO}$, etc.)
- Second-order self-model states to control feed forward exchanges from System Owner, Department, and EHS department to the Organisation ($\mathbf{W}_{W_SW_O}$, $\mathbf{W}_{W_DW_O}$, $\mathbf{W}_{W_EW_O}$)
- Second-order self-model states to control in feedback exchanges from the organization back to the System Owner, Department, and EHS department ($\mathbf{W}_{W_OW_S}$, $\mathbf{W}_{W_OW_D}$, $\mathbf{W}_{W_OW_E}$)
- Two extra context states to activate $\mathbf{W}_{\mathbf{W}}$-states by site Leadership: Lead FB and Lead FF.

Note that the organization is added to the model creating a place of shared knowledge. The design was specified in detail by role matrices used as input into the dedicated software environment (Treur 2020a, b, c; Chap. 9; Bogenrieder 2002). These role matrices can be found in Appendix 1.

5 Scenario Variations

This section describes the variations to the base scenario.

Scenario 1: Leadership does not allow for the organization to implement the EHS standards robustly: In this scenario the organization does acknowledge that it is important to have the EHS standards implemented but does not allow the implementation team to take the time to properly document and institutionalize the findings. Computationally this would be simulated by not enabling the feed-forward and feedback context states.

Scenario 2: EHS professional does not have the time to provide regular input. The EHS professional has to communicate with the Global organization, local and national authorities and has a role in incident investigations on a local level communicating with the site. When recourses are scarce, the tendency is to use them for immediate threats/risks and postpone activities like Standard implementation, reverting to only reactively answering the questions an SO has for the EHS expert. Computationally this shows by adding a Steponce in the second order state, called the 'EHS manager' which replaces the regular FF/FB loop by the leadership. The EHS manager has influence on the resources spent and postpones the attribution of the EHS professional.

Scenario 3: System Owner does not get resources from own manager. In this scenario the System Owner has the task of standard implementation next to his/her regular activities within the company. In this way the employee is always divided in his/her work. A departmental manager who has a SO in its team will need to provide guidance on how many time can be spent. Computationally this is simulated by adding a Stepmod in the second order state, called the 'SO manager' enabling the states connected to the SO only 20% of the times using the Stepmod function.

Scenario 4: EHS professional leaves the company and an interim external contractor is instated. Nowadays switching jobs is a regularly occurring phenomenon, especially in the EHS world. Companies take al lot of effort to keep personnel in place but the temptation to switch is always luring. In event of such a job change by the EHS professional, there will be a temporary stop in information flow. If the company is lucky they soon find a replacement or are able to hire an interim EHS consultant to keep shop. A similar situation could occur when an employee is on long term sick leave. The afore mentioned consultant will not be as embedded in the organization as the permanent employee. Due to practical reasons (access to folders/ company laptop), temporal reasons (i.e. only available 3 days a week), and social reasons (only communicates for 1 project with the necessary people). Computational this would be simulated by reducing the connection weights from the EHS expert (which in this case is an external consultant) to all the other parts of the organization.

Scenario 5: EHS expert is appointed to also be the System Owner. Some of the EHS standards are essential or require a level of expertise which is not found on the shop floor. In those cases it could be decided to have the EHS expert also function as the System Owner. In that case both expert input and process guidance would come from the same person. Computational this would be simulated by adding direct connections between the EHS expert and the System owner on both the base level as on the first order. This way of portraying is chosen because both activities are role's and not persons.

6 Results

In our base scenario the SO and organization have mental models for all process steps. The department has a mental model for all steps except 'Having subject matter input' and the EHS expert has a mental model for all steps except 'Having department specific information' (Fig. 2).

Where knowledge in the organization is present, an initial value of 1 in Ww states was assigned. This was done for: $\mathbf{W}_{cD}$, $\mathbf{W}_{bE}$, $\mathbf{W}_{bdO}$, $\mathbf{W}_{cdO}$, $\mathbf{W}_{eO}$, $\mathbf{W}_{fO}$. Practically this means that the SO has no actions or starting information except the need to start implementing, the department only knows its own situation, the EHS expert has knowledge of the standard and legislation, and the organization knows how to make operating procedures, train employees and do internal auditing. Without organizational learning ($t - 0$–50) this knowledge is not shared. After $t = 50$ feedback ($t = 50$) and feed forward ($t = 100$) learning kick in and we see all mental models rising with a baseline of at least 0.6 showing stability.

In Fig. 3 only the base states are shown, in general the SO and department have levels closer to each other. EHS follows the same trend but stays behind in its mental models. This shows that the EHS expert does not have the capacity to align fully with all mental models but needs to provide a solid basis for every implementation team in Fig. 4.

Looking at the second and third level iterations it is shown that after starting the organizational FB and FF learning the learning levels rise and show maximum efficiency.

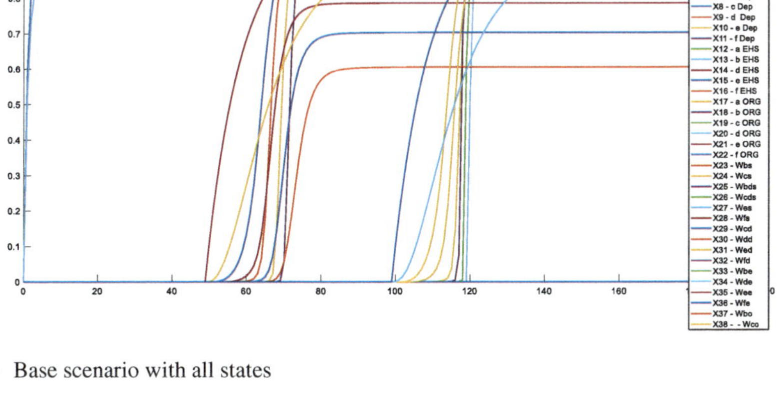

Fig. 2 Base scenario with all states

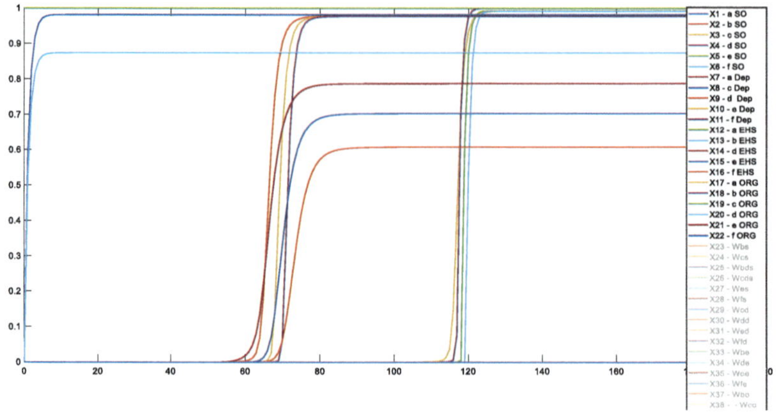

Fig. 3 Base states of base scenario

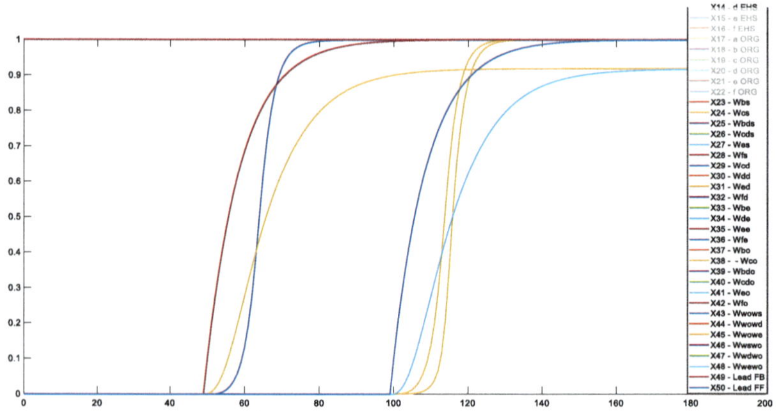

Fig. 4 Higher iterations of base scenario

Alternative Scenarios

Scenario 1: Leadership does not allow for the organization to implement the EHS standards robustly. In this scenario the Leadership of the organization wants to 'get implementation over with' as soon as possible and not learn during the process. Teams are formed but processes are not well documented. In this way every team needs to start from scratch and cannot profit from previous teams. Not sharing the information with other streams is shown in the model. The SO only starts the process, the EHS expert is only having subject matter information and the department only having its own information. Organizational actions of drafting the operating procedure, training and doing an audit are all 0 because there is no documentation/standard process in place (Fig. 5).

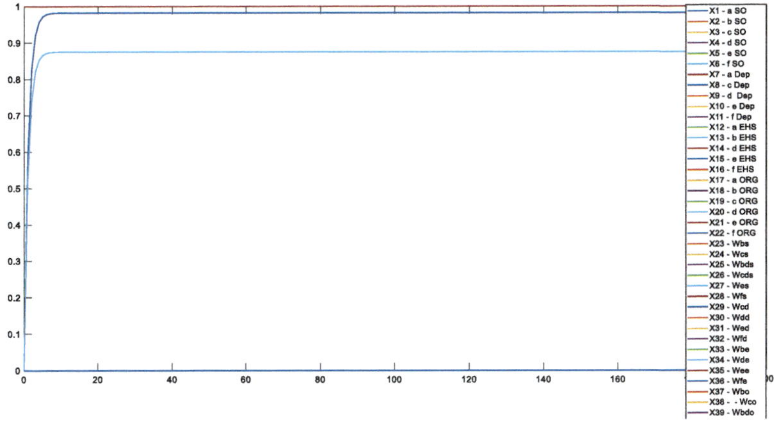

Fig. 5 Feedback and feed forward learning disabled

Scenario 2: EHS professional does not have the time to provide input. To visualize the EHS department not having time to give input for the standard implementation it is shown that they do not give input until $t = 150$ where it is decided they do need to make time to help. A Steponce connected to the EHS experts **W**-states was introduced and named the 'EHS manager' who would usually decide where the resources of the department are spent (Fig. 6).

Note that after the 'go for implementation' by leadership, the EHS expert still does not join the mental model formed by SO, department and organization. In practice this would mean that there is a standard implemented without subject matter input. The EHS Expert will need to spend more energy influencing the mental model as the team already had time to form an initial one. Below the input of the EHS expert is shown by only showing base and **W**-states for the B-node. The EHS expert shows single knowledge on subject matter information until he/she is allowed to spend

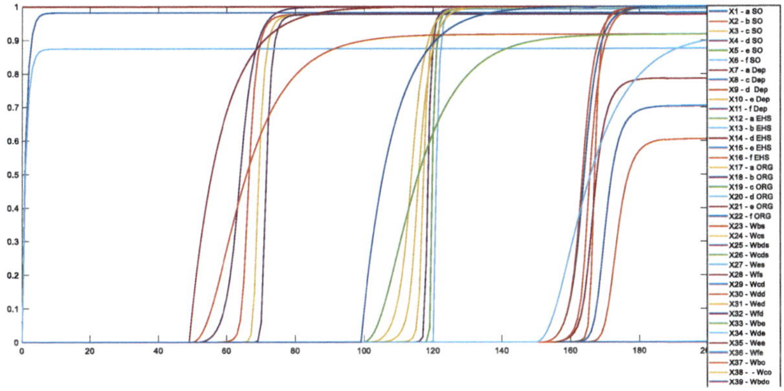

Fig. 6 Adding the EHS manager

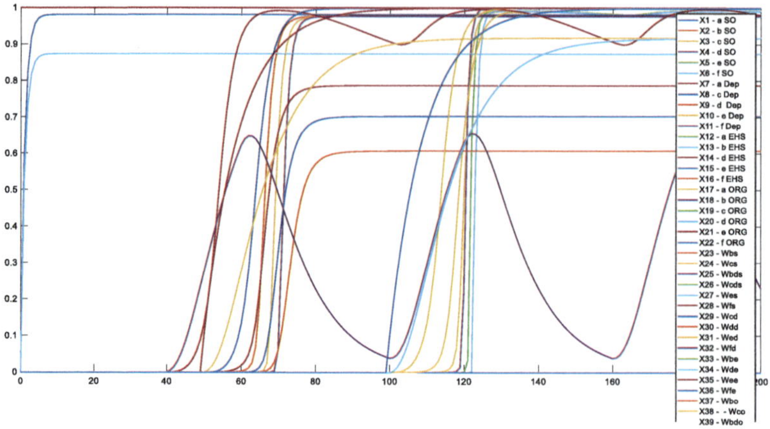

Fig. 7 SO manager restricts time spent

time at $t = 150$ where he/she starts interacting with the other team members and organization.

Scenario 3: System Owner does not get resources by own manager. In this scenario the SO spends only 20% of the required time for standard implementation which is shown by adding a stepmod function peaking between $t = 40$ and 60 of every cycle (Fig. 7).

It is interesting to see that this modularity of the SO does not influence the mental models of others much. In practice this could not be well received as the other part of the organization is doing their part and the SO only spends little time. Note that it does not by definition need to be 20% of the FTE (Full Time Equivalent), but rather is 20% of the time needed to perform tasks.

Scenario 4: EHS professional leaves the company and an interim external contractor is instated. When the EHS expert is not available and needs to be replaced by a consultant this results in less efficient communication. To visualize this, the speed factors for the EHS expert were cut in half. The trend in mental models is secured, however it takes more time to reach the same level, showing that with a weaker connection the organization does not learn as well. In the case of a consultant, the SO also learns at a (marginal) slower rate than in the base scenario. The next page shows both the situations with only the base states for EHS and SO are displayed to better show the slack created with a consultant. The W and Ww states for EHS learning are also shown (Figs. 8, 9 and 10).

Scenario 5: EHS expert is appointed to also be the System Owner. Computational this would be simulated by adding direct connections between the EHS expert and the System owner on both the base level as on the first order. This way of portraying is chosen because both activities are role's and not persons. In the figure it seems

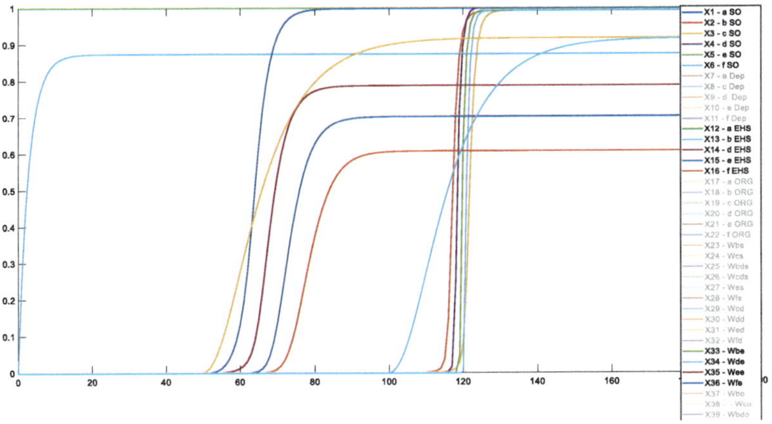

Fig. 8 Base states EHS consultant and SO combined with W and Ww states for EHS

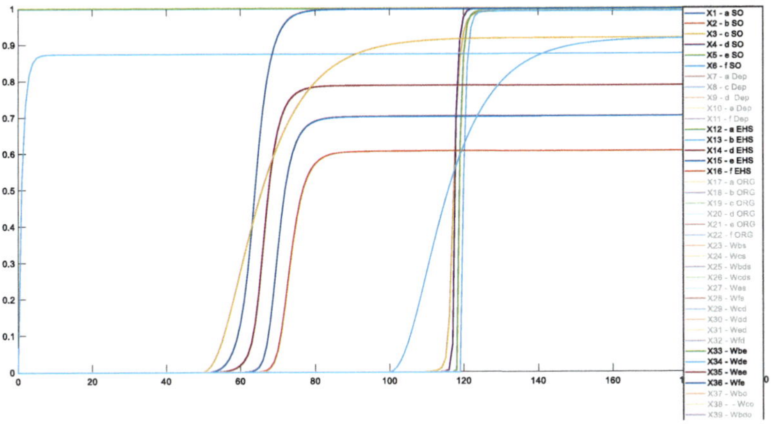

Fig. 9 Base states EHS fixed employee and SO combined with W and Ww states for EHS

that there are less lines/mental models, but this is just a visual trait of the SO and EHS mental models overlapping as they are the same person. While this seems like a very efficient solution it is only shown for 1 EHS standard. In practice it would not be possible to have this done for every standard as it would lead to very large and costly EHS departments.

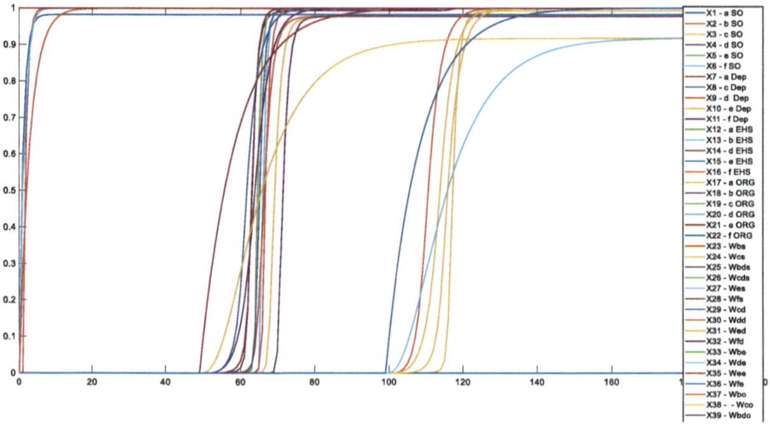

Fig. 10 The EHS expert is appointed to be SO

7 Conclusions

This study shows the possibility of using a self-modelling social network in the execution of EHS activities. It demonstrates the importance of cooperation and open communication between the SO, EHS expert and department. This is needed to be able to learn about processes and efficiently safeguard the organization and its members. The material is based on Wiewiora et al. (2020).

The Scenarios show that:

- It is important to enable feed forward and feedback learning to ensure that the organization keeps learning. Remember: Mistakes can be made but should not happen more than once.
- It is important to provide proper resourses (time, money etc.) to enable the implementation team to gather the information needed and ensure compliance to the standard.
- External support can help a organization in interim or egress situations but to be able to fully wield an external's knowledge, and organizational learning: consultants should be integrated in to the organization as much as possible.
- EHS experts having a double role (SO and EHS expert) work very efficiently.

Discussion

This study continues the previous work in the field of computational modeling and organizational learning (Canbaloğlu et al. 2021, 2023; Iftikhar and Wiewiora 2021; Kim 1993; Mathieu et al. 2000; McComb and Simpson 2014; Robinson et al. 2016; Rouse and Morris 1986; Treur 2020a, b, c) by applying it specifically to the field of EHS. A field which always has a challenge to connect to the organization it is in. The modeling approach used is based on multi-order adaptive self-modeling networks, which in another domain has turned out to be useful to model plasticity and metaplasticity of neural and mental processes (Bogenrieder 2002; Bouma et al. 2023; Canbaloğlu et al. 2021; Hurck et al. 2022; Wiewiora et al. 2019). In the current study this modelling approach has been successfully applied to a real-world case in the social computation domain addressing computational organisational learning (Abraham and Bear 1996; Canbaloğlu et al. 2022a; Kim 1993; Mathieu et al. 2000; McComb and Simpson 2014; Robinson et al. 2016).

In some organizations, the EHS department is not seen as a partner in operations but rather as an audit or inspection party. In that context, it is often said that EHS departments exceed their targets and unnecessarily slow down production with 'their rules'. This leads to departments not sharing all the information they have, fearing judgement and possible retributions by the EHS department. The social factors involved in this specific scenario where out of scope in this study. Alternative scenarios show that with enough resources, the implementation of EHS standards would benefit, but does not properly describe the consequences. Work on EHS is often a 'side issue' to regular production. Spending too much resources on supporting processes like this is often seen only as c osts. It is important to also show that EHS helps improve processes on the shopfloor and makes people work more efficient, longer and healthier. This also translates into the external support and making EHS experts to be dedicated SO's. Both of these scenario's will require extra resources, either financial or in time spent (which in the end is also financial). The easy option of 'making the EHS expert SO for all standards' only works if you hire enough EHS experts to do so. As the industry often has between 30 and 50 EHS standards. This model, as all models, is a simplification of the reality only changing one specific aspect of a social model at the time to be able to research the impact. The conclusions in this study are a logical response to situations changing. The fact that we were able to model and predict them helps us fundamentally in understanding computational social modelling and being able to predict more complicated scenario's. Further research is advised into the more complex EHS implementation issues like installing engineering controls or performing risk assessment studies. One could also expand research into multi-disciplinary and transcending or cross-departmental issues like medical surveillance or site expansion.

Appendix Role Matrices

mb	Base connectivity	1	2	3	4
X_1	a_S	X_1			
X_2	b_S	X_1			
X_3	c_S	X_1			
X_4	d_S	X_2	X_3		
X_5	e_S	X_4			
X_6	f_S	X_5			
X_7	a_D	X_7			
X_8	c_D	X_7			
X_9	d_D	X_8			
X_{10}	e_D	X_9			
X_{11}	f_D	X_{10}			
X_{12}	a_E	X_{12}			
X_{13}	b_E	X_{12}			
X_{14}	d_E	X_{13}			
X_{15}	e_E	X_{14}			
X_{16}	f_E	X_{15}			
X_{17}	a_O	X_{17}			
X_{18}	b_O	X_{17}			
X_{19}	c_O	X_{17}			
X_{20}	d_O	X_{18}	X_{19}		
X_{21}	e_O	X_{20}			
X_{22}	f_O	X_{21}			
X_{23}	$\mathbf{W}_{bS}$	X_{23}	X_{37}		
X_{24}	$\mathbf{W}_{cS}$	X_{24}	X_{38}		
X_{25}	$\mathbf{W}_{bdS}$	X_{25}	X_{39}		
X_{26}	$\mathbf{W}_{cdS}$	X_{26}	X_{40}		
X_{27}	$\mathbf{W}_{eS}$	X_{27}	X_{41}		
X_{28}	$\mathbf{W}_{fS}$	X_{28}	X_{42}		
X_{29}	$\mathbf{W}_{cD}$	X_{29}	X_{38}		
X_{30}	$\mathbf{W}_{dD}$	X_{30}	X_{40}		
X_{31}	$\mathbf{W}_{eD}$	X_{31}	X_{41}		
X_{32}	$\mathbf{W}_{fD}$	X_{32}	X_{42}		
X_{33}	$\mathbf{W}_{bE}$	X_{33}	X_{37}		
X_{34}	$\mathbf{W}_{dE}$	X_{34}	X_{39}		
X_{35}	$\mathbf{W}_{eE}$	X_{35}	X_{41}		
X_{36}	$\mathbf{W}_{fE}$	X_{36}	X_{42}		
X_{37}	$\mathbf{W}_{bO}$	X_{37}	X_{23}	X_{33}	
X_{38}	$\mathbf{W}_{cO}$	X_{38}	X_{24}	X_{29}	
X_{39}	$\mathbf{W}_{bdO}$	X_{39}	X_{25}	X_{34}	
X_{40}	$\mathbf{W}_{cdO}$	X_{40}	X_{26}	X_{30}	
X_{41}	$\mathbf{W}_{eO}$	X_{41}	X_{27}	X_{31}	X_{35}
X_{42}	$\mathbf{W}_{fO}$	X_{42}	X_{28}	X_{32}	X_{36}
X_{43}	$\mathbf{W}_{\mathrm{W_O W_S}}$	X_{49}			
X_{44}	$\mathbf{W}_{\mathrm{W_O W_D}}$	X_{49}			
X_{45}	$\mathbf{W}_{\mathrm{W_O W_E}}$	X_{49}			
X_{46}	$\mathbf{W}_{\mathrm{W_S W_O}}$	X_{50}			
X_{47}	$\mathbf{W}_{\mathrm{W_D W_O}}$	X_{50}			
X_{48}	$\mathbf{W}_{\mathrm{W_E W_O}}$	X_{50}			
X_{49}	Lead FB	X_{49}			
X_{50}	Lead FF	X_{50}			

mcw	Connection weights	1	2	3	4
X_1	a_S	1			
X_2	b_S	X_{23}			
X_3	c_S	X_{24}			
X_4	d_S	X_{25}	X_{26}		
X_5	e_S	X_{27}			
X_6	f_S	X_{28}			
X_7	a_D	1			
X_8	c_D	X_{29}			
X_9	d_D	X_{30}			
X_{10}	e_D	X_{31}			
X_{11}	f_D	X_{32}			
X_{12}	a_E	1			
X_{13}	b_E	X_{33}			
X_{14}	d_E	X_{34}			
X_{15}	e_E	X_{35}			
X_{16}	f_E	X_{36}			
X_{17}	a_O	1			
X_{18}	b_O	X_{37}			
X_{19}	c_O	X_{38}			
X_{20}	d_O	X_{39}	X_{40}		
X_{21}	e_O	X_{41}			
X_{22}	f_O	X_{42}			
X_{23}	$\mathbf{W}_{bS}$	1	X_{43}		
X_{24}	$\mathbf{W}_{cS}$	1	X_{43}		
X_{25}	$\mathbf{W}_{bdS}$	1	X_{43}		
X_{26}	$\mathbf{W}_{cdS}$	1	X_{43}		
X_{27}	$\mathbf{W}_{eS}$	1	X_{43}		
X_{28}	$\mathbf{W}_{fS}$	1	X_{43}		
X_{29}	$\mathbf{W}_{cD}$	1	X_{44}		
X_{30}	$\mathbf{W}_{dD}$	1	X_{44}		
X_{31}	$\mathbf{W}_{eD}$	1	X_{44}		
X_{32}	$\mathbf{W}_{fD}$	1	X_{44}		
X_{33}	$\mathbf{W}_{bE}$	1	X_{45}		
X_{34}	$\mathbf{W}_{dE}$	1	X_{45}		
X_{35}	$\mathbf{W}_{eE}$	1	X_{45}		
X_{36}	$\mathbf{W}_{fE}$	1	X_{45}		
X_{37}	$\mathbf{W}_{bO}$	1	X_{46}	X_{48}	
X_{38}	$\mathbf{W}_{cO}$	1	X_{46}	X_{47}	
X_{39}	$\mathbf{W}_{bdO}$	1	X_{46}	X_{48}	
X_{40}	$\mathbf{W}_{cdO}$	1	X_{46}	X_{47}	
X_{41}	$\mathbf{W}_{eO}$	1	X_{46}	X_{47}	X_{48}
X_{42}	$\mathbf{W}_{fO}$	1	X_{46}	X_{47}	X_{48}
X_{43}	$\mathbf{W}_{w_O w_S}$	1			
X_{44}	$\mathbf{W}_{w_O w_D}$	1			
X_{45}	$\mathbf{W}_{w_O w_E}$	1			
X_{46}	$\mathbf{W}_{w_S w_O}$	1			
X_{47}	$\mathbf{W}_{w_D w_O}$	1			
X_{48}	$\mathbf{W}_{w_E w_O}$	1			
X_{49}	Lead FB	1			
X_{50}	Lead FF	1			

mcfw	Combination function weights	alogistic	steponce
X_1	a_S	1	
X_2	b_S	1	
X_3	c_S	1	
X_4	d_S	1	
X_5	e_S	1	
X_6	f_S	1	
X_7	a_D	1	
X_8	c_D	1	
X_9	d_D	1	
X_{10}	e_D	1	
X_{11}	f_D	1	
X_{12}	a_E	1	
X_{13}	b_E	1	
X_{14}	d_E	1	
X_{15}	e_E	1	
X_{16}	f_E	1	
X_{17}	a_O	1	
X_{18}	b_O	1	
X_{19}	c_O	1	
X_{20}	d_O	1	
X_{21}	e_O	1	
X_{22}	f_O	1	
X_{23}	$\mathbf{W}_{bS}$	1	
X_{24}	$\mathbf{W}_{cS}$	1	
X_{25}	$\mathbf{W}_{bdS}$	1	
X_{26}	$\mathbf{W}_{cdS}$	1	
X_{27}	$\mathbf{W}_{eS}$	1	
X_{28}	$\mathbf{W}_{fS}$	1	
X_{29}	$\mathbf{W}_{cD}$	1	
X_{30}	$\mathbf{W}_{dD}$	1	
X_{31}	$\mathbf{W}_{eD}$	1	
X_{32}	$\mathbf{W}_{fD}$	1	
X_{33}	$\mathbf{W}_{bE}$	1	
X_{34}	$\mathbf{W}_{dE}$	1	
X_{35}	$\mathbf{W}_{eE}$	1	
X_{36}	$\mathbf{W}_{fE}$	1	
X_{37}	$\mathbf{W}_{bO}$	1	
X_{38}	$\mathbf{W}_{cO}$	1	
X_{39}	$\mathbf{W}_{bdO}$	1	
X_{40}	$\mathbf{W}_{cdO}$	1	
X_{41}	$\mathbf{W}_{eO}$	1	
X_{42}	$\mathbf{W}_{fO}$	1	
X_{43}	$\mathbf{W}_{w_O w_S}$	1	
X_{44}	$\mathbf{W}_{w_O w_D}$	1	
X_{45}	$\mathbf{W}_{w_O w_E}$	1	
X_{46}	$\mathbf{W}_{w_S w_O}$	1	
X_{47}	$\mathbf{W}_{w_D w_O}$	1	
X_{48}	$\mathbf{W}_{w_E w_O}$	1	
X_{49}	Lead FB		1
X_{50}	Lead FF		1

mcfp	Combination function / parameters	1 alogistic 1 σ	1 alogistic 2 τ	2 steponce 1 α	2 steponce 2 β
X_1	a_S				
X_2	b_S	10	0.5		
X_3	c_S	10	0.5		
X_4	d_S	10	0.5		
X_5	e_S	10	0.5		
X_6	f_S	10	0.5		
X_7	a_D	10	0.6		
X_8	c_D	10	0.6		
X_9	d_D	10	0.6		
X_{10}	e_D	10	0.6		
X_{11}	f_D	10	0.6		
X_{12}	a_E	5	0.6		
X_{13}	b_E	5	0.6		
X_{14}	d_E	5	0.6		
X_{15}	e_E	5	0.6		
X_{16}	f_E	5	0.6		
X_{17}	a_O	5	0.6		
X_{18}	b_O	2	0.7		
X_{19}	c_O	2	0.7		
X_{20}	d_O	2	0.7		
X_{21}	e_O	2	0.7		
X_{22}	f_O	2	0.7		
X_{23}	$\mathbf{W}_{bS}$	5	0.5		
X_{24}	$\mathbf{W}_{cS}$	5	0.5		
X_{25}	$\mathbf{W}_{bdS}$	5	0.5		
X_{26}	$\mathbf{W}_{cdS}$	5	0.5		
X_{27}	$\mathbf{W}_{eS}$	5	0.5		
X_{28}	$\mathbf{W}_{fS}$	5	0.5		
X_{29}	$\mathbf{W}_{cD}$	5	0.5		
X_{30}	$\mathbf{W}_{dD}$	5	0.5		
X_{31}	$\mathbf{W}_{eD}$	5	0.5		
X_{32}	$\mathbf{W}_{fD}$	5	0.5		
X_{33}	$\mathbf{W}_{bE}$	5	0.5		
X_{34}	$\mathbf{W}_{dE}$	5	0.5		
X_{35}	$\mathbf{W}_{eE}$	5	0.5		
X_{36}	$\mathbf{W}_{fE}$	5	0.5		
X_{37}	$\mathbf{W}_{bO}$	5	0.5		
X_{38}	$\mathbf{W}_{cO}$	5	0.5		
X_{39}	$\mathbf{W}_{bdO}$	5	0.5		
X_{40}	$\mathbf{W}_{cdO}$	5	0.5		
X_{41}	$\mathbf{W}_{eO}$	5	0.5		
X_{42}	$\mathbf{W}_{fO}$	5	0.5		
X_{43}	$\mathbf{W}_{W_OW_S}$	5	0.5		
X_{44}	$\mathbf{W}_{W_OW_D}$	5	0.5		
X_{45}	$\mathbf{W}_{W_OW_E}$	5	0.5		
X_{46}	$\mathbf{W}_{W_SW_O}$	5	0.5		
X_{47}	$\mathbf{W}_{W_DW_O}$	5	0.5		
X_{48}	$\mathbf{W}_{W_EW_O}$	5	0.5		
X_{49}	Lead FB			100	500
X_{50}	Lead FF			100	500

ms	speed factors	1
X_1	a_S	0.7
X_2	b_S	0.7
X_3	c_S	0.7
X_4	d_S	0.7
X_5	e_S	0.7
X_6	f_S	0.6
X_7	a_D	0.6
X_8	c_D	0.6
X_9	d_D	0.6
X_{10}	e_D	0.6
X_{11}	f_D	0.6
X_{12}	a_E	0.6
X_{13}	b_E	0.6
X_{14}	d_E	0.6
X_{15}	e_E	0.6
X_{16}	f_E	0.6
X_{17}	a_O	0.3
X_{18}	b_O	0.3
X_{19}	c_O	0.3
X_{20}	d_O	0.3
X_{21}	e_O	0.3
X_{22}	f_O	0.3
X_{23}	$\mathbf{W}_{bS}$	0.3
X_{24}	$\mathbf{W}_{cS}$	0.3
X_{25}	$\mathbf{W}_{bdS}$	0.3
X_{26}	$\mathbf{W}_{cdS}$	0.3
X_{27}	$\mathbf{W}_{eS}$	0.3
X_{28}	$\mathbf{W}_{fS}$	0.3
X_{29}	$\mathbf{W}_{cD}$	0.3
X_{30}	$\mathbf{W}_{dD}$	0.3
X_{31}	$\mathbf{W}_{eD}$	0.3
X_{32}	$\mathbf{W}_{fD}$	0.3
X_{33}	$\mathbf{W}_{bE}$	0.3
X_{34}	$\mathbf{W}_{dE}$	0.3
X_{35}	$\mathbf{W}_{eE}$	0.3
X_{36}	$\mathbf{W}_{fE}$	0.3
X_{37}	$\mathbf{W}_{bO}$	0.3
X_{38}	$\mathbf{W}_{cO}$	0.3
X_{39}	$\mathbf{W}_{bdO}$	0.3
X_{40}	$\mathbf{W}_{cdO}$	0.3
X_{41}	$\mathbf{W}_{eO}$	0.3
X_{42}	$\mathbf{W}_{fO}$	0.3
X_{43}	$\mathbf{W}_{w_O w_S}$	0.1
X_{44}	$\mathbf{W}_{w_O w_D}$	0.1
X_{45}	$\mathbf{W}_{w_O w_E}$	0.1
X_{46}	$\mathbf{W}_{w_S w_O}$	0.1
X_{47}	$\mathbf{W}_{w_D w_O}$	0.1
X_{48}	$\mathbf{W}_{w_E w_O}$	0.1
X_{49}	Lead FB	0.1
X_{50}	Lead FF	0.1

References

Abraham, W.C., Bear, M.F.: Metaplasticity: the plasticity of synaptic plasticity. Trends Neurosci. **19**(4), 126–130 (1996)

Bogenrieder, I.: Social architecture as a prerequisite for organizational learning. Manag. Learn. **33**(2), 197–212 (2002)

Bouma, D., Canbaloğlu, G., Treur, J., Wiewiora, A.: adaptive network modeling of the influence of leadership and communication on learning within an organization; Bouma, D., Canbaloğlu, G., Treur, J., Wiewiora, A. (eds.).: Adaptive network modeling of the influence of leadership and communication on learning within an organization. Cogn. Syst. Res. **79**, 55–70 (2022)

Canbaloğlu, G., Treur, J., Roelofsma, P.H.M.P.: Computational modeling of organisational learning by self-modeling networks. Cogn. Syst. Res. **73**, 51–64 (2022b)

Canbaloğlu, G., Treur, J., Wiewiora, A.: Computational modeling of the role of leadership style for its context-sensitive control over multilevel organisational learning. In: Proceedings of the 7th International Congress on Information and Communication Technology, ICICT'22. Lecture Notes in Networks and Systems, vol. 447, pp. 223–239. Springer Nature Switzerland AG (2021). https://www.researchgate.net/publication/355186556

Canbaloğlu, G., Treur, J., Wiewiora, A.: Multilevel organisational learning in a project-based organisation: computational analysis based on a third-order adaptive network model. Procedia Comput. Sci. **213**, 71–82. Elsevier (2022)

Canbaloğlu, G., Treur, J., Wiewiora, A.: computational modeling of multilevel organisational learning: from conceptual to computational mechanisms. In: Computational Intelligence: Select Proceedings of InCITe 2022. Lecture Notes in Electrical Engineering, vol. 968, pp. 1–17. Springer Nature, Cham (2022). https://www.researchgate.net/publication/354617515

Canbaloğlu, G., Treur, J., Wiewiora, A. (eds.).: Computational Modeling of Multilevel Organisational Learning and its Control Using Self-modeling Network Models. Springer Nature Switzerland AG (2023). https://www.researchgate.net/publication/355582177

Crossan, M.M., Lane, H.W., White, R.E.: An organizational learning framework: from intuition to institution. Acad. Manag. Rev. **24**, 522–537 (1999)

Iftikhar, R, Wiewiora, A.: Learning processes and mechanisms for interorganizational projects: insights from the islamabad–rawalpindi metro bus project. IEEE Trans. Eng. Manag. (2021). https://doi.org/10.1109/TEM.2020.3042252

Kim, D.H.: The link between individual and organisational learning. sloan management review. In: Fall, pp. 37–50 (1993). Reprinted In: Klein, D.A. (ed.) The Strategic Management of Intellectual Capital. Routledge-Butterworth-Heinemann, Oxford (1993)

Mathieu, J.E., Heffner, T.S., Goodwin, G.F., Salas, E., Cannon-Bowers, J.A.: The influence of shared mental models on team process and performance. J. Appl. Psychol. **85**(2), 273 (2000)

McComb, S., Simpson, V.: The concept of shared mental models in healthcare collaboration. J. Adv. Nurs. **70**(7), 1479–1488 (2014)

Robinson, B.L., Harper, N.S., McAlpine, D.: Meta-adaptation in the auditory midbrain under cortical influence. Nat. Commun. **7**, e13442 (2016)

Rouse, W.B., Morris, N.M.: On looking into the black box: prospects and limits in the search for mental models. Psychol. Bull. **100**(3), 349 (1986)

Treur, J.: Modeling higher-order adaptivity of a network by multilevel network reification. Netw. Sci. **8**, 110–144 (2020a)

Treur, J.: Network-Oriented Modeling for Adaptive Networks: Designing Higher-Order Adaptive Biological, Mental and Social Network Models. Springer Nature, Cham (2020b)

Treur, J.: Modeling multi-order adaptive processes by self-modeling networks. In: Tallón-Ballesteros, A.J., Chen, C.-H. (eds.) Proceedings of the 2nd International Conference on Machine Learning and Intelligent Systems, MLIS'20. Frontiers in Artificial Intelligence and Applications, vol. 332, pp. 206–217. IOS Press (2020c)

Van den Hurck, H., Treur, J., Roelofsma, P.H.M.P: Adaptive modelling of the implementation of environment health and safety standards. In: Proceedings of the 15th International Conference on Social Computing and Networking, SocialCom'22 (2022)
Wiewiora, A., Smidt, M., Chang, A.: The 'how'of multilevel learning dynamics: a systematic literature review exploring how mechanisms bridge learning between individuals, teams/projects and the organization. Eur. Manag. Rev. **16**(1), 93–115 (2019)
Wiewiora, A., Chang, A., Smidt, M.: Individual, project and organizational learning flows within a global project-based organization: exploring what, how and who. Int. J. Project Manag. **38**, 201–214 (2020)

Results, Further Perspectives, and Discussion

The sixth part of this volume focuses on summarizing significant findings and outlining future research directions in the realm of safety and security in cyberspace. The chapters in this part explore innovative approaches and insights into addressing cybersecurity challenges within healthcare organizations, emphasizing the role of adaptive network-oriented modeling and AI-driven solutions.

Chapter "Addressing Organizational Cybersecurity Challenges in Healthcare Using Adaptive Network-Oriented Modeling", by van den Hout et al., presents a comprehensive overview of current technological trends in healthcare and the cybersecurity risks accompanying these advancements. The chapter underscores the prevalence of phishing attacks as a primary threat vector and discusses the implementation of phishing training simulations as a behavioral intervention strategy. Furthermore, it highlights a multi-level adaptive network model, integrating elements from the COM-B framework, to enhance cybersecurity resilience within healthcare settings. The chapter concludes with insights into future opportunities for refining and expanding the model to address emerging cybersecurity threats effectively.

Chapter "Towards an AI Coach for Risk Management and Cybersecurity in Health Systems", authored by Roelofsma and Treur, delves into the role of AI-driven solutions in bolstering cybersecurity and risk management in healthcare environments. The chapter outlines a network-oriented computational analysis that explores how an AI Coach can support medical teamwork and contribute to enhancing safety and cybersecurity measures. Through formalized computational models, the chapter illustrates various scenarios where AI coaches can intervene to mitigate risks effectively. Moreover, the authors propose a Cybersecurity Simulation Lab project as a crucial next step to further advance network-oriented modeling in cybersecurity risk management.

What Will be addressed in This Part

Addressing Organizational Cybersecurity Challenges in Healthcare
In this chapter, the authors highlight the critical cybersecurity challenges faced by healthcare organizations amidst evolving technological landscapes. The focus on phishing attacks as a prevalent threat underscores the need for proactive strategies

to educate and train healthcare professionals. The chapter's approach of using adaptive network-oriented modeling and the COM-B framework provides a structured methodology to analyze and mitigate cybersecurity risks. By integrating behavioral interventions such as phishing training simulations, healthcare organizations can enhance their resilience against cyber threats.

A key contribution of the chapter lies in its identification of future research opportunities. The authors emphasize the ongoing evolution of cybersecurity threats and the need for adaptive models that can dynamically respond to emerging risks. The discussion on enhancing the COM-B framework for network-oriented modeling suggests avenues for refining predictive capabilities and improving decision-making processes in cybersecurity management. Overall, the chapter advocates for a proactive approach to cybersecurity in healthcare, emphasizing continuous learning and adaptation to safeguard sensitive patient data and operational integrity.

Towards an AI Coach for Risk Management and Cybersecurity in Health Systems

This chapter explores the transformative potential of AI-driven solutions in healthcare cybersecurity. By employing network-oriented computational analysis, the authors point out how AI coaches can function as proactive agents in identifying and mitigating cyber risks. The chapter's focus on formalized computational models demonstrates the feasibility and efficacy of AI coaches in enhancing safety protocols and cybersecurity measures within health systems.

The practical examples and scenarios presented in the chapter underscore the versatility of AI coaches in supporting medical teams and improving organizational resilience against cyber threats. The proposal for a Cybersecurity Simulation Lab project represents a forward-looking initiative to advance research and development in cybersecurity risk management. This project, as outlined in the chapter, promises to integrate cutting-edge technologies and methodologies to simulate and analyze complex cybersecurity scenarios, thereby preparing healthcare organizations to respond effectively to real-world threats.

In conclusion, the chapter advocates for continued investment and research into AI-driven solutions in healthcare cybersecurity. By leveraging network-oriented modeling and AI technologies, healthcare organizations can fortify their defenses, mitigate vulnerabilities, and uphold patient confidentiality and safety in an increasingly digitized healthcare landscape.

Conclusion

The sixth part of this volume encapsulates critical insights and future directions in safety and security within cyberspace, as explored through adaptive network-oriented modeling and AI-driven approaches. Chapters "Addressing Organizational Cybersecurity Challenges in Healthcare Using Adaptive Network-Oriented

Modeling" and "Towards an AI Coach for Risk Management and Cybersecurity in Health Systems" underscore the importance of proactive cybersecurity strategies in healthcare, highlighting innovative methodologies to address evolving threats. By advancing research in adaptive modeling and AI technologies, organizations can strengthen their cybersecurity posture, ensuring robust protection of sensitive data and maintaining operational continuity in healthcare settings. The chapters' findings and proposals lay a foundation for future research endeavors aimed at enhancing safety, security, and resilience in cyberspace.

Addressing Organizational Cybersecurity Challenges in Healthcare Using Adaptive Network-Oriented Modeling

Niek Jan Van den Hout, Debby Bouma, Charlotte Hoffmans, Natalia Zwarts, Jan Treur, and Peter Roelofsma

Abstract This chapter introduces the reader to current technological trends that are shaping the healthcare sector and the pressing cyber security risks associated with these trends. The importance of phishing as a common attack vector is highlighted and the execution of phishing training simulations as a behavioral intervention is described. In the discussion section some challenges regarding the use of the COM-B framework as a basis for network-oriented models are discussed and potential future improvement opportunities are highlighted. Following the description of a phishing training simulation scenario, a multi-level adaptive network model based on key elements of the COM-B framework is described.

N. J. Van den Hout · D. Bouma · C. Hoffmans · N. Zwarts · J. Treur (✉) · P. Roelofsma
Risk Management and Cybersecurity Group, Centre of Expertise Cyber Security, The Hague University of Applied Sciences, Hague, The Netherlands
e-mail: j.treur@vu.nl; j.treur@hhs.nl

N. J. Van den Hout
e-mail: n.j.vandenhout@hhs.nl

D. Bouma
e-mail: d.bouma@hhs.nl

C. Hoffmans
e-mail: c.c.a.hoffmans@hhs.nl

N. Zwarts
e-mail: n.h.zwarts@hhs.nl

P. Roelofsma
e-mail: p.h.m.p.roelofsma@hhs.nl

D. Bouma · C. Hoffmans · J. Treur
Social AI Group, Department of Computer Science, Vrije Universiteit Amsterdam, Amsterdam, The Netherlands

P. Roelofsma
Department of Neonatal and Pediatric Intensive Care, Division Neonatology, Erasmus University Medical Center, Rotterdam, The Netherlands

P. H. M. P. Roelofsma et al. (eds.), *Using Shared Mental Models and Organisational Learning to Support Safety and Security Through Cyberspace: A Computational Analysis Approach*, Studies in Systems, Decision and Control 570,
https://doi.org/10.1007/978-3-031-72075-8_12

Keywords Cyber security · Healthcare · COM-B framework · Adaptive network model

1 Introduction

Because of multiple socio-demographic trends the worldwide demand for medical care is growing fast (Liu et al. 2017; Prince et al. 2015). Already today many healthcare systems are under economic and workload pressure (Liu et al. 2017; World Health Organization. Regional Office for Europe: Health and Care Workforce in Europe 2022; British Medical Association 2023). To handle this growing workload healthcare organizations are implementing a variety of novel information technology (IT) solutions. The growing reliance on information technology, however, introduces a wide range of new cybersecurity risks (Coventry and Branley 2018; Bhuyan et al. 2020; European Union Agency for Cybersecurity 2023). These risks include the unavailability of critical data, such as electronic health records, the unintended modification of data, such as the settings of implanted medical devices and the publication or theft of data, such as an individual's medical history.

A high-profile scenario that illustrates the impact of these risks on healthcare organizations is the WannaCry ransomware incident at the UK National Health Service (NHS) that took place in 2017. As a result of the WannaCry ransomware, data of thousands of patients were encrypted and therefore not accessible to healthcare professionals (Martin et al. 2018). As a consequence surgeries and other types of patient appointments were cancelled which in turn negatively influenced the health of the patients (Ghafur et al. 2019).

A cohort study of 364 ransomware incidents at US healthcare organizations between January 2016 to December 2021 showed that this incident was not an exception and that clinics (57,8%), hospitals (21,9%) and ambulatory surgical centers (15,0%) are all targets of this type of attack (Neprash et al. 2022). Two factors that set apart healthcare organizations specifically, are the vulnerability of their clients (patients) and the widespread adoption of innovative digital technology (Jalali and Kaiser 2018; Kraus et al. 2021). The combination of having substantial security requirements and the implementation of new information technology results in a variety of pressing cybersecurity challenges.

2 Current Trends in Healthcare

The global healthcare sector is changing rapidly and is facing a wide variety of new cybersecurity risks. This chapter briefly describes some trends in healthcare and addresses some of the associated cybersecurity risks in more detail.

As mentioned earlier the worldwide demand for healthcare is growing. The average global life expectancy has been increasing rapidly in the past decades. Age and lifestyle related diseases, such as chronic non-communicable diseases (NCDs), continue to cause the highest disease burden worldwide, and their impact has worsened in the past two decades. Treatment of these diseases can be difficult and relatively expensive adding extra pressure to healthcare systems and leading to a global healthcare spending increase. Global healthcare spending is estimated to have grown more than 40% between 2018 and 2022, reaching $12 trillion and this increase is predicted to continue (World Economic Forum 2023). Recent data from the EU also shows a 35% increase in EU healthcare expenditure between 2014 and 2021 (Eurostat 2023). Another factor that has increased pressure on healthcare organizations and associated organizational actors is the long-term effect of the COVID-19 pandemic. During the pandemic many healthcare organizations experienced a sharp increase of leading to an overworked workforce and a wave of professionals leaving the sector (Hassamal et al. 2021; Poon et al. 2022; Ornell et al. 2020). In many countries there remains a shortage of healthcare professionals, keeping pressure high for existing practitioners.

Another current trend in healthcare is the advancement of personalized diagnoses, treatment and medicine (Hamburg and Collins 2010). Personalized (or precision) medicine considers the patients unique clinical, genetic, genomic, and environmental information. Emerging technologies, such as DNA sequencing, proteomics, AI assisted medical imaging and personal health monitoring devices take into account the (inter-)individual variation in disease processes (Goetz and Schork 2018). Since these technologies leverage advanced data acquisition and analysis technology in order to provide personalized treatment, the role of medical data storage solutions is as important as ever. The increased use of Electronic Health Records (EHRs) or Electronic Patients Records (EPRs) further illustrates the growing importance of digital medical storage solutions (Menachemi and Collum 2011; Kierkegaard 2011).

Lastly, another major development within the healthcare sector is the increasing importance of care provision outside of the hospital. In fact, growth in expenditure on health provision and care at home is expected to outpace healthcare spend in nearly all other areas (World Economic Forum 2023). This investment will be used to further home diagnostics, home administered drug delivery systems, patient monitoring devices and a variety of other eHealth solutions, sometimes called the Internet of Medical Things (IoMT). This shift to care at home is further illustrated by sectoral agreements, such as in the Integral Care Agreement in which Dutch healthcare organizations agreed that, in order to meet future workload demands, healthcare should take place at home when possible. This new perspective is summarized as *"yourself if possible, at home if possible and digital if possible"* (Dutch Government 2022). This transition is also highlighted in sectoral reports from the United Kingdom (National Health Service 2019), the United States (American Medical Association 2023) and the European Union (European Commission 2021).

3 Cybersecurity Risks in Healthcare

The described trends in healthcare all significantly depend on the technological advancements involving the improved acquisition, analysis and application of data. It is therefore not surprising that in 2021, digital investments in the healthcare sector nearly doubled to $57 billion (World Economic Forum 2023). A number of countries have recognized the transformative potential of digital health technology and have invested significantly in the digital transformation of healthcare (World Economic Forum 2023). In 2021, the EU launched the EU4Health Programme 2021–2027, which, among other goals, will be focusing on *"reinforcing health data, digital tools and services, [and the] digital transformation of healthcare"* (European Commission 2021). Meanwhile, the UK has begun a major initiative towards the digital transformation of the NHS and also within the US many healthcare organizations are changing to digital service delivery (American Medical Association 2022).

It is clear that a wide range of digital eHealth solutions will play an irreplaceable role in the future of healthcare. This increase in digital dependency highlights the relevance of tackling the cybersecurity risks that the healthcare industry is facing.

Cybersecurity risks are often categorized based on their impact on three information security properties, namely: confidentiality, integrity and availability, generally known as the CIA-triad (Samonas and Coss 2014). The CIA triad requirements are presented in the table below.

Confidentiality	The information (system) is only accessible to authorized entities or processes
Integrity	The information (system) is accurate and complete
Availability	The information (system) is accessible and usable on demand by authorized entities

The disclosure of personal medical data to unauthorized parties (a breach of confidentiality) can lead to a wide variety of negative financial, operational, legal and safety related consequences. Sectoral reports highlight the continuing threat of ransomware and associated data disclosure (European Union Agency for Cybersecurity 2023; UK National Cyber Security Centre 2023). Although ransomware traditionally mostly led to the unavailability of systems and data, cybercriminals have adapted their modus operandi by also stealing large amounts of sensitive (patient) data and requesting a ransom to be paid to prevent the publication of the data. The earlier described cohort study of 374 ransomware attacks on US healthcare delivery organizations indicates that Personal Health Information (PHI) of nearly 42 million patients was exposed (Neprash et al. 2022). Between 2021 and March 2023 ENISA reported nearly 100 data related incidents at European healthcare organizations (European Union Agency for Cybersecurity 2023). In addition to ransomware there are also instances of sensitive medical information disclosure because of successful targeted attacks, through the use of social engineering, including phishing, or through the exploitation of software vulnerabilities. The theft of sensitive medical information can lead a variety of financial and legal consequences, including the payment of ransom to cybercriminals and closer scrutiny from a variety of regulators, such as data protection authorities. Additionally, it can lead to a diminished trust of patients, identity theft and insurance fraud.

In addition to this, the manipulation of medical data and devices (a breach of integrity) can also lead to a wide variety of negative financial, operational, legal and safety related consequences. A concrete example of this is the manipulation of electronic implants with a therapeutic function, such as pacemakers, implantable cardioverter defibrillators (ICDs) and insulin pumps. Most of these implants are nowadays connected to the internet, making them susceptible to a variety of cyberattacks (Groeneveld et al. 2019).

In 2017 the US Food and Drug Administration (FDA) issued an alert for a voluntary recall of approximately 500,000 pacemakers citing the potential exploitation of security vulnerabilities (Kuehn 2018). In 2018 security researchers demonstrated the exploitation of multiple vulnerabilities in pacemakers developed by Medtronic Kuehn 2018. The researchers were able to manipulate the pacemaker's programmer which controlled the electronic impulses that are send to the heart potentially causing heart failure. In 2019 the FDA issued a mandatory recall of the company's insulin pumps, again citing cybersecurity concerns, which could enable malicious parties to manipulate (from distance) insulin delivery to patients, leading to hypoglycemia or high blood sugar and potentially life-threatening diabetic ketoacidosis (DKA) (FDA 2021). Another risk that could affect the integrity of hospital data is the defacement of healthcare websites by hacktivists, due to worldwide increasing geopolitical tensions (European Union Agency for Cybersecurity 2023; Romagna and Hout 2017). The defacement of these websites could lead to potentially misinformed patients and a decrease of trust in healthcare organizations.

Moreover, deleting or otherwise blocking access to medical information systems and data (a breach of availability) can also lead to a wide variety of negative financial, operational, legal and safety related consequences. As previously mentioned, ransomware remains a critical threat which can cause medical data to become inaccessible. The inaccessibility of data by ransomware can cause financial damage, because of the ransom that has to be paid (ranging from a few thousand to millions of Euro's), but more importantly it could lead to significant operational and safety related impact. The previously described review of ransomware incidents at US healthcare organizations reports operational disruptions, such as ambulance diversion, cancelled appointments and surgeries, unavailable equipment such as X-ray and CT scanners and disrupted systems, including EHR systems (Neprash et al. 2022). Of all organization types hospitals were the ones that reported the most disrupting attacks (Neprash et al. 2022). Other studies also report increased waiting times and increased patient mortality (McGlave et al. 2023). Another type of attack which directly impacts the availability of data is a Distributed Denial of Service (DDoS) attack. Although not as prevalent as ransomware attacks, recent geopolitical tensions have caused an increase in hacktivist activity, including attacks against hospitals (European Union Agency for Cybersecurity 2023). Reports suggest however, that the financial and operational impact of these attacks is often limited (European Union Agency for Cybersecurity 2023).

4 Phishing as a Common Attack Factor

With the increasing use of information technology in healthcare organizations, the cyber security risks threatening the confidentiality, integrity and availability (CIA) of (patient) information (systems) are of unprecedented significance. As described earlier the risks associated with ransomware attacks are considered especially critical. Before ransomware is activated and starts encrypting files, the attacker first has to gain access to the victim's computer network (Brewer 2016). Although healthcare organizations typically have deployed a variety of technological and organizational controls, such as an Information Security Management System (ISMS), to protect the network against, among other things, unauthorized access, healthcare organizations are still vulnerable to network breaches. The most common technique to gain access to an organizations computer network and bypass any security controls is the use of phishing (Alkhalil et al. 2021). Phishing can be described as *"scam emails or text messages that contain links to websites which may contain malware, or may trick users into revealing sensitive information (such as passwords) or transferring money"* (UK National Cyber Security Centre 2024) and is often used to gather user credentials which then can be used to access the organizations network. An example of this could be a healthcare employee receiving a phishing e-mail from an attacker which attempts to convince the employee to click on a link which redirects to a malicious website. On this website the employee is tricked to login with his organizational username and password, sending these credentials to a server under the attacker's control. As a consequence, the attacker can now utilize these credentials to access sensitive information and privileged network systems and perform malicious actions on the organizations network. The desired behavior in this case would be that the employee notices the phishing attempt and disregards or, even better, reports the e-mail to the IT security department.

In order to reinforce secure employee behavior, organizations can conduct a wide range of interventions. One of such interventions is the execution of phishing training simulations during which the organizations security department recreates phishing scenarios by sending out artificial phishing e-mails to employees (Gordon et al. 2019). Instead of visiting a malicious website the employee that clicks on the link in the e-mail is redirected to a notification page with a message that the employee took part in a phishing simulation and that the employee failed to report the phishing e-mail. The desired situation in such a scenario would be that the employee reports the phishing e-mail to the IT security department. Often this is facilitated by adding a *'report phishing e-mail'* button to the employee's e-mail client. Recent experiments have shown that phishing simulations in healthcare institutions improve employee behaviour, especially when the intervention is systematically repeated (Gordon et al. 2019). To fully understand the effects of phishing training simulations on employee behavior and, as a consequence, on the organizations cyber security posture, several behavioral analysis methods could be used.

5 The COM-B Framework for Behavioral Analysis

A widely used approach to examine employee (security) behavior and design behavioral change interventions is the COM-B framework. The COM-B framework was created in 2011 by Michie et al. and focusses on how capability (C), opportunity (O) and motivation (M) factors influence behavior (B) (Michie et al. 2011). The COM-B framework has been used frequently in the field of cyber security to understand and change employee security behavior (Alshaikh et al. 2019; Kleij et al. 2020, 2021; Steen 2023; Bouma et al. 2024).

Within the COM-B framework capability is defined as *"the individual's psychological and physical capacity to engage in the activity"* and includes having the *"necessary knowledge and skills"* (Michie et al. 2011). Motivation is defined as *"all those brain processes that energize and direct behavior"* (Michie et al. 2011). It includes *"habitual processes, emotional responding, as well as analytical decision-making"* (Michie et al. 2011). Opportunity is defined as *"all the factors that lie outside the individual that make the behavior possible or prompt it"* (Michie et al. 2011). Figure 1 highlights the individual components and the relationship between components in a simplified manner.

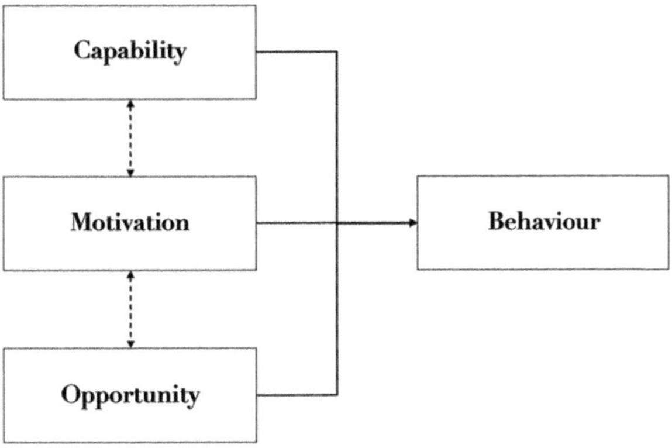

Fig. 1 Simplified version of the relationship between the different elements of the COM-B framework for behavioral change as described by Michie et al. (2011)

Scenario

The COM-B framework can be applied to the previously described phishing training scenario as previously described by Bouma et al. (2024). In this case the capability, opportunity and motivation of targeted employees are taken into account, resulting in behaviour of the employee that includes either clicking on the phishing link or reporting the phishing e-mail (Bouma et al. 2024).

In the earlier described scenario the employee receives feedback on his behavior in two instances. When clicking on the malicious link the employee receives a notification indicating that they fell for the phishing e-mail and that they failed to notice and report it. When the employee clicks the *"report phishing e-mail"* button the employee receives a notification that they succeeded in noticing and reporting the phishing email. In both instances a learning process is initiated which has a certain effect on the capability, motivation and opportunity factors. As also noted by Bouma et al. (2024), the behavior of the employee could be different in a subsequent phishing training because of this learning process.

To illustrate the effects of this learning process, the effectiveness of interventions in changing employee behaviour and to simulate this through network-oriented modeling, Bouma et al. 2024 propose a realistic scenario taking place in a hospital environment. They consider an individual starting to work as a secretary at a hospital: during a workday the secretary receives many emails from different recipients. In order to improve the hospitals organizational cyber security, phishing simulation training is implemented immediately upon the secretary's start of employment (Bouma et al. 2024). The employee is characterized by two individual traits: capability and motivation, and by external conditions (opportunity) provided by the company, based on the COM-B framework for behavioral analysis as earlier described. In Bouma et al. 2024 scenario the employee is motivated to learn, but has limited capability and opportunity to recognize and respond properly to phishing e-mails. Throughout employment, the employee encounters simulated phishing e-mails, as part of a training intervention, offering the options to either click on the link or report it. As previously highlighted, feedback is provided in both cases.

6 The Self-modelling Network Approach

The technique to simulate effects of behavioral interventions that is presented in this chapter is dynamic computational modelling through self-modelling networks as described by Treur 2020. In the following paragraphs the self-modeling network approach is explained in more detail and the model that was presented by Bouma et al. (2024) is discussed and expanded upon.

- Connectivity characteristics: Network characteristics describing which nodes X connect to which nodes Y, and with what connection weight $\boldsymbol{\omega}_{X,Y}$ indicating how strongly nodes are connected.
- Aggregation characteristics: Network characteristics which describe the way in which causal impacts have combined effects on node *Y* by assigning a combination function $\mathbf{c}_Y$ to Y, see Table 1 for the combination functions used in this paper. Stepmod and Stepmodopp are used to model repetitive external context factors.
- Timing characteristics: Network characteristics specifying speed factors $\boldsymbol{\eta}_Y$ for all nodes Y by which certain changes of node activations take place, for example used for timing of enabling or disabling certain mechanisms in the model; moreover, initial values are needed for all of the nodes.

Standard difference equation: Based on these network characteristics, the following difference equation is used for simulation and analysis. Here the X_i are all states from which state Y has incoming connections:

$$Y(t + \Delta t) = Y(t) + \boldsymbol{\eta}_Y\left[\mathbf{c}_Y\left(\boldsymbol{\omega}_{X_1,Y}X_1(t), \ldots, \boldsymbol{\omega}_{X_k,Y}X_k(t)\right) - Y(t)\right]\Delta t \qquad (1)$$

The network characteristics are specified in a standard table format called role matrices. These are input for the dedicated software environment, after which the above difference equation is automatically generated and simulated within the software environment (Treur 2020).

Table 1 Examples of combination functions

Name	Notation	Formula	Parameters
Advanced logistic sum	$\mathbf{alogistic}_{\sigma,\tau}(V_1, \ldots, V_k)$	$\left[\frac{1}{1+e^{-\sigma(V_1+\cdots+V_k-\tau)}} - \frac{1}{1+e^{\sigma\tau}}\right](1 + e^{-\sigma\tau})$	p(1) = steepness σ p(2) = threshold τ
Complemental advanced logistic sum	$\mathbf{compa}_{\sigma,\tau}(V_1, \ldots, V_k)$	$1-\left[\frac{1}{1+e^{-\sigma(V_1+\cdots+V_k-\tau)}} - \frac{1}{1+e^{\sigma\tau}}\right](1 + e^{-\sigma\tau})$	p(1) = steepness σ p(2) = threshold τ
Stepmod	$\mathbf{stepmod}_{\rho,\delta}$	1 if t mod $\rho > \delta$ else 0 (time t)	p(1) = repetition period ρ p(2) = tipping point δ
Stepmodopp	$\mathbf{stepmodopp}_{\rho,\delta}$	0 if t mod $\rho > \delta$ else 1 (time t)	p(1) = repetition period ρ p(2) = tipping point δ

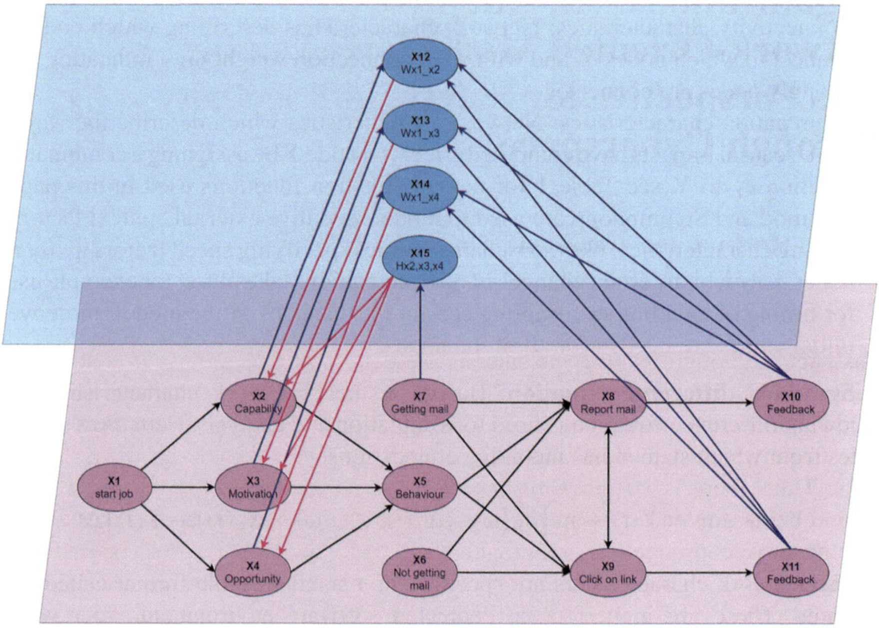

Fig. 2 A graphical representation of the phishing simulation model as described by Bouma et al. 2024

Self-model nodes for adaptation: To model adaptivity, self-model nodes are added. For example, if a connection weight $\boldsymbol{\omega}_{X,Y}$ changes over time due to learning or forgetting, then a self-model node $\mathbf{W}_{X,Y}$ is introduced to represent the dynamic value of $\boldsymbol{\omega}_{X,Y}$. During processing, the activation value of $\mathbf{W}_{X,Y}$ is utilized for $\boldsymbol{\omega}_{X,Y}$. This approach enriches causal modeling, allowing not only node activations but also the underlying causal relations to evolve over time. Self-model nodes are depicted at a higher level with a conceptual 3D structure, as shown in Fig. 2. They are integrated into the network structure by specifying connections to and from them, along with other network characteristics. The majority of first-order network adaptation principles known in the literature consider the adaptation of connection weights over time, with additional extensions for changes in structure, for example adaptive excitability thresholds or processing speed Treur 2020.

7 An Example Adaptive Network Model

As described by Bouma et al. (2024), through computationally modelling phishing simulations, it is possible to show that the elements of the COM-B framework are influenced by the learning process using an adaptive layer to adjust the weights of the COMB elements over time. By adopting this approach insights could be gathered into the effectiveness of these interventions over time. Bouma et al. (2024) used a simplified version of the COM-B framework to show the effects of feedback learning on responding to phishing e-mails.

Assumption

To represent the previously described scenario as a model, certain simplifications and assumptions are made, as described by Bouma et al. (2024).

Bouma et al. (2024) describe that all states, except for the two email states, exist on a spectrum. This means that they range from 0 (completely off) to 1 (completely on). They describe that this can be adjusted for each state to enhance realism and versatility. The emails arrive at regular intervals, and the employee only learns when an email arrives and receives feedback. Bouma et al. (2024) assume that the elements outlined in the COM-B framework, as applied in the context of their paper, are defined as:

- **Capability**: *Does the employee have the physical and psychological capacity (such as competences, skills, and knowledge) to notice and report the simulated phishing email?*
- **Opportunity**: *Does the working environment (including policies, procedures, and social environment) allow and support the employee to report the simulated phishing email?*
- **Motivation**: *Is the employee motivated (e.g., are the beliefs, habits, and emotions supportive) to report the simulated phishing email?*
- **Behaviour**: *The behavior of the employee is either clicking on the simulated malicious link or reporting the phishing email by pressing the"report phishing" button in the e-mail client.*

In addition, Bouma et al. (2024) mentioned several assumptions that are made about the employee in the scenario regarding motivation, capability, and opportunity: since the employee is motivated to learn, the initial value of X13, representing the weight between the start state and motivation, starts the highest of the three W-states (X12, X13, and X14). Bouma et al. (2024) describe that the person struggles with capability; therefore, X12 has a lower initial value than X13. Lastly, X14 (learning associated with opportunity) has the lowest initial value since training within the organization has not yet begun Bouma et al. (2024).

Bouma et al. (2024) describe that for all states, a combination function is utilized and that it is assumed that the chosen functions represent the states accurately:

- **Alogistic**: Used in states X1-X5, X8, and X10-X15.
- **Compalogistic**: Used in X9, since it happens opposite to when X8 occurs.
- **Stepmod**: Used in X6 to represent when an email is sent.
- **Stepmodopp**: Used in X7 to represent when there is no email sent.

8 Enhancement of the Adaptive Network Model

In line with the work of Michie et al. (2011), West and Michie (2020) describe the COM-B framework in more detail. They elaborate on the sub-elements of capability, opportunity and motivation, namely: physical capability, psychological capability, reflective motivation, automatic motivation, physical opportunity and social opportunity. In Fig. 4 the relationship between the sub-elements are defined based on the description provided by West and Michie (2020).

The capability element is divided into physical capability and psychological capability. The motivation element is divided into reflective motivation and automatic motivation. The opportunity element is divided into physical opportunity and social opportunity. In Table 2 the sub-elements are defined based on the description provided by West and Michie (2020) (Figs. 3 and 4).

State nr	State name	Explanation	Level
X1	Start job	Start of simulation, someone starts a new job	
X2	Capability	Level of employee's capability	
X3	Motivation	Level of employee's motivation	
X4	Opportunity	Level of employee's opportunity to click a link in a phishing email	
X5	Behaviour	Likeliness of the employee to behave safely	
X6	Not getting mail	Period where there is no email send	
X7	Getting mail	Period where there is an email send	
X8	Report mail	Action of not opening the link and sending a report to the IT department	
X9	Click on link	Action of clicking on the phishing	Base level
X10	Feedback	When reporting the phishing email, employee receives a praising email due to their safe behaviour	
X11	Feedback	In case of an employee clicking the phishing link, they receive feedback on their unsafe behaviour	
X12	Wx1_x2	Self-model state for the weight of the connection between start job and capability, learned by feedback of X10 and X11 and the initial values of capability	
X13	Wx1_x3	Self-model state for the weight of the connection between start job and motivation, learned by feedback of X10 and X11 and the initial values of motivation	
X14	Wx1_x4	Self-model state for the weight of the connection between start job and opportunity, learned by feedback of X10 and X11 and the initial values of opportunity	First-order
X15	Hx2,x3,x4	Self-model state for the speed of the state's capability, motivation and opportunity, when there is no e-mail there is no learning.	self-model level

Fig. 3 Explanation of the different states of the phishing simulation model as described by Bouma et al. (2024)

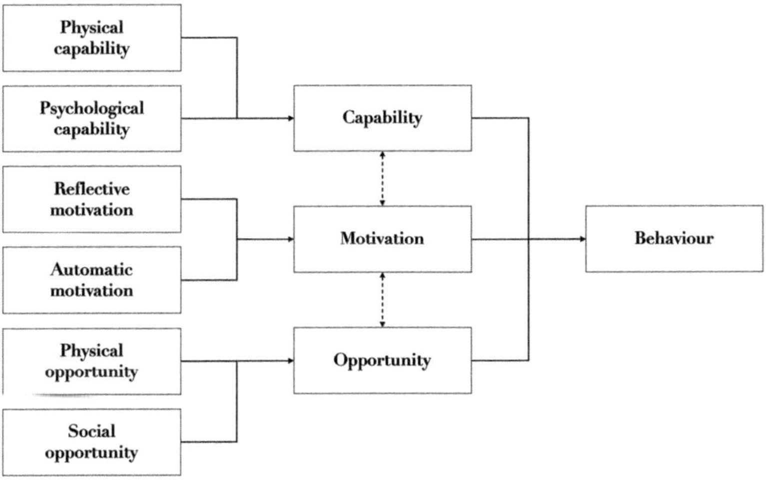

Fig. 4 A simplified overview of the relationships between the different elements, including subelements of capability, motivation and opportunity, of the COM-B framework for behavioral change as described by West and Michie (2020)

Physical capability	Capability that involves a person's physique, and musculoskeletal functioning (e.g., balance and dexterity)
Psychological capability	Capability that involves a person's mental functioning (e.g., understanding and memory)
Reflective motivation	Motivation that involves conscious thought processes[11] (e.g., plans and evaluations)
Automatic motivation	Motivation that involves habitual, instinctive, driverelated, and affective processes (e.g., desires and habits)
Physical opportunity	Opportunity that involves inanimate parts of the environmental system and time (e.g., financial and material resources)
Social opportunity	Opportunity that involves other people and organizations (e.g., culture and social norms)

Using the specification of the sub-elements, the initial model presented by Bouma et al. (2024) can be expanded. Instead of three factors used as input for behavior all six sub-elements could be considered in the model. This would result in the addition of three base-level and three second-level (learning) nodes. An expanded version of the network-oriented model can be seen in Fig. 5. By adding the sub-elements to the model, it becomes possible to individually change the values of these sub-elements potentially resulting in a more precise and realistic simulation (Fig. 6).

A concrete example of taking into account the individual differences of different scenario and organizational characteristics, could be the distinction between physical opportunity and social opportunity. An employee could be part of a close-knit team in which colleagues stimulate each other to behave in a secure manner and

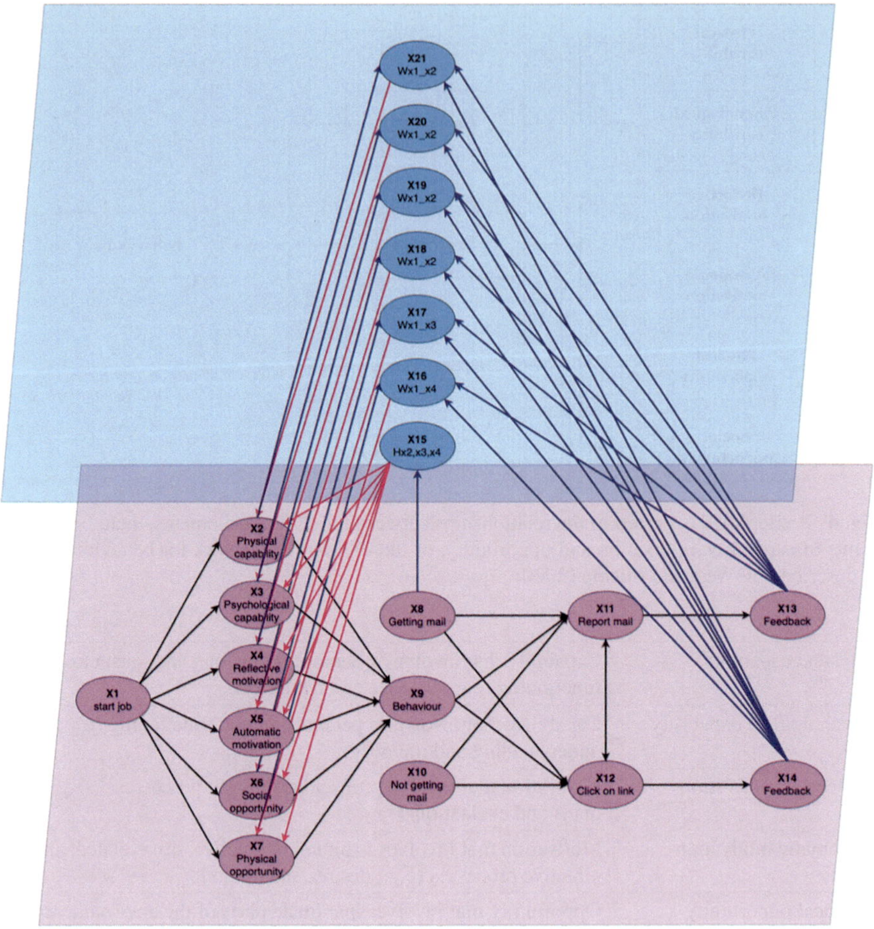

Fig. 5 A graphical representation of an enhanced phishing simulation model

report all phishing e-mails (the social opportunity value would be higher) while the working environment could be less conducive of secure behavior, because of high work pressure and the lack of policies and procedures regarding the reporting of phishing e-mails (the physical opportunity value would be lower). Due to the addition of the extra COM-B sub-elements the described scenario could be modeled more accurately.

Another example that highlights the benefits of using an enhanced model is taking into account the difference between reflective motivation and automatic motivation. An employee could be relatively unmotivated to consciously recognize and report (simulated) phishing e-mails (the reflective motivation value would be lower), but because of the employees previous training the employee could have formed habits to recognize and report (simulated) phishing e-mails (the automatic motivation value

State nr	State name	Explanation	Level
X1	Start job	Start of simulation, someone starts a new job	Base level
X2	Physical capability	Level of employee's physical capability	
X3	Psychological capability	Level of employee's psychological capability	
X4	Reflective motivation	Level of employee's reflective motivation	
X5	Automatic motivation	Level of employee's automatic motivation	
X6	Social opportunity	Level of social opportunity	
X7	Physical opportunity	Level of physical opportunity	
X8	Behaviour	Likeliness of the employee to behave safely	
X9	Not getting mail	Period where there is no email send	
X10	Getting mail	Period where there is an email send	
X11	Report mail	Action of not opening the link and sending a report to the IT department	
X12	Click on link	Action of clicking on the phishing	
X13	Feedback	When reporting the phishing email, employee receives a praising email due to their safe behaviour	
X14	Feedback	In case of an employee clicking the phishing link, they receive feedback on their unsafe behaviour	
X15	Wx1_x2	Self-model state for the weight of the connection between start job and physical capability, learned by feedback of X13 and X14 and the initial values of physical capability	First-order self-model level
X16	Wx1_x3	Self-model state for the weight of the connection between start job and psychological capability, learned by feedback of X13 and X14 and the initial values of psychological capability	
X17	Wx1_x4	Self-model state for the weight of the connection between start job and reflective motivation, learned by feedback of X13 and X14 and the initial values of reflective motivation	
X18	Wx1_x5	Self-model state for the weight of the connection between start job and automatic motivation, learned by feedback of X13 and X14 and the initial values of automatic motivation	
X19	Wx1_x6	Self-model state for the weight of the connection between start job and social opportunity, learned by feedback of X13 and X14 and the initial values of motivation	
X20	Wx1_x7	Self-model state for the weight of the connection between start job and physical opportunity, learned by feedback of X13 and X14 and the initial values of opportunity	
X21	Hx2,x3,x4,x5,x6,x7	Self-model state for the speed of the state's physical capability, psychological capability, reflective motivation, automatic motivation, social opportunity and physical opportunity, when there is no e-mail there is no learning.	

Fig. 6 Explanation of states of the enhanced model

would be higher). In this scenario the enhanced model could take into account this difference and support a more precise simulation, highlighting another use-case for the enhanced model.

Further Enhancement of the Adaptive Network Model

An additional consideration that could be taken into account when modeling employee behavior during phishing training simulations is the content of the phishing e-mail itself. While the enhanced model takes into external opportunity factors, this might not be sufficient to account for different types of phishing e-mails, such as generic phishing e-mails or spear-phishing e-mails and the use of specific techniques such as e-mail address spoofing. The nature and content of the phishing e-mail is likely to have an effect on the behavior of the employee. Future research could focus on how these factors could be incorporated into the adaptive network model in addition to the (sub-) elements of the COM-B framework.

9 Conclusion and Discussion

In this chapter the reader is introduced to technological trends that are shaping the healthcare industry and the pressing cyber security risks that come along with it. The importance of phishing as an attack vector is highlighted and the application of phishing training simulations as an intervention to reduce the risks associated with phishing is described. Based on earlier work by Bouma et al. (2024) an adaptive network model is presented that is based on a phishing training simulation scenario and on some of the elements of the COM-B framework. Afterwards an enhanced model is presented, which includes additional sub-elements of the COM-B framework, allowing for greater customization and more accurate results. In addition to this some practical customization scenarios are provided and the potential role of the nature of the phishing e-mail is highlighted.

One challenge that remains when applying adaptive network modeling to the types of scenarios described in this chapter is determining the correct values of all (sub-) elements of the COM-B framework. In order to properly determine the values of all factors associated with the different (sub-) elements, the elements have to be precisely defined in order to support the quantification of these elements. While the COM-B framework includes descriptions of the different (sub-) elements (see Table 2), the elements could be further defined. Some example considerations include the presence of the *"report phishing e-mail"* button as part of the physical opportunity factor and the effect of periodic phishing simulation training as part of the automatic motivation factor. Bouma et al. (2024) also highlight the significance of these considerations, they write that during the creation of the initial adaptive network model several simplifications were made, such as that *"the influence of the employee's colleagues and existing organizational security policies and guidelines have not been taken into account"* (as part of the opportunity value). While the enhanced adaptive network model presented in this chapter incorporates extra sub-elements as described by West and Michie (2020) and thereby allowing more precise simulations to be executed, the correct translation of these elements into quantifiable factors and, more importantly, the empirical validation of these factors, remains a topic for future research.

References

Alkhalil, Z., Hewage, C., Nawaf, L., Khan, I.: Phishing attacks: a recent comprehensive study and a new anatomy. Front. Comput. Sci. **3** (2021) https://doi.org/10.3389/fcomp.2021.563060

Alshaikh, M., Naseer, H., Ahmad, A., Maynard, S.B.: Toward Sustainable Behaviour Change: An Approach for Cyber Security Education Training and Awareness (2019)

American Medical Association: Future of Health Closing the Digital Health Disconnect: A Blueprint for Optimizing Digitally Enabled Care (2022)

American Medical Association: The State of Health at Home Models: Key Considerations and Opportunities (2023)

British Medical Association: An NHS under pressure (2023)

Bhuyan, S.S., Kabir, U.Y., Escareno, J.M., Ector, K., Palakodeti, S., Wyant, D., Kumar, S., Levy, M., Kedia, S., Dasgupta, D., Dobalian, A.: Transforming healthcare cybersecurity from reactive to proactive: current status and future recommendations. J. Med. Syst. **44**(5), 98 (2020). https://doi.org/10.1007/s10916-019-1507-y

Bouma, D., Hoffmans, C., Van den Hout, N.J., Zwarts, N., Treur, J., Roelofsma, P.: Cyber Security in Hospitals: A Network-Oriented Model for Behavioural Learning of Employees During Phishing Simulations (2024)

Brewer, R.: Ransomware attacks: detection, prevention and cure. Netw. Secur. **2016**(9), 5–9 (2016). https://doi.org/10.1016/S1353-4858(16)30086-1

Coventry, L., Branley, D.: Cybersecurity in healthcare: a narrative review of trends, threats and ways forward. Maturitas **113**, 48–52 (2018). https://doi.org/10.1016/j.maturitas.2018.04.008

Dutch Government: Integraal Zorg Akkoord—Samen werken aan gezonde zorg (2022)

European Commission: EU4Health programme 2021–2027 – a vision for a healthier European Union (2021)

European Union Agency for Cybersecurity: ENISA Threat Landscape (2023)

FDA: Medtronic Recalls Remote Controllers Used with Paradigm and 508 MiniMed Insulin Pumps for Potential Cybersecurity Risks (2021)

Ghafur, S., Kristensen, S., Honeyford, K., Martin, G., Darzi, A., Aylin, P.: A retrospective impact analysis of the wannacry cyberattack on the nhs. npj Dig. Med. **2**(1), 98 (2019). https://doi.org/10.1038/s41746-019-0161-6

Goetz, L.H., Schork, N.J.: Personalized medicine: motivation, challenges, and progress. Fertil. Steril. **109**(6), 952–963 (2018)

Gordon, W.J., Wright, A., Aiyagari, R., Corbo, L., Glynn, R.J., Kadakia, J., Kufahl, J., Mazzone, C., Noga, J., Parkulo, M., Sanford, B., Scheib, P., Landman, A.B.: Assessment of employee susceptibility to phishing attacks at US health care institutions. JAMA Netw. Open **2**(3), 190393 (2019)

Groeneveld, S.A., Jongejan, N., Winter, B.J.S.A.A.F., Fiolet, A.T.L.: [hacking into a pacemaker; risks of smart healthcare devices]. Ned Tijdschr Geneeskd **163** (2019)

Hamburg, M.A., Collins, F.S.: The path to personalized medicine. New England J. Med. **363**(4), 301–304 (2010). https://doi.org/10.1056/NEJMp1006304, https://doi.org/10.1056/NEJMp1006304

Hassamal, S., Dong, F., Hassamal, S., Lee, C., Ogunyemi, D., Neeki, M.M.: The psychological impact of COVID-19 on hospital staff. West. J. Emerg. Med. **22**(2), 346–352 (2021)

Health care expenditure by financing scheme. Technical report, Eurostat (2023)

Jalali, M.S., Kaiser, J.P.: Cybersecurity in hospitals: a systematic, organizational perspective. J. Med. Internet Res. **20**(5), 10059 (2018). https://doi.org/10.2196/10059

Joinson, A., Steen, T.: Human aspects of cyber security: behaviour or culture change? Cyber Secur. Peer-Rev. J. **1**(4), 351–360 (2018)

Kierkegaard, P.: Electronic health record: wiring europe's healthcare. Comput. Law Secur. Rev. **27**(5), 503–515 (2011). https://doi.org/10.1016/j.clsr.2011.07.013

Kleij, R., Wijn, R., Hof, T.: An application and empirical test of the capability opportunity motivation-behaviour model to data leakage prevention in financial organizations. Comput. Secur. **97**, 101970 (2020)

Kleij, R., Hoff-De Goede, S., Weijer, S., Leukfeldt, R.: How safely do we behave online? an explanatory study into the cybersecurity behaviors of dutch citizens. In: International Conference on Applied Human Factors and Ergonomics, pp. 238–246. Springer, Berlin (2021)

Kraus, S., Schiavone, F., Pluzhnikova, A., Invernizzi, A.C.: Digital transformation in healthcare: Analyzing the current state-of-research. J. Bus. Res. **123**, 557–567 (2021). https://doi.org/10.1016/j.jbusres.2020.10.030

Kuehn, B.M.: Pacemaker recall highlights security concerns for implantable devices. Circulation **138**(15), 1597–1598 (2018) https://doi.org/10.1161/CIRCULATIONAHA.118.037331, https://doi.org/10.1161/CIRCULATIONAHA.118.037331

Liu, J.X., Goryakin, Y., Maeda, A., Bruckner, T., Scheffler, R.: Global health workforce labor market projections for 2030. Hum. Resour. Health **15**(1), 11 (2017). https://doi.org/10.1186/s12960-017-0187-2

Martin, G., Ghafur, S., Kinross, J., Hankin, C., Darzi, A.: Wannacry—a year on. BMJ **361**(2018). https://doi.org/10.1136/bmj.k2381, https://www.bmj.com/content/361/bmj.k2381.full.pdf

McGlave, C., Neprash, H., Nikpay, S.: Hacked to Pieces? The Effects of Ransomware Attacks on Hospitals and Patients (2023). Available at SSRN: https://ssrn.com/abstract=4579292 or https://doi.org/10.2139/ssrn.4579292

Menachemi, N., Collum, T.H.: Benefits and drawbacks of electronic health record systems. Risk Manag. Healthcare Policy **4**, 47–55 (2011). https://doi.org/10.2147/RMHP.S12985, https://doi.org/10.2147/RMHP.S12985. PMID: 22312227

Michie, S., van Stralen, Maartje M., West, R.: The behaviour change wheel: a new method for characterising and designing behaviour change interventions. Implementation Sci. **6**(1), 42 (2011) https://doi.org/10.1186/1748-5908-6-42

National Health Service: The NHS Long Term Plan (2019)

Neprash, H.T., McGlave, C.C., Cross, D.A., Virnig, B.A., Puskarich, M.A., Huling, J.D., Rozenshtein, A.Z., Nikpay, S.S.: Trends in ransomware attacks on US hospitals, clinics, and other health care delivery organizations, 2016–2021. JAMA Health Forum **3**(12), 224873 (2022)

Ornell, F., Halpern, S.C., Kessler, F.H.P., Narvaez, J.C.D.M.: The impact of the covid-19 pandemic on the mental health of healthcare professionals. Cadernos de Sau´de Pu´blica **36**(4), 00063520 (2020). https://doi.org/10.1590/0102-311X00063520

Poon, Y.-S.R., Lin, Y.P., Griffiths, P., Yong, K.K., Seah, B., Liaw, S.Y.: A global overview of healthcare workers' turnover intention amid COVID-19 pandemic: a systematic review with future directions. Hum. Resour. Health **20**(1), 70 (2022)

Prince, M.J., Wu, F., Guo, Y., Gutierrez Robledo, L.M., O'Donnell, M., Sullivan, R., Yusuf, S.: The burden of disease in older people and implications for health policy and practice. Lancet **385**(9967), 549–562 (2015). https://doi.org/10.1016/S0140-6736(14)61347-7

Romagna, M., Van den Hout, N.J.: Hacktivism and website defacement: Capabilities, motivations and potential threats. In: Proceedings of the 27th International Virus Bulletin Conference (2017)

Samonas, S., Coss, D.L.: The cia strikes back: Redefining confidentiality, integrity and availability in security (2014). https://api.semanticscholar.org/CorpusID:215838643

Steen, T.: Measuring behavioural cybersecurity: an overview of options. In: International Conference on Human-Computer Interaction, pp. 460–471. Springer, Berlin (2023)

Treur, J.: Network-oriented Modeling. Springer, Berlin (2016)

Treur, J.: Network-Oriented Modeling for Adaptive Networks: Designing Higher-order Adaptive Biological, Mental and Social Network Models. Springer Nature, Cham (2020)

UK National Cyber Security Centre: Glossary (2024)

UK National Cyber Security Centre Ransomware: extortion and the cyber crime ecosystem (2023)
West, R., Michie, S.: A Brief Introduction to the Comb Model of Behaviour and the Prime Theory of Motivation. Qeios (2020) https://doi.org/10.32388/WW04E6.2
World Economic Forum: Global Health and Healthcare Strategic Outlook: Shaping the Future of Health and Healthcare—Insight Report (2023)
World Economic Forum: Global Health and Healthcare Strategic Outlook: Shaping the Future of Health and Healthcare (2023)
World Health Organization. Regional Office for Europe: Health and Care Workforce in Europe: Time to Act, p. 190. World Health Organization. Regional Office for Europe (2022)

Towards an AI Coach for Risk Management and Cybersecurity in Health Systems

Peter H. M. P. Roelofsma and Jan Treur

Abstract This chapter discusses the network-oriented computational analysis that was used to obtain insight on how an AI coach in cyberspace can provide support to medical teamwork and contribute to safety and security. Several options were analyzed. Many of these options were illustrated by a several realistic examples that were formalized computationally by adaptive network models. Other options are offered as interesting options for future research. As a first step to this future, we are discussing setting up a Cybersecurity Simulation Lab and a Living Lab. In these labs network-oriented modeling and AI coaching in cyberspace will be a basis to handle cyber risk management problems. This chapter describes the conclusions and further envisioned impact of this work.

Keywords Risk management · Cybersecurity · Network-oriented modelling · Virtual AI coach · Mental model · Medical team

---

P. H. M. P. Roelofsma · J. Treur (✉)
Risk Management and Cybersecurity Group, Centre of Expertise on Cybersecurity, The Hague University of Applied Sciences, The Hague, The Netherlands
e-mail: j.treur@vu.nl; j.treur@hhs.nl

P. H. M. P. Roelofsma
e-mail: p.h.m.p.roelofsma@hhs.nl

P. H. M. P. Roelofsma
Department of Pediatrics, Division Neonatology, Erasmus MC, University Medical Center-Sophia Children's Hospital, Rotterdam, The Netherlands

J. Treur
Social AI Group, Department of Computer Science, Vrije Universiteit Amsterdam, Amsterdam, The Netherlands

P. H. M. P. Roelofsma et al. (eds.), *Using Shared Mental Models and Organisational Learning to Support Safety and Security Through Cyberspace: A Computational Analysis Approach*, Studies in Systems, Decision and Control 570,
https://doi.org/10.1007/978-3-031-72075-8_13

1 Introduction

In this chapter the integrated results of separate scientific contributions are evaluated that focus on the development of an AI coach for health safety and cybersecurity. The integrated contribution is considered, and the future potential is examined. In health systems, risk management of cybersecurity directly also implies health safety management. The AI coach solution is a crucial future venue for supporting such risk management processes. Several related knowledge gaps are assessed, and a Cybersecurity Simulation Lab and Living Lab approach are suggested as a main method of future research for safety and cybersecurity in health systems. The expected social impact of this approach is described, and our main conclusion is presented.

2 An AI Coach Based Approach for Safety and Cybersecurity Health Risk Management

Each of the preceding chapters in this book has demonstrated successful input for an AI coach for health safety and security in cyberspace.

The integrated AI coach that is described introduces a platform-based solution that includes a user-centred AI communication and cooperation system which:

(1) interacts in a natural and ambient way with user and organisations
(2) supports/transforms this group into a concerned user community that updates and adapts the users' and organisations' shared mental models
(3) identifies safety, security and comfort challenges and solutions for needs to be solved.

The research group has contributed by joining forces from multiple disciplines and sectors to design models that may change the way we understand and respond to health safety and security in cyberspace. The models include technical subsystems, social subsystems, sociotechnical systems as well macro level ecosystem visions. We foresee a long-lasting impact.

3 Knowledge Gaps

The AI coach project addresses, specifically, the gap and unmet challenge that current guidelines exceed the cognitive bandwidth of the average safety and cybersecurity officer, especially where multidisciplinary and multi-unit, multi-stakeholder attention is provided. Furthermore, achieving and confirming that all involved workers have a shared mental model and are speaking up without fear is very challenging in settings, for which no adequate intervention is currently available. On top of that, cultural and country-specific differences add complexity to implementation and evaluation of cybersecurity. Our proposed AI-Coach platform enables workers to access context-relevant information, to use/implement quality assured procedures and to adhere to harmonised and standardised patient security guidelines.

4 What Has Been Achieved

The chapters' contributions cover a unique combination of experts in AI and cyber risk management, software development, ICT commercialisation, management science, and social and organisational psychology. The proposed AI-supported solution offers a novel, multi-facetted, and intriguing approach to organisational resilience and (cyber) risk management. The researchers composition and the work performed promises a carefully designed, thoroughly validated, and widely marketable adaptive ICT solution that undoubtedly constitutes a huge step forward in employing modern AI technology to successfully and cost-effectively address societal security challenges. Founded on computational formalisation of the use of shared mental models (Van Ments et al. 2025) and organisation learning (Canbaloğlu et al. 2025, this volume), different case studies for an AI coach to support safety and security in the neonatal domain have been undertaken as described in Mokadem et al. (2025), Chawla et al. (2025), Sebahi et al. (2025), Xu et al. (2025, see this volume). Moreover, in a similar manner in Doornkamp et al. (2025), Weigl et al. (2025, this volume), in particular the role of an AI coach to support safety and security for communication for speaking up between health care practitioners and communication with parents to avoid risks on postpartum depression was analysed.

Following the conceptual framework of three-level cognitive architectures for mental models and organisational learning outlined in Roelofsma and Treur (2025a, b), where protocols are modeled as (shared) mental models, the developed architecture of an AI coach includes the capabilities summarized in Table 1. Here, adaptive self-modeling network modelling first-order self-model **W**-states have been used to represent connections within a mental model in an adaptive manner. Moreover, adaptive self-modeling network modelling second-order self-model $\mathbf{W}_{\mathbf{WW}}$-states have been used to model adaptive connections between **W**-states of different agents to control the exchange of knowledge between the AI coach and health care practitioners.

Further variations have been described addressing learning from mistakes (Hosseini et al. 2025, this volume), and implementation of safety standards within a large organisation (Van den Hurk et al. 2025, this volume). Moreover, in Van den Hout et al. (2025, this volume), future perspectives for cybersecurity are discussed.

Table 1 Capabilities of the AI coach

	Base level: Prediction by internal simulation of mental models and discrepancy notification	First-order self-model level: learning mental models by exchange of knowledge	Second-order self-model level: control of communication of mental models
Monitoring	• Predicting next steps of a protocol by internal simulation of the own mental model • Observing the real world • Checking for a discrepancy between prediction and observation		
Intervening when noticing a discrepancy	• Communicating that a discrepancy was noticed	• Communicating missing knowledge of the mental model (feedback) from AI coach to team member (from **W**-states of the AI coach to **W**-states of team members)	• Control of (feedback) communication of from **W**-states of the AI coach to **W**-states of team members ($\mathbf{W}_{\mathbf{WW}}$-states)
Sharing mental models by organisational learning		• Communicating mental models (feed forward) from team members to AI coach (from **W**-states of team member to **W**-states of AI coach) • Communicating mental model (feedback) from AI coach to team members (from **W**-states of AI coach to **W**-states of team members)	• Control of (feed forward) communication from **W**-states of team member to **W**-states of AI coach ($\mathbf{W}_{\mathbf{WW}}$-states) • Control of (feedback) communication from **W**-states of AI coach to **W**-states of team members ($\mathbf{W}_{\mathbf{WW}}$-states)

5 The Cybersecurity Living Lab: Co-Creating Cyber Security

We suggest as a next step for the future research the further development of the AI coach in a Cyber Living Lab approach focusing on organisational cybersecurity and specific security pathways. In the Living Lab multidisciplinary co-creation is performed with all stakeholders, knowledge partners and end users. The developed framework and AI-Coach includes the personal, team and organisation level, and provides an adaptive evaluation of executed processes. This focused approach allows us to develop the scientific framework and AI-Coach as a technology with already built-in aspects of implementation and scalability aspects.

In the Living Lab participating organisations and consortium partners will develop strategies for stakeholder engagement, dissemination and communication strategy, which provide a route for engaging with national organisations and policymakers. In the Living Lab a Security Operation Center (SOC) will be established. Here 24/7 monitoring and response management will take place. Work, learning and research takes place at the same time so that co-creation of cybersecurity is an ongoing activity.

Due to open access/source outputs such as the scientific framework, the open source licences for AI-Coach and the AI-Coach training, organisations can easily test and adopt our approach and AI platform. The scaling-up of our approach depends on the success of the system, which we will safeguard during the project. Positive feedback from participating organisations will help us to deliver a functioning system. As mentioned before, partners can offer paid support and training, respectively, to interested parties. Furthermore, cooperation with other projects can be achieved so as to join forces on cybersecurity and propose synergistic solutions, closing knowledge gaps and addressing challenges. In this way further significant impact can be achieved.

The living lab integrates off-the-shelf technology (e.g. sensors that monitor the organizations endpoints and network devices) and builds on existing structures, thus permitting an individually tailored approach. The integrated AI coach approach is based upon the following four principles:

1. *Active end user involvement*: Users associated with the Living Labs are directly involved in the iterative process of AI Coach solution design, installation, operation, and evaluation. This user-centred approach using simulation, gaming and gamification climaxes in a large-scale validation research.
2. *Behavioural focus*: The AI Coach prompts users directly to beneficial and sustained levels of team communication and coordination for behavioural change rather than relying on knowledge or attitude driven changes. This will be accomplished by applying advanced motivational frameworks which also encompass the users' perceptions of autonomy, technology acceptance and positive rewards.
3. *Improvement science*: The AI coach component enables and encourages the learning process and the implementation of the innovation in the organisation throughout the project offering many benefits and safety options that make the use of it a rewarding experience.
4. *Adaptivity, scalability, and open-source*: The AI coach builds on existing standards, features open interfaces and employs open-source technology to ensure high levels of adaptivity (e.g., changing end user needs) and scalability (e.g., additional sensors, network segments, endpoint devices, languages, and combinations with existing and future hardware and software applications).

6 The Cybersecurity Simulation Lab

As a first step to this, we suggest to examine the effect of AI on cybersecurity more focusedly. More specifically, we suggest to substantiate the success of our research on network oriented modelling with a focus on cyber risk management. The venue that would be most promising is the apply the network oriented modelling approach to standard and non standard cyber risk management problems. Such research work has not yet been performed and would be a necessary logical next development of our current work. If that would be successful, the models may be applied in de Living Lab using real data from several field settings.

7 Social Impact

We expect the following impact of the suggested research next steps.

Health policy and systems adopt a holistic approach:

The solutions like the AI-Coach and hospital quality safety and security modelling allow for monitoring and evaluation of the interventions and security in a holistic way. In this way we can measure identifiable changes e.g., error registry, in and beyond our participating hospitals. By monitoring the number of organisations using the new shared mental model-framework and the AI-Coach we will evaluate implementation and the cultural and organisational changes across organisations in order to measure our impact.

Health care providers are trained and equipped with the skills and competences:

The open-access framework description, including methodology like, Cybersecurity Simulation Lab and the Living Lab learning community, the open source licences for the AI-Coach, and AI-coach training materials allows every health care provider to test and use our approach and train their Safety and Cybersecurity professionals. Our partners can provide training and support to interested hospitals.

Citizens can be supported to play a key role in managing their own risks:

Further development of the AI-Coach envisions inclusion of the patient or other users of ICT as a key stakeholder. Professionals interact daily with users, which allows them to directly feedback their input and perspective into pathway design.

In sum: Our AI coach approach for safety and security in cyberspace have shown significant progress with strong lasting social impact. The next step would be to develop a research program that explicitly focuses on the application of our current work that is performed in this book on AI and cybersecurity risk management.

8 Discussion

The initiative to develop an AI coach for risk management and cybersecurity in health systems stands as a testament to the potential of artificial intelligence to revolutionize healthcare. This AI coach project, detailed extensively in this chapter, is not merely an isolated technological endeavor; it embodies a comprehensive, multi-disciplinary effort that integrates insights from AI, cyber risk management, software development, organizational psychology, and management science. Central to this project is the use of network-oriented computational analysis. This method allows for the simulation and evaluation of complex interactions within healthcare environments, providing critical insights into how AI can support medical teamwork and ensure safety and security. Through adaptive network models, realistic examples are formalized computationally, offering a dynamic approach to understanding and managing cybersecurity risks. These models encompass technical subsystems, social subsystems, sociotechnical systems, and macro-level ecosystem visions, thus ensuring a holistic view of healthcare security.

The AI coach introduces a platform-based solution characterized by its user-centered AI communication and cooperation system. This system:

- Interacts in a natural and ambient manner with users and organizations, ensuring seamless integration into daily routines.
- Supports and transforms groups into a concerned user community that continuously updates and adapts shared mental models.
- Identifies safety, security, and comfort challenges, proposing solutions to address these needs effectively.

This user-centered approach is pivotal as it ensures that the AI coach is not just a tool but an integral part of the healthcare environment, fostering a culture of safety and proactive risk management.

One of the significant achievements of the AI coach is its ability to bridge cognitive gaps that exist within current safety and cybersecurity protocols. The complexity and volume of guidelines often exceed the cognitive bandwidth of the average safety officer, particularly in multidisciplinary and multi-stakeholder environments. The AI coach mitigates this by providing context-relevant information and standardizing patient security guidelines, making them more accessible and actionable.

A core element of the AI coach's success is its emphasis on shared mental models and open communication within medical teams. Ensuring that all team members have a shared understanding of protocols and are encouraged to speak up without fear is critical for effective risk management. This is particularly challenging in settings where hierarchical structures and cultural differences may inhibit open communication. The AI coach fosters an environment where feedback is actively solicited and valued, promoting a culture of continuous improvement and learning.

As a foundational step, the establishment of a Cybersecurity Simulation Lab will provide a focused environment for examining the impact of AI on cybersecurity. This lab will enable researchers to apply network-oriented modeling to standard and non-standard cyber risk management problems, providing a controlled setting for rigorous testing and validation.

The proposed Cybersecurity Living Lab represents a significant leap forward in the AI coach project. This Living Lab will facilitate multidisciplinary co-creation with all stakeholders, including knowledge partners and end users. The framework developed within the Living Lab will address personal, team, and organizational levels, providing an adaptive evaluation of executed processes.

1. **Active End User Involvement**: Users associated with the Living Labs will be directly involved in the iterative process of AI coach solution design, installation, operation, and evaluation. This user-centered approach, utilizing simulation, gaming, and gamification, will culminate in large-scale validation research.
2. **Behavioral Focus**: The AI coach will prompt users towards beneficial and sustained levels of team communication and coordination, focusing on behavioral change rather than relying solely on knowledge or attitude-driven changes.
3. **Improvement Science**: The AI coach will enable and encourage learning and innovation implementation throughout the organization, offering numerous safety options and making its use a rewarding experience.
4. **Adaptivity, Scalability, and Open-Source**: The AI coach will build on existing standards, feature open interfaces, and employ open-source technology to ensure high levels of adaptivity and scalability.

The AI coach's holistic approach allows for comprehensive monitoring and evaluation of interventions and security measures. This enables healthcare systems to identify and measure changes, such as error registries, across participating hospitals. By monitoring the adoption of shared mental models and the AI coach, the project can evaluate implementation and cultural shifts within organizations, providing valuable insights into its impact.

Healthcare providers will be equipped with the necessary skills and competences through the AI coach's open-access framework, including methodologies like the Cybersecurity Simulation Lab and the Living Lab learning community. The availability of open-source licenses and training materials ensures that any healthcare provider can adopt and test the AI coach, with partners providing training and support as needed.

The project envisions including patients and other ICT users as key stakeholders in the AI coach framework. This inclusion ensures that professionals can directly incorporate patient feedback into pathway design, enhancing the relevance and effectiveness of the AI coach.

Conclusion

The development of an AI coach for risk management and cybersecurity in health systems marks a significant advancement in integrating artificial intelligence with healthcare. This project leverages network-oriented computational analysis and interdisciplinary collaboration to address critical knowledge gaps and provide a scalable, user-friendly solution. The establishment of the Cybersecurity Simulation Lab and

Cybersecuity Living Lab will further advance research and development, ensuring the AI coach remains adaptive and effective in real-world settings.

The potential social impact of the AI coach is substantial. By adopting a holistic approach to health safety and cybersecurity, healthcare systems can effectively monitor and evaluate interventions, equipping providers with the necessary skills and competences, and involving patients as key stakeholders. As the project progresses, continuous evaluation and feedback from participating organizations will be crucial in refining and scaling the system. Ultimately, the AI coach promises to transform risk management and cybersecurity in healthcare, leading to safer, more resilient health systems.

References

Canbaloğlu, G., Treur, J., Roelofsma, P.H.M.P.: Learning of safety and security management through cyberspace: an adaptive self-modeling network model for multilevel organizational learning. In: Roelofsma, P.H.M.P., Jabeen, F., Taal, H.R., Treur, J. (eds.) Using Shared Mental Models and Organisational Learning to Support Safety and Security Through Cyberspace: A Computational Analysis Approach. Springer Nature, Cham, Switzerland (this volume) (2025)

Chawla, R., Jabeen, F., Treur, J., Taal, H.R., Roelofsma, P.H.M.P.: Supporting risk management through cyberspace: an adaptive network model simulating AI coach effects by inducing adherence to guidelines in neonatal medical protocols . In: Roelofsma, P.H.M.P., Jabeen, F., Taal, H.R., Treur, J. (eds.) Using Shared Mental Models and Organisational Learning to Support Safety and Security Through Cyberspace: A Computational Analysis Approach. Springer Nature, Cham, Switzerland (this volume) (2025)

Doornkamp, S., Jabeen, F., Treur, J., Taal, H.R., Roelofsma, P.H.M.P.: Optimizing speaking up behavior for safety and security through cyberspace: a controlled adaptive computational network model of a virtual coach. In: Roelofsma, P.H.M.P., Jabeen, F., Taal, H.R., Treur, J. (eds.) Using Shared Mental Models and Organisational Learning to Support Safety and Security Through Cyberspace: A Computational Analysis Approach. Springer Nature, Cham, Switzerland (this volume) (2025)

Hosseini, M., Treur, J., Kucharska, W..: Learning from mistakes within organizations: an adaptive network-oriented model for a double bias perspective for safety and security through cyberspace. In: Roelofsma, P.H.M.P., Jabeen, F., Taal, H.R., Treur, J. (eds.) Using Shared Mental Models and Organisational Learning to Support Safety and Security Through Cyberspace: A Computational Analysis Approach. Springer Nature, Cham, Switzerland (this volume) (2025)

Mokadem, N., Jabeen, F., Treur, J., Taal, H.R., Roelofsma, P.H.M.P.: Increasing safety and security through cyberspace by an adaptive network model for AI-assisted risk management of neonatal respiratory distress. In: Roelofsma, P.H.M.P., Jabeen, F., Taal, H.R., Treur, J. (eds.) Using Shared Mental Models and Organisational Learning to Support Safety and Security Through Cyberspace: A Computational Analysis Approach. Springer Nature, Cham, Switzerland (this volume) (2025)

Roelofsma, P.H.M.P., Treur, J.: Computational analysis to support safety and security through cyberspace using shared mental models and organisational learning. In: Roelofsma, P.H.M.P., Jabeen, F., Taal, H.R., Treur, J. (eds.) Using Shared Mental Models and Organisational Learning to Support Safety and Security Through Cyberspace: A Computational Analysis Approach. Springer Nature, Cham, Switzerland (this volume) (2025a)

Roelofsma, P.H.M.P., Treur, J.: Towards an AI coach for risk management and cybersecurity in health systems. In: Roelofsma, P.H.M.P., Jabeen, F. Taal, H.R. Treur, J. (eds.) Using Shared Mental

Models and Organisational Learning to Support Safety and Security Through Cyberspace: A Computational Analysis Approach. Springer Nature, Cham, Switzerland (this volume) (2025b)
Sebahi, Y., Jabeen, F., Treur, J. Taal, H.R., Roelofsma, P.H.M.P.: Improving risk management through cyberspace: optimizing neonatal respiratory support through network oriented modeling. In: Roelofsma, P.H.M.P., Jabeen, F., Taal, H.R., Treur, J. (eds.) Using Shared Mental Models and Organisational Learning to Support Safety and Security Through Cyberspace: A Computational Analysis Approach. Springer Nature, Cham, Switzerland (this volume) (2025)
Van den Hout, N.J., Bouma, D.. Hoffmans, C., Zwarts, N., Treur, J.. Roelofsma, P.: Addressing organizational cyber security challenges in healthcare using adaptive network-oriented modeling. In: Roelofsma, P.H.M.P., Jabeen, F., Taal, H.R., Treur, J. (eds.) Using Shared Mental Models and Organisational Learning to Support Safety and Security Through Cyberspace: A Computational Analysis Approach. Springer Nature, Cham, Switzerland (this volume) (2025)
Van den Hurk, N., Treur, J., Roelofsma, P.H.M.P.: Organisational learning for safety and security through cyberspace: adaptive modelling of the implementation of environment health and safety standards. In: Roelofsma, P.H.M.P., Jabeen, F., Taal, H.R., Treur, J. (eds.) Using Shared Mental Models and Organisational Learning to Support Safety and Security Through Cyberspace: A Computational Analysis Approach. Springer Nature, Cham, Switzerland (this volume) (2025)
Van Ments, L., Treur, J., Klein, J. Roelofsma, P.H.M.P.: Developing a safety and security AI coach: a second-order adaptive network model of shared mental models in hospital teamwork. In: Roelofsma, P.H.M.P., Jabeen, F., Taal, H.R., Treur, J. (eds.) Using Shared Mental Models and Organisational Learning to Support Safety and Security Through Cyberspace: A Computational Analysis Approach. Springer Nature, Cham, Switzerland (this volume) (2025)
Weigl, L., Jabeen, F., Treur, J., Taal, H.R., Roelofsma, P.H.M.P.: Learning for a better safety and security culture within an organization: Reducing the risk in communication with AI coaching for security communication through cyberspace. In: Roelofsma, P.H.M.P., Jabeen, F., Taal, H.R., Treur, J. (eds.) Using Shared Mental Models and Organisational Learning to Support Safety and Security through Cyberspace: A Computational Analysis Approach. Springer Nature, Cham, Switzerland (this volume) (2025)
Xu, Y., Jabeen, F., Treur, J., Taal, H.R., Roelofsma, P.H.M.P.: Supporting medical team decision in the neonatal domain: how safety and security risk management through cyberspace can be supported by adaptive agent network models with internal mental models. In: Roelofsma, P.H.M.P., Jabeen, F., Taal, H.R., Treur, J. (eds.) Using Shared Mental Models and Organisational Learning to Support Safety and Security through Cyberspace: A Computational Analysis Approach. Springer Nature, Cham, Switzerland (this volume) (2025)

GPSR Compliance

The European Union's (EU) General Product Safety Regulation (GPSR) is a set of rules that requires consumer products to be safe and our obligations to ensure this.

If you have any concerns about our products, you can contact us on ProductSafety@springernature.com

In case Publisher is established outside the EU, the EU authorized representative is:

Springer Nature Customer Service Center GmbH
Europaplatz 3
69115 Heidelberg, Germany

Batch number: 10755079

Printed by Printforce, the Netherlands